D1259279

Genesis and Propagation of Cosmic Rays

NATO ASI Series

Advanced Science Institutes Series

A Series presenting the results of activities sponsored by the NATO Science Committee, which aims at the dissemination of advanced scientific and technological knowledge, with a view to strengthening links between scientific communities.

The series is published by an international board of publishers in conjunction with the NATO Scientific Affairs Division

A Life Sciences B Physics	Plenum Publishing Corporation London and New York
C Mathematical and Physical Sciences	D. Reidel Publishing Company Dordrecht, Boston, Lancaster and Tokyo
D Behavioural and Social Sciences E Applied Sciences	Martinus Nijhoff Publishers Dordrecht, Boston and Lancaster
F Computer and Systems Sciences G Ecological Sciences H Cell Biology	Springer-Verlag Berlin, Heidelberg, New York, London, Paris, and Tokyo

Series C: Mathematical and Physical Sciences Vol. 220

Genesis and Propagation of Cosmic Rays

edited by

Maurice M. Shapiro

University of Maryland,
College Park, Maryland, U.S.A.

and

John P. Wefel

Department of Physics and Astronomy,
Louisiana State University, Baton Rouge,
Louisiana, U.S.A.

D. Reidel Publishing Company

Dordrecht / Boston / Lancaster / Tokyo

Published in cooperation with NATO Scientific Affairs Division

Proceedings of the NATO Advanced Study Institute on
Genesis and Propagation of Cosmic Rays
Erice, Sicily, Italy
June 1-9, 1987

Library of Congress Cataloging in Publication Data

NATO Advanced Study Institute on Genesis and Propagation of Cosmic Rays (1986 : Erice, Sicily)
Genesis and propagation of cosmic rays.

(NATO ASI series. Series C, Mathematical and physical sciences; vol. 220)
"Published in cooperation with NATO Scientific Affairs Division."
"Proceedings of the NATO Advanced Study Institute on Genesis and Propagation of Cosmic Rays, Erice, Sicily, Italy, June 1–9, 1986"—T.p. verso.
Includes indexes.
1. Cosmic rays—Congresses. 2. Nuclear astrophysics—Comgresses. I. Shapiro, Maurice M. (Maurice Mandel), 1915– . II. Wefel, J. P. III. North Atlantic Treaty Organization. Scientific Affairs Division. IV. Title. V. Series: NATO ASI series. Series C, Mathematical and physical sciences; no. 220.
QC484.8.N42 1986 523.01′97223 87–26330
ISBN 90–277–2628–0

Published by D. Reidel Publishing Company
P.O. Box 17, 3300 AA Dordrecht, Holland

Sold and distributed in the U.S.A. and Canada
by Kluwer Academic Publishers,
101 Philip Drive, Assinippi Park, Norwell, MA 02061, U.S.A.

In all other countries, sold and distributed
by Kluwer Academic Publishers Group,
P.O. Box 322, 3300 AH Dordrecht, Holland

D. Reidel Publishing Company is a member of the Kluwer Academic Publishers Group

---

Printed in The Netherlands

Dedicated

to

INEZ

in loving memory

--an exemplary life

ever to be cherished

--M. M. S.

TABLE OF CONTENTS

1. J. Grunsfeld
2. D. Averna
3. N. Smith
4. N. Mascarenhas
5. J. Beatty
6. M. Machin
7. M. J. Rogers
8. C. Koch
9. J. Isbert
10. N. Porter
11. J. Perko
12. U. Heinbach
13. A. Monk
14. W. Dröge
15. B. Dingus
16. P. Watson
17. K. Lesko
18. C. Setti
19. C. Cattadori
20. H. Völk
21. M. M. Shapiro
22. K. Weiler
23. K. Oschlies
24. M. A. Miah
25. M. Potgieter
26. N. Rana
27. W. Markiewicz
28. C. Akujor
29. D. Green
30. M. Lawrence
31. D. Breitschwerdt
32. S. Sutton
33. K. Levenson
34. J. Szabelski
35. J. F. Valdes-Galicia
36. E. Eleftheriou
37. K. Doutsi
38. W. Stamm
39. J. Mitchell
40. P. P. Maggioli
40b. C. Belardi
41. M. Gottardi
42. K. O. Thielheim
43. M. Samorski
44. A. Tylka
45. A. Drukier
46. J. Engelage
47. V. Pirronello
48. G. Lanzafame
49. F. Ashton
50. D. de Martino
51. I. Nastari
52. G. Pizzichini
53. E. Palazzi
54. R. Binns
55. C. P. Vankov
56. T. Montmerle
57. P. Chadwick

PREFACE

The fifth Course of the International School of Cosmic-Ray Astrophysics "Genesis and Propagation of Cosmic Rays," which gave rise to this volume, was held in June, 1986 at the Ettore Majorana Center for Scientific Culture (CCSEM) in Erice, Sicily. Its recognition and co-sponsorship as an Advanced Study Institute (ASI) by NATO contributed much to the success of the Course. Valuable support was also rendered by the European Physical Society,, the Italian Ministry of Education, the Italian Ministry of Scientific and Technological Research,the Sicilian Regional Government, and the National Science Foundation of the United States.

For a decade the International School of Cosmic-Ray Astrophysics has consistently aimed to reflect the interdisciplinary character of this field of research. This Course was no exception; it extended the astrophysical horizons of students and senior participants. While the sessions were mainly pedagogical, they also served as a workshop, dealing with current and controversial research issues. Nearly all participants were active investigators in cosmic radiation and related fields.

The Editors, who were co-directors of the Institute, owe thanks to the sponsors and to many colleagues: notably to Lecturers and other participants who made this volume possible; to Professor Antonino Zichichi, Director of CCSEM; to Assistant Secretary General Professor Henry Durand and Dr. Craig Sinclair of NATO's Scientific Affairs Division; to the distinguished members of NATO's Scientific Committee; to Dr. Alberto Gabriele, Dr. Pinola Savalli and Ms. Sandra Wefel for unstinting administrative support. We acknowledge the patient editorial help of Ms. N. M. Pols-v.d. Heijden and her associates at the D. Reidel Publishing Company.

Members of the Scientific Advisory Committee of the School have given timely advice: Professors Pierre Auger, G.P.S. Occhialini, Bruno Rossi, Maurice Shapiro (Director of the School), Rein Silberberg, John Simpson, James A. Van Allen, and Antonino Zichichi.

This volume helps celebrate a noteworthy milestone in the history of cosmic rays--the seventy-fifth anniversary of their discovery. In 1912 no one could have predicted that the balloon flights of Victor Hess would engender new disciplines--particle astronomy, high-energy astrophysics, and elementary particle physics. Is it too sanguine to expect that further treasure-troves await discovery in these fertile fields?

Maurice M. Shapiro*

Department of Physics and Astronomy
University of Maryland
College Park, MD, USA
and
Laboratory for Cosmic-Ray Physics (emeritus)
Naval Research Laboratory
Washington, D. C., USA

John P. Wefel

Department of Physics and Astronomy
Louisiana State University
Baton Rouge, LA 70803, USA

---

*Address for correspondence:

205 Yoakum Pkwy. # 1720
Alexandria, VA 22304, USA

AN OVERVIEW OF COSMIC RAY RESEARCH: COMPOSITION, ACCELERATION AND PROPAGATION

John P. Wefel
Department of Physics and Astronomy
Louisiana State University
Baton Rouge, LA 70803-4001

ABSTRACT. An overview of Cosmic Ray Research and its relation to other areas of High Energy Astrophysics is presented. Particular attention is paid to providing a summary of current results on the composition, acceleration and propagation of the cosmic radiation -- topics which will be discussed in greater detail in other papers in this volume.

1. WHAT IS COSMIC RAY RESEARCH?

The question "What is Cosmic Ray Research?" is frequently asked by both students and scientists from other fields, and has direct relevance to this 5th Course of the International School of Cosmic Ray Astrophysics. It is often difficult to give a unique answer to this question since the definition may vary among the many practitioners engaged in cosmic ray research and since this area of study has encompassed, and been the midwife for, many separate research fields during its long history.

Cosmic ray research goes back to the beginning of this century, since its origin is usually ascribed to the pioneering manned balloon flights of Victor Hess, around 1911-12. As inscribed on a plaque on the wall of the Secretariat room in the Ettore Majorana Centre in Erice, Italy:

"Here in the Erice maze
Cosmic Rays are the craze
and this because a guy named Hess
ballooning up found more not less."

This little rhyme indicates the nature of Hess's discovery, namely that the radiation was extraterrestrial in origin! Thus, one definition of cosmic ray research would be the study of "Rays from Outer Space."

From our present vantage point, we might list the possible components of these rays from outer space as:

M. M. Shapiro and J. P. Wefel (eds.), Genesis and Propagation of Cosmic Rays, 1–40.

(i) electrically charged particles, both positive and negative
(ii) electromagnetic quanta
(iii) electrically neutral particles
(iv) dust particles

Each of these components has been found above the Earth's protective atmosphere and all have been involved, to one degree or another, in cosmic ray research. Cosmic dust, for example, has been studied on high flying aircraft, balloons and space missions -- most recently on the Halley's Comet missions -- and shown to consist of some grains of interplanetary origin and others, possibly, of interstellar origin (e.g. Brownlee, 1985). The study of the morphology and the element/isotope patterns in these grains will provide new insights into the formation of our solar system and processes in the Interstellar Medium (ISM). These investigations generally fall under the area of Cosmochemistry and are, today, not considered a part of cosmic ray research.

Under "neutral" particles we would include neutrons, neutrinos, neutral mesons and hyperons, and, possibly, the magnetic monopole and predicted new particles such as the supersymmetric photino. In addition, the elusive "graviton" is being actively pursued with a variety of gravitational wave detector systems. There has been considerable progress in the last decade in detecting the decay products or in observing the interactions of neutral particles, and some of these will be summarized in later papers in this volume. As examples of such observations, one might point to the discovery in interplanetary space of a proton pulse produced by the decay of the neutrons emitted during a solar flare (e.g. Evenson et al. 1983) or to the new studies of neutrino interactions becoming possible with the large underground proton decay instruments (see Cherry et al., 1987).

The electromagnetic spectrum, of course, forms the basis of astronomical research. Ground based observations are, however, limited to photons with frequency below the optical or near ultra-violet region of the spectrum, so high energy photons (far-UV, x-rays and γ-rays) must be studied near the top of the Earth's atmosphere using balloons, rockets or space vehicles. High energy photons were first discovered in the cosmic radiation, and their investigation remains interwoven with cosmic ray research.

A brief return to the history of cosmic rays is in order. Following Hess's discovery, many researchers became involved in trying to discover the nature of this extraterrestrial radiation, performing experiments from sea level to mountain tops and at a wide variety of latitudes on long sea voyages. By the 1930's, it had been shown that the radiation arriving at the top of the Earth's atmosphere was, mainly, positively charged and was of high energy, as shown by its great penetrating power. By the late 1930's - early 1940's, the radiation had been confirmed to be protons, not electrons, and the importance of secondary particles produced by interactions in the atmosphere or within detectors had been realized. The discovery of mesons at about the same time ushered in an era spanning 10-20 years during which the cosmic radiation was used to provide the high energy particle "beam" for investigations of elementary particle production. This new field later diverged from cosmic ray research with the development of particle accelerators and their controlled beams of high energy particles; however, the cosmic radiation remains the only current source of extremely high energy (>TeV) particles and is still employed today to search for new states of matter formed in ultrahigh energy interactions (e.g. Burnett et al., 1986).

The next milestone in cosmic ray research occurred in 1948 with the discovery of helium and heavier elements ($Z \lesssim 28$) among the cosmic rays (Freier et al., 1948). This opened the possibility of studying, at first hand, a sample of matter from elsewhere in our galaxy. This was extended in the mid-60's with the discovery of ultra-heavy (UH) cosmic rays with atomic numbers extending from $Z > 28$ through the rest of the periodic table, up to and including uranium (Fleischer et al., 1967; Fowler et al., 1967). By the late 1970's, the experimental techniques had improved to such a degree that the isotopes of various elements below the iron peak could be measured. These developments turned part of the focus of cosmic ray research towards "Particle Astrophysics," the study of astrophysical environments using the information in the charge, mass and energy spectra of the cosmic rays.

There were several other important discoveries made in the 1960's. Primary electrons, and, shortly thereafter, positrons, were confirmed in the cosmic rays. These components provide a powerful tracer of the matter and magnetic field distributions encountered by the cosmic rays in the galaxy. At about the same time, extra-terrestrial x-rays and γ-rays were discovered near the top of the Earth's atmosphere. These high energy photons have instigated a virtual "revolution" in astronomy, opening up many high energy processes for detailed astrophysical investigation. Yet, it is high energy charged particles which are responsible for many of the high energy photons observed, and, as mentioned above, this connection provides the bond between particle astrophysics and x-ray and γ-ray astronomy, which together compose the area of research known as High Energy Astrophysics.

Thus, currently, cosmic ray research involves the study of astrophysical systems, with scales ranging from the solar system, to our galaxy, to groupings of galaxies, focusing on the high energy processes that occur in these environments. Information is gained from photon astronomy, both classical and the newer high energy astronomy, from neutrinos and other radiations, and from charged particles. Central to this research is the study of the high energy cosmic ray particles to determine their origin, nucleosynthesis, acceleration, history and their role in the dynamics of our galaxy.

Some of these questions are illustrated in schematic form in Figure 1. On the right is shown the origin of cosmic ray particles from the matter that is to become cosmic rays through acceleration to high energy and their subsequent propagation in the galaxy and in the heliosphere (the region of space around our sun controlled by the outflowing solar wind) before being observed at the Earth. In this figure the question of the "origin" of cosmic ray matter involves the contributions of several different sources -- individual stars, the interstellar medium, supernovae — to the matter that is to be accelerated. Resolving this question requires working backwards from the composition and energy measurements made in the vicinity of Earth, through the propagation and acceleration processes, to study the composition of matter in the sources. Outlining these areas will be the topic of the remainder of this paper.

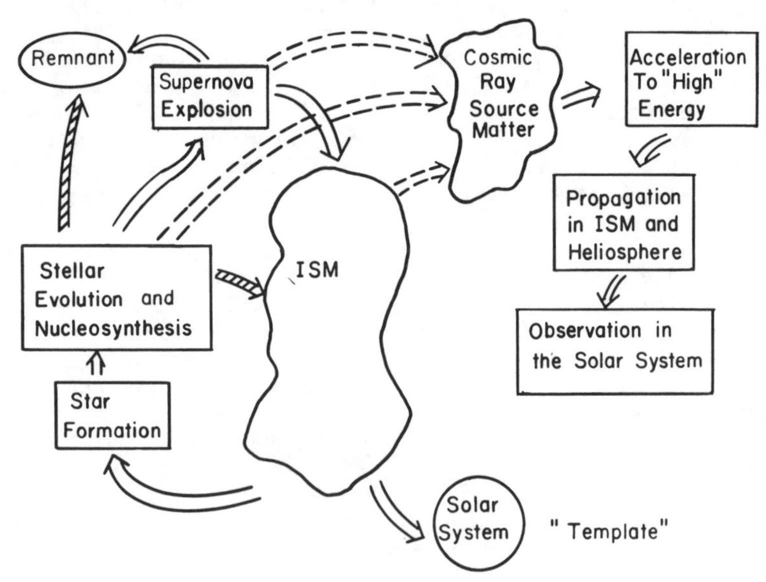

Figure 1. Illustration of the relationships between cosmic ray processes (right) and astrophysical processes which may contribute matter to the cosmic radiation.

2. COMPOSITION

2.1. The Solar System "Template"

As Figure 1 indicates, the interstellar medium (ISM) is the largest reservoir of matter in the galaxy, and we might expect that the composition of cosmic ray material would be similar to that of the ISM. We cannot, unfortunately, measure the detailed composition of the ISM directly, although much information is available from spectroscopy applied to different types of stars and from observations of HI regions, HII regions and molecular clouds. However, the best known sample of the ISM is our solar system, which withdrew from active processes in the galaxy about 4.6×10^9 years ago as the early solar nebula formed and began the process of the formation of the solar system. Thus, we can use solar system material as a "template" against which spectroscopic or cosmic ray composition measurements can be compared.

The study of element and isotope abundances in solar system material has a long history and involves measurements not only of the composition of the sun but also of relative abundances in the planets, the moon and, most importantly, in meteorites. There is one class of meteorites, the carbonaceous chrondrites, which show little differentiation (unlike the planets) and so may represent the most primitive sample of the material from which the solar system formed. There have been many compilations of solar system abundances, among the more recent of which are Meyer (1979), Cameron (1982) and Anders and Ebihara (1982). Figure 2 shows the relative abundances for the first

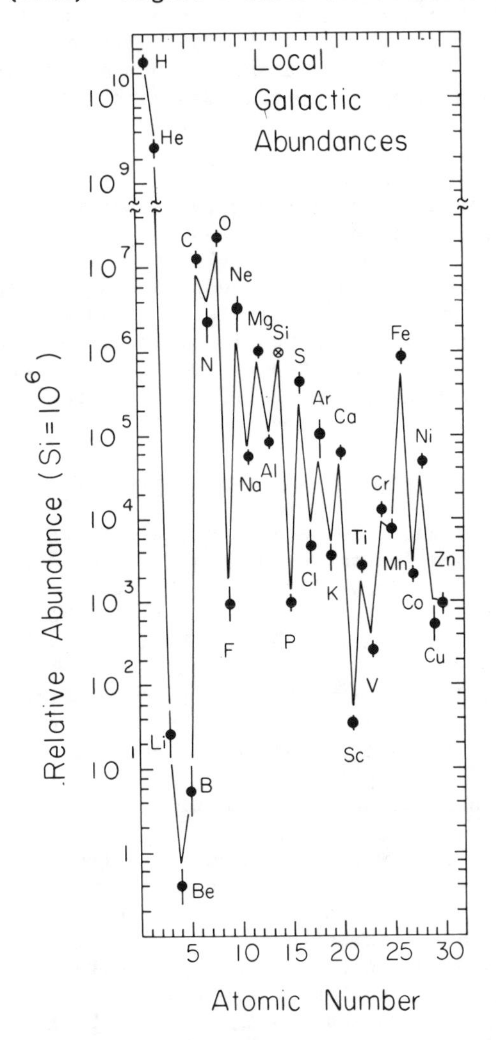

Figure 2. The relative abundances of the elements $1 \leqslant Z \leqslant 30$ in the solar neighborhood (Meyer, 1979) relative to silicon.

thirty elements, adopted by Meyer (1979) as "Local Galactic" abundances from a study of both solar system material and nearby HI and HII regions and unevolved stars. The uncertainty limits are determined by Meyer from the ranges of the measurements. Starting from H and He, the most abundant elements, the relative abundance plummets by 10 orders of magnitude to the light elements, Li, Be and B, which are clearly very rare in nature. Abundance peaks appear at the medium elements, C, N and O, the intermediate elements, Ne-S, and the iron peak. Just below the iron peak is a valley encompassing the rare sub-Fe elements (Sc, Ti, V, Cr, Mn), and the odd-Z elements are significantly less abundant than their even-Z neighbors. Explaining the origin of these patterns in the element abundances (plus the remaining 2/3 of the periodic table up to uranium) is the goal of the study of stellar evolution and heavy element nucleosynthesis.

2.2. Abundances at Earth ($Z \leq 30$)

Figure 3 shows the measured abundances in the galactic cosmic rays (GCR) at the top of the Earth's atmosphere. The data are from the review of Simpson (1983), and three points are plotted for most elements (He - Zn) corresponding to low, intermediate and high energy data. Overall, the agreement among the results at different energies is good, indicating that the cosmic ray composition is largely energy independent. There are, however, some notable exceptions to this general trend. The relative abundances of He, C and O decrease with increasing energy well outside the measurement uncertainties. Also, the odd-Z elements show more scatter than their even-Z neighbors, but this may be a reflection of experimental difficulties in making the measurements.

A comparison of Figures 2 and 3 shows marked differences between the galactic cosmic ray and local galactic abundances. The deep valley at the light elements and the valley below the iron peak in Figure 2 have both been filled in Figure 3. In addition, the even-Z/odd-Z ratio is much smaller in Figure 3 than in Figure 2. These effects are the result of cosmic ray propagation through the ISM between the source regions and the Earth. During their confinement, the primary cosmic rays (the particles that left the source regions) interact with interstellar H and He and undergo nuclear fragmentation reactions, such as $^{12}C + p \rightarrow {}^{10}B + 2p + n$. These produce secondary nuclei at approximately the same velocity as the primary particle, and what is observed at Earth is a mixture of primary and secondary components. Nuclear fragmentation would be expected to take the peaks in Figure 2 and move some of the particles into the valleys. In the large valleys such as the light elements and some of the sub-Fe elements, the cosmic rays observed at Earth are effectively all secondaries, and the relative abundances and energy spectra of these secondary elements can be used to provide information on the conditions of cosmic ray confinement and propagation in the galaxy.

There are several peculiarities in Figure 3 that are not explainable by propagation in the galaxy and indicate a fundamental difference between cosmic rays and local galactic matter. The ratio of carbon to oxygen in the cosmic rays is approximately unity rather than a value of 0.6 as observed in the solar system, and sulphur appears to be underabundant by a factor of two in the cosmic rays compared to the local galactic abundances. In addition, neon appears underabundant in cosmic rays, although the element neon is very difficult to interpret in solar system materials, and, consequently, this element is assigned a large uncertainty in the local galactic abundance compilation. These peculiarities in the cosmic ray

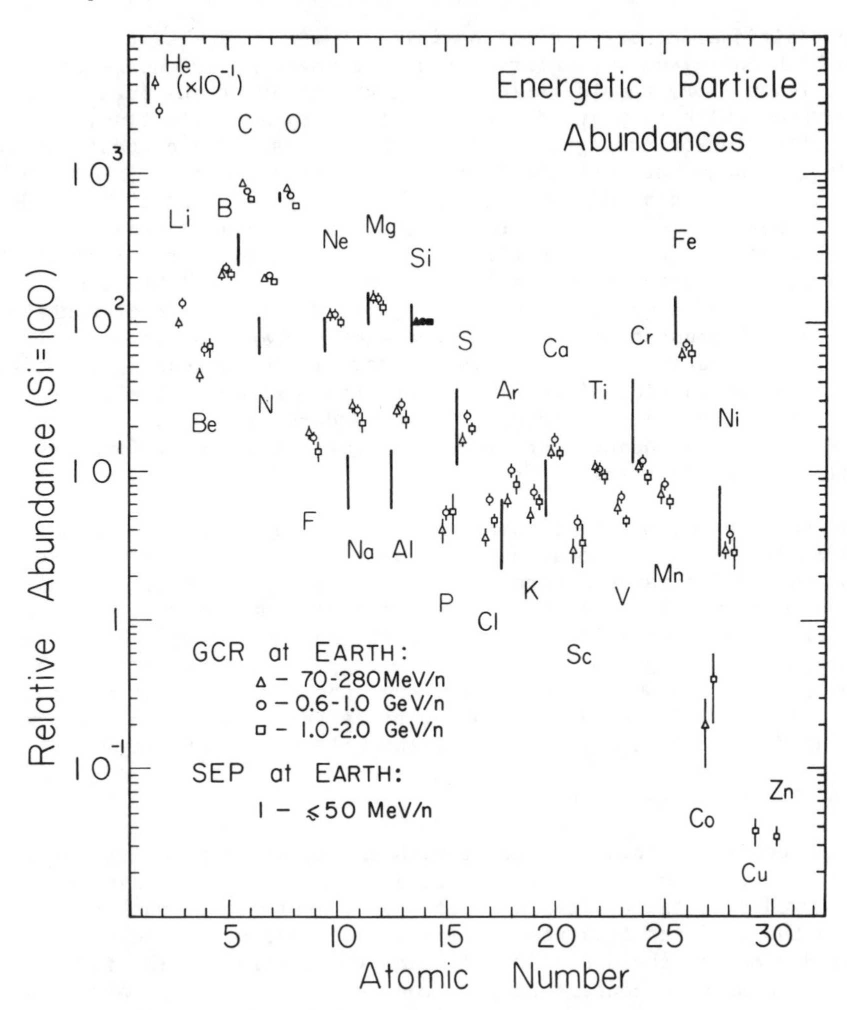

Figure 3. A summary of relative abundances for the elements helium to zinc measured at Earth. The galactic cosmic ray (GCR) abundances are shown for three energy intervals and are compared to abundances observed among solar energetic particles (SEP).

composition may indicate a difference in the composition of the source material for cosmic rays, or they may be due to chemical selection prior to the particle acceleration process. There may well be other differences between cosmic rays and solar system abundances, but to study these it is necessary to unfold the secondary particle contributions and extrapolate the abundances back to the source regions.

2.3. Solar Energetic Particles

An additional comparison is possible. Our sun accelerates particles to high energy during solar flares, and these solar energetic particles (SEP) should reflect the composition of the regions on the sun from which they are derived, modified by any selection, acceleration or propagation introduced biases. The composition of SEP's is somewhat variable from flare to flare, but it is possible to extract an "unbiased" composition from the large body of flare data (e.g. Meyer, 1985a and references therein). These unbiased SEP abundances are plotted in Figure 3 as solid bars to the left of the GCR points for the elements for which there are reliable data. Note that for the secondary elements (Li, Be, B, the sub-Fe elements) there are no SEP abundances shown. These elements are not observed in SEP's, indicating that there is no appreciable nuclear fragmentation between the particles' release at the flare site and their observation at Earth. This is consistent with the odd-Z elements, for which the SEP relative abundances are generally well below those found among the galactic cosmic rays.

The similarity between SEP and cosmic ray abundances is strongest for the even-Z, mainly primary, elements. Note that carbon observed in SEP's favors the local galactic value, but the SEP result for neon and sulphur shows an underabundance. This implies that the "peculiarity" at Ne and S in GCR is probably due to an effect of the acceleration process rather than to the source matter. Note, however, the possibly large abundance of Cr among SEP's, an effect which is not observed in the GCR abundances. Thus, the comparison of SEP and GCR data can provide insights into the processes responsible for both solar flare particle emission and the galactic cosmic rays.

2.4. The Elements Beyond Zinc

The preceding section has dealt with the first third of the periodic table, but, as mentioned in the first section, cosmic rays are observed over the entire periodic table. For atomic numbers greater than those of the iron peak elements, the absolute abundances drop rapidly making these $Z > 30$ (UH) nuclei quite rare. The first experiments to observe contemporary UH cosmic rays flew large area (many m^2) balloon packages composed of nuclear emulsion and etchable plastic detector sheets. These experiments confirmed the existence of UH nuclei up to the actinide region and obtained the first rough charge spectrum. Techniques for studying UH particles evolved rapidly, and culminated in two satellite experiments, HEAO-C3 and Ariel-VI, both of which achieved a long exposure above the Earth's atmosphere.

The results from these satellite experiments are compared to the solar system abundances (solid histogram) for Z ⩾ 30 in Figure 4. Note that between iron and the plateau just above iron (32 ⩽ Z ⩽ 40) the abundance has dropped by a factor of 10^4 and declines by another 2-3 orders of magnitude over the rest of the spectrum. The highest statistics data, therefore, are available in the 32 ⩽ Z ⩽ 40 and the 50 ⩽ Z ⩽ 60 charge ranges. The overall agreement among the two satellite experiments is reasonably good with Ariel-VI observing perhaps somewhat more UH events, relative to iron, than HEAO-C3. The charge resolution displayed by the HEAO instrument is, however, superior to that of the Ariel-VI apparatus. Given the limited statistics and the resolution, only even-Z (or even-Z plus neighboring odd-Z) abundances have been determined reliably.

The striking feature of Figure 4 is the relative lack of structure in the charge spectrum. The peaks at Z = 50-56 and Z = 76-82 are observed in the cosmic rays, but the valleys below these peaks appear to have been filled completely. These massive nuclei have a larger interaction cross section than iron, so more nuclear fragmentation is expected, but the observed abundances of secondaries still appear larger than can be explained easily by propagation effects.

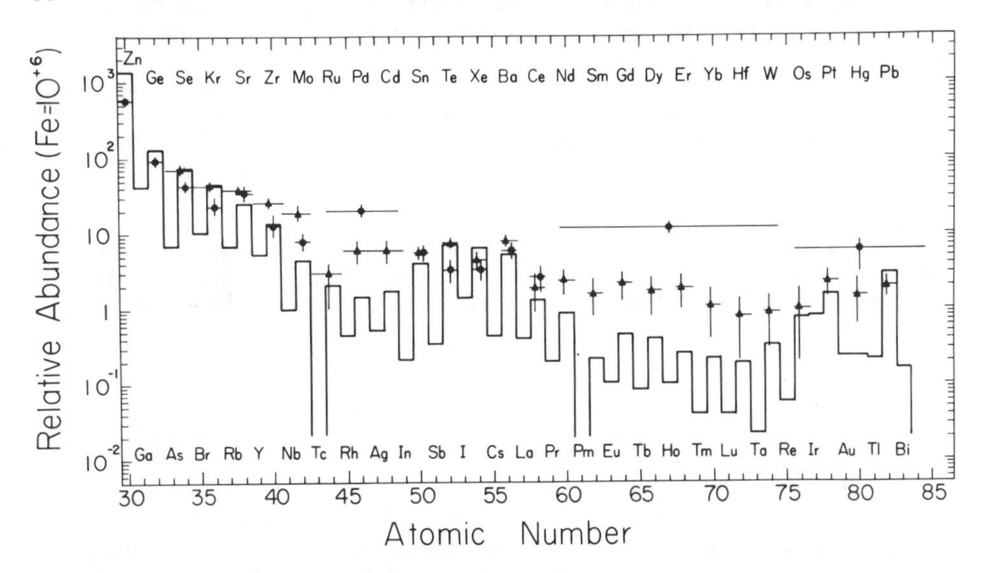

Figure 4. Measured abundances of UH cosmic rays from HEAO-C3; filled circles: Binns et al., 1981; 1983; 1984 and from Ariel-VI; filled triangles: Fowler et al., 1981; 1985a; 1985b (see also Simpson, 1983), compared to solar system abundances (Cameron, 1982) shown as the histogram. The lateral bars indicate the element range included in the points.

There are some interesting features in Figure 4 such as the underabundance of Zn and the apparent overabundances at Sr, Zr and Mo. The somewhat low Te and Xe abundance relative to Ba and Ce may be a statistical fluctuation, but could indicate a nucleosynthesis effect. The goal here is to understand these effects in terms of contributions from propagation, acceleration and/or neutron capture nucleosynthesis in a manner consistent with the more abundant and better studied nuclei at $Z < 30$. In this regard, the UH nuclei provide, almost, a new dimension for cosmic ray studies.

2.5. Electrons, Positrons and Anti-nuclei

In addition to the nucleonic component, the cosmic radiation observed at Earth contains electrons (~1%), both primary and secondary, and positrons believed to be predominantly of secondary origin. While the nucleonic component interacts strongly with the matter in the galaxy, the electronic component also experiences losses due to interactions with the magnetic field (synchrotron emission) and with the photon fields (inverse compton emission) in the confinement region. Thus, the electrons provide information on particle confinement and transport which is complementary to that derived from the nucleonic component (e.g. Webber, 1983).

The positrons in the cosmic rays are derived from proton-ISM interactions which produce π^+ mesons that subsequently decay to positrons. Once the pion production spectrum is known from accelerator studies, the positron source spectrum can be calculated. In like manner, we can calculate the expected production of other anti-matter, in particular anti-protons, in high energy interactions. A predicted ratio for a leaky box model (discussed in Section 4.2) with a mean pathlength of 7 g/cm^2 gives $\bar{p}/p \sim 2 \times 10^{-4}$ at 10 GeV, and this ratio decreases rapidly with decreasing energy. Anti-protons have been observed in the cosmic rays, and between 3-12 GeV the $\bar{p}/p$ ratio is found to be $5\text{-}7 \times 10^{-4}$ (Golden et al., 1979). This high abundance calls into question the propagation models used in the calculation or suggests the presence of primary anti-protons. No anti-helium or heavier anti-nucleus has been observed to a level of 10^{-4} - 10^{-5}.

An even more surprising result was the report of $\bar{p}/p \sim 2 \times 10^{-4}$ at 150-300 MeV, well below the cut-off energy for secondary anti-proton production (Buffington et al., 1981). This observation is still unconfirmed and may be due to an instrumental problem. However, should this low energy flux of anti-protons be substantiated by new experiments planned in the next few years, then the question of a primary origin for anti-protons will have to be investigated anew (see Ahlen et al., 1982).

2.6. Isotopic Composition

Isotopic, as opposed to elemental, abundances can provide additional information on the composition of the cosmic radiation. Isotopic ratios are, often, more accurately known than elemental ratios, and isotopic ratios can be directly related to the processes of nucleosynthesis responsible for the isotopes. Improvements in technology in the 1970's now make it possible to measure the isotopic composition of all elements up through the iron peak, provided a sufficiently long exposure time can be achieved. To date, detailed isotopic ratios are available for only a limited number of elements, over a limited energy range, due to the lack of opportunities for long space exposures for the instruments.

The isotopes in the cosmic rays can be divided into those that are mainly secondary in origin, those that are principally primary and those that are mixed, having both a primary source component and a substantial secondary component produced by nuclear interactions. Table I gives an approximate classification of the isotopes of the first thirty elements. Even those elements listed as primary may contain a small component of secondary nuclei which must be unfolded to obtain accurate source abundance ratios.

The secondary isotopes are ones for which the solar system abundance is essentially negligable in comparison to the abundance observed at the Earth. These species provide information on the propagation process and are employed to determine the propagation model used to unfold the secondary components from the measured distribution.

The isotopes labeled with an asterick (*) on Table I are naturally radioactive, decaying via beta particle emission or electron capture processes. These latter species can be particularly important since, at high energy, these isotopes will be completely stripped of electrons and, therefore, will be effectively stable in the cosmic rays. As the energy decreases, the probability of electron attachment to the nucleus increases and consequently so does the probability of decay by the electron capture process. Thus, the energy dependence of the abundance of these radioactive isotopes can provide a measure of the density of material traversed by the cosmic rays.

Secondary isotopes decaying by beta emission provide a method for determining the average confinement time of the cosmic rays or the average density in the confinement region. Figure 5 shows the status of isotopic data for two of these isotopes: ^{10}Be with a half-life of ~1.5 million years and ^{26}Al with a half life of ~0.7 million years. The parameter labeling the curves is the mean matter density in the confinement region, and the results displayed in Figure 5 are consistent with a mean matter density of ~0.2 atom/cm^3. For relativistic particles, this corresponds to a mean confinement time of about 10-20 million years (Garcia-Munoz et al., 1977). One should note that the calculated curves in Figure 5 involve the nuclear excitation functions (production cross section as a function of energy) for these isotopes but do not include the uncertainties in the nuclear physics data (in some cases, still rather large).

The conclusion to be drawn from Figure 5 is that the cosmic rays are relatively young, compared to the material of our solar system, and therefore provide a sample of <u>recent</u> galactic matter for study. The mean density indicated in Figure 5 is substantially below the average density for the disk of our galaxy. This implies that cosmic rays spend a considerable part of their lifetime either in the low

TABLE I. Classification of the Cosmic Ray Isotopes

Element	Primary	Secondary	Mixed
Hydrogen	^{1}H	^{2}H	
Helium	^{4}He	^{3}He	
Lithium		^{6}Li, ^{7}Li	
Beryllium		^{7}Be*, ^{9}Be, ^{10}Be*	
Boron		^{10}B, ^{11}B	
Carbon	^{12}C	^{14}C*	^{13}C
Nitrogen		^{15}N	^{14}N
Oxygen	^{16}O	^{17}O	^{18}O
Fluorine		^{19}F	
Neon	$^{20,22}Ne$	^{21}Ne	
Sodium			^{23}Na
Magnesium	^{24}Mg, ^{25}Mg, ^{26}Mg		
Aluminum		^{26}Al*	^{27}Al
Silicon	^{28}Si, ^{29}Si, ^{30}Si		
Phosphorous		^{31}P	
Sulphur	^{32}S, ^{34}S	^{33}S	^{36}S
Chlorine		^{36}Cl*	^{35}Cl, ^{37}Cl
Argon	^{36}Ar, ^{38}Ar	^{37}Ar*, ^{40}Ar	
Potassium		^{40}K*	^{39}K, ^{41}K
Calcium	^{40}Ca, ^{44}Ca	^{41}Ca*, ^{43}Ca, ^{46}Ca	^{42}Ca, ^{48}Ca
Scandium		^{45}Sc	
Titanium		^{44}Ti*, ^{46}Ti, ^{47}Ti, ^{49}Ti, ^{50}Ti	^{48}Ti
Vanadium		^{49}V*, ^{50}V*, ^{51}V	
Chromium		^{50}Cr, ^{51}Cr*, ^{53}Cr, ^{54}Cr	^{52}Cr
Manganese		^{53}Mn*, ^{54}Mn*	^{55}Mn
Iron	^{56}Fe, ^{57}Fe, ^{58}Fe	^{55}Fe*, ^{60}Fe*	^{54}Fe
Cobalt	^{59}Co	^{57}Co*	
Nickel	^{58}Ni, ^{60}Ni, ^{62}Ni, ^{64}Ni	^{56}Ni*, ^{59}Ni*	^{61}Ni
Copper			^{63}Cu, ^{65}Cu
Zinc	^{64}Zn, ^{66}Zn, ^{68}Zn	^{67}Zn	^{70}Zn

*Long-lived Radioactive Isotope

density, hot portions of the ISM or beyond the galactic disk in a low density galactic halo. Refining such conclusions involves three steps. First, the nuclear excitation functions which are important

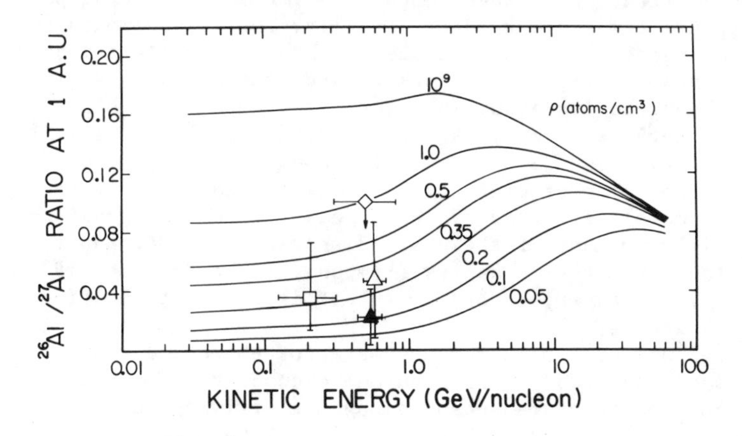

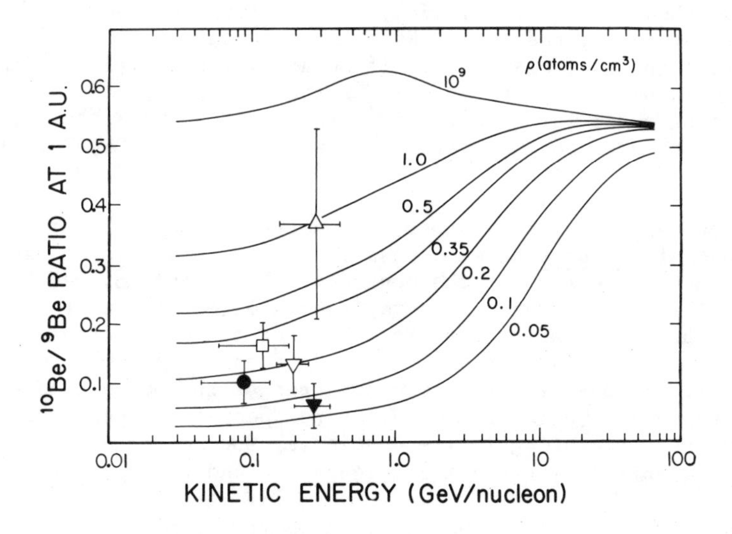

Figure 5. Comparison of experimental measurements of the radioactive isotopes ^{26}Al and ^{10}Be to predictions calculated for different mean densities in the confinement region (Guzik et al., 1985). The data points for ^{26}Al are; open square: Wiedenbeck (1983); filled triangle: Webber et al. (1985); open triangle: Webber (1982); open diamond: Young et al. (1979). The data points for ^{10}Be are; closed circle: Garcia-Munoz et al. (1981a); open square: Wiedenbeck and Greiner (1980); open inverted triangle: Webber et al. (1977); open triangle: Hagen et al. (1977); filled inverted triangle: Webber and Kish (1979).

for calculating the curves in Figure 5 must be measured experimentally. Second, the uncertainties in the cosmic ray measurements must be improved and, since these uncertainties are mainly due to statistical limitations, experiments with larger collecting power and longer exposure times are needed. Finally, the measurements must be extended to higher energies to trace out experimentally the shape of the curves shown in Figure 5. Work towards these goals is in progress at several laboratories around the world.

Considering the primary isotopes listed in Table I, significant measurements have been made for the isotopes of H, He, C, O, Ne, Mg and Si and pioneering studies have been performed for heavier isotopes. A complete review of these measurements is beyond the scope of this article -- many of these are discussed in other papers in this volume or in recent review papers (Simpson, 1983; Mewaldt, 1983; Wiedenbeck, 1984; Meyer, 1985b) -- but it is useful to mention the significant results.

The most striking measurements were for the element neon, which has three isotopes, two primary and one secondary. The heavy isotope ^{22}Ne was shown to be overabundant among the arriving cosmic rays (Fisher et al., 1976; Garcia-Munoz et al., 1979a; Mewaldt et al., 1980; Wiedenbeck and Greiner, 1981). Correcting for the small secondary component, the ^{22}Ne/^{20}Ne ratio in cosmic ray source matter is 3-4 times the equivalent ratio in solar system material. Smaller excesses have been found for ^{25}Mg/^{24}Mg and ^{26}Mg/^{24}Mg, and there was an indication of a similar effect for ^{29}Si/^{28}Si (but see Webber et al., 1985). The results for heavier elements (e.g. S, Ar, Ca, Fe, Ni) are still statistically limited and no conclusions can be drawn for these elements.

The current results, however, demonstrate that there are significant <u>differences</u> between cosmic ray and solar system material. The pattern of excesses for the neutron rich isotopes of Ne, Mg and possibly Si show that the cosmic ray source material has its own special composition and/or history. Whether these differences are due to evolution of the ISM between the time of solar system formation and the present, or to a mixture between matter of the ISM and material from other sources (c.f. Figure 1), or to specialized processes of nucleosynthesis operating in the cosmic ray source regions remains a major experimental/theoretical challenge for the future.

3. THE COSMIC RAY ENERGY SPECTRUM AND ANISOTROPY

In addition to composition, another distinguishing feature of the galactic cosmic radiation is its energy spectrum, which extends over 14 orders of magnitude and can be divided into four characteristic regions:

(a) below several GeV/nucleon, where solar modulation processes are important,
(b) from several GeV/nucleon to ~10^6 GeV/particle, where a single power law dominates,

(c) between 10^6 and $\sim 10^{10}$ GeV/particle, where a steeper spectrum is evident, and
(d) above 10^{10} GeV/particle, a region most likely dominated by extragalactic cosmic rays.

Figure 6 shows a compilation of measurements of the differential energy spectrum for cosmic ray protons, extending over six decades in energy from 0.01 - 10^4 GeV/nucleon. Over the most of this range the spectrum has the form of a power law in energy with an index of ~2.7. The highest energy points shown are the emulsion chamber results of Burnett et al. (1983a), which demonstrate that the power law extends to beyond 10^4 GeV. Below several GeV/nucleon the spectrum rolls over, peaking at several hundred MeV/nucleon and decreasing to lower energies. These features appear to be common to the spectra measured for all of the primary elements, although the experimental coverage in energy space is most complete for protons.

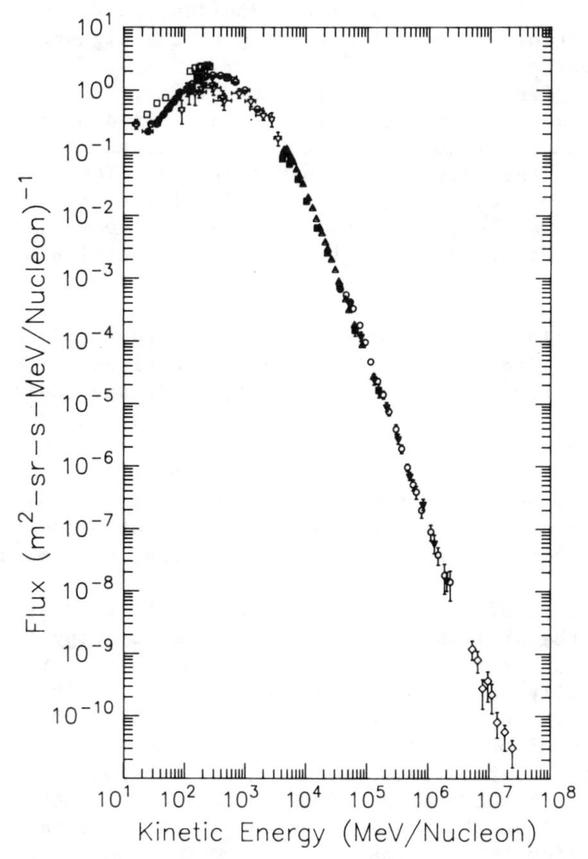

Figure 6. The differential energy spectrum for galactic cosmic ray protons compiled from measurements reported in the literature.

The observed turn-over in the differential energy spectrum below several GeV/nucleon is understandable as a combination of the intrinsic spectral parameter (e.g. total energy, momentum, rigidity) and the influence of the solar wind (solar modulation). The cosmic ray flux obtained below several GeV/nucleon is variable and depend upon the phase of the solar cycle. The largest flux is obtained during periods of solar minimum conditions (solar quiet times) decreasing to a minimum flux at solar maximum (solar active times). Figure 6 shows data from solar minimum conditions.

At energies above those shown in Figure 6, current instruments for measuring individual elements do not have sufficient collecting power. Instead, investigations must utilize the Earth's atmosphere as a target and measure the products produced in the resulting air showers. Parameters such as the shower size, total number of particles, light emitted from the atmosphere or hadron/lepton abundances are studied in ground-based arrays or atmospheric optical systems to determine the total energy of the incident particles. This results in an "all-particle" spectrum extending to the highest energies, $\sim 10^{11}$ GeV per particle.

Figure 7, adapted from Fichtel and Linsley (1986), presents a summary of the measurements. Here the flux is multiplied by $E^{2.5}$, in order to reduce the number of decades to be displayed, and the data are plotted in terms of the total energy per particle. The lower energy portion of Figure 7 overlaps, for several decades, the high energy part of Figure 6 and shows the tail of the $E^{-2.7}$ power law. The obvious feature of Figure 7 is the break in the power-law spectrum at $\sim 10^6$ GeV with the spectrum continuing, with a much steeper slope, up to $\sim 10^{10}$ GeV where an apparent flattening occurs for the highest energy events. The region around 10^6 GeV, often called the "knee", may be due to a propagation effect related to the confinement conditions for these ultrahigh energy particles, or it may indicate a transition between one source/acceleration mechanism and another. Above $\sim 10^{10}$ GeV, the particles are usually attributed to an extra-galactic source, but below this energy they are probably of galactic origin (Hillas, 1984). Note that the region around the knee may not show a smooth transition but rather an increased flux at a few times 10^6 GeV per particle, and this region needs to be studied in more detail.

An additional source of information is provided by measurements of the anisotropy of the arriving particles, where anisotropy is defined as the deviation from an isotropic distribution of arrival directions and is usually analyzed by binning the data in right ascension or in galactic coordinates. Below $\sim 10^6$ GeV allowance must be made for the motion of our solar system through the galaxy.

The anisotropy is observed to be small ($\sim 0.07\%$) below 10^4 GeV, gradually increasing with increasing energy to beyond 10^{10} GeV, with an apparent change in slope at about the region of the spectral "knee". This correlation between the energy spectrum and the amplitude of the anisotropy has been interpreted as evidence that confinement or "leakage" from the galaxy is responsible for the

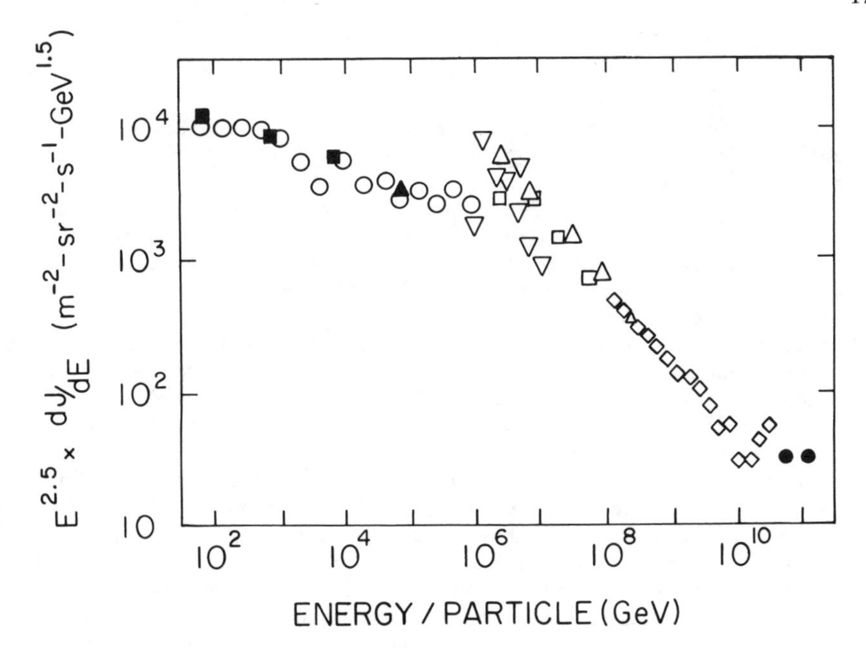

Figure 7. All particle energy spectrum for very high energy cosmic rays, adapted from Fichtel and Linsley (1986). The ordinate is multiplied by $E^{2.5}$ and the abscissa is the total energy per particle. Data are from; filled squares: Ryan et al. (1972), Simon et al. (1980), Burnett et al. (1983a); filled triangle: Burnett et al. (1983b); open circle: Grigorov et al. (1971); inverted open triangle: Khristiansen et al. (1983); open square: Efimov and Sokurov (1983); open triangle: LaPointe et al. (1968); open diamond: Cunningham et al. (1980); filled circle: Bower et al. (1983).

knee (Hillas, 1984); however, alternative explanations have also been given (e.g. Watson, 1984).

The direction of the observed anisotropy also provides some surprising results. Analyzed in galactic coordinates, the particles below about 10^8 GeV tend to be consistent with arrival from the galactic plane. Above this energy, however, there appears to be an excess of particles arriving from the southern galactic hemisphere. This effect changes abruptly to a northern excess at ~10^{10} GeV. There is general agreement that above 10^{10} GeV the particles are extragalactic, coming from the direction of the Virgo supercluster, but below this energy there are several possible interpretations of the results.

4. ACCELERATION, CONFINEMENT AND PROPAGATION

A complete model of the cosmic radiation must explain the acceleration, confinement and transport to Earth of the particles from the four major portions of the energy spectrum discussed in Section 3. Such a model must be consistent with the experimental constraints imposed by the measured composition (both primary and secondary), energy spectrum and anisotropy of the particles, and it must indicate a source of high energy particles of sufficient power to maintain the cosmic ray energy density approximately constant for at least the last several billion years. This is a large order! While much progress has been made in recent years in understanding various parts of the problem, we are still far from having this complete model.

One difficulty is that the questions of acceleration, confinement and propagation are coupled. In the past, models for these processes have tended toward the assumption either of "weak" coupling, in which the particles are accelerated in specific astrophysical objects and subsequently injected into the confinement/propagation region, or of "strong" coupling, in which particles are continuously accelerated throughout the confinement/propagation region (Lingenfelter, 1979). Transitions between these extreme views were prompted by new experimental observations and by more refined theoretical analyses of the effects/problems. Current thinking in this area, however, seems to be converging, at least for the bulk of the cosmic rays below 10^6 GeV, toward models in which both discrete source acceleration and continuous acceleration play a role. This is due in part to the realization that shock waves are efficient particle accelerators (e.g. Axford, 1981; Achterberg, 1984) and in part to the realization that the ISM is a very turbulent environment driven to a large degree by the action of supernovae and their shock waves (e.g. McCray and Snow, 1979). Studying the implications of these two ideas for the acceleration and propagation of cosmic rays is, today, one of the most active areas of investigation in particle astrophysics.

4.1. Acceleration Processes

Particle acceleration is commonplace in astrophysical environments ranging from the Earth's magnetosphere, to interplanetary space, to the sun and other stellar systems, to the magnetospheres of other planets (Jupiter, Saturn), pulsars and neutron stars, to supernova remnants and to radio galaxies and active galactic nuclei. Observations of particle acceleration are made directly (in situ by spacecraft) within our solar system (e.g. Scholer, 1984 and references therein) and indirectly (through the associated emission of radio, optical, x-ray or γ-ray photons) for objects/regions beyond the solar system (e.g. Bignami, 1986; Trumper, 1986; Ramaty and Lingenfelter, 1982; Weiler, 1987).

There are many mechanisms which can accelerate particles, including mass motions, electric fields, magnetic field irregularities or the interaction of particles with electromagnetic waves, all of which can be observed within the solar system. Mass motions produce adiabatic energy changes in particles in the Earth's magnetosphere or trapped in the expanding solar wind. Impulsive acceleration by magnetic field reconnection is suggested for some solar flares and is observed in the Earth's geomagnetic tail. Induced electric fields accelerate particles rapidly in the auroral regions of the magnetosphere, and more gradual acceleration due to wave-particle interactions is observed throughout planetary magnetospheres. Acceleration due to magnetic fields in shock waves is observed at the bow shock of planetary magnetospheres, on the sun and in interplanetary space. Many of these processes find counterparts elsewhere in the galaxy, and may be involved, to one degree or another, in the acceleration of the galactic cosmic rays.

Fermi (1949) was the first to propose a model for the acceleration of GCR based on particle scattering in a turbulent, magnetized conducting medium. With this model, he showed that acceleration was quite efficient for ordered motion of the scattering centers (i.e. particles trapped between converging centers -- first order Fermi acceleration -- see below) and that acceleration was also possible, though less efficient, for random motion of the scatters (second order Fermi acceleration, often called "stochastic" acceleration). Although Stochastic acceleration may be realized in the ISM through large scale mass motions combined with magnetohydrodynamic waves which can interact with the particles, it is difficult to evaluate its relative importance since the degree of turbulence (on both small and large scales) and the spectrum of waves are unknown. In the ISM stochastic acceleration generally involves a long time scale, probably longer than the 10-20 million year residence time of the cosmic rays, but in more localized objects the acceleration time can be much shorter. Thus, stochastic acceleration is probably not the major mechanism responsible for the GCR, but it may be important in other environments, for example in maintaining the high energy electron population in supernova remnants (Achterberg, 1984).

First order Fermi acceleration, however, is important to GCR through its realization in particle acceleration by shock waves in a scattering medium. Charged particles are trapped near the shock front by scattering on irregularities in the magnetic field and move back and forth across the shock boundary. Due to the difference in velocity of the scattering centers on each side of the shock front, the particles observe converging scattering centers upon each passage through the shock and are accelerated by the first order process. Such shock acceleration is likely to occur in the expanding envelope following a supernova explosion. These SN shocks are quite strong (high compression ratios) in the early stages and gradually become weaker as the supernova remnant expands. For a supernova rate of one per 30 years, only ~1% of the supernova energy need be converted to particle energy in order to maintain the cosmic ray energy density. Thus, SN shocks provide a means for "tapping" the reservoir of energy available in supernovae.

An earlier suggestion to tap the energy of supernovae for GCR, reviewed recently by Colgate (1984), involves accelerating matter in the supernova explosion itself. The explosion expells most of the mass of the pre-supernova star with a small fraction of the mass reaching extremely high velocities. The main difficulty here is that this matter may lose much of its energy in the expansion and subsequent formation of the supernova remnant.

Supernovae are not the only sources of shock waves in the ISM. Novae, expanding HII regions, and stellar winds, especially those connected with O and B or Wolf-Rayet stars (Casse and Paul, 1982), all produce shock waves. Although less energy is available in these shocks, compared to SN shocks, nevertheless, they may play an important role in the acceleration of charged particles in particular regions of the galaxy.

Shock wave acceleration appears, currently, to be the most promising mechanism for the acceleration of GCR, but can it meet the constraints of the experimental data? Answering this question involves constructing detailed models of the astrophysical environment in which the shock waves exist and investigating the resulting acceleration. One remarkable feature of most of the models is that shock acceleration produces a single power-law energy spectrum covering many decades in energy between the minimum effective energy, E_{min}, and E_{max}, the upper limit to the energy obtainable from the model, determined by the relative velocity of the scattering centers, the scattering mean free path and the time available for the acceleration. Taking optimistic parameters for a typical SN shock, Lagage and Cesarsky (1983) have shown that E_{max} is $< 10^5$ GeV. For stellar wind shocks containing large terminal velocities and a large magnetic field strength, the upper limit can be extended by about an order of magnitude. Thus, shock acceleration in galactic environments can provide a single power-law energy spectrum, as observed for the bulk of the cosmic rays, up to about the knee in the energy spectrum.

Shock acceleration requires the existence of suprathermal particles above some minimum energy, E_{min}, before the acceleration becomes effective, thus requiring some form of "injection." In addition, the particles probably suffer some energy loss (e.g. by ionization) between shock crossings. In general then, E_{min} must be several times the downstream thermal energy, and as E_{min} increases, the ionization losses decrease and the total time for acceleration to relativistic energy also decreases. The "injection" mechanism must also supply sufficient suprathermal particles to counteract losses (e.g. escape or nuclear interaction) and must supply the correct mixture of elements to reproduce the cosmic ray composition.

Another question for shock acceleration theories is whether the cosmic rays gain most of their energy in a single encounter with a strong shock wave or through many repeated encounters with weaker shocks over their lifetime. The latter case involves acceleration intermixed with propagation and is constrained by the energy dependence of the measured secondary to primary ratios which (as discussed in Sec. 4.3) decrease with increasing energy above a few GeV/nucleon. This leads to the conclusion that the GCR's probably gain most of their energy during a single acceleration event and do not encounter repeated strong acceleration during their lifetime

(Blandford and Ostriker, 1980; Cowsik, 1980; Fransson and Epstein, 1980; Eichler, 1980; but see also Lerche and Schlickeiser, 1985). Nevertheless, it would seem likely that cosmic rays encounter a number of weak shock waves during their propagation through the ISM. This may lead to a small amount of distributed acceleration which, as suggested by Silberberg et al. (1983a), may have consequences for the energy dependence of element and isotope ratios observed at Earth.

The SN shock waves, which may be responsible for the acceleration of the lower energy cosmic rays, do not appear to account for the highest energy particles, those above the knee in the spectrum. At these high energies attention has been focused mainly upon discrete sources such as pulsars (e.g. Michel, 1984; Arons, 1981) or binary systems containing a magnetized neutron-star or black hole which is undergoing accretion of matter from its companion (e.g. Kundt, 1984). In these environments, strong electric fields are generated and plasma/field instabilities can develop that result in the acceleration of high energy particles. Such processes may have been observed with the discovery of a number of objects, such as Cygnus X-3, (Samorski and Stamm, 1983; see also Turver, 1985; Hillas, 1985) emitting ultrahigh energy γ-rays (10^3 - 10^7 GeV). The high photon energies observed imply the existence of even higher energy charged particles within these systems, and only a small "leakage" of particles from a number of such sources would be required to maintain the cosmic ray density between 10^6 - 10^{10} GeV.

At still higher energies, above $\sim 10^{10}$ GeV, acceleration becomes more difficult to understand because energy loss processes involving photon and magnetic fields become quite severe. Models for these highest energy particles (believed to be extragalactic) have generally focussed on active galactic nuclei or the extended lobes of radio galaxies (Hillas, 1984), but no fully convincing model has yet appeared, due in part to the lack of experimental constraints on the astrophysical environments or the expected particle spectrum.

4.2. Particle Confinement

The fundamental questions in this area are "where" and "how". We know from γ-ray observations and radio synchrotron studies that cosmic rays pervade the disk of our galaxy (e.g. Hermsen and Bloemen, 1984), but are these particles also confined for some part of their lifetime to regions around the source/acceleration sites and/or do they spend a significant amount of time in a low density halo around the galaxy? In this regard, the possibility of the existence of a galactic wind, constantly expelling gas from the galaxy (Jokipii, 1976), and its convective effect on the cosmic rays must be considered.

The basic interactions of a high energy cosmic ray "gas" with the ISM involve energy loss and nuclear spallation in the matter of the ISM (the cosmic ray propagation problem discussed in the next section) and scattering by magnetic field irregularities. The latter interaction limits the cosmic ray streaming velocity and provides the means to confine the particles in the galaxy. In the case of very strong scattering, the cosmic rays can be viewed as diffusing rather than streaming, and diffusion models have often been applied to the problem. The magnetic field irregularities (or waves) responsible for scattering the particle can come from a number of sources: the cosmic rays themselves can generate magnetohydrodynamic waves and be scattered by them, while other turbulent motions (SN shocks, stellar winds, cloud-cloud collisions, etc.) also generate a spectrum of waves which interact with the particles. In addition, the waves interact with the ambient medium, so the basic process involves the interchange of energy between waves and particles and between mass motions and waves in the ISM. A critical review of these processes has been provided by Cesarsky (1980).

There have been a number of "models" proposed to describe the confinement and propagation of the GCR. Among them are:

(i) Leaky-box Model (Cowsik et al., 1968; Gloeckler and Jokipii, 1969) -- Cosmic rays are confined in a region with partially reflecting boundaries from which they slowly "leak-out".

(ii) Nested Leaky-box Model (Cowsik and Wilson, 1973; 1975; Simon, 1977) -- There are two confinement regions, one around the sources and the other corresponding to the galaxy. The GCR leak from both regions but at different rates.

(iii) Closed Galaxy Model (Rasmussen and Peters, 1975) -- The boundaries of the confinement region are perfectly reflecting so particles cannot escape. They are destroyed by nuclear interactions and energy loss processes.

(iv) Dynamical Halo Model (Jokipii, 1976) -- Our galaxy has a galactic wind (much like the solar wind) which expels gas at a constant rate. The GCR and magnetic fields are tied to the gas and are, therefore, being convected out of the system.

(v) Diffusion Models (e.g. Ginzburg et al., 1980) -- The cosmic ray particles are tied tightly to the magnetic field lines, which random walk, due to irregular/turbulent motion, to the boundary of the confinement region where the particles can escape. With appropriate boundary conditions, diffusion models reduce to some of the models mentioned above and, in addition, can easily incorporate a galactic halo.

(vi) Combination Models -- These involve combinations of the above models (e.g. Three Tier Model: Stephens, 1981; Multi-Cloud-Model: Silberberg et al., 1983b).

One of the key parameters to be considered in comparing these models, is the nature of the boundary to the confinement region, i.e. do particles escape and, if so, how fast. For galactic confinement and sources in the galactic disk, an equally important question is how the particles are transported and arrive at the boundary of the disk and whether this is really a boundary or just an access point to a larger cosmic ray halo.

Deciding between these models involves a combination of theoretical arguments and experimental data. The secondary components in the cosmic rays (fragmentation products, positrons, anti-protons) are most sensitive to the differences in the models, and measurements of these components already provide some experimental constraints, but more work, especially on the energy dependence of the secondaries, is required before a firm distinction between various models will be possible.

4.3. Propagation

Cosmic ray propagation encompasses the changes induced in the cosmic ray "beam" during confinement in the galaxy (or other regions). These changes are of two types. First, the composition of the GCR is modified due to nuclear interactions in the ISM which fragments heavy nuclei into lighter species and creates secondary particles such as positrons and anti-protons. Second, the particle energies are modified through ionization energy loss in the ISM, synchrotron radiation in the galactic magnetic fields, inverse Compton collisions with the photon fields (starlight, infrared 2.7 K radiation) and, possibly, energy gain due to stochastic acceleration or to passage through interstellar shocks. (The energy loss processes, of course, are responsible for the production of γ-ray and radiosynchrotron emission, and this connection allows the photons to be used as tracers of the GCR spatial distribution.) Propagation, therefore, connects the GCR source/acceleration regions to the energy spectrum and composition that exists in local interstellar space (LIS), just beyond our solar cavity.

The final propagation step, to move from LIS to the orbit of Earth where the actual observations are made, involves considering the effects of the outflowing solar wind on the GCR particles, which must move upstream to reach the Earth. This results in a rigidity dependent change in the particle energy, called adiabatic deceleration (typically of the order of 200 - 500 MeV/nucleon for $A/Z = 2$ particles, varying from solar minimum to solar maximum conditions), which affects both the measured relative composition and energy spectrum. This solar modulation can be described in a spherically symmetric, steady state model (Fisk, 1979) which can be used to unfold the modulation effects for data taken at different times during the solar cycle. Solar modulation is discussed elsewhere in this volume, and a recent review of measurements and model comparisons can be found in Garcia-Munoz et al., 1986.

Propagation studies cannot, in general, deal with the micro-structure of the particle-matter or particle-field interactions but are confined to studying the average process, since what we observe at Earth is the product of the histories of many individual particles. What are measured at Earth are the composition and energy of the particles, and one of the goals of propagation studies is to utilize this data to derive information about the average history of the cosmic rays. In general, this involves assuming a model for cosmic ray confinement and propagation, and the derived information is model dependent. Some of the derived quantities that have typically been discussed include: (i) the mean amount of matter traversed by the particles, (ii) the distribution of pathlengths followed by the particles, (iii) the mean density in the confinement region, (iv) the average energy loss for the different species of particles, (v) the mean confinement time , (vi) the distribuiton of interstellar shock waves encountered, (vii) the loss or "leakage" rate from the confinement volume, (viii) the mean intensity of the photon fields, and (ix) the average magnetic field strength in the confinement region. These average history parameters are, in general, not independent from one another and information on one parameter is obtained by experimentally constraining or making assumptions about the values of other parameters. Propagation is, in addition, an energy dependent process, and the parameters cited above are a function of energy. (For a general review of the leaky-box model applied to cosmic ray propagation to derive average history parameters see Ormes and Freier, 1978).

A realistic propagation calculation must include all of the modes of particle loss and gain and all of the energy changing processes, and the calculation must be performed as a function of energy. The important processes affecting the particle species are nuclear fragmentation reactions and radioactive decay. If all of the stable or long lived radioactive isotopes between ^{4}He and ^{64}Ni are to be considered, the calculations must contain over 80 individual isotopes as well as the cross sections, as a function of energy, for fragmentation of these species. Ideally, one would like to have cross sections for all these reactions from measurements in nuclear physics laboratories, but this is a monumental and, as yet, uncompleted task. Thus, it is necessary to rely upon cross sections determined from semi-empirical formulae (e.g. Silberberg and Tsao, 1973) to predict values for the unmeasured cross sections.

The most important energy changing processes for nuclear species are ionization energy loss and distributed acceleration. For the electrons, however, energy losses by synchrotron and inverse Compton processes must also be included. Schlickeiser (1986) has given a detailed theoretical treatment of the different processes involved and has applied the analysis to several different astrophysical environments. Silberberg et al. (1987) discuss many of the outstanding problems in propagation and the interpretation of the existing cosmic ray database.

As an example of the information that can be obtained from propagation investigations, we consider the cosmic ray pathlength distribution (PLD). The PLD is one of the parameters cited above that can be determined experimentally, and it constrains the different models described in Section 4.2. The path length distribution is the function describing the relative probability for cosmic rays to pass through different amounts of matter before being observed at Earth, and is determined experimentally from measurements of secondary to primary ratios as a function of energy.

Early observations of significant abundances of light elements in the cosmic rays were interpreted in terms of propagation of all particles through a "slab" of interstellar material (e.g. Ginzburg and Syrovatskii, 1964). As the measurements improved, it became clear that a single slab of matter could not reproduce the secondary to primary ratios for both the light elements and the iron peak elements (Shapiro et al., 1970) and that this delta function PLD must be generalized to a continuous function. Various forms were investigated, and an exponential function was found to best reproduce the available data. An exponential PLD results from models in which the cosmic rays are assumed to be in equilibrium with a small, constant probability of leaving the confinement region, and therefore, corresponds to the leaky-box model. Figure 8, part A, shows a sketch of an exponential PLD.

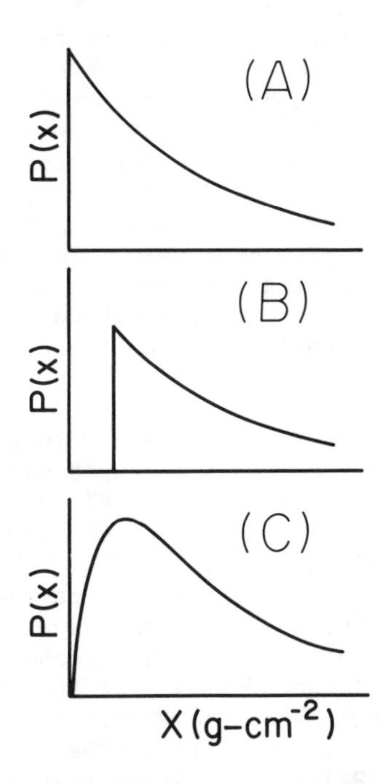

Figure 8. Cosmic ray path-length distributions. P(x) is the probability for a particle to encounter an amount of matter x (g/cm^2) during propagation.

The cosmic ray propagation process is very complex, and the PLD is probably not purely exponential in form. There may well be more than one confinement region involved and, if so, this will alter the shape of the PLD. For example, suppose there were a shell of matter around the source/acceleration regions through which all of the particles must pass before being released into the galaxy (e.g. Garcia-Munoz and Simpson, 1970), then the PLD would have the form shown in Figure 8, part B, with no particles having a possibility of passing through less than some minimum amount of matter. Alternatively, the source regions may represent a confinement volume from which the particles escape with an exponential probability into the larger galaxy from which they also escape with an exponential probability (different from the source region probability). This situation is the nested leaky-box model and yields a PLD such as shown in Figure 8, part C. In addition, the propagation process is energy dependent, so Figure 8 represents the PLD's only at one particular energy, with the distribution being narrower or broader at other energies.

The task, then, is to trace out the shape and the energy dependence of the PLD using secondary to primary ratios. For this analysis, the weighted-slab technique (Fichtel and Reames, 1968) is employed. Starting with a set of relative abundances of elements and isotopes and a common energy spectrum at the source/acceleration regions, the different species are propagated through slabs of interstellar matter, and the results are then integrated over a PLD to obtain the particle densities in LIS. These densities are then transported to the orbit of Earth, using a model of solar modulation, and compared to the experimental data. The procedure is iterated with different PLD's until a good fit to the data is obtained. In this technique, the PLD enters as an empirical parameter which combines the effects due to leakage, convection and the distribution of source/ acceleration sites. Each theoretical model for cosmic ray confinement and propagation (Sec. 4.2) can be used to derive a path length distribution which can be compared to the PLD determined empirically from the secondary to primary ratios.

Figure 9 shows a compilation of data for the secondary to primary ratios B/C and sub-Fe/Fe, extending from the lowest measured energies to just below 100 GeV/nucleon. While there are still significant uncertainties on many of the experimental results, the accumulated data trace the energy dependence of these ratios quite well. The B/C ratio (representative of light/medium secondary to primary ratios -- Li/C, Be/C, F/Ne would suffice as well, but none of these have been as well measured as B/C) has been analyzed in a model with an exponential PLD to determine the energy dependence of the mean, X_o, of the PLD. Correcting for solar modulation and allowing for the errors in the B/C data and the uncertainties in the nuclear fragmentation cross sections, Garcia-Munoz et al., (1979b, 1981b, 1987) demonstrated that the PLD must be energy dependent over the full range of energies covered by the data in Figure 9.

The energy dependence of X_o above ~1 GeV/nucleon was discovered about fifteen years ago (Juliusson et al., 1972; Smith et al., 1973) and was interpreted as a decreasing lifetime against escape from the galaxy with increasing energy. This energy dependence provides a constraint on models for the confinement and propagation of GCR, and was cited above as an argument against repeated accelerations of GCR throughout the galaxy. The new feature from the analysis of the complete energy range is that X_o also decreases with decreasing energy below about 1 GeV/nucleon. This decrease implies that the physical process responsible for the energy dependence at high energies cannot be the controlling mechanism at the lower energies. One possibility is that escape via convection in a galactic wind dominates at low energies. The energy dependence of X_o below 1 GeV/nucleon is consistent with such a dynamical halo model (Garcia-Munoz et al., 1981b).

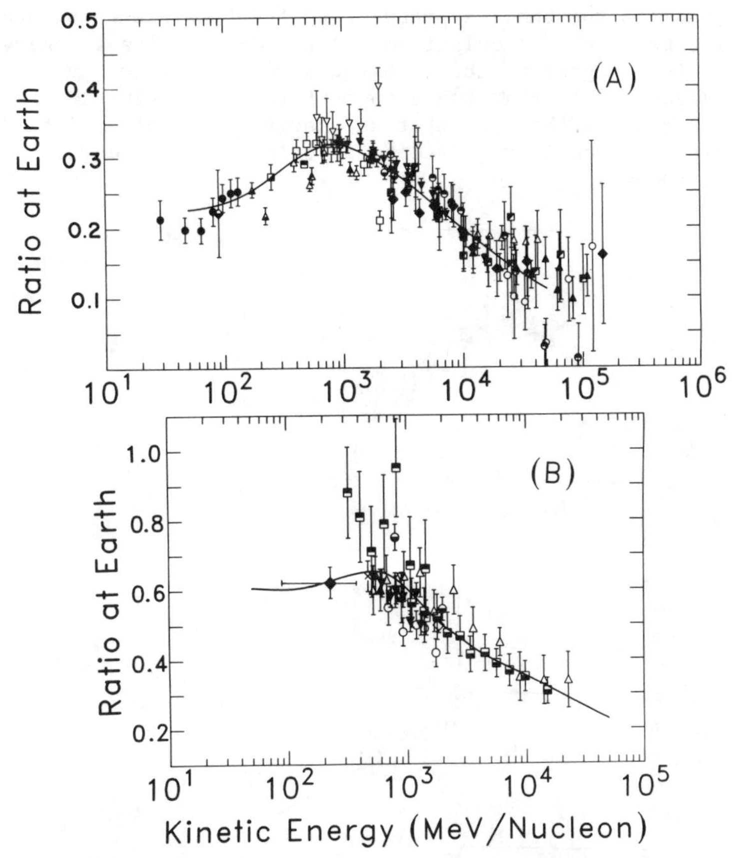

Figure 9. Compiled data at Earth for (A) the B/C ratio and (B) the sub-Fe/Fe ratio as a function of energy (see Garcia-Munoz et al., 1984 for references to the experimental data).

The question of the overall shape of the PLD can only be addressed by comparing several secondary to primary ratios which are sensitive to different regions of the PLD. Iron has a total interaction cross section which is about three times larger than the equivalent cross section for carbon, so the production of iron secondaries (the sub-Fe elements) is sensitive to the short path lengths (1-5 g/cm^2) in the PLD whereas the B/C ratio is determined mainly by the longer paths (5-10 g/cm^2). Thus, by comparing the sub-Fe/Fe and B/C ratios, a measure of the relative amounts of short and long pathlengths is obtained. (This procedure can be extended by using UH secondary to primary ratios to investigate even shorter paths or by using, for example, $^3He/^4He$ to investigate longer paths.) Applying an exponential PLD (Figure 8-A) with an energy dependent X_o to calculate the sub-Fe/Fe ratio yields the curves shown in Figure 10. The energy dependence of X_o selected to reproduce the B/C ratio in Figure 10 predicts too small a production of sub-Fe elements, and the discrepancy between the calculations and the data is itself energy dependent, being largest at the lowest energies. The conclusion to be drawn from Figure 10 is that the exponential PLD contains too large a proportion of short paths, and that an energy dependent depletion of the short path lengths is required to describe the cosmic ray propagation process.

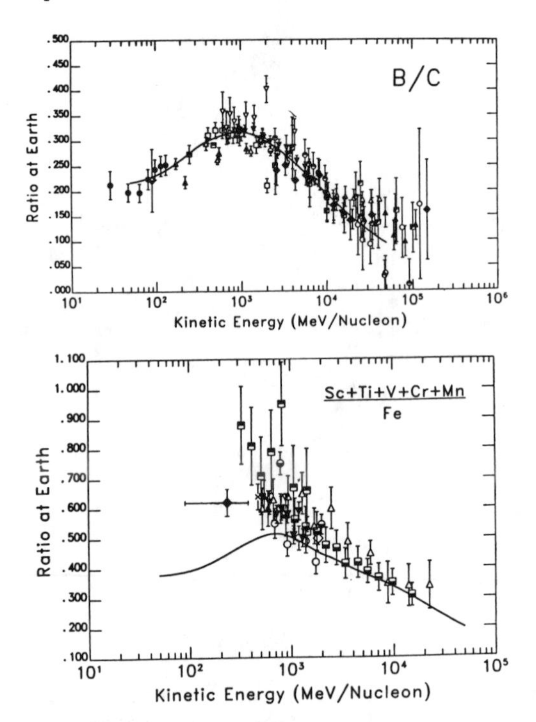

Figure 10. Compiled measurements of the B/C and sub-Fe/Fe ratios compared to propagation calculations using an energy dependent exponential PLD.

The suggestion of a depletion of short path lengths, or truncation of the PLD, is not new (Shapiro and Silberberg, 1970) and has been the subject of considerable controversy between different groups comparing different sets of data covering different energy ranges. The energy dependent truncation suggested by Figure 10 resolves these problems and allows the calculations to reproduce the data over the full range of the measurements, as demonstrated by the curves displayed in Figure 9.

There are many models which can be employed to obtain a depletion of short pathlengths. As an example, consider the PLD of Figure 8-B. This is reproduced as the inset (B) in Figure 11, where the cutoff point is labeled X_c and represents the upper limit to the short path lengths which are completely removed. This PLD is described by two parameters: X_c and X_o, the mean of the exponential part. Requiring the calculations to reproduce, simultaneously, the energy dependence of both the B/C and sub-Fe/Fe ratios determines the energy dependence of these two parameters as shown (along with the mean path length) in part (A) of Figure 11. The energy dependence of X_o in Figure 11 is determined mainly by the B/C data and shows the energy dependence both above and below 1 GeV/nucleon, as discussed previously. The new feature here is the energy dependent X_c parameter, which may be interpreted as a second confinement region (possibly around the sources) from which the particles escape in an energy dependent

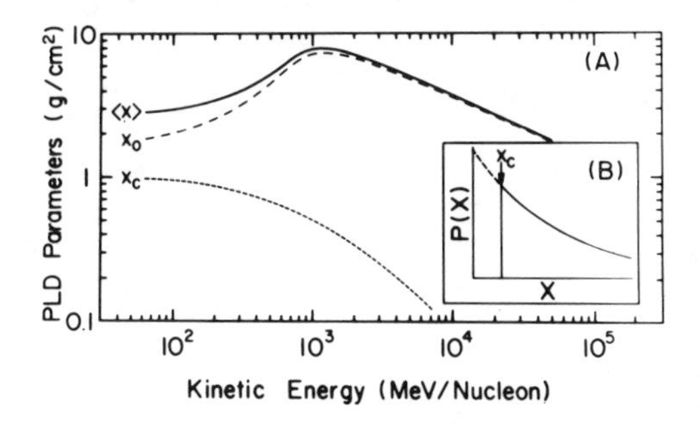

Figure 11. Energy dependence of the parameters (see inset) of the PLD of Figure 8, part B plus the energy dependence of the average path-length encountered by the GCR.

fashion with the lowest energy particles remaining trapped for the longest times.

Two conclusions can be drawn from this discussion. First, the purely exponential PLD as represented by the leaky-box model does not reproduce the full range of the available data and must be superceded by a more complex model. Second, the need for energy dependent truncation in the PLD suggests the existence of two separate confinement regions for the GCR (Guzik and Wefel, 1984), such as are incorporated in the nested leaky-box model. However, this analysis shows that the particles are confined in these regions in an energy dependent manner, and this represents a new constraint which has not been included in previous models for cosmic ray confinement and propagation.

Investigations of cosmic ray propagation, using a fully developed, energy dependent calculation, depend for success upon accurate cosmic ray elemental or isotopic abundance measurements and knowledge of the relevant nuclear physics parameters. For some elements, such as boron, these conditions are satisfied reasonably well, but for other elements and for most of the isotopes, this is not the case. One example is the element helium, which in the cosmic rays consists of both the primary isotope ^{4}He and the secondary isotope ^{3}He. The ^{3}He/^{4}He ratio was one of the first isotopic ratios to be measured (e.g. O'Dell et al., 1965) and remains an important tracer of the propagation history of the lightest cosmic rays. Figure 12 shows a compilation of recent measurements of the ^{3}He/^{4}He ratio compared to the prediction of the propagation calculations (solid line), using the fully energy dependent model just described. In addition, a prediction using a PLD with a constant X_o below 1 GeV/nucleon is shown by the dashed line. In both cases, the calculations fall below most of the data by almost a factor of two; although, to the degree that it can be determined, the data follow the shape of the calculated curve. This may be interpreted as showing the need for more ^{4}He fragmentation in the calculations, which could imply that a larger number of long pathlengths are needed in the PLD. However, such a conclusion may be premature. Most of the data points in Figure 12

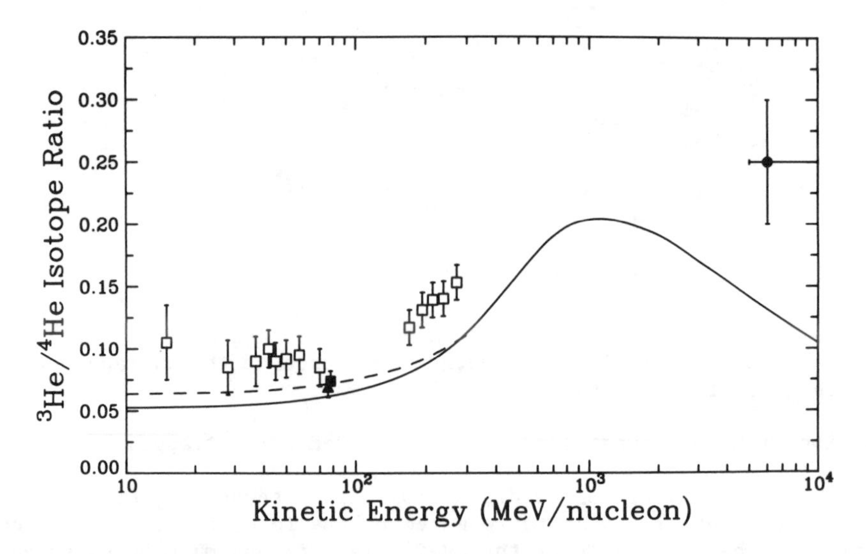

Figure 12. Measurements of the ^{3}He/^{4}He ratio as a function of energy compared to the predictions of propagation calculations (see text for details). The data points are: open squares, Webber and Yushak (1983); filled square, Beatty (1986); filled triangle, Kroeger (1986); filled circle, Jordan (1985).

come from a single experiment, and, where there is overlap at ~80 MeV/nucleon, other experiments show smaller ^{3}He/^{4}He ratios. This suggests a possible experimental problem (e.g. Mewaldt, 1986) and points to the need for accurate experimental results, from several different experiments, covering as broad an energy range as possible. Secondly, the cross sections used in the calculations for ^{3}He production have not been measured uniformly. If these cross sections are systematically underestimated, the calculated $^3He/^4He$ ratio would fall below the cosmic ray data. This emphasizes the need for precise accelerator measurements of the nuclear fragmentation cross sections before astrophysical conclusions can be drawn from the comparison of propagation calculations and cosmic ray experimental data. In fact, there are many outstanding cosmic ray problems whose resolution requires improved nuclear physics data (Meyer, 1985b), and the $^3He/^4He$ question shown in Figure 12 is representative of the challenging problems remaining for future work in GCR propagation.

5. Source Abundances

One of the major goals of cosmic ray propagation studies is to tie the measurements made at Earth to the source/acceleration regions by unfolding the secondary component in the elemental or isotopic abundances measured at Earth. Once a calculational model for propagation is developed, as described in the previous section, the relative element or isotope abundances in the source/acceleration regions can be determined, and these source abundances can be used to provide constraints on the injection, acceleration or nucleosynthesis of the cosmic rays. The GCR source abundances depend, of course, upon the detailed propagation model employed and upon the nuclear cross sections used in the analysis. There have been many calculations of source abundances, and, for the elements which are mainly primary, the different determinations are in substantial agreement and show no energy dependence to within ~20% (Mewaldt, 1981).

Figure 13 compares the derived composition of the matter at the cosmic ray source to the local galactic abundances (LG -- see Figure 2) for the elements ($Z \leqslant 30$) for which reliable GCR source abundances have been obtained. The first feature to note in Figure 13 is the trend to increasing abundance with increasing atomic number, i.e. cosmic ray source matter is enriched in heavy elements or, conversely, depleted in light elements relative to the local galactic abundances, with the transition occurring between neon and sodium ($Z = 10$-11). This pattern, however, is broken by the intermediate elements sulphur and argon as well as by zinc ($Z = 30$). Also shown in Figure 13 are the equivalent ratios for solar energetic particle (SEP) abundances plotted as the bars to the left of the cosmic ray points. The agreement between the ratios for GCR and SEP is striking, with the exception of helium and nitrogen. Thus, if the GCRS source matter is depleted in the elements below sodium, the regions on the sun that give rise to solar flare particles must be as well. In fact, except for He and N, a plot of the GCRS/SEP ratio verses element is consistent with unity (Mewaldt, 1981).

The agreement between the GCRS and SEP relative abundances shown in Figure 13 implies that the cosmic rays are derived from local galactic matter in much the same way as solar flare particles are derived from the solar composition (which is, basically, LG). The elements that appear underabundant (He,C,N,O,Ne,S,Ar) are those elements that are difficult to ionize whereas the elements showing ratios of about unity in Figure 13 are relatively easy to ionize. This correlation with ionization implies that the amount of a given element that is accelerated to high energy (in both SEP and GCR) depends upon a chemical property of the element and is not related to nucleosynthesis, which depends upon the properties of the nucleus.

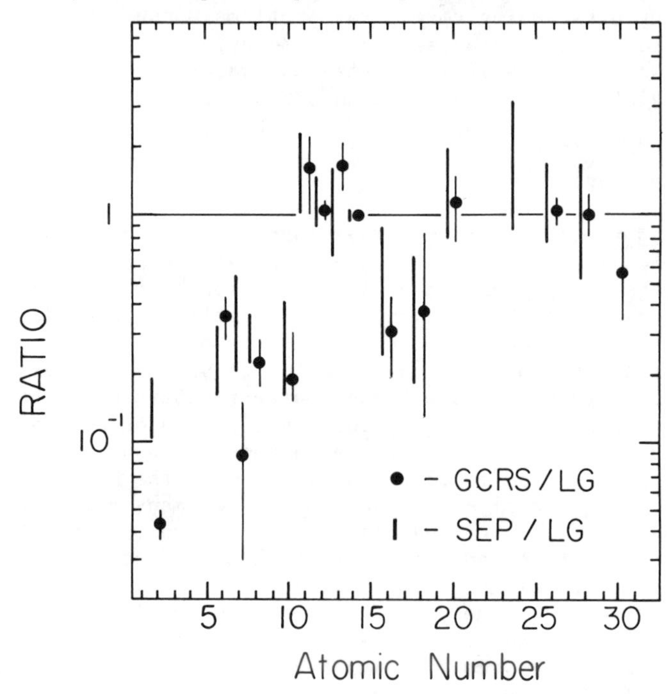

Figure 13. Ratios to the local galactic (LG) abundances of galactic cosmic ray source (GCRS) abundances and solar energetic particle (SEP) abundances (all relative to silicon) versus the atomic number of the element.

The correlation of abundance with ionization potential has been known for about a decade and has been investigated in detail for solar flare events (e.g. Breneman and Stone, 1985; Meyer, 1985a). Figure 14, from the review of Guzik (1987), shows the ratio of SEP to LG abundances (open symbols) plotted versus the first ionization potential of the elements. The solid error bars include only the uncertainties in the SEP measurements, while the dashed error bars also include the uncertainties in the local galactic abundances (c.f. Figure 2). The ratios in Figure 14 are near unity for the low ionization potential elements and begin to decrease at an ionization potential of 9 - 10 eV. Whether this decrease is a step function or a more gradual change is not well determined from the data. The difficulty in understanding this correlation is that the decrease at 9 - 10 eV implies an ionization equilibrium at a temperature of

$\sim 10^4$K, characteristic of the solar photosphere, while the solar flares occur higher in the solar atmosphere, in the corona, which is several orders of magnitude hotter. The higher temperatures of the corona should remove the ionization potential selection effect shown in Figure 14.

Recent gamma ray line observations of solar flares are relevant to this problem. During a solar flare many of the energetic particles are directed downward and interact with the ambient matter in the

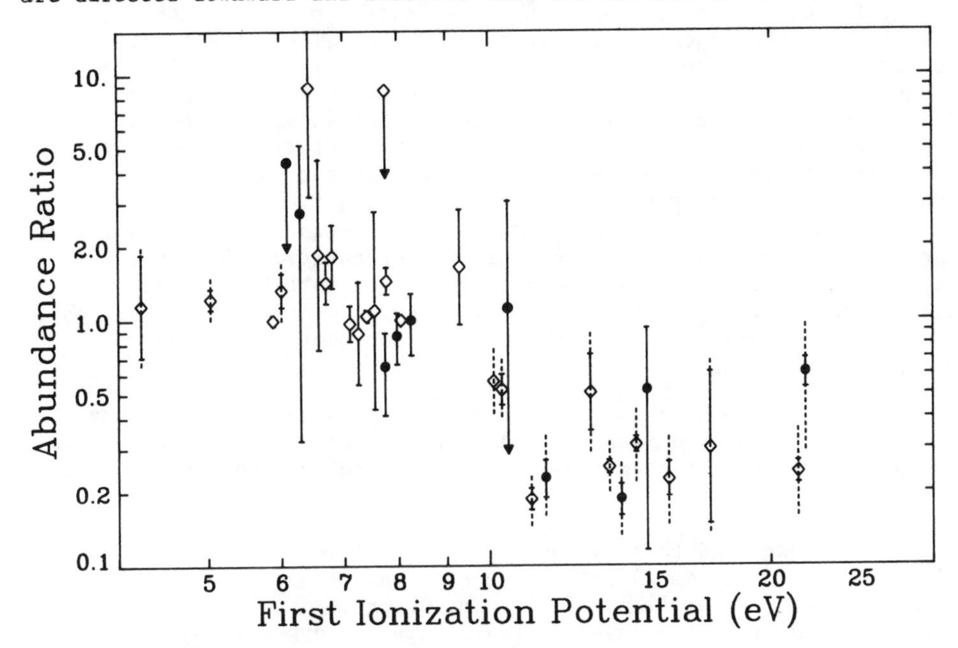

Figure 14. Abundance ratio with respect to local galactic (normalized to silicon) for SEP abundances (open diamonds) and gamma-ray line derived abundances (filled circles) as a function of the first ionization potential of the elements.

chromosphere, producing excited nuclei which decay by gamma ray emission. From the intensities of the lines observed, Murphy et al. (1985) have been able to determine the relative abundances of the ambient matter in the chromosphere. These abundances (divided by the LG abundance) are plotted in Figure 14 as the filled symbols and show the same correlation with first ionization potential as do the SEP results.

The agreement between the relative abundances derived from γ-ray lines and those obtained from solar fares may be interpreted (e.g. Guzik, 1987) as showing that the matter in the chromosphere has been chemically selected by first ionization potential from the normal (LG composition) matter in the photosphere by some as yet unknown process. During solar flare outbursts, this differentiated matter is accelerated and released as high energy particles. Thus, the ionization potential correlation is not related to the particle acceleration mechanism but is the result of a chemical separation in the matter that is to become SEP's.

For the galactic cosmic rays, which show the same ionization potential selection, this implies that either the matter originates in a region of temperature $\sim 10^4$K or that a differentiation process similar to that occurring on the sun must take place prior to particle acceleration. This leads naturally to models for sources of GCR in stellar flares, which provide both the selection by ionization potential and the injection at suprathermal energies. Additional acceleration, for example by SN shocks, is required to explain the measured energy spectrum, and the difficulty with such models is in finding the proper astrophysical environment which allows the stellar flare particles to get to the SN shocks without significant energy loss and/or nuclear fragmentation and in sufficient numbers to maintain the density of the galactic cosmic rays.

The conclusion to be drawn from this discussion of the cosmic ray source element abundances is that most of the observed abundance pattern can be explained by processes acting on matter of local galactic composition, so that GCR could originate anywhere in the ISM or in stars of normal composition. This is in contrast to the isotopic observations. SEP events show no enhancement of ^{22}Ne or ^{25,26}Mg over the local galactic values, but the galactic cosmic ray source (as discussed in Section 2.6) shows an enhancement of these isotopes. Although enhancements of neutron rich isotopes would be expected in highly evolved, massive stars or in metal rich stars, these objects are not an adequate source for the GCR since such objects would not be expected to show local galactic element abundances. One solution, of course, would be to take some material from evolved stars and mix it with matter from normal stars prior to acceleration, and such scenarios are being investigated actively (Meyer, 1985b and reference therein).

Thus, the original problem sketched on Figure 1, namely the origin of the matter that is to be accelerated to form the bulk of the galactic cosmic rays, remains largely unresolved. Much information has been obtained from the source elemental composition and from the pioneering isotopic abundance measurements that have been performed, but much remains to be done. The uncertainties on the comparisons shown in Figures 13 and 14 must be reduced through more precise cosmic ray measurements, more reliable propagation calculations and better measurements of the local galactic or solar system "template". The isotopic measurements must be extended to additional elements to look for other examples of isotope effects, down to the 10-20% level -- a significant experimental challenge. Finally, detailed models of realistic astrophysical environments must be developed and analyzed to determine if the proposed sources can, indeed, meet the constraints imposed by our knowledge of the source abundances.

6. SUMMARY

The cosmic ray particles span ~14 decades in energy, consist of nuclei of all of the elements in the periodic table plus electrons, positrons and anti-protons, and provide the only sample of matter from beyond our solar system available for direct study. As such, the cosmic rays are the key to understanding the recent history of nucleosynthesis in the galaxy — since the time of the formation of the solar system. In broad outline, the cosmic rays appear to be just another sample of local galactic matter, but when studied in detail, through the isotopes of the elements, for example, small but significant differences from local galactic abundances emerge. These subtle differencies show that either the ISM has undergone changes in the last 4.6 billion years or the source of cosmic ray matter involves contributions from stars, in various stages of evolution, and/or from supernovae.

The energization of the cosmic rays also appears to involve supernovae since SN shock waves are efficient in accelerating particles and there is sufficient total energy available. Stochastic acceleration occurring throughout the ISM may play a minor role as do individual objects/sources for the bulk of the particles at energies below 10^{14} eV. However, at $E > 10^{14}$ eV, individual objects/sources such as Cygnus X-3 may be the main contributors. Gamma-ray observations have shown that there are high energy particles in some pulsars, x-ray sources and binary systems, but whether or not these particles can overcome losses in these objects and escape to contribute to the galactic cosmic ray population remains an unanswered question.

The confinement and propagation of the cosmic rays have been studied through investigation of the secondary components among the particles observed at Earth and reveal that the cosmic rays are relatively young (10-20 million years old), that they have traversed a relatively large amount of matter (~9-10 g/cm^2 at ~1 GeV/nucleon), that the mean matter density in the confinement region is 0.2-0.3 atom/cm^3, that the amount of matter traversed decreases both with increasing energy above and with decreasing energy below a GeV/nucleon and that there is an energy dependent depletion of short pathlengths in the PLD. Results such as these provide the basis for constructing improved models for particle confinement and transport in the galaxy.

The investigation of galactic cosmic rays continues to provide fundamental information which, combined with results from x-ray and γ-ray observations and radio and optical astronomy, yields a fuller understanding of the high energy processes operating in our galaxy. Cosmic ray physics has a long history, spanning three-quarters of a century, and has evolved considerably in breadth and scope. Today, particle astrophysics confronts fundamental questions covering a range of topics from nucleosynthesis to galactic dynamics, to acceleration processes, to compact objects, to extragalactic matter, and to the basic enigma of the cosmic rays themselves — the question of their "Genesis" — the answer to which remains an elusive but tantalizing goal.

7. ACKNOWLEDGEMENTS

Thanks are due to T. G. Guzik and J. W. Mitchell for useful discussions and for a critical review of the manuscript and to G. Sutton for her excellent work in manuscript preparation. This work was supported, in part, by NASA under grant NAGW-550, by the Office of Naval Research under contract N-00014-83-K-0365 and by DOE under grant FG05-84ER40147.

8. REFERENCES

Achterberg, A., 1984, Adv. Space Res., 4, 193.

Ahlen, S. P., Price, P. B., Salamon, M. H. and Tarle, G., 1982, Ap. J., 260, 20.

Anders, E. and Ebihara, M., 1982, Geochim. et Cosmochim. Acta, 46, 2363.

Arons, J., 1981, in Origin of Cosmic Rays, eds. G. Setti, G. Spada, A. W. Wolfendale, (Dordrecht, Holland: D. Reidel Co.), p. 175.

Axford, W. I., 1981, in Origin of Cosmic Rays, eds. G. Setti, G. Spada, A. W. Wolfendale, (Dordrecht, Holland: D. Reidel Co.), p. 339.

Beatty, J. J., 1986, Ap. J., 311, 425.

Bignami, G. F., 1986, in Cosmic Radiation in Contemporty Astrophysics, ed. M. M. Shapiro, (Dordrecht, Holland: D. Reidel Co.), p. 175.

Binns, W. R., Fixsen, D. J., Garrard, T. L., Israel, M. H., Klarmann, J., Stone, E. C. and Waddington, C. J., 1984, Adv. Space Res., 4, 25.

Binns, W. R., Fickle, R. K., Garrard, T. L., Israel, M. H., Klarmann, J., Krombel, K. E., Stone, E. C. and Waddington, C. J., 1983, Ap. J. Lett., 267, L93.

Binns, W. R., Fickle, R. K., Garrard, T. L., Israel, M. H., Klarmann, J., Stone, E. C. and Waddington, C. J., 1981, Ap. J. Lett., 247, L115.

Blandford, R. D. and Ostriker, J. P., 1980, Ap. J., 237, 793.

Bower, A. J., Cunningham, G., Linsley, J., Reid, R. J. O. and Watson, A. A., 1983, J. Phys. G, 9, L53.

Breneman, H. H. and Stone, E. C., 1985, Ap. J. Lett., 299, L57.

Brownlee, D. E., 1985, Ann. Rev. Earth Planet. Sci., 13, 147.

Buffington, A., Schindler, S. M and Pennypacker, C. R., 1981, Ap. J., 248, 1179.

Burnett, T. H., Dake, S., Fuki, M., Gregory, J. C., Hayashi, T., Holynski, R., Iwai, J., Jones, W. V., Jurak, A., Lord, J. J., Miyamura, O., Ogata, T., Parnell, T. A., Saito, T., Strausz, S., Tabuki, T., Takahashi, Y., Tominaga, T., Wilczynska, B., Wilkes, R. J., Wolter, W. and Wosiek, B., 1986, Phys. Rev. Lett., 57, 3249.

Burnett, T. H., Dake, S., Fuki, M., Gregory, J. C., Hayashi, T., Holynski, R., Huggett, R. W., Hunter, S. D., Iwai, J., Jones, W. V., Jurak, A., Lord, J. J., Miyamura, O., Oda, H., Ogata, T., Parnell, T. A., Saito, T., Tabuki, T., Takahashi, Y., Tominaga, T., Watts, J. W., Wilczynska, B., Wilkes, R. J., Wolter, W. and Wosiek, B., 1983a, Phys. Rev. Lett., 51, 1010.

__________, 1983b, 18th ICRC Conference Papers, 2, 105.

Cameron, A. G. W., 1982, in Essays in Nuclear Astrophysics, eds. C. A. Barnes, D. D. Clayton and D. N. Schramm, (New York: Cambridge University Press), p. 23.

Casse, M. and Paul, J. A., 1982, Ap. J., 258, 860.

Cesarsky, C. J., 1980, Ann. Rev. Astron. Astrophys., 18, 289.

Cherry, M. L., Corbato, S., Kieda, D., Lande, K. and Lee, C. K., 1987, in Genesis and Propagation of Cosmic Rays, eds. M. M. Shapiro and J. P. Wefel (Dordrecht, Holland: D. Reidel Co.) this volume.

Colgate, S. A., 1984, Adv. Space Res., 4, 367.

Cowsik, R., 1980, Ap. J., 241, 1195.

Cowsik, R., Pal, Y., Tandon, S. N. and Verma, R. P., 1968, Canadian J. Phys., 46, S646.

Cowsik, R. and Wilson, L. W., 1975, 14th ICRC Conference Papers, 2, 659.

______, 1973, 13th ICRC Conference Papers, 1, 500.

Cunningham, G., Lloyd-Evans, J., Pollock, A. M. T., Reid, R. J. O., and Watson, A. A., 1980, Ap. J. Lett., 236, L71.

Efimov, N. M. and Sokurov, V. F., 1983, 18th ICRC Conference Papers, 2, 123.

Eichler, D., 1980, Ap. J., 237, 809.

Evenson, P., Meyer, P. and Pyle, K. R., 1983, Ap.J., 274, 875.

Fermi, E., 1949, Phys. Rev., 75, 1169.

Fichtel, C. E. and Linsley, J., 1986, Ap. J., 300, 474.

Fichtel, C. E. and Reames, D. V., 1968, Phys. Rev., 175, 1564.

Fisher, A. J., Hagen, F. A., Maehl, R. C., Ormes, J. F. and Arens, J. F., 1976, Ap. J., 205, 938.

Fisk, L. A., 1979, in Solar System Plasma Physics, eds. E. N. Parker, C. F. Kennel and L. J. Lanzerotti, (Amsterdam: North Holland Publ. Co.), p. 179.

Fleischer, R. L., Price, P. B., Walker, R. M., Maurette, M. and Morgan, G., 1967, J. Geophys. Res. 72, 355.

Fowler, P. H., Masheder, M. R. W., Moses, R. T., Walker, R. N. F., Worley, A. and Gay, A. M., 1985a, 19th ICRC Conference Papers, 2, 115.

___________, 1985b, 19th ICRC Conference Papers, 2, 119.

Fowler, P. H., Walker, R. N. F., Masheder, M. R. W., Moses, R. T and Worley, A., 1981, Nature, 291, 45.

Fowler, P. H., Adams, R. A., Cowen, V. G. and Kidd, J. M., 1967, Proc. Roy. Soc. A. 301, 39.

Fransson, C. and Epstein, R. I., 1980, Ap. J., 242, 411.

Freier, P. S., Lofgren, E. J., Oppenheimer, F., Bradt, H. L. and Peters, B., 1948, Phys. Rev., 74, 213.

Garcia-Munoz, M., Simpson, J. A., Guzik, T. G., Wefel, J. P. and Margolis, S. H., 1987, Ap. J. Suppl., 64, 269.

Garcia-Munoz, M., Meyer, P., Pyle, K. R., Simpson, J. A. and Evenson, P. A., 1986, J. Geophys. Res., 81, 2858.
Garcia-Munoz, M., Guzik, T. G., Simpson, J. A. and Wefel, J. P., 1984, Ap. J. Lett., 280, L13.
Garcia-Munoz, M., Simpson, J. A. and Wefel, J. P., 1981a, 17th ICRC Conference Papers, 2, 72.
Garcia-Munoz, M., Guzik, T. G., Margolis, S. H., Simpson, J. A. and Wefel, J. P., 1981b, 17th ICRC Conference Papers, 9, 195.
Garcia-Munoz, M., Simpson, J. A. and Wefel, J. P., 1979a, Ap. J. Lett., 232, L95.
Garcia-Munoz, M., Margolis, S. H., Simpson, J. A. and Wefel, J. P., 1979b, 16th ICRC Conference Papers, 1, 310.
Garcia-Munoz, M., Mason, G. M. and Simpson, J. A., 1977, Ap. J., 217, 859.
Garcia-Munoz, M. and Simpson, J. A., 1970, Acta Phys. Acad. Sci. Hungaricae, 29, Suppl. 1, 325.
Ginzburg, V. L., Khazan, Ya. M. and Ptuskin, V. S., 1980, Ap. Space Sci., 68, 295.
Ginzburg, V. L. and Syrovatskii, S. I., 1964, The Origin of Cosmic Rays, trans. H. S. W. Massey, ed. D. ter Haar (New York: McMillan), p. 426.
Gloeckler, G. and Jokipii, J. R., 1969, Phys. Rev. Lett., 22, 1448.
Golden, R. L., Horan, S., Mauger, B. G., Badhawar, G. D., Lacy, J. L., Stephens, S. A., Daniel, R. R. and Zipse, J. E., 1979, Phys. Rev. Lett., 43, 1196.
Grigorov, N. L., Mamontova, N. A., Rapoport, I. D., Savenko, I. A., Akimov, V. V. and Nesterov, V. E., 1971, 12th ICRC Conference Papers, 5, 1746.
Guzik, T. G., 1987, submitted to Solar Physics.
Guzik, T. G. and Wefel, J. P., 1984, Adv. Space Res., 4, 215.
Guzik, T. G., Wefel, J. P., Garcia-Munoz, M. and Simpson, J. A., 1985, 19th ICRC Conference Papers, 2, 76.
Hagan, F. A., Fisher, A. J. and Ormes, J. F., 1977, Ap. J., 212, 262.
Hermsen, W. and Bloemen, J. B. G. M., 1984, Adv. Space Res., 4, 393.
Hillas, A. M., 1984, Ann. Rev. Astron. Astrophys., 22, 425.
______, 1985, 19th ICRC Conference Papers, 9, 407.
Jokipii, J. R., 1976, Ap. J., 208, 900.
Jordan, S. P., 1985, Ap. J., 291, 207.
Juliusson, E., Meyer, P. and Muller, D., 1972, Phys. Rev. Lett., 29, 445.
Khristiansen, G. B., Fomin, Yu. A., Aliev, N. A., Alimov, T. A., Khakimov, N. Kh., Kakhkharov, M. K., Machmudov, B. M., Rakhimova, N. R. and Tashpulatov, R. T., 1983, 18th ICRC Conference Papers, 9, 195.
Kroeger, R., 1986, Ap. J., 303, 816.
Kundt, W., 1984, Adv. Space Res., 4, 381.
Lagage, P. O. and Cesarsky, C. J., 1983, Astron. Astrophys., 125, 249.
LaPointe, M., Kamata, K., Gaebler, J., Escobar, I., Domingo, V., Suga, K., Murakami, K., Toyoda, Y. and Shibata, S., 1968, Canadian J. Phys., 46, S68.
Lerche, I. and Schlickeiser, R., 1985, Astron. Astrophys., 151, 408.

Lingenfelter, R. E., 1979, 16th ICRC Conference Papers, 14, 135.
McCray, R. and Snow, T. P., 1979, Ann. Rev. Astron. Astrophys., 17, 213.
Mewaldt, R. A., 1986, Ap. J., 311, 979.
Mewaldt, R. A., 1983, Rev. Geophys. Space Phys., 21, 295.
Mewaldt, R. A., 1981, 17th ICRC Conference Papers, 13, 49.
Mewaldt, R. A., Spalding, J. D., Stone, E. C. and Vogt, R. E., 1980, Ap. J. Lett., 235, L95.
Meyer, J. P., 1985a, Ap. J. Suppl., 57, 151.
Meyer, J. P., 1985b, 19th ICRC Conference Papers, 9, 141.
Meyer, J. P., 1979, in Les Elements et leurs Isotopes dans l'Univers, (Liege: University of Liege, Institute of Astrophysics), p. 153.
Michel, F. C., 1984, Adv. Space Res., 4, 387.
Murphy, R. J., Ramaty, R., Forrest, D. J. and Kozlovsky, B., 1985, 19th ICRC Conference Papers, 4, 249.
O'Dell, F. W., Shapiro, M. M., Silberberg, R. and Stiller, B., 1965, 9th ICRC Conference Papers, 1, 412.
Ormes, J. F. and Freier, P. S., 1978, Ap. J., 222, 471.
Ramaty, R. and Lingenfelter, R. E., 1982, Ann. Rev. Nucl. Part. Sci., 32, 235.
Rasmussen, I. L. and Peters, B., 1975, Nature, 258, 412.
Ryan, M. J., Balasubrahmanyan, V. K. and Ormes, J. F., 1972, Phys. Rev. Lett., 28, 1497.
Samorski, M. and Stamm, W., 1983, Ap. J. Lett., 268, L17.
Schlickeiser, R., 1986, in Cosmic Radiation in Contemporary Astrophysics, ed. M. M. Shapiro (Dordrecht, Holland: D. Reidel Co.), p. 27.
Scholer, M., 1984, Adv. Space Res., 4, 419.
Shapiro, M. M. and Silberberg, R., 1970, Ann. Rev. Nucl. Sci., 20, 323.
Shapiro, M. M., Silberberg, R. and Tsao, C. H., 1970, Acta Phys. Acad. Sci. Hungaricae, 29, Suppl. 1, 471.
Silberberg, R., Tsao, C. H., Letaw, J. R. and Shapiro, M. M., 1987, in Genesis and Propagation of Cosmic Rays eds. M. M. Shapiro and J. P. Wefel (Dordrecht, Holland: D. Reidel Co.), this volume.
__________, 1983a, Phys. Rev. Lett., 51, 1217.
Silberberg, R., Tsao, C. H., Shapiro, M. M. and Letaw, J. R., 1983b, 18th ICRC Conference Papers, 2, 179.
Silberberg, R. and Tsao, C. H., 1973, Ap. J. Suppl., 25, 315.
Simon, M., 1977, Astron. Astrophys., 61, 833.
Simon, M., Spiegelhauer, H., Schmidt, W. K. H., Siohan, F., Ormes, J. F., Balasubrahmanyan, V. K. and Arens, J. F., 1980, Ap. J., 239, 712.
Simpson, J. A., 1983, Ann. Rev. Nucl. Part. Sci., 33, 323.
Smith, L. H., Buffington, A., Smoot, G. F., Alvarez, L. W. and Wahlig, M. A., 1973, Ap. J., 180, 987.
Stephens, S. A., 1981, 17th ICRC Conference Papers, 9, 199.
Trumper, J., 1986, in Cosmic Radiation in Contemporary Astrophysics, ed. M. M. Shapiro (Dordrecht, Holland: D. Reidel Co.), p. 217.
Turver, K. E., 1985, 19th ICRC Conference Papers, 9, 399.
Watson, A. A., 1984, Adv. Space Res., 4, 35.

Webber, W. R., 1983, in Composition and Origin of Cosmic Rays, ed. M. M. Shapiro (Dordrecht, Holland: D. Reidel Co.), p. 83.
Webber, W. R., 1982, Ap. J., 252, 386.
Webber, W. R. and Yushak, S. M., 1983, Ap. J., 275, 391.
Webber, W. R., Kish, J. C. and Schrier, D. A., 1985, 19th ICRC Conference Papers, 2, 88.
Webber, W. R. and Kish, J. C., 1979, 16th ICRC Conference Papers, 1, 389.
Webber, W. R., Lezniak, J. A., Kish, J. C. and Simpson, J. A., 1977, Astro. Lett., 18, 125.
Weiler, K. W., 1987, in Genesis and Propagation of Cosmic Rays, eds. M. M. Shapiro and J. P. Wefel (Dordrecht, Holland: D. Reidel Co.), this volume.
Wiedenbeck, M. E., 1984, Adv. Space Res., 4, 15.
Wiedenbeck, M. E., 1983, 18th ICRC Conference Papers, 9, 147.
Wiedenbeck, M. E. and Greiner, D. E., 1981, Phys. Rev. Lett., 46, 682.
Wiedenbeck, M. E. and Greiner, D. E., 1980, Ap. J. Lett., 239, L139.
Young, J. S., Freier, P. S. and Waddington, C. J., 1979, 16th ICRC Conference Papers, 1, 442.

PROPAGATION AND TRANSFORMATIONS OF COSMIC RAYS: FROM SOURCES TO EARTH

R. Silberberg and C. H. Tsao,
Naval Research Laboratory, Hulburt Center for Space Research, Washington, D.C. 20375
J. R. Letaw, Severn Communications Corporation, Severna Park, MD 21146
M. M. Shapiro, Alexandria, VA

The particles that are accelerated to cosmic ray energies show selection effects of an injection (or pre-injection) environment of 10^4 K, i.e., nuclei that are neutral near 10^4 K, due to high first ionization potentials, are reduced in abundance by a factor of about 6. Such temperatures (with corresponding energies of about 1 eV) are characteristic of stellar photospheres. A similar selection effect (related to migration of particles from the solar photosphere or chromosphere into the corona) affects the composition of the corona, the solar wind and solar flare particles. (Typical solar wind speeds of 400 km/sec correspond to energies of 10^3 eV.) The annual Galactic energy input into cosmic rays is about 10^{60} eV/year, (calculated from the cosmic ray energy density, galactic radio disk volume and cosmic ray confinement time therein). The corresponding number of cosmic rays accelerated per year is about 10^{51} with a mean energy near 10^9 eV. The above huge power output and number of particles accelerated is likely to be provided by shock waves of expanding supernova remnants ploughing into stellar wind zones. After acceleration, cosmic rays are confined for about 10^7 years in the Galactic radio disk (and probably even longer in the halo) suffering nuclear spallation reactions with the interstellar gas, mainly in the clouds. These clouds thus also serve as sites of meson production with π^0-generated gamma rays. A modest amount of cosmic-ray reacceleration takes place in interstellar space by old, weak shock waves. The cosmic rays that reach the solar system suffer solar modulation by the outflowing solar plasma that results in adiabatic deceleration and reduction of the intensity of low energy particles. The particles that reach the earth undergo geomagnetic modulation; lower energy cosmic rays are deflected back by the earth's magnetic field. Close to the geomagnetic equator, even protons with energies of 10^{10} eV are prevented entry.

I. INTRODUCTION

In this article we explore the changes in composition of cosmic rays induced by nuclear collisions and changes in energy and diffusive

M. M. Shapiro and J. P. Wefel (eds.), Genesis and Propagation of Cosmic Rays, 41–70.

scattering and deflections induced by magnetic fields, and plasma flow. We shall now consider various problems about cosmic-ray origin and propagation, and show how experimental and theoretical investigations of cosmic rays and of astrophysics permits us to find answers to these questions.

What is the origin of the material that is accelerated to cosmic ray energies: Is it the hot component of the interstellar gas, or stellar wind particles and stellar flare particles in young OB stellar associations that have frequent supernovae, or is it the accreting gas in a binary system? What is the degree of admixture of recently nucleosynthesized material in the cosmic rays and how does it differ from the solar system composition? Investigations of the cosmic ray source composition provide answers to these problems. How much time has passed since the nucleosynthesis of the cosmic ray nuclides? The abundance of long-lived radioactive primary nuclei permits an investigation of this problem. How much time has passed between nucleosynthesis and the acceleration of cosmic ray nuclei? The abundance of primary nuclides that decay only by electron capture and hence become stable when stripped of electrons and accelerated to cosmic ray energies provides clues to this problem.

How much material have cosmic rays traversed since their acceleration? The relative abundance of the stable secondary nuclei permits the resolution of this problem. Is most of the matter traversed by cosmic rays in clouds (either at the source regions or in interstellar space)? The energy dependence of the abundance of the relatively long-lived secondary nuclides that decay by electron capture permits an investigation of this problem; (the lifetime has to be sufficiently long, so that the nucleus is re-stripped of its electron in the relatively dense medium before it can capture this electron.) Do higher energy cosmic rays leak out at a faster rate from their confinement region? The relative abundance of secondary-to-primary cosmic rays, measured as a function of energy, permits a resolution of this question. Is the confinement volume where most of the traversal of matter takes place near the source regions (e.g. the O-B stellar associations with clouds and frequent supernovae) or in the interstellar gas of the Galaxy? A comparison of the energy dependence of the relative abundance of stable secondaries-to-primaries vs. the energy dependence of the directional anisotropy of cosmic rays helps to explore this problem. Also the gamma ray flux, including that of gamma-ray lines from these objects, permits an investigation of this problem. Do cosmic rays escape from confinement volume where most material is traversed by a diffusive process (scattering by magnetic field structures) or a convective process (associated with bursting of the confining magnetic field)? The relative abundance of various secondary cosmic rays, with different nuclear interaction mean path lengths helps to resolve this problem; the distribution of path lengths is broad and exponential in shape in the case of diffusive escape, and narrow in the case of convective escape. Are cosmic rays confined in two successive confinement volumes, one near the source regions, and then in the Galaxy as a whole? An investigation of the relative abundances of sets of secondary nuclei permits the determination of the structure of the distribution of path lengths; a dearth of short path lengths implies a double confinement volume, (or possibly the absence of a relatively close-by sources).

Are cosmic rays reaccelerated to some extent in the interstellar space after their main acceleration? What is the extent of adiabatic deceleration by the outflowing solar plasma? These problems are explored by investigating the relative abundance of various secondary cosmic rays and the energy dependence of their production cross sections. It is also explored from the energy dependence of the abundance of secondary nuclei that decay only by electron capture; decay occurs at low energy where the capture cross section is large, with acceleration, these nuclei in the depleted energy interval are shifted to higher energies. Weak reacceleration also distorts the energy dependence of the secondary-to-primary ratios, and displaces the peaks of the primary energy spectra.

How long are cosmic rays confined by the magnetic fields in the Galaxy? Long-lived radioactive secondary nuclei provide clues to this problem. Are cosmic rays confined principally to the Galactic disc, or are there magnetic structures in the low-density Galactic halo that confine the particles? The mean path length (derived from the stable secondaries) combined with the confinement time helps to resolve this problem. Is the escape of cosmic rays from the Galaxy diffusive or convective? The energy dependence of the abundance of long-lived secondary nuclides permits an investigation of this problem.

The intensity variations of cosmic ray energy spectra provide information on solar wind interaction with the cosmic rays. The magnetic field of the earth results in the cutoff (deflection) of lower energy cosmic rays.

The composition of cosmic rays will be discussed by Prof. J. Wefel at this Course. Associated interpretations will be presented by him, as well as in this paper.

2. ORIGIN OF COSMIC RAYS

Prior to acceleration to cosmic-ray energies there is an injection phase, at a temperature near 10^4 °K, since cosmic rays exhibit a dependence on the first ionization potential, as shown in Figure 1. There are two sites where such temperatures occur: (a) the outer portions of the interstellar clouds, close to the interface with the hot, low-density phase of the interstellar medium and (b) the potospheres and chromospheres of stars. However, in case (a) the composition of the gas is unlike the cosmic-ray composition, since the non-volatile elements are largely in grains. In case (b), there are two alternative subcases; injection of flare star particles to energies near 1 MeV, or stellar wind particles, at energies near 1

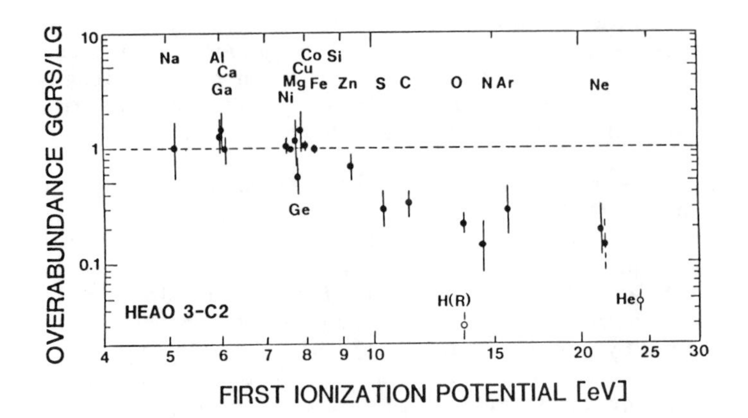

Figure 1: The ratios of the elemental abundances in the cosmic radiation and in the Local Galactic matter are plotted as a function of the first ionization potential. Normalization is at silicon.

keV. The number of particles in the former subcase is somewhat lower (Gorenstein, 1979), compared to the galactic cosmic-ray requirements of $\sim 10^{-3}$ cosmic rays/m^{-3}, and a confinement time of 10^7 years. This corresponds to the generation of 10^{51} cosmic rays per year. The scenario we prefer hence is the acceleration of stellar wind particles by shock waves of supernova remnants. The material in this scenario is largely derived from stars, while the energy is derived from supernovae.

Since supernovae of type II are formed at the final stage of the young, rapidly burning O and B stars, many of the cosmic rays are likely to originate in young O, B stellar associations though not necessarily from the O and B stars themselves. Since stars form by collapse of gas clouds, young stellar associations (e.g. the Orion nebula) have gas clouds associated with them. The origin of cosmic rays in O, B stellar associations and of many gamma rays in the associated clouds has been proposed by Montmerle (1979).

There are a couple of anomalies in the isotopic composition of cosmic rays (when compared to the composition of the solar system) that imply the contribution of some special nucleosynthesis to the cosmic ray material. Wiedenbeck and Greiner (1981) and Webber et al (1985) find that the source abundance of $^{22}Ne/^{20}Ne$ is 3.2 ± 0.5 times higher in cosmic rays and $^{25,26}Mg/^{24}Mg$ is 1.4 ± 0.3 times higher.

Meyer (1985) has proposed an explanation in terms of some contribution from the very massive (10 to 50 $M_{\odot}$) Wolf-Rayet stars. Since there are only about 10^3 in the Galaxy, if cosmic rays originate within a few hundred parsecs of the sun, fluctuating (over a period of 10^6 years) contributions from such stars are possible. In the Wolf-Rayet stars the CNO nucleosynthesis cycle yields ^{14}N, which in turn, by helium burning yields ^{18}O, and then by helium burning yields ^{22}Ne. Some of the latter burns into ^{25}Mg and ^{26}Mg. Furthermore, the 3α reaction yields some enhancement of ^{12}C. Meyer (1985) showed that a 2% contribution of the Wolf-Rayet stars, to the nuclei from which cosmic rays are accelerated, is sufficient to resolve the above composition anomalies. Since the stellar wind material outflow from these stars exceeds that of solar-like stars by more than a factor of 10^6, the contribution from a couple of such stars can be significant.

Prantzos et al (1985) find that Wolf-Rayet stars yield a significant enhancement of carbon in cosmic rays (factor of 2.9), and also of oxygen, by a factor of 1.6. We note that with this enhancement, the experimental C and O abundances (see Figure 1) yield a significantly better fit to the first ionization potential relationship. Furthermore, the N/O source abundance anomaly ceases to be an anomaly if the contribution of Wolf-Rayet stars to oxygen is considered. Guzik et al. (1980) find the low energy N/O source ratio to be 0.03 to 0.06. Distributed acceleration (Silberberg et al. 1983) raises it slightly to about 0.06, which is also the value of Byrnak et al. (1983) and Goret et al. (1983) at high-energies ($\sim$ 3 GeV/nucleon). One can thus have an N/O ratio of 0.1 for the bulk of the stars that contribute material to cosmic ray sources, [in agreement with the Galactic abundance estimate of Meyer (1985)] and have the ratio reduced to 0.06 by the contribution of Wolf-Rayet stars to oxygen.

There is an other abundance "anomaly" in cosmic rays that may be spurious. The abundance ratio (Pt-group)/(Pb-group) is high in cosmic rays. Margolis and Blake (1985) have proposed a modification of the s-process nucleosynthesis to explain the observations. However, Grevesse and Meyer (1985) have another explanation, based on recently measured solar spectra: the abundance of lead in the solar system, i.e., in "general abundances" is less by a factor of 2 than the previously adopted value, based on elemental abundances in meteorites. However, there is an abundance anomaly that is not yet fully explained: the lightest elements H and He are underabundant in cosmic rays by a factor of 4. H was earlier believed to be more underabundant, but using momentum per nucleon for comparison (based on shock wave acceleration) rather than the rigidity interval, it, too, is a factor of 4, as shown by Meyer (1986). One possible explanation is the preferential acceleration of heavier nuclei ($Z > 2$ or $Z \geq 6$) in cosmic rays. If cosmic rays originate in regions of new stellar associations with frequent supernovae, a modest amount of recently nucleosynthesized material could be present in cosmic rays.

A test would be provided by the ratios Pu/U and Cm/U, i.e. is there a slight admixture of $\sim 10^7$ year old material mixed in with $> 10^9$ year old material? Figure 2, from Blake and Schramm (1974) gives the relative abundances of the long-lived actinides as a function of time after nucleosynthesis; this figure, combined with actinide abundance measurements in cosmic rays, will permit a measurement of the time after nucleosynthesis or a test of some admixture of ancient and recently nucleosynthesized material. The nuclide ^{244}Pu has a half life of 8 x 10^7 years, and ^{247}Cm of 1.5 x 10^7 years. The ratio U/Th changes as ^{235}U and ^{236}U decay.

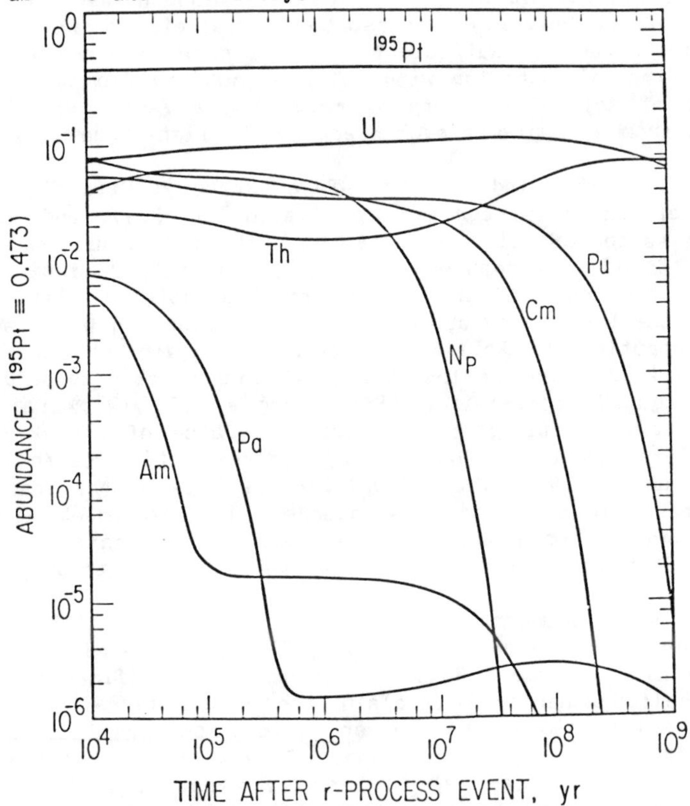

Figure 2: The relative abundances of the individual actinides as a function of time after their nucleosynthesis in an r-process event.

The nuclides that decay only by electron capture and originate in cosmic ray sources permit a determination of the time T_a between nucleosynthesis and acceleration of cosmic-ray nuclei, Soutoul et al. (1975). In the synthesis of elements certain nuclides are produced that normally would decay by electron capture.

If, however, they are accelerated prior to decay, they will survive in cosmic rays. These nuclides can thus be used to determine whether the cosmic ray "material" originates in supernovae or in old interstellar gas. Methods for determining the time T_a were proposed by Silberberg et al (1976); they are outlined in Table 1.

TABLE 1

The Ratio Co/Ni and the Abundance of ^{44}Ti versus the Time T_a between Nucleosynthesis and Acceleration

T_a (years)	Conditions determining Co and ^{44}Ti abundances	Expected* Ratio Co/Ni	Expected Ratio $^{44}Ti/^{46}Ti$
$\ll 1$	Original ^{57}Co survives	0.4	0.10
$1 \ll T_a \ll 50$	^{57}Co decays, secondary Co dominates, ^{44}Ti survives	0.06	0.10
$50 \ll T_a \ll 10^5$	Co and ^{44}Ti predominantly	0.06	0.04
$\gg 10^5$	^{59}Co from decay of ^{59}Ni, some secondary Co	0.10	0.04

*The "expected" abundances of ^{57}Co and ^{44}Ti at the sources are based on Cameron's (1982) values of ^{57}Fe and ^{44}Ca.

Juliusson and Meyer (1978) measured the ratio Co/Ni and found it to be 0.15 and two standard deviations below 0.4; Ormes et al (1975) found it to be < 0.14. Hence Soutoul et al. (1975) and Shapiro and Silberberg (1975) concluded that ^{57}Co has decayed, and that cosmic rays must have been accelerated at least one year after synthesis.

More recently, Koch-Miramond (1981) based on analytic work of Soutoul, Cassé and Juliusson (1978) explored the relative abundances of Fe, Co, Ni measured on HEAO 3-2C, shown in Figure 3. These data imply that ^{59}Ni has decayed into ^{59}Co, and hence at least 10^5 years have passed between nucleosynthesis in supernovae and the acceleration of cosmic ray nuclei. Since the strength of the evidence is about 2 standard deviations, a further test, outlined by Adams et al (1979) is suggested: If ^{59}Ni has not decayed the ratio of fluxes would be $J(^{57}Co) = 8$; if Ta > 10^5 years, i.e. if ^{59}Ni has decayed $J(^{57}Co)/J(^{59}Co) = 1$.

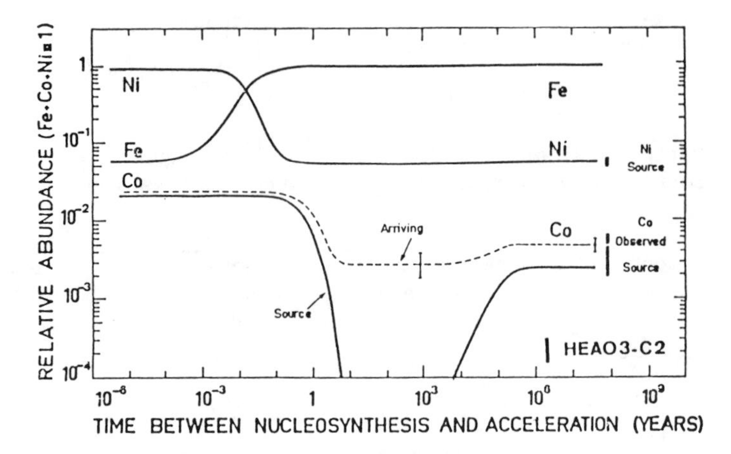

Figure 3: The expected relative abundance of Fe, Co, Ni at the galactic cosmic ray source vs. time elapsed between nucleosynthesis and GCR acceleration.

4. COSMIC RAY PROPAGATION

The ratios of groups of stable secondary-to-primary nuclei, e.g., L/M or $(3 \leq Z \leq 5)/(6 \leq Z \leq 8)$ permit the determination of the amount of material traversed by cosmic rays between acceleration and escape from the galaxy. The ratio L/M was first measured reliably by O'Dell, Shapiro, and Stiller (1962) who obtained 0.25 ± 0.04 at energies ≥ 1.5 GeV/nucleon, above a geomagnetic rigidity cutoff. (Above an energy cutoff, the value reduces to 0.23 ± 0.04). The corresponding path length traversed is about 6 g/cm^2 of interstellar H and He, and about half as much if there is some distributed acceleration.

A comparison of the ratios L/M and $(17 \leq Z \leq 25)$/Fe permits the calculation of the distribution of path lengths. (Iron nuclei have a cross section about 3 times that of the M nuclei, hence are depleted faster). An exponential distribution of path lengths (i.e. many short path lengths and some long ones) fits both of the above abundance ratios [Shapiro, Silberberg and Tsao (1969)]. Davis (1959) explored the theoretical bases for various path length distributions; Davis noted that a steady state, with uniform injection throughout the volume, rapid diffusion and slow leakage at the surface leads to an exponential distribution.

The ratio of secondary to primary nuclei is energy dependent, implying a rigidity (momentum/charge) dependent leakage from a confinement region; however, weak distributed reacceleration can simulate this energy dependence. This energy dependence was noted by Juliusson, Meyer, Muller (1982), Smith et al. (1983), Webber et al. (1983) and Ormes and Balasubrahmanyan (1972).

Figure 4, based on Ormes and Protheroe (1983) shows the energy dependence of the B/C ratio. We note here the decrease of the ratio below 1 GeV/nucleon and shall explore it later in terms of the hypothesis of distributed acceleration.

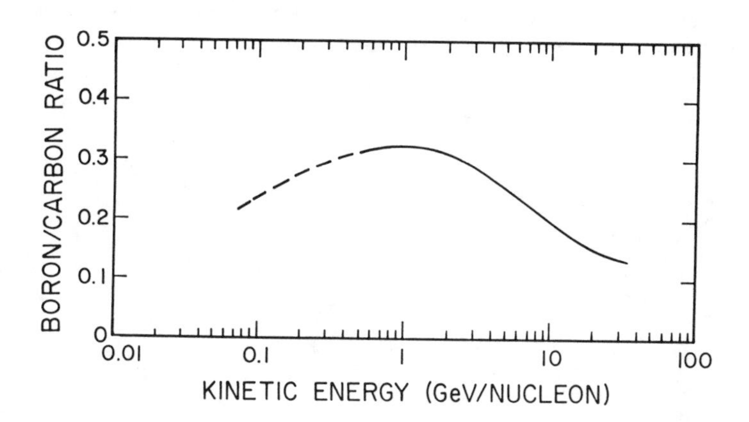

Figure 4: Experimental B/C ratio as a function of energy.

Meyer (1986) in his Figure 11, based on the work of Brewster et al. (1983) Klarmann et al. (1985) and Fowler (1985) find that the secondary/primary ratios of the ultraheavy nuclei with $Z > 60$ also have an energy dependence, probably rather similar to that of the previous paragraph.

Klarmann et al. (1985) find that they can fit the ultraheavy secondary/primary ratios with the standard leaky box model, using the cross sections of Kertzman et al. (1985). However, the latter were measured near 1 GeV/nucleon, while the cosmic rays fitted are at $E > 5$ GeV/nucleon. Hence, this actually means that the cross sections appropriate to the distributed acceleration hypothesis [Silberberg et al. (1983)] fit the data.

The decreasing ratio of secondary to primary nuclei at high energies has been interpreted in terms of a faster diffusion rate from the Galaxy at high energies. However, the "nested leaky box" model of Cowsik and Wilson (1973, 1975) with rigidity dependent leakage from the source region (e.g. a young stellar association with recent supernovae and clouds) is another plausible hypothesis. The latter is even supported by the near absence of the energy dependence of anisotropy of cosmic rays between 10^{10} and 3×10^{13} eV, shown in Figure 5. The nested leaky box has been further explored by Silberberg et al. (1983, 1985), Tang and Muller (1985), and Stephens (1985).

Normally, the young stellar associations with supernovae and clouds are considered as the inner volume. However, different versions, where the solar system is inside the inner volume have been proposed: (a) Peters and Westergaard (1977) according to whom the galactic arms constitute the inner confinement volume, and the Galaxy with the halo the outer volume and (b) the superbubble model of Kafatos et al. (1981), Streitmatter et al. (1983). Furthermore, weak distributed reacceleration generates a decreasing ratio of secondary to primary nuclei that decreases between 1 and 100 GeV/nucleon.

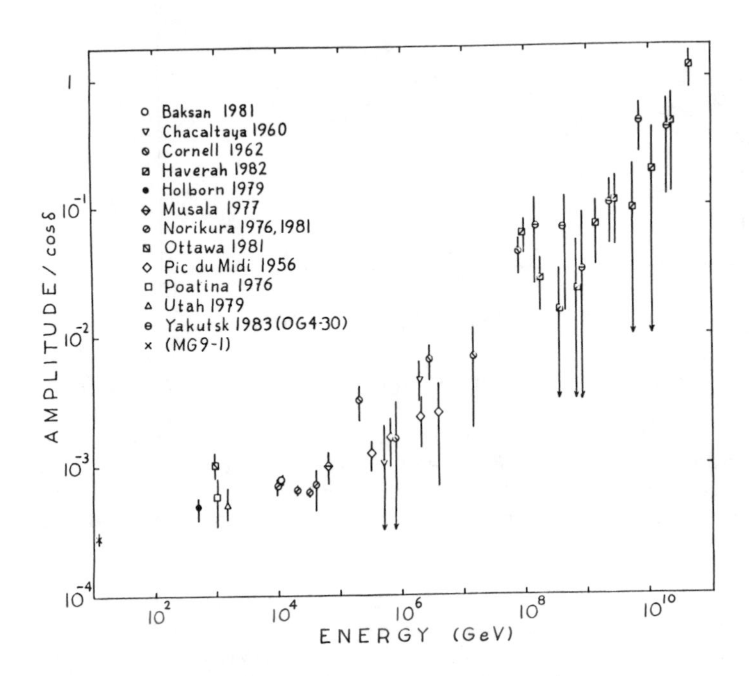

Figure 5: Cosmic Ray anisotropy as a function of energy.

Over 90% of cosmic-ray spallation reactions take place in interstellar clouds. These should hence be significant gamma-ray sources, as has also been observed. We shall now justify the above value of over 90%. Blandford and Ostriker (1980) discuss 3 components of the interstellar medium: (1) a hot component that is maintained by repeated supernova explosions, having a very low density $\rho \sim 3 \times 10^{-3}$ atoms/cm^3, that fill ~ 75% of interstellar space, (2) fluffy warm clouds with $\rho \sim 0.25$ atoms/cm^3, that fill ~ 20% of space, and (3) cool clouds (cores within warm regions, that fill 2 to 5% of space, with $\rho \sim 40$ atoms/cm^3). For a nearly uniform cosmic ray density, the number of interactions [$\propto \rho \times$ volume]; an inspection of the above values shows that component (3) i.e., the dense cores of clouds will dominate.

The nuclides that decay by electron capture and are formed by spallation during propagation are ^{7}Be, ^{37}Ar, ^{41}Ca, ^{44}Ti, ^{49}Vi, ^{51}Cr, ^{53}Mn, ^{55}Fe and about 50 more that are heavier than iron. These isotopes can be employed to measure adiabatic deceleration in the solar system [Raisbeck, et al. (1973)]. They then appear to survive at low energies where they would have decayed by electron capture. These isotopes [Silberberg et al. (1983)] can test the hypothesis of distributed reacceleration. Then these nuclides would be depleted at intermediate energies as they would have captured the electrons at lower energies before their final reacceleration. Also the element Eu can be used to explore distributed acceleration, since the abundance of Eu depends on electron capture, which is energy dependent, as illustrated in Figure 6, from Letaw et al. (1985). In addition, long-lived nuclei that decay by electron capture, e.g. ^{44}Ti, ^{93}Mo and ^{157}Tb can test whether cosmic ray spallation reactions occur mainly in clouds, and explore the density of these clouds, as illustrated in Figure 7 from Letaw et al. (1985).

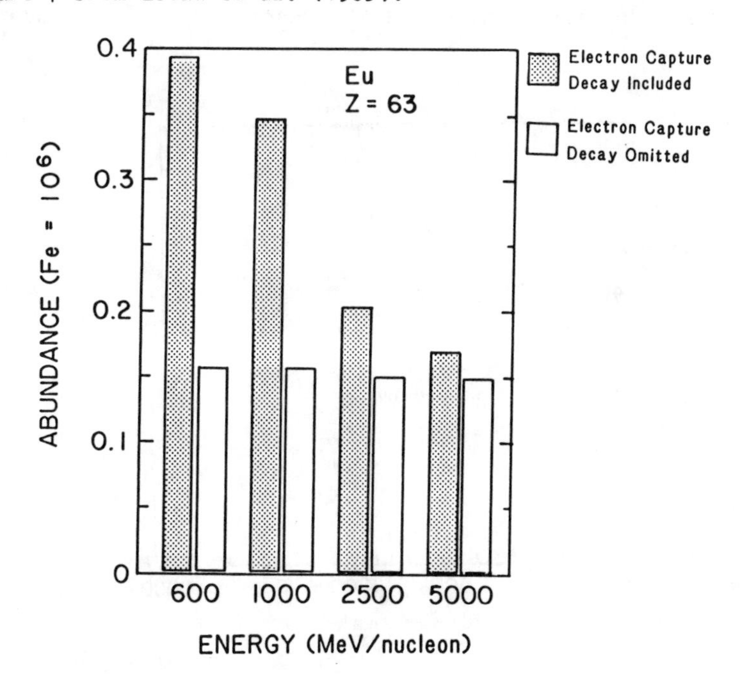

Figure 6: The abundances of Eu (Z = 63) calculated with and without electron capture decay effect as a function of energy.

Long-lived radioactive secondary nuclei that are formed by spallation during propagation permit an estimate of the confinement time of cosmic rays in the Galaxy. Radionuclides useful for the determination of this time include ^{10}Be (half-life 1.5 x 10^6 y), ^{26}Al (7.4 x 10^5y), ^{36}Cl (3 x 10^5y) and ^{54}Mn (~2 x 10^6y). The best estimates of the confinement time are based on the measurement of ^{10}Be relative to ^{9}Be and ^{7}Be.

Using the publications since 1980 that also have the best isotopic resolution, those of Garcia-Munoz, Simpson and Wefel (1981) and Wiedenbeck and Greiner (1980), measured at 60 to 200 MeV/nucleon, one can conclude that only about 20% or 25% of ^{10}Be has survived decay. The confinement time is 10^7 years, within a factor of 2. Combining this with the measurement of the path length traversed yields an average density of 0.2 or 0.3 atoms/cm^3. This is smaller than the mean density in the galactic disk, ~ 1 atom/cm^3; hence the particles we observe in the galactic disk spend part of their confinement time in the inner regions of the halo. Already earlier, and at higher energies, near 3 GeV/nucleon, O'Dell et al. (1975) measured and explored the Be/B ratio, and concluded that at these higher energies, where the Lorentz factor $\gamma \gg 1$, most ^{10}Be survives. Silberberg, Tsao, and Shapiro (1976) estimated that the confinement time is $< 10^7$ years within one standard deviation and $< 5 \times 10^7$ years within two standard deviations.

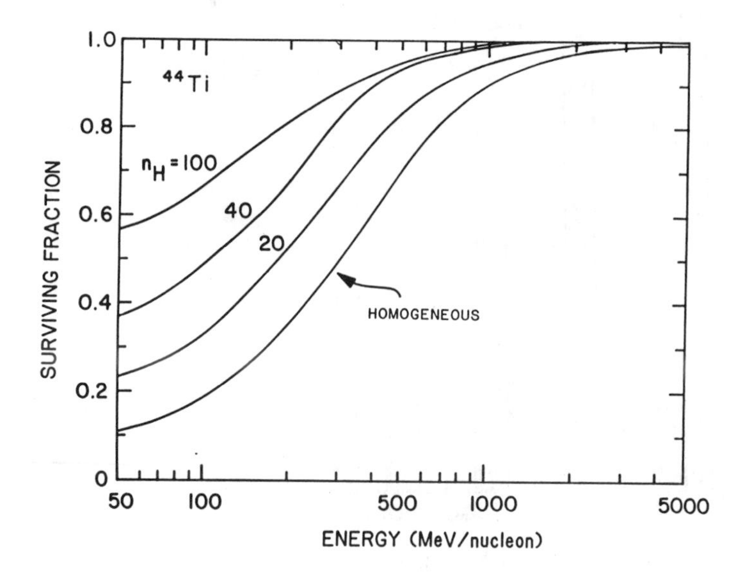

Figure 7: The surviving fraction of ^{44}Ti as a function of energy, calculated for various densities in interstellar clouds.

Koch et al. (1981) have reached a similar conclusion regarding ^{54}Mn; below 1 GeV/nucleon most appears to have decayed, while at values of several GeV/nucleon, most ^{54}Mn seems to survive. Measurements of ^{26}Al and ^{36}Cl by Wiedenbeck (1983) and Wiedenbeck (1985) are also consistent with a confinement time of 10^7 years. A test for a moderate degree of distributed reacceleration is provided by secondary/primary ratios, if the production cross section is highly energy dependent, e.g. single-nucleon stripping, and by the energy at which the electron capture effects are observed for nuclides that can decay only by electron capture (Silberberg et al., 1983). These effects then appear to be shifted to higher energies than otherwise expected.

The shape of the cosmic-ray energy spectrum at low energies (below 1 GeV/nucleon) can be explained in terms of distributed reacceleration. Figure 8 shows the calculated spectrum of carbon outside the solar cavity, assuming a strong shock-wave acceleration that yields a power law in momentum, and subsequent distributed reacceleration by weak shocks.

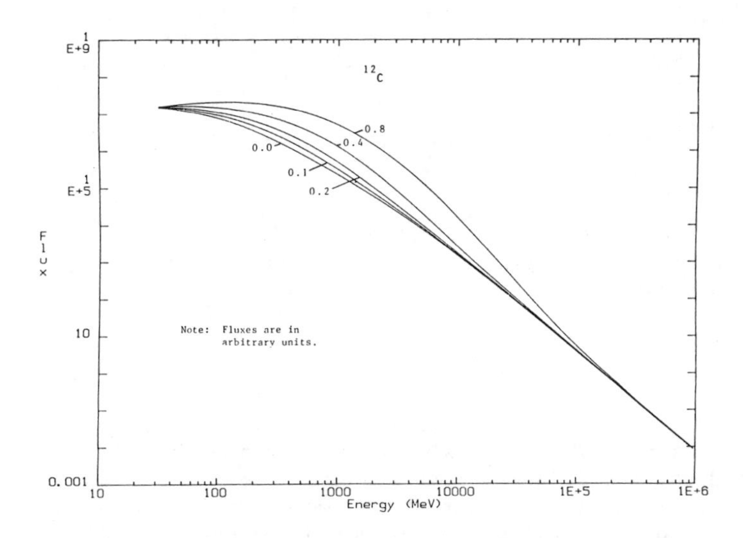

Figure 8: Calculated spectrum of carbon outside the heliosphere. The number of weak shocks per g/cm^2 are given in the figure.

Figure 9 shows the calculated B/C ratio, assuming a power law for the escape path $\lambda_e \propto R^{-0.6}$, (also a smaller exponent, even -0.1 provides a satisfactory fit), and distributed reacceleration. The bending over of the ratio below 1 GeV/nucleon (as is observed experimentally) is a consequence of distributed reacceleration.

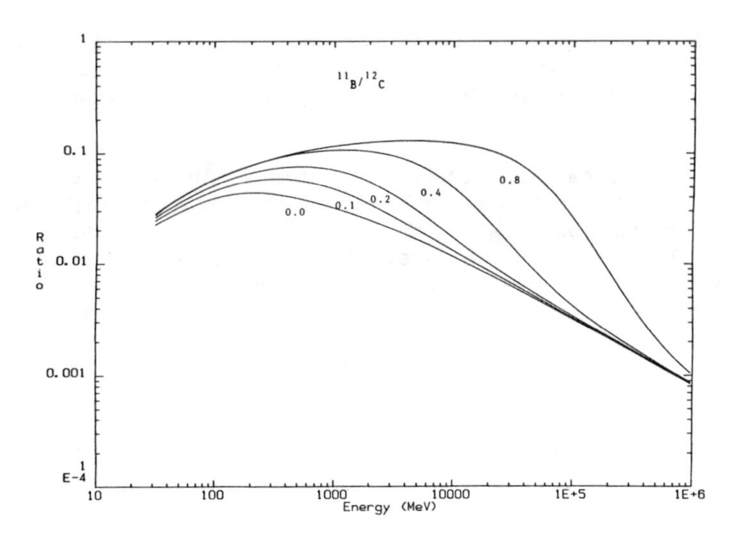

Figure 9: Calculated $^{11}B/^{12}C$ ratio (^{11}B derived from ^{12}C only) as a function of energy. The number of weak shocks per g/cm^2 are also shown.

5. GAMMA RAY LINES

Gamma ray astrophysics is closely associated with cosmic rays and nuclear reactions. Cosmic ray protons (especially those above ~700 MeV) have an appreciable collision cross section for π^0 production, which upon decay yields gamma rays (usually two). At lower energies, gamma ray line production is important, when cosmic rays collide with the nuclei of interstellar gas or clouds or accretion disks of binary neutron stars. Radioactive decay (associated with nucleosynthesis in supernovae) is a major but sporadic source of gamma ray lines (Clayton, 1973). E.g. the formation of ^{56}Ni that decays into ^{56}Fe via ^{56}Co could yield a strong gamma ray line for a few months; (^{56}Co has a half life of 58 days). Also cosmic ray electrons contribute to gamma rays via the inverse Compton effect, (i.e. electron collision with a less energetic photon), and by bremsstrahlung.

Unlike the charged cosmic rays that are scattered in the interstellar magnetic fields, the directions of gamma rays are associated with localized sources where particle acceleration of propagation takes place: the sun, supernova remnants, pulsar-neutron stars, accreting neutron stars, the Galactic Center, and active galactic nuclei and quasars, which probably have ultra-massive black holes, with accretion disks.

A gamma ray line due to nucleosynthesis--that of the radioactive nuclide ^{26}Al--has already been observed at a flux level of 5×10^{-4}/ cm^2-sec. rad (Mahoney et al. 1984 and Share et al. 1985). The latter line is probably due to nucleosynthesis in novae (Clayton, 1984).

In nuclear collisions, there are three processes that give rise to gamma ray lines: (1) Quasi-elastic collisions at low energies that generate excited nuclei. The relevant cross sections for these reactions have been reviewed by Ramaty et al. (1979). The main uncertainty here is the abundance of the low energy nuclei (near 10 MeV). (2) Nuclear spallation reactions, with the spallation products in various excited states, that de-excite promptly by gamma-ray line emission. Here the uncertainty is due to lack of information of the population of the various excited states. (3) Gamma rays produced upon the decay of various radioactive nuclides generated in nuclear spallation reactions. The associated gamma-ray energies and emission probabilities for the latter case are sufficiently well known. In this section we shall concentrate on processes (2) and (3).

The rate at which a nuclide of type j is created per unit volume in the interstellar medium from proton interactions (for protons in a given energy interval) is:

$$R_j \text{ (nuclides/cm}^3 \text{ sec)} = 10^{-27} \, J \, n_H \, F_j$$

Here 10^{-27} is the conversion factor from mb to cm^2, J = cosmic ray protons/cm^2-sec, in a given energy interval. (We assume that the contribution of nuclei heavier than protons results in a second-order correction, though at energies of a couple of MeV/u, the contribution of alpha-induced reactions is important). n_H is the number of hydrogen atoms in the medium, per cm^3, and

$$F_j = \Sigma_i \, \sigma_{ji} f_i,$$

where σ_{ji} is the cross section for protons with nuclide i, in units of mb yielding nuclide j, and f_i is the abundance of i per H atom in the collision medium.

Figure 10 shows the values of F_j as a function of cosmic ray proton energy E, using the elemental and isotopic abundances of Cameron (1982). Assuming that most of the prompt de-excitations pass to the ground state via the first excited state, the most prominent lines induced by high energy cosmic ray protons would be 4.44 MeV from ^{12}C, 2.12 MeV from ^{11}B, 5.27 MeV from ^{15}N, 5.18 MeV from ^{15}O. In addition, the radioactive decay of ^{15}O and ^{11}C yields positrons.

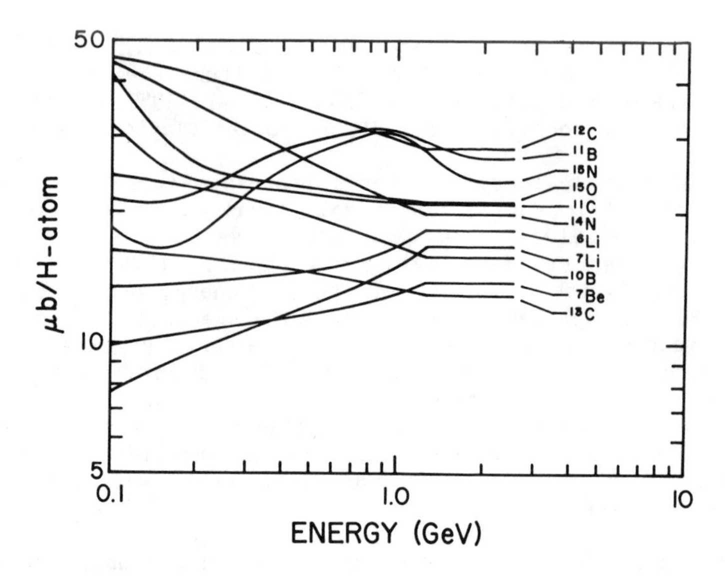

Figure 10: The energy dependence of F_j, the production rate function for nuclides whose secondary yields are largest.

We shall now estimate the gamma-ray line flux from clouds, the Galactic disk and from accreting pulsars. The flux from a large cloud is calculated, using the properties discussed by Issa et al. (1981) i.e. we assume a cloud mass of about 10^5 solar masses, or 10^{62} hydrogen atoms. A distance of 400 pc is assumed. If one adopts Cameron's abundances for the calculations, F = 0.03 for the ^{12}C 4.44 MeV line. J, for the local interstellar cosmic ray intensity is about 10/cm^2-sec. With the most conservative assumptions, the flux of the 4.4 MeV carbon line is about 2 x 10^{-9}/cm^2-sec. However, Issa et al.` (1981) noted that the cosmic ray flux is higher by a factor of 5, especially in clouds associated with O-B stars, where frequent supernovae occur. Furthermore, such O-B regions should have nucleosynthetic enhancement of elements heavier than helium. Adopting a factor of 5 for this enhancement, the flux is 5 x 10^{-8}. Reeves (1978), based on the general abundances of isotopes of B, has concluded that there is a very large flux of low-energy particles. A large low energy cosmic ray flux coupled to the large low-energy cross sections would further increase the 4.4 MeV gamma ray flux from a large cloud by an order of magnitude, to values near 5 x 10^{-7}/cm^2-sec. With a large proton flux between 5 and 30 MeV, such as considered by Ramaty et al. (1979), the contribution of nuclear excitations to the gamma ray line flux becomes dominant; the 4.4 MeV gamma ray line flux could then be boosted further by a factor of about 10.

With the "optimistic" latter assumptions, the flux from the galactic disk (near 4 MeV) due to spallation is about $3 \times 10^{-6}/cm^2$-sec, and with nuclear excitation included, about $3 \times 10^{-5}/cm^2$-sec (Ramaty et al. 1979). The gamma ray background from the galactic disk (due largely to bremsstrahlung) is about 10^{-4} photons/cm^2-sec MeV at 4 MeV and about 10^{-2} at 1 MeV (O'Neil et al. 1983). This background renders the lines from the galactic disk (near 4 MeV) difficult to observe if the fluxes are less than $10^{-5}/cm^2$-sec, unless the line is narrow (i.e. if the recoil nuclei are produced in grains and come to rest therein) and the instrument has an excellent resolution (about 5 keV).

Consider a pulsar beam incident on an accretion disk that is sufficiently thick for most particles to collide. Adopting (a) Cameron's (1982) abundances for the accretion disk material, (b) power input of 10^{37} ergs/sec per decade of energy interval, and (c) a distance of 1 kpc, we estimate a flux of about 10^{-6} f/cm^2-sec for the 4.4 MeV ^{12}C spallation-induced gamma rays. (The flux ^{12}C is about 10 times larger if low-energy nuclear excitation reactions are considered.) Here f is the gamma ray suppression factor due to opacity of the accretion disk: f < 1.

Table 2 summarizes the 4.4 MeV ^{12}C fluxes from the sources discussed above. The 5.1-5.3 MeV fluxes of ^{15}O, ^{15}N, ^{14}N are nearly as abundant.

Table 2. The 4.4 MeV ^{12}C Flux (cm^{-2} sec^{-1}) from Various Sources

Dominant Reaction	Large cloud	Inner Galactic Disk	Accreting Pulsar[a]
Spallation	5×10^{-7}	3×10^{-6}	10^{-6} f
Excitation[b]	5×10^{-6}	3×10^{-5}	10^{-5} f

[a] The assumed proton fluxes and properties of these sources are given in the text.
[b] Excitation dominates over spallation if low-energy cosmic rays dominate.

Gamma ray lines due to nuclear spallation reactions should be detectable with a detector that has a threshold of $10^{-6}/cm^2$-sec from sources like the galactic disk (if the low-energy 20-50 MeV cosmic ray flux is high) and from pulsars and active galactic nuclei that beam into accretion disks. The above detection threshold is about 10 times lower than that of GRO/OSSE. However, if the flux between 5-30 MeV is high, as suggested by Reeves et al. (1978) and Ramaty et al. (1979), nuclear excitation lines will be observable from the above sites (with the GRO/OSSE detector) as well as from supernova remnants in clouds (Morfill et al. 1981).

6. SOLAR MODULATION OF COSMIC RAYS

There are several types of solar modulations: (a) Transient modulations where the cosmic ray intensity is reduced and then recovers after some weeks or a month. These so called Forbush decreases are associated with solar flares propagating into interplanetary space, and with prominences, which upon breaking off from the sun move out as plasma clouds. (b) Co-rotating solar wind flows generate also transient modulations. Such flow is illustrated in Figure 11, from Burlaga (1983).

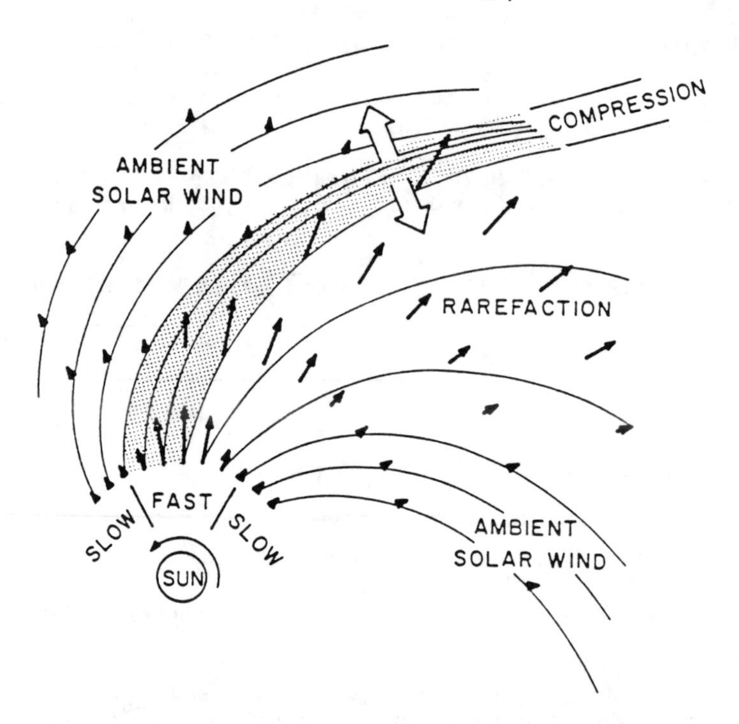

Figure 11: Corotating stream structure. (From Burlaga).

The fast solar wind flow from coronal holes results in rarefaction, but on overtaking the normal velocity component, generates a compression with a higher magnetic field intensity. Such as outward moving wave suppresses temporarily the cosmic ray intensity. (c) Recurrent solar wind flows. The sun rotates with a period of 27 days.

The corotating flow thus can repeat itself. An example is shown in Figure 12 from Iucci et al. (1979) where we note the 27-day periodicity; times with increased flow velocity and higher magnetic field are seen to be anti-correlated with cosmic ray intensity as measured by neutron monitors. (d) Series of Forbush decreases. Before the recovery of the cosmic-ray intensity decrease, another Forbush decrease occurs, resulting in a long-term step-wise reduction of cosmic ray intensity. (e) The 11-year solar cycle. The decreasing

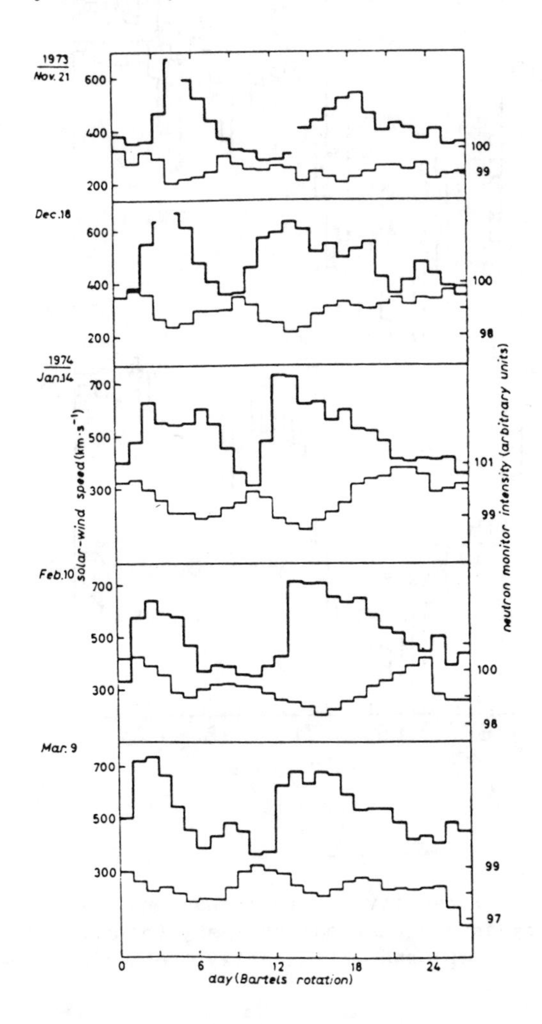

Figure 12: Recurrent solar wind streams (heavy lines) and recurrent Forbush decreases (light lines). (From Iucci et al.).

phase of cosmic ray intensity appears correlated with series of Forbush decreases and frequent flares. This correlation is illustrated in Figure 13, from Hatton (1980) and Figure 14 from McDonald et al. (1981).

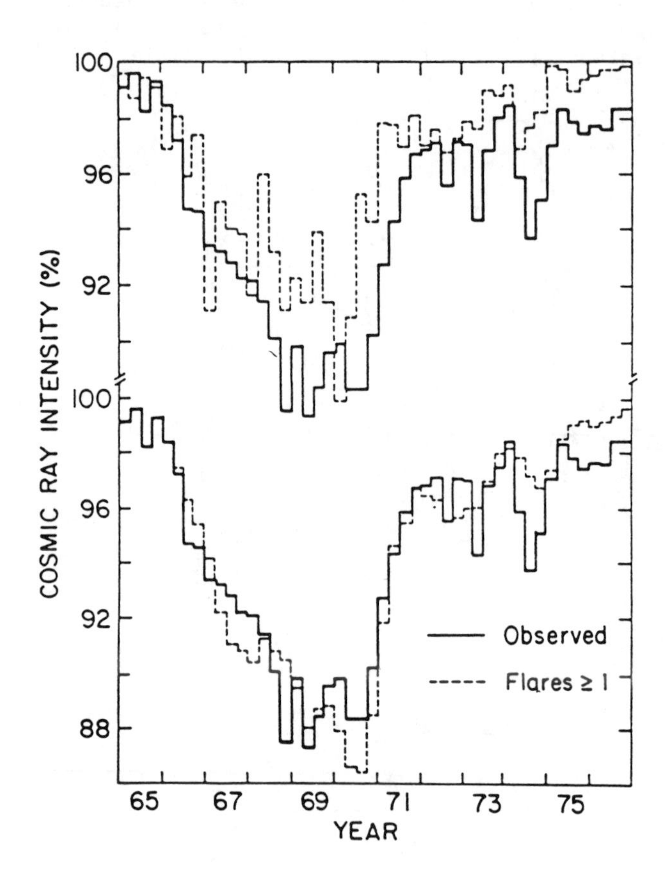

Figure 13: Cosmic ray intensity (solid lines) and the number of flares of importance > 1 plotted inversely (dashed lines) without a time delay (top) and a time delay (bottom). (From Hatton).

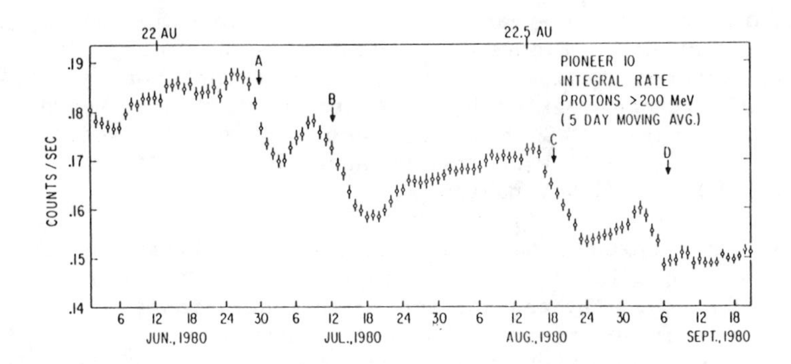

Figure 14: Pioneer data for 4 large Forbush decreases in mid-1980. (From McDonald et al.)

We note that the fit is improved if we adopt a decay that corresponds to the solar wind speed. The cosmic ray intensity variation over 2.5 solar cycles is illustrated in Figure 15 from McKibben et al. (1982).

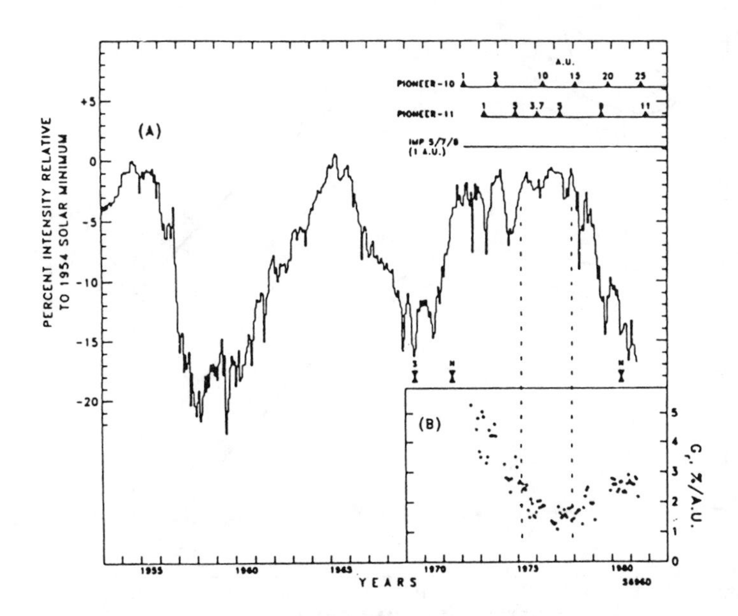

Figure 15: (A) Solar cycle variations of cosmic ray intensity. The insert (B) shows measurements of the radial gradient G_r. (From McKibben et al.)

Solar modulation (the 11-year cycle) affects more low-energy cosmic rays as illustrated in Figure 16, which shows the energy spectrum of cosmic-ray hydrogen, helium and iron. We note that near 100 MeV, the intensity of hydrogen varies by an order of magnitude between solar maximum and minimum. Cosmic rays are furthermore adiabatically decelerated by the outflowing solar plasma; at solar minimum, the loss in energy is about 200 MeV/nucleon.

There are also longer term solar activity variations, occurring at intervals of a couple of hundred years. During these periods, the sunspots are practically absent, and solar modulation is reduced, i.e. the cosmic ray intensity is increased. Such a period, the Maunder minimum, occurred in the late 17th century. Figure 17 from Beer et al. (1983) shows the corresponding increase in the production of ^{10}Be by cosmic-ray generated spallation reactions in the atmosphere.

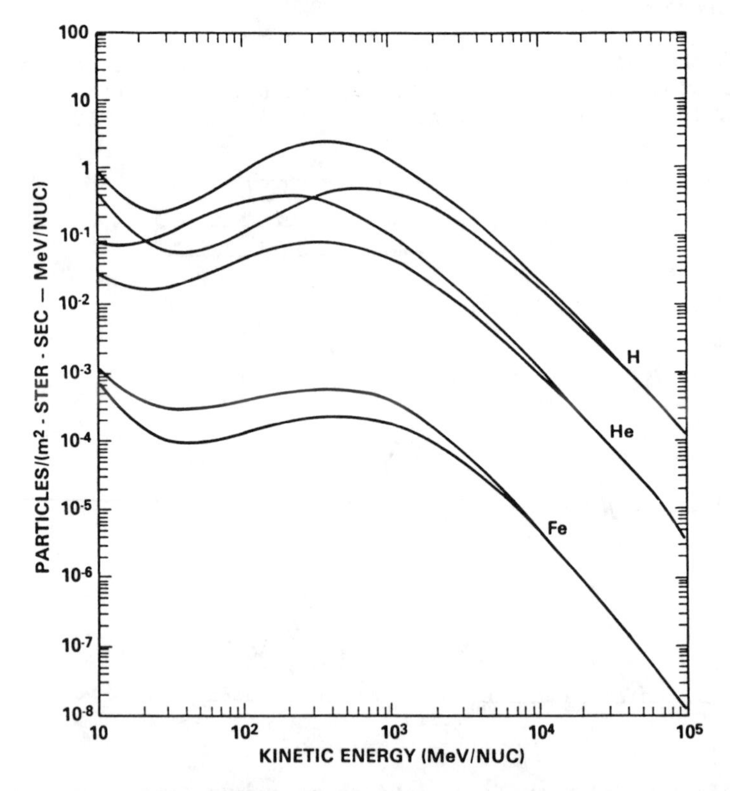

Figure 16: The differential energy spectra of cosmic-ray hydrogen, helium and iron, near solar minimum and maximum.

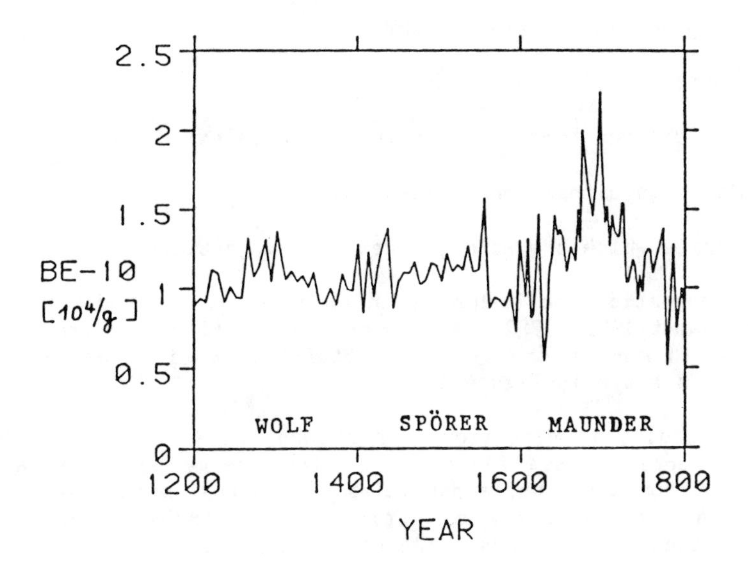

Figure 17: ^{10}Be concentration in units of 10000 atoms/g ice as a function of age in the Milcent ice core (Greenland). (From Beer et al.)

7. GEOMAGNETIC MODULATION

The geomagnetic field acts as a natural spectrometer, suppressing the arrival of cosmic rays below a given rigidity threshold--these particles are deflected away. (Rigidity is defined as the momentum per unit charge of the cosmic ray particle). Regions in the outer magnetosphere and near the magnetic poles can be reached by particles with much lower rigidities than is required near the equator. For each point on the earth's surface in the magnetosphere there exists a magnetic rigidity below which cosmic rays cannot arrive. This value is the geomagnetic cutoff. The fluxes recorded by neuton monitors or muon telescopes at the earth's surface thus also depend on the geomagnetic cutoffs at their location. The geomagnetic cutoff was first calculated by C. Stormer (1930) using a dipole approximation for the earth's magnetic field. He showed that the cutoff rigidity at the earth's surface is given by:

$$p = \frac{60}{r^2} [1-(1-\cos\gamma \cos^3\lambda)^{1/2}]^2/[\cos\gamma \cos\lambda]^2$$

for positively charged particles, where

p = magnetic rigidity in GeV/ec,

r = radial distance from the dipole center in earth radii

λ = latitude in dipole coordinates and

γ = the angle which trajectory makes with magnetic west

Vertical geomagnetic cutoffs have been computed at altitudes of 20 km by Shea and Smart(1975, 1983) using a precise model of the geomagnetic field. A world map of the vertical cutoff rigidities, expressed in units of GV, is shown in Figure 18.

The cutoff rigidity in directions other than the vertical depends on the angle of arrival (see Figure 19). Low energy particles arrive predominantly from west, while particles with energy much greater than the cutoff arrive from any direction. Exact arrival angles are described by the Stormer cones. The relation between the Stormer cone angle, γ, and cutoff is approximately:

$$R = \frac{4R_v}{[1 + (1 - \cos\gamma \cos^3\lambda)^{1/2}]^2}$$

where

$$\cos^4\lambda = \frac{R_v}{17.6}\left[1 + \frac{\text{altitude (km)}}{6371}\right]^2$$

Plots of the ratio of cutoff to vertical cutoff for various vertical cutoffs and Stormer cone angles are displayed in Figure 19. The Stormer cone angle, γ, is related to the azimuthal and zenith angles by:

$$\cos\gamma = \cos\phi \sin\theta$$

The above equations allow the geomagnetic cutoff for any arriving angle to be determined from the vertical cutoff and altitude. Further discussion of the geomagnetic field effect on cosmic rays is contained in Adams et al. (1983). Figure 19 shows the ratio of cutoff to vertical cutoff as a function of vertical cutoff and Stormer cone angles.

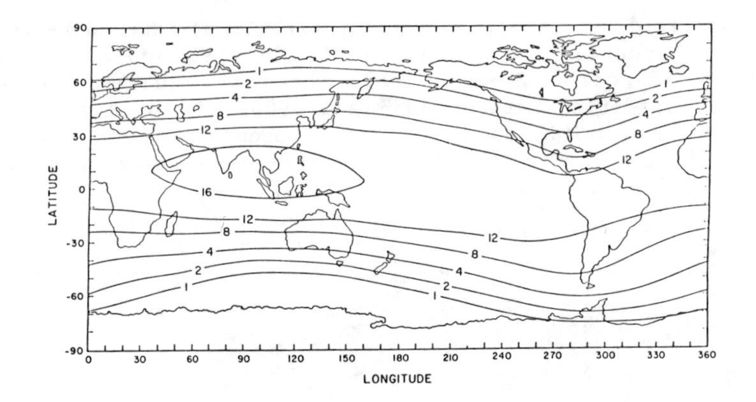

Figure 18: World map of vertical cutoff rigidities.

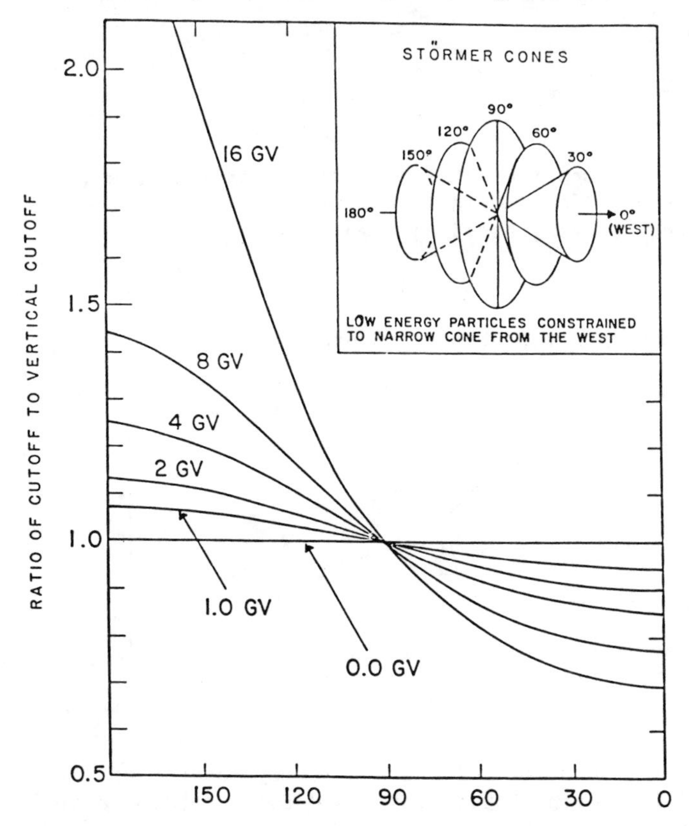

Figure 19: Geomagnetic cutoffs as a function of Störmer cone angle.

The geomagnetic field varies to some extent with time; there have been times of field reversals, with epochs of field disappearance in between. Figure 20, from Stoker (1983) shows the changes in the vertical cutoff rigidities between 1955 and 1980. We note that the regions of high cutoffs (13 to 17 GV) have become reduced, e.g. in Central Brazil there was a region at the geomagnetic equator where the vertical cutoff exceeded 13 GV, but is now less.

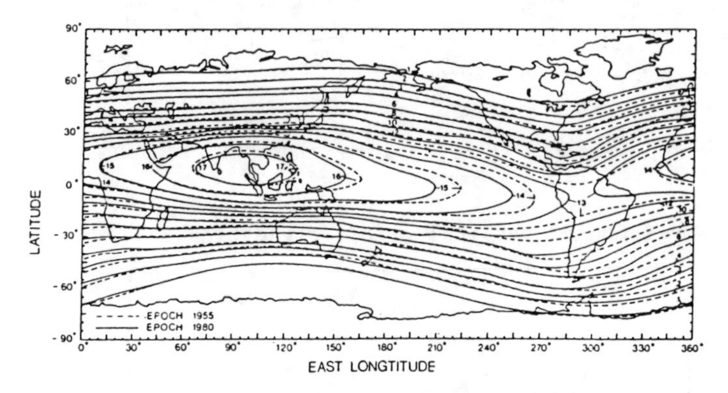

Figure 20: Contours for cutoff rigidities as indicated, comparing epoch 1980 (solid lines) with epoch 1955 (broken lines). (From Stoker, based on Shea et al.)

Figure 21 from Shea et al (1983) shows that between 1955 and 1980 the geomagnetic equator has shifted north by two or three degrees over South America and over the Atlantic Ocean.

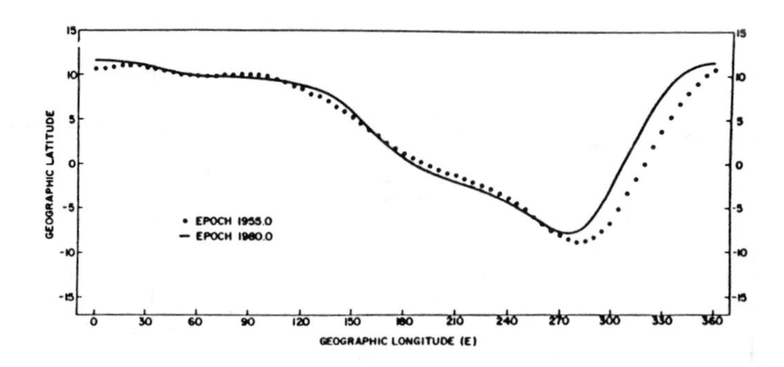

Figure 21: The locations of the cosmic ray equator as calculated for Epoch 1955.0 (dots) and Epoch 1980.0 (solid line). (From Shea et al.)

REFERENCES

Adams, J. H. Jr., Shapiro, M. M., Silberberg, R. and Tsao, C. H. 1979, 17th ICRC, Paris 2, 256.

Adams, J. H., Jr., Bellingham, J. and Graney, P.E., 1983, NRL Memo Report.

Beer, J., et al. 1983, 18th ICRC, Bangalore, 9, 312.

Blake, J. B. and Schramm, D.N., 1974, Ap. Space Sci. 30, 275.

Blankford, R. D. and Ostriker, J. P., 1980, Ap. J., 237, 793.

Brewster, N. R., Freier, P. S. and Waddington, D. J., 1983, Ap. J. 264, 324.

Burlaga, L. F., Invited Paper, 18th ICRP, Bangalore, 12, 21.

Byrnak, B., et al., 1983, 18th ICRC,, Bangalore, 9, 135.

Cameron, A.G.W., 1982, Ap. Space Sci., 82, 123.

Clayton, D. D., 1973, Explosive Nucleosynthesis,, ed. D. Schramm and W. D. Arnett, Univ. Texas Press, Austin, P. 263.

Clayton, D. D., 1984, Ap. J. 280, 144.

Cowsik, R. and Wilson. L. W., 1973, 13th ICRC, Denver 1, 500.

Cowsik, R. and Wilson, L. W., 1975, 14th ICRC, Munich, 2, 659.

Davis, L., Jr., 1959, 6th ICRC, Moscow, 3, 220.

Fowler, P. H., et al, 1985, private communication to J.-P. Meyer. See his 1986 ref.

Garcia-Munoz, M., Simpson, J. A. and Wefel, J. P., 1981, 17th ICRC, Paris 2, 72.

Gorenstein, P., 1981, 17th ICRC, Paris, 12, 99.

Grevesse, N. and J.-P. Meyer, 1985, 19th ICRC, La Jolla, 3, 5.

Guzik, T. G., Wefel, J. P., Crawford, H. J., Greiner, D. E., Lindstrom, P. J., Schimmerling, W. and Symons, T.J.M., 1985, 18th ICRC, LaJolla, 2, 80.

Hatton, C.J., 1980, Solar Phys., 66, 159.

Issa, M. R., Riley, P. A., Li Ti Pei and Wolfendale, A. W., 1981, 17th ICRC, Paris, 1, 150.

Iucci, N., Parisi, M., Storini, M. and Villoresi, G., 1979, Lett. Nuovo-Cimento 24, 225.

Juliusson, E., Meyer, P. and Muller, D., 1972, Phys. Rev. Letters, 29, 445.

Juliusson, E. and Meyer, P., 1975, Ap. J. 201, 76.

Kafatos, M., Bruhweiler, S. and Sofia, S., 1981, 17th ICRC, Paris, 2, 222.

Kertzman, M. P., Klarmann, J., Newport,, B. J., Stone, E. C., Waddington, C. J., Binns, W. R., Garrard, T. L. and Israel, M. H., 1985, 19th ICRC, La Jolla, 3, 95.

Klarmann, J., Margolis, S. H., Stone, E. C., Waddington, C. J., Binns, W. R., Garrard, T. L., Israel, M. H. and Kertzman, M. P., 1985, 19th ICRC, La Jolla, 2, 127.

Koch-Miramond, L., 1981, 17th ICRC, Paris, 12, 21.

Koch, L., Perron, C., Goret, P., Cesarsky, C. J., Juliusson, E., Soutoul, A. and Rasmussen, J. L., 1981, 17th ICRC, Paris, 2, 18.

Letaw, J.R., Silberberg, R. and Tsao, C. H., 1985, Ap. Space Sci. 114, 365.

Mahoney, W. A., Ap. J., 1984, 286, 578.

Margolis, S. H. and Blake, J. B., 1985, 19th ICRC, La Jolla, 3, 21.

McDonald, F. B., Trainor, J. H. and Webber, W. R., 1981, 17th ICRC, 10, 147.

McKibben, R. B., Pyle, K. R. and Simpson, J. A., 1982, Ap. J. 254, L23.

Meyer, J.-P., 1985, Ap. J. Suppl. 57, 173.

Meyer, J.-P., 1986, Rapporteur Papers, 19th ICRC, LaJolla, (to be publ).

Montmerle, T., 1979, Ap. J., 231, 95.

Morfill, G. E. and Meyer, P., 1981, 17th ICRC, Paris, 9, 56.

Muller, D. and Tang, J., 1985, 19th iCRC, La Jolla, 2, 378.

O'Dell, H. F., Shapiro, M. M. and Stiller, B. 1962, J. Phys. Soc. Japan Suppl. A-III, 17, 23.

O'Dell, F. W., Shapiro, M. M., Silberberg, R. and Tsao, C. H., 1975, 14th ICRC, Munich, 2, 526.

O'Neill, T., Dayton, B., Long, J., Zanrosso, E., Zych, A. and White, R. S., 1983, 18th ICRC, Bangalore, 9, 45.

Ormes, J. F. and Balasubrahmanyan, V. K., 1972, Nature Phys. Sci., 241, 95.

Ormes, J. F., Fisher, A., Hagen, F., Maehl, R. and Arens, J. F., 1975, 14th ICRC, Munich, 1, 245.

Ormes, J. F. and Protheroe, R. J., 1983, Ap. J. 272, 756.

Peters, B. and Westergaard, N. J., 1977, Ap. Space Sci., 48, 21.

Prantzos, N., Arnould, M., Arioragi, J. P. and Casse, M., 1985, 19th ICRC, La Jolla 3, 167.

Raisbeck, G. M., Perron, C. Touissaint, J., and Yiou, F., 1975, 13th ICRC, Denver, 1,

Ramaty, R., Kozlovsky, B. and Lingenfelter, R. E., 1979, Ap. J. Suppl., 40, 487.

Reeves, H. and Meneguzzi, M., 1978, Gamma Ray Spectroscopy in Astrophysics, Ed. T. L. Cline and R. Ramaty, NASA Tech. Memo 79619, p. 283.

Shapiro, M. M., Silberberg, R. and Tsao, C. H., 1969, 11th ICRC, Budapest, paper OG-87.

Shapiro, M. M. and Silberberg, R., 1975, 14th ICRC, Munich, 2, 538.

Share, G. H., Kinzer, R.L., Kurfess, J. D., Forrest, D. J., Chupp. E. L. and Rieger, E., 1985, Ap. J. (Lett.), 292, L61.

Shea, M.A. and Smart, D. F., 1975, Report No. AFCRL-TR-75-0185, Hanscom AFB, Ma.

Shea, M. A. and Smart, D. F., 1983, 18th ICRC, Bangalore, 3, 415.

Shea, M. A., Smart, D. F. and Gentile, L. C., 1983, 18th ICRC, Bangalore, 3, 423.

Silberberg, R., Tsao, C. H. and Shapiro, M. M., 1976, in Spallation Nuclear Reactions and Their Applications, p. 49, ed. S. P. Shen and M. Merker, Publ. Reidel, Dordrecht.

Silberberg, R., Tsao, C. H., Shapiro, M. M. and Letaw, J. R., 1983, 18th ICRC Bangalore, 2, 179.

Silberberg, R., Tsao, C.H., Letaw, J. R. and Shapiro, M. M., 1983, Phys. Rev. Letters 51, 1217.

Silberberg, R. Tsao, C. H., Letaw, J. R. and Shapiro, M. M., 1985, 19th ICRC LaJolla 3, 238.

Smith, L. H., Buffington, A., Smoot, G. F., Alvarex, L. W. and Wahlig, W. A., 1973, Ap. J. 180, 987.

Soutoul, A., Casse, M. and Juliusson, E., 1975, 14th ICRC, Munich, 2, 455.

Soutoul, A., Casse, M. and Juliusson, E., 1978, Ap. J., 219, 753.

Stephens, S. A., 1985, 19th ICRC, LaJolla, 2, 350.

Stoker, P. H., Rappporteur Paper, 1983, 18th ICRC, Bangalore, 12, 425.

Stormer, C, 1930, Z. Astrophys. 1, 237.

Streitmatter, R. E., Balasubrahmanyan, V. K., Ormes, J. F. and Protheroe, R. J., 1983, 18th ICRC, Bangalore, 2, 183.

Webber, W. R., Kish, J. C. and Schrier, D. A., 1985, 19th ICRC, LaJolla, 2, 88.

Wiedenbeck, M. E., 1983, 18th ICRC, Bangalore, 9, 147.

Wiedenbeck, M. E., 1985, 19th ICRC, La Jolla, 2, 84.

Wiedenbeck, M. E. and Greiner, D. E., 1980, Ap. J. Letters, 239, L139.

Wiedenbeck, M. E. and Greiner, D. E., 1981, Ap. J. Letters, 247, L122.

ULTRA HEAVY NUCLEI IN THE COSMIC RADIATION

W. Robert Binns
Department of Physics, and the McDonnell Center for the Space Sciences, Washington University
St. Louis, Missouri 63130, U.S.A.

ABSTRACT. This paper describes the measurements of the Ultra-Heavy cosmic ray abundances obtained by the Heavy Nuclei Experiment aboard the NASA High Energy Astronomy Observatory-3. A subset of the HEAO data which has been recently determined to give improved resolution in the charge region from Z = 40 to 60 is described and preliminary abundances from these data are given. In general we find that the cosmic ray abundances are in broad agreement with solar system abundances with a step-FIP fractionation model applied although in detail there are some differences. In particular Ge and Pb appear to be underabundant in the cosmic radiation. However if the recent solar photospheric abundance estimates of Grevesse and Meyer for Ge and Pb are used, solar system abundances are brought into agreement with our data. Although the platinum/lead ratio and the actinides are consistent with some r-process enhancement, we see that the cosmic ray source is not dominated by the r-process up through the 50's as evidenced by the Sr/Rb ratio and by the abundance of Sn and Ba. In addition the actinides are not greatly enhanced, ruling out freshly synthesized r-process production as the primary source of the heavy cosmic rays.

1. INTRODUCTION

The study of Ultra Heavy (UH) nuclei in the cosmic radiation (nuclei with Z > 30) is significant in Astrophysics for several reasons. The first is that the measured cosmic ray abundances can be compared to specific nucleosynthesis process predictions to look for the signature of these production processes, or mixes of these processes [J. Wefel, this conference]. In particular we look for abundance peaks in the element distribution which are characteristic of a particular process. Secondly, the measured abundances can be studied for evidence of fractionation effects which may have modified the cosmic ray source abundances in some systematic way. For example, solar cosmic ray abundances have been shown to be fractionated by a mechanism which modifies the abundance of elements depending on the first ionization potential of the element. Finally UH cosmic rays are important in the study of the propagation and history of cosmic radiation.

M. M. Shapiro and J. P. Wefel (eds.), Genesis and Propagation of Cosmic Rays, 71–89.

The standard of comparison which we have in nature is the abundances of elements in our solar system. These abundances have been compiled by several authors [1-5] and are based on measurements of element abundances in carbonaceous chondrite meteorites and, for some elements, upon solar photospheric measurements. Since galactic cosmic ray abundances have been shown to be broadly similar to these "solar system abundances" it is instructive to plot them as shown in Figure 1 where the even and odd charge abundances are plotted separately to more clearly show the charge spectrum features. For a cosmic ray source with similar abundances we would expect to have good "source visibility" for elements which are more abundant than heavier elements, so that most of the nuclei detected at earth are source nuclei and not products of heavier nuclei which have interacted in the interstellar medium. The elements which are expected to be substantially primary are plotted as filled data points and those which would be expected to be mostly secondary are plotted as open points. To see this more quantitatively I have plotted in Figure 2 the fraction of elements with charge between 30 and 62 detected at earth which are primary, assuming Anders and Ebihara source abundances modified by a step first ionization potential (FIP) model [6] and propagated using a leaky box model with an escape pathlength of $7 g/cm^2$ of hydrogen [7]. We see that there are a number of elements in the $30 \leq Z \leq 40$ and $50 \leq Z \leq 60$ charge regions that are mostly primary in their origin while those in the $41 \leq Z \leq 49$ are mostly secondary in origin. Not shown in this figure is the $Z \geq 62$ region in which the elements with $Z = 76, 77, 78$, and 82 (Os, Ir, Pt, and Pb) are mainly primary with the other elements having significant secondary components. Thus source abundances can be studied most directly by focusing on those elements which are mainly primary in composition and the propagation of these elements can be studied by looking at those which are mainly secondary in origin.

The cosmic ray abundances can also be examined for evidence of nucleosynthesis by the "slow" addition of neutrons to a seed nucleus, the s-process, and by the "rapid" addition of neutrons to a seed nucleus, the r-process. Slow and rapid here refer to the rate of neutron addition to the seed nuclei present in the region of active nucleosynthesis which is slow or rapid compared to the radioactive decay rate of beta unstable nuclei. The s-process is believed to occur, for example, in normal stars in the red giant stage with the r-process occurring in explosive processes such as supernovae. The solar system abundances have been decomposed into an r- and s- process composition by Cameron [4] and Binns, *et al.* [8]. These two decompositions give nearly the same results for the elemental composition of the r- and s-processes and the Cameron decomposition is shown in Figure 3 for the charge region from $Z = 30$ to 92. From the figure we see that the solar system mix is predominately s-process in the $Z = 30–40$ charge region. In the 50-60 region the elements $_{50}Sn$, $_{56}Ba$, $_{58}Ce$, and $_{60}Nd$ are primarily s-process with elements $_{52}Te$ and $_{54}Xe$ being r-process peaks. For the very heavy elements

$_{92}$U, with the only s-process element being $_{82}$Pb. Thus experimental data on element abundances in the cosmic radiation can be studied for evidence of the occurrence of these specific processes.

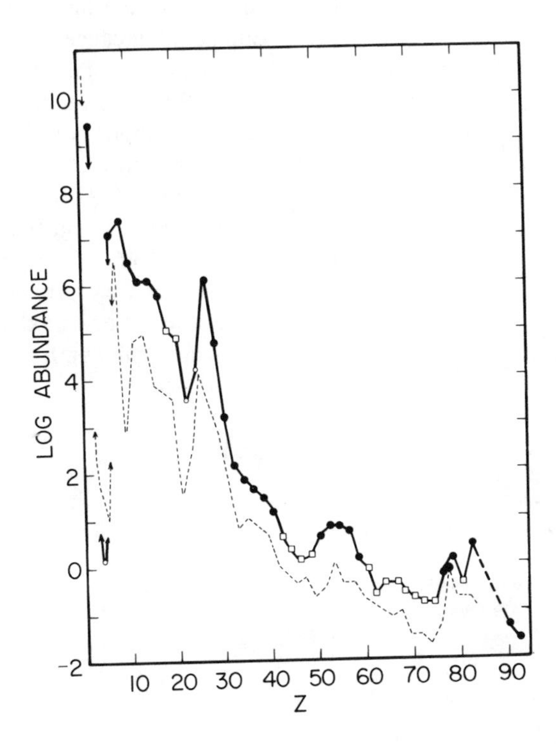

Figure 1: Relative abundances of the elements [4-5]. The solid line connects even-Z elements and dashed line connects odd-Z elements. The points plotted as filled circles are elements which are expected to be mostly primary after propagating to earth from the source and those plotted as open circles are expected to be mostly secondary interaction products from propagation through the interstellar medium.

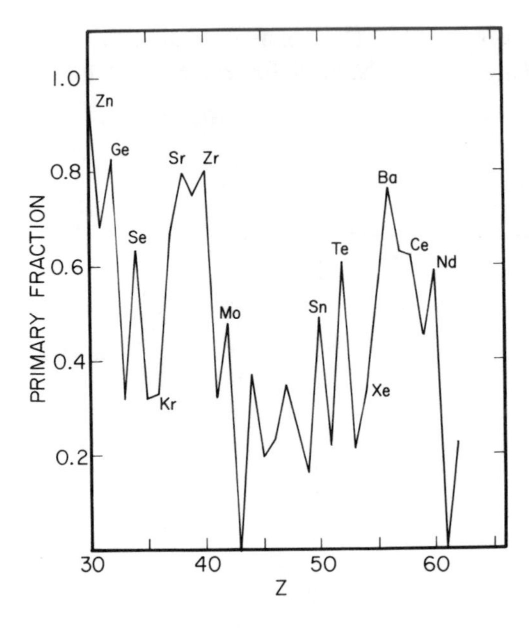

Figure 2: The primary fraction of cosmic ray nuclei arriving at earth after propagation through the interstellar medium is plotted. The assumed source abundances are the Anders and Ebihara abundances [1] with a step FIP bias applied. The FIP function that was used gave a multiplier of 12.6 for elements with FIP < 7 eV, of 8.37 exp(−0.27FIP) for 7 < FIP < 13.6, and 0.21 for FIP > 13.6 eV. These abundances were propagated using a leaky box model with an escape pathlength of 7 g/cm^2 of hydrogen.

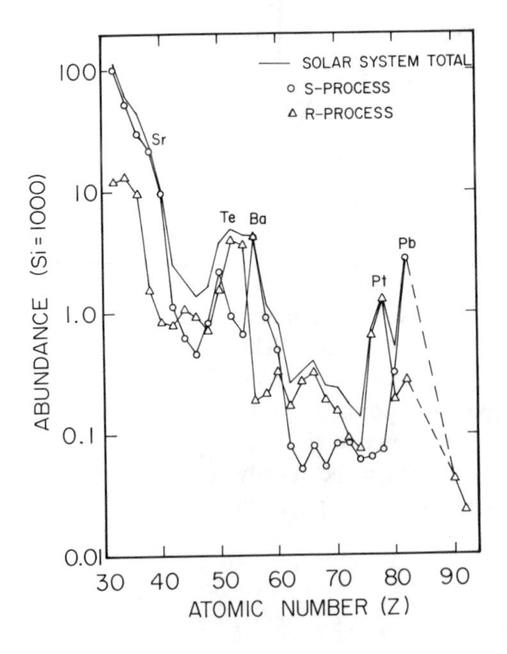

Figure 3: Contribution of the r-process (triangles) and the s-process (circles) to the total solar system abundance of each even-Z element (upper line) [4].

2. THE HEAVY NUCLEI EXPERIMENT INSTRUMENT

The Heavy Nuclei Experiment (HNE) was launched aboard the NASA High Energy Astronomy Observatory (HEAO-3) satellite in 1979 for the purpose of studying the abundances of elements heavier than iron in the cosmic radiation. This instrument as shown in Figure 4 consists of six independently pulse height analyzed ionization chambers for dE/dx measurements, a Cherenkov counter with eight photomultipliers independently pulse height analyzed, and wire ionization hodoscope counters for pathlength corrections to detector response [9]. Both the ionization and Cherenkov signals are proportional to the charge of the nucleus (Z) squared, to first order, but depend differently upon the particle velocity. Figure 5 gives a plot of ionization signal versus Cherenkov with the energy as a parameter. Nuclei with kinetic energy less than 1 GeV/amu fall along the charge contours to the left of the diagonal line and give unambiguous charge identification. Cosmic rays with rigidity greater than about 8 GV fall along the upward rising part of each contour on the right, also providing unambiguous charge identification. The remaining nuclei which have energy greater than about 1 GeV/amu or are detected when the geomagnetic rigidity cutoff is less than 8 GV fall in the minimum part of the hook and some of these overlap nearby contours of other charges. Thus we have in the past chosen the low energy and high geomagnetic cutoff regions as our best regions for obtaining unambiguous charge resolution. Figure 6 shows a histogram of nuclei in the iron region of the charge spectrum which is a combination of nuclei from the "low energy" and "high rigidity" data subsets. We observe that the element peaks for the even charge elements are clearly resolved and we see the dramatic fall off in abundance as we go to charges heavier than iron. A resolution in charge of 0.34 c.u. was obtained for these data. Very recent analysis indicates that excellent charge resolution can also be obtained for a large data set in the "ambiguous" energy region for portions of the charge spectrum. This will be discussed in more detail below.

3. COSMIC RAY OBSERVATIONS IN THE $32 \leq Z \leq 42$ CHARGE REGION

In Figure 7 we show a histogram of nuclei in the low energy and high rigidity data sets combined for the charge range extending from $Z = 32$ to 42 selected in the same way as the iron events in Figure 6. Clear peaks are evident at the even charges extending from charge 32 through 38 with the statistics becoming poor for 40 and 42. This data set corresponds to 2.57×10^6 iron nuclei and is roughly 15% of the total number of nuclei detected. We have fit this data with a gaussian maximum liklihood algorithm to derive element abundances and obtained our best fit for a distribution sigma of 0.40 c.u. at charge 32, with the sigma increasing linearly by 0.016 c.u./charge unit as we go up the charge scale. The best fit is shown as the smooth curve in Figure 7.

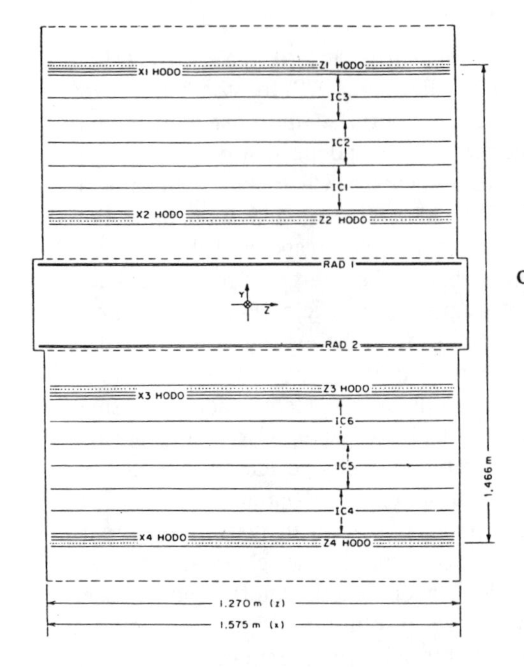

Figure 4: Schematic cross-section of the HNE instrument on HEAO-3.

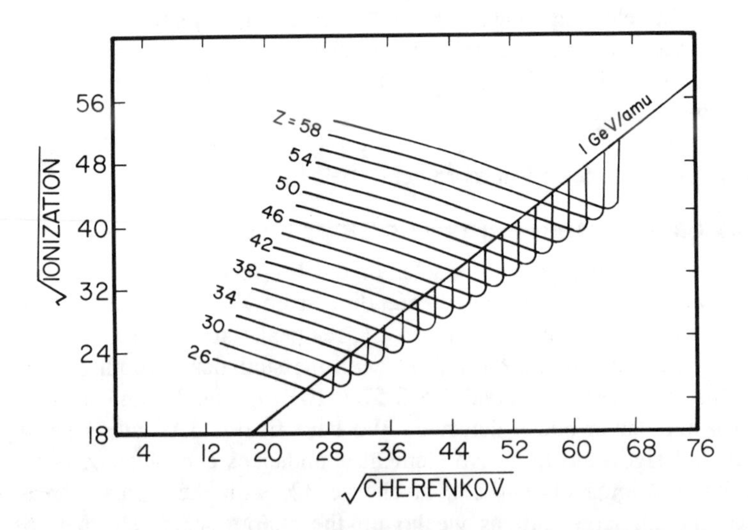

Figure 5: The square root of the ionization and Cherenkov signals are plotted for even charges with energy as a parameter. Nuclei with energy less than 1 GeV/amu lie to the left of the diagonal line and those with a greater energy to the right of the line.

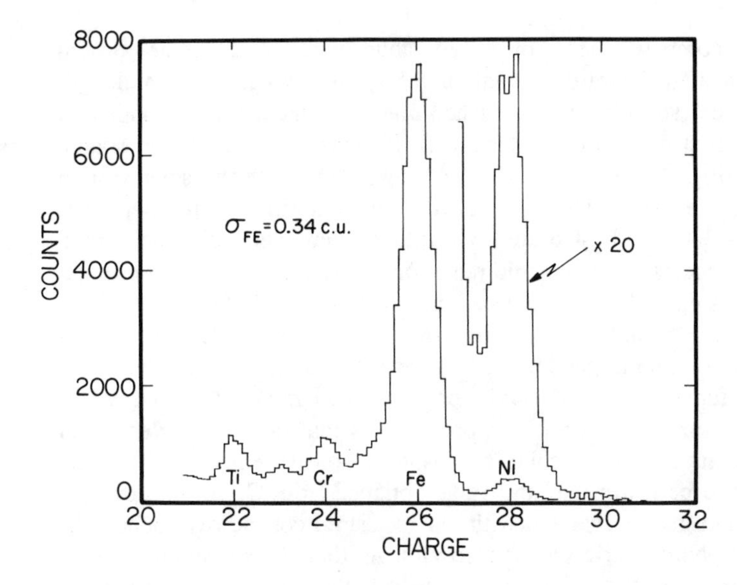

Figure 6: Histogram of nuclei in the iron region of the charge spectrum chosen from the low energy and high rigidity data sets. The histogram labeled x 20 has the number of counts multiplied by 20 to more easily show the shape of the nickel peak.

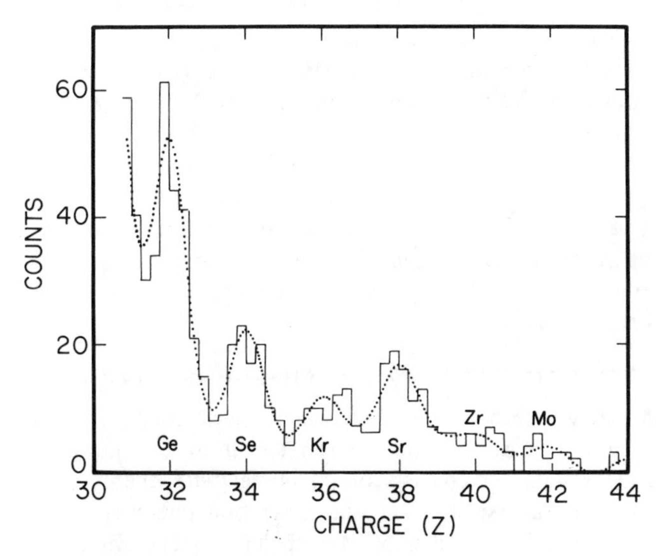

Figure 7: Histogram of nuclei selected in the same way as the iron region events shown in Figure 6.

In Figure 8(a) we compare these measured abundances with solar system abundances [5] propagated forward through a 5.5 g/cm^2 exponential pathlength distribution of Hydrogen (solid line). The dashed line gives these same abundances with an exponential first ionization potential (FIP) bias applied at the source (f = 9.31exp(–0.288FIP)). We see generally good agreement with the solar system abundances with the exponential-FIP applied although in detail there are significant differences. $_{32}$Ge and $_{37}$Rb are both volatile elements [10] which could account for their deficiencies in the cosmic radiation. However Rb can be brought into agreement with a step-FIP model such as that proposed by Letaw *et al.* [6] or Meyer [10-13]. Also the Sr agreement is improved using a step-FIP bias instead of the exponential form. The dotted line in Figure 8(a) results from applying the Letaw *et al.* step-FIP function. Ge cannot be brought into agreement with any FIP model since it has the same first ionization potential as that of Fe. Grevesse and Meyer [14] have examined recent solar photospheric measurements of the Ge abundance and find it to be lower than the value obtained from C1 meteorites by a factor 0.61 which would give agreement with our measured cosmic ray abundance. However these recent photospheric results also indicate that Fe is overabundant in the photosphere by a factor of 1.45 compared to C1 abundances. Although the photospheric Fe and Ge measurements are decoupled it appears that we have reached a level of accuracy in our cosmic ray measurements which is better than our current understanding of the proper standard for comparison. A similar comparison with the Cameron Solar System r-process [5] with and without a FIP bias is shown in Figure 8(b). Our measured abundances are almost a factor of 5 greater than the r-process part of the SS abundances. If we permit an r-process enhancement by multiplying the predicted r-process abundances by a factor of 4.5 as shown in Figure 8(c), our even Z abundances roughly follow those from an r-process source with FIP. However the Cameron SS r-process with or without a FIP bias has more Rb than Sr for any conventional FIP function. Our data as shown in both the histogram and the fitted values clearly show that there is more Sr than Rb in the cosmic radiation. This is a strong indicator that the Cameron SS r-process is not dominant in this charge region. Thus we see that in the 30's the data show good agreement with solar system abundances with a step-FIP fractionation model applied.

4. COSMIC RAY OBSERVATIONS IN THE $40 \leq Z \leq 60$ CHARGE REGION

We have recently found a new subset of our data which gives greatly improved resolution and statistics in the region of the charge spectrum extending from Z = 40 through 60 [7]. This data set consists of all nuclei with energy greater than about 1 GeV/amu which traverse at least one ionization chamber on both sides of the Cherenkov counter. The number of iron nuclei corresponding to this data set is 8.0×10^6. To put this data set in the context of the charge region from 26 to 40 just discussed we show charge histograms of this data set from charge 20 through 30 (Figure 9) and charges 30 through 42 (Figure 10) where the charge is identified on the basis of the Cherenkov signal alone. We see that by comparison with the previous data set the resolution is considerably worse. In Figure 9 we see that the iron peak has a long tail extending down the charge scale. This is not surprising since this data set includes a significant number of nuclei with energy low enough so that their Cherenkov signal has not quite reached the plateau value, thus giving a low apparent charge. In Figure 10 we see peaks evident at charge 34 and 38 but with a much worse peak to valley ratio than observed previously in Figure 7. The iron spectrum has been scaled and plotted in Figure 10 as a pure Z = 38 particle beam should appear if it had an identical shape. Using this resolution function we have fit the charge histogram and we see that there is good agreement between the resulting fit and the data.

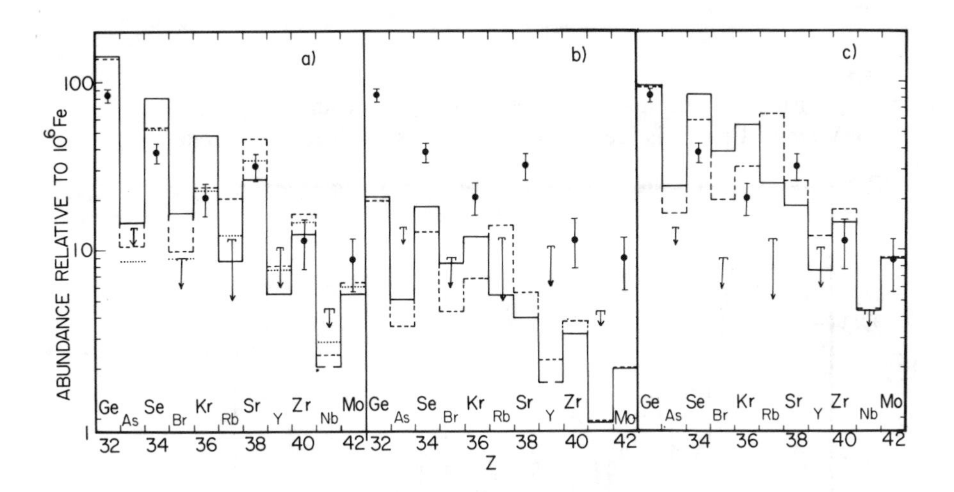

Figure 8: Our data compared with a) Cameron solar system abundances propagated through 5.5 g/cm^2 exponential path length distribution of hydrogen (Solid line); the same but with an exponential-FIP bias applied (Dashed line); the same but with a step-FIP bias applied (Dotted line). b) Cameron r-process abundances propagated (Solid line); the same but with an exponential-FIP bias applied (Dashed line). c) Cameron s-process abundances propagated (Solid line); the same but with an exponential-FIP bias applied (Dashed line).

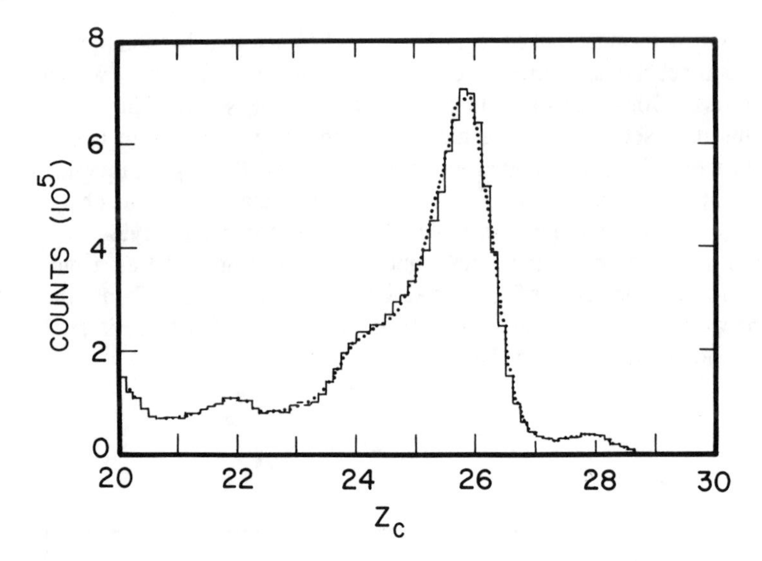

Figure 9: Histogram for nuclei in the iron region with energy greater than 1 GeV/amu. Charge is identified on the basis of Cherenkov alone.

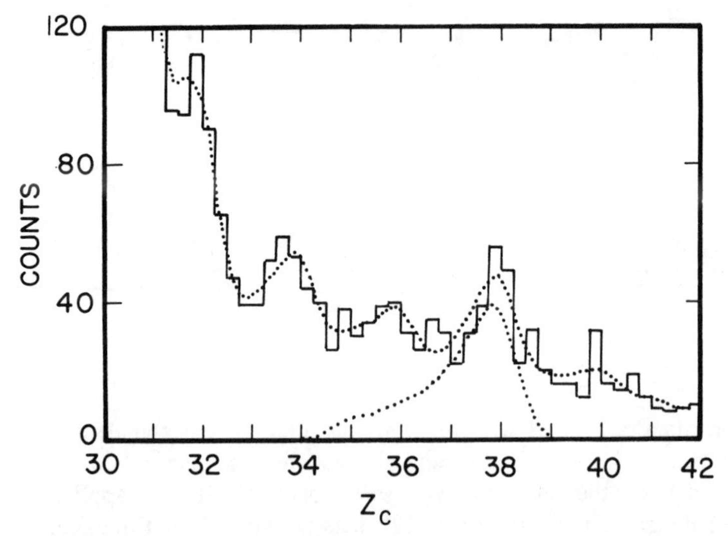

Figure 10: Histogram of nuclei in the charge range from Z = 31 to 42. The upper dotted curve is the fit curve using the iron spectrum scaled by Z. The lower dotted curve is the iron distribution scaled by Z to charge 38. Charge is identified on the basis of Cherenkov alone.

Figure 11 extends the histogram from charge 40 through 62 and we see that we obtain peaks at every even element from charge 40 through 60 and that they are remarkably sharp compared to those in the 30's. This is the first data set for which we have resolved peaks in the 40's region. If we try to fit this data with the same fitting function (the iron fitting function) that was used in the 30's we see that the fit consistently underestimates the element peaks and over estimates the valleys (Fig. 11-smooth curve). However if we try fitting this data with a gaussian resolution function we obtain very good agreement between the data and the fit as shown in Figure 12. This result was totally unexpected since we have used the resolution at iron where abundant statistics were available to serve as our guide for finding good resolution data sets for the lower statistics, high charge data. This apparent transition from poor resolution at iron and in the 30's to the good resolution in the 40-60 range suggests that the energy spectra of nuclei with $Z > 40$ might be deficient at low energies (< several GeV/amu) compared to iron, or that a systematic instrumental or selection bias against low energy, high charge nuclei is present, but we have been unable to find any mechanism which might give this result. A third possibility is that the Cherenkov emission dependence on charge and velocity is different than we have assumed. However we have performed Bevalac calibrations with nuclei ranging from Neon through Uranium and have found no evidence for significant deviations from the expected Cherenkov response for energies up to about 1.8 GeV/amu. Thus although we have obtained much improved resolution in this charge region, we do not yet understand the reason for this good resolution and are actively studying this at the present time.

The preliminary abundances derived from this gaussian fitting procedure are plotted in Figure 13 in charge pairs, the abundance of each even charge being combined with the odd charge immediately below it, with the exceptions of 39-41 which are combined and plotted as a single point and 42 which is plotted by itself. The abundances are presented in this fashion since there may still be a residual tail extending down from the more abundant even charges. The data analysis and fitting procedure are described in detail in Newport's thesis [7]. For comparison we show our previously published abundances in Figure 13 [15-16] and see that we obtain very good agreement with this earlier, but smaller, data set. Figure 14 shows these new abundances compared with the Ariel-6 abundances [17 and 18] and again we see good agreement over this entire charge region.

To compare these abundances with those of the solar system we have taken the ratios of our measured abundances to the Anders and Ebihara abundances fractionated according to the Letaw *et al.* step-FIP model and propagated them to earth using a rigidity dependent path length leaky box model [7 and 19]. This ratio is shown in Figure 15 plotted versus charge and we see that there is good agreement over the charge spectrum from charge 34 to 60. Similar comparisons with an exponential-FIP fractionation gives poorer agreement as do comparisons with pure s- or r-process abundances propagated to earth [7].

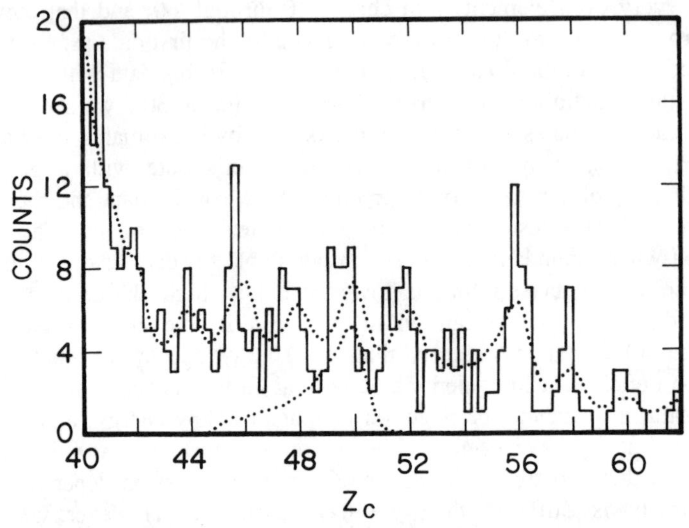

Figure 11: Histogram of nuclei in the charge range from Z = 40 to 62. The upper dotted curve is the fit using the iron spectrum scaled by Z. The lower dotted curve is the iron distribution scaled by Z to charge 50. Charge is identified on the basis of Cherenkov alone.

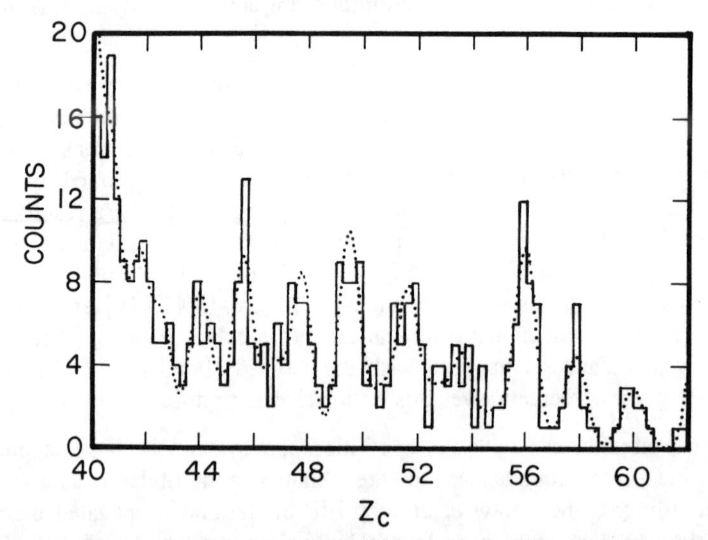

Figure 12: The same data is plotted as in Figure 11 but with a gaussian fit. It is seen that the peak to valley ratio of the fit curve agrees well with the histogram.

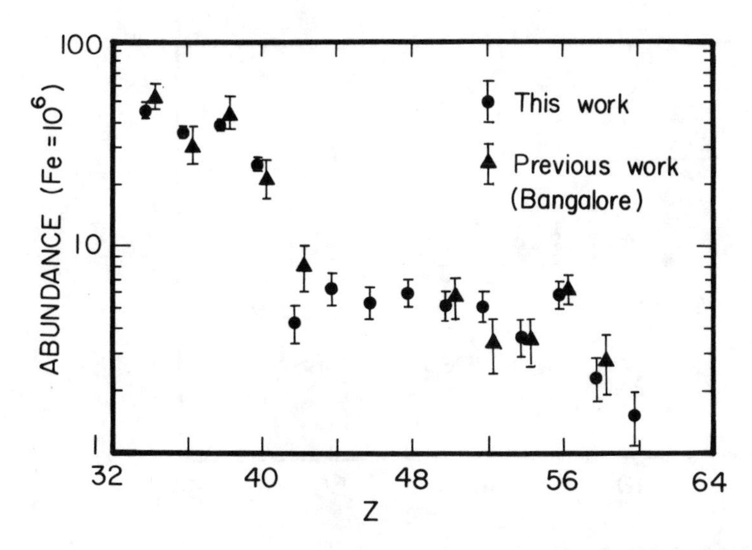

Figure 13: The abundances obtained from the gaussian fit are plotted and compared to our previously published abundances in the 30's and 50's charge regions.

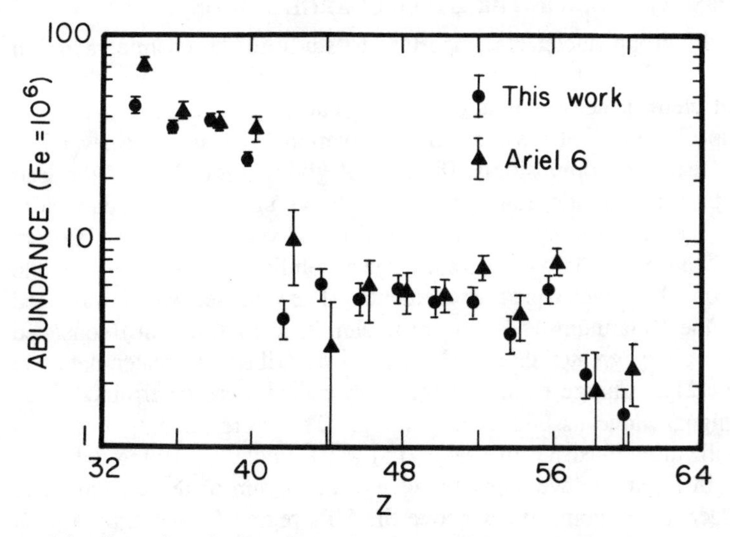

Figure 14: Our abundances from the gaussian fit compared to the Areal-6 abundances.

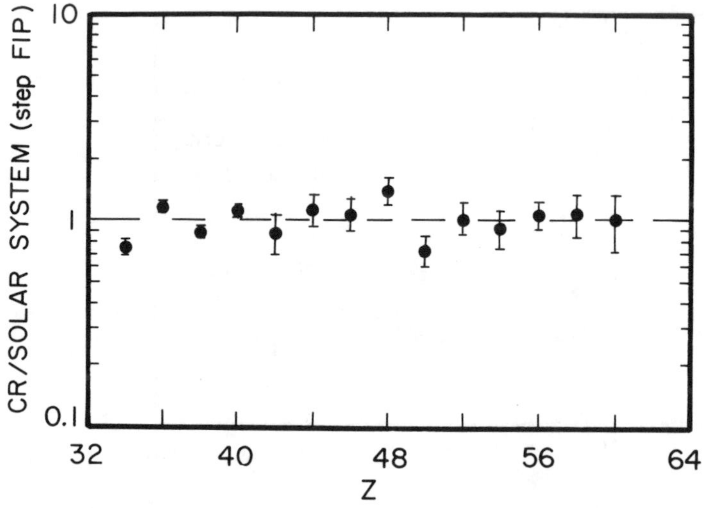

Figure 15: The ratio of our measured abundances to the Anders and Ebihara abundances fractionated using a step-FIP model is plotted versus charge.

5. COSMIC RAY OBSERVATIONS IN THE Z > 60 CHARGE REGION

The "platinum-lead" group nuclei are important to study in the cosmic radiation since $_{76}$Os, $_{77}$Ir, and $_{78}$Pt are primarily r-process peaks with $_{82}$Pb being the only relatively abundant element heavier than charge 70 that has a substantial s-process component. In the Cameron solar system decomposition [5] lead is nearly all s-process with more recent decompositions [8 and 20] giving about half of the lead produced in the s-process. In addition there is a 4 charge unit gap between charge 78 and 82 with the intervening elements having a very low solar system abundance. Thus even with relatively poor charge resolution it should be possible to separate lead from the lower r-process elements. The data set which was used in the analysis of the "platinum-lead" group nuclei (*i.e.,* $74 \leq Z \leq 86$) consisted of all events with rigidity greater than 5 GV and is described in greater detail in Waddington *et al.* [21]. Charge estimates for these nuclei were determined from their Cherenkov signal alone assuming Z^2 scaling. The corresponding iron data set used for normalization consisted of $(9.6 \pm 0.5) \times 10^6$ nuclei satisfying the same selection criteria. In Figure 16 we show the charge histogram of these events and note the general decrease in abundances above the 50's region followed by a peak in the 74-82 charge region. Although our decreased resolution at these high charges prevents us from identifying individual element peaks, we can obtain an estimate of the ratio of "lead" to "platinum" where we define lead and platinum as those nuclei given charge estimates $81 \leq Z \leq 86$ and $74 \leq Z \leq 80$ respectively. From this data we identify 10 nuclei as "lead" and 42 as "platinum". Figure 17 shows this abundance ratio compared to r-process and solar system abundances with and without a FIP fractionation. The Ariel-6 value [18] is also plotted and within the accuracy of the measurements is in agreement with our ratio. Our HEAO data point is distinctly lower than the solar system or r-process abundance estimates. This result might suggest that, unlike the cosmic rays with Z < 60, the cosmic rays with Z near 80 come from a source with a distinctly different nucleosynthesis history than do the solar system elements. However alternatives to this conclusion are that the Pb abundance in the cosmic ray source may be suppressed by some form of source fractionation which depends upon a different parameter than FIP (*e.g.* volatility), or that the Pb abundances assumed in model calculations are not really representative of the solar system or of the r- or s-process contributions to the solar system. Specifically Grevesse and Meyer [14] have re-examined the spectroscopic data on both Pb and Ge (which is the other prominent element which is underabundant in the cosmic radiation). They concluded that the best photospheric abundance for Pb is about 0.63 of the standard (C1 meteorite) abundance which brings the solar system abundances into agreement with our measurement. It should be noted that this would reduce the r-process ratio by an even greater amount since the r-process abundances are obtained by subtracting the calculated s-process abundance from the assumed solar system value. At the present time the interpretation of this measurement is not completely resolved. Waddington *et al.* [21] and Meyer [13] present a more detailed discussion of the significance of this result.

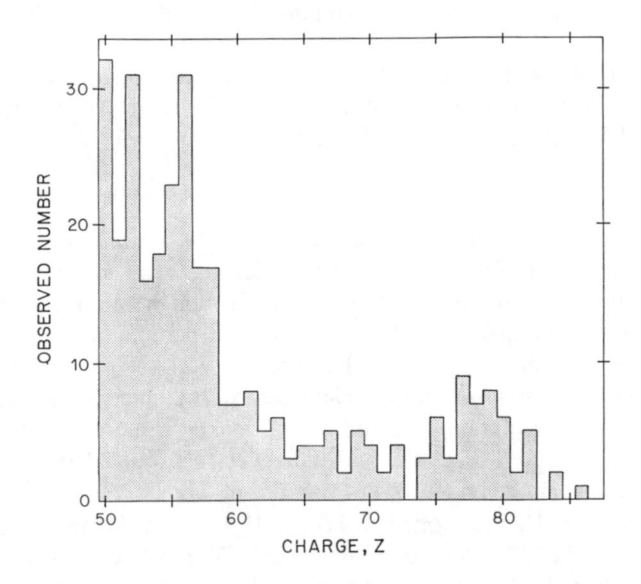

Figure 16: Histogram of nuclei in the 50 to 90 charge region.

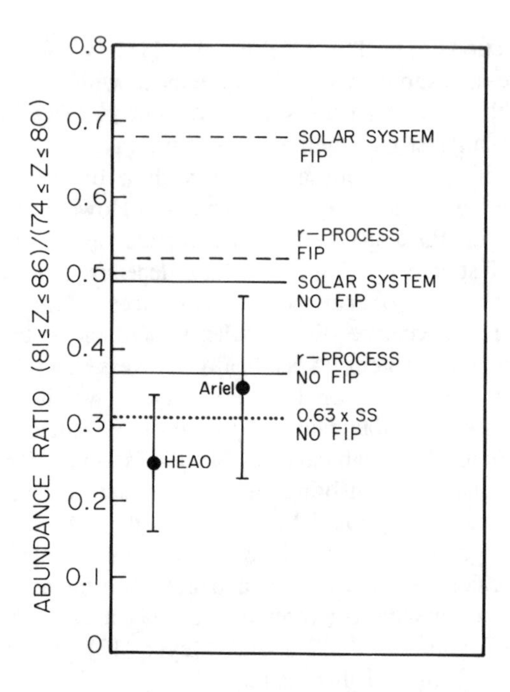

Figure 17: The lead ($81 \leq Z \leq 86$) to platinum ($74 \leq Z \leq 80$) abundance ratio compared to solar system and r-process abundances. The solid lines have no FIP fractionation applied, with the dashed lines having a FIP fractionation applied at the source. The dotted line is the solar system ratio if the Grevesse and Meyer [14] lead abundance is used.

To study the abundance of the heaviest nuclei in the cosmic radiation, the actinides, the selection requirements were significantly relaxed to avoid any possible selection bias against these rare nuclei, even though this meant that the charge resolution was degraded. We detected only one particle with an assigned charge in the charge region $88 \leq Z$ and that nucleus was assigned a charge of $Z = 89$. The short half-lives of all the elements in the interval $84 \leq Z \leq 89$, combined with the poor resolution of this data set, makes it more likely that this was a nucleus of $_{90}$Th or $_{92}$U. Together with this event we found 101 events with $74 \leq Z \leq 87$ giving us an actinide to Pt-Pb ratio of about 1%, with an 84% confidence level of 3%. This is plotted in Figure 18 along with the Areal-6 data and earlier balloon flight [22] and skylab data [23]. The balloon flight data has recently been reevaluated [24] and is labeled "balloons revised" in Figure 18. With the very low actinide statistics in these experiments, both the HEAO-3 and the Ariel-6 results are consistent with a result formed by combining the two, 2.4% (+1.9% — 1.2%). This result is consistent with solar system abundances although with such low statistics we cannot rule out a significant enhancement of r-process actinides in the cosmic ray source. However, as can be seen from the figure, our measurement is inconsistent with freshly synthesized r-process material.

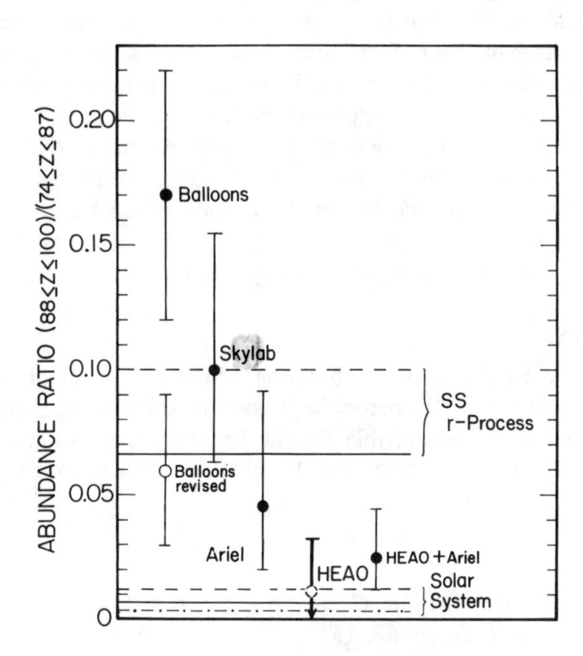

Figure 18: The actinide ($88 \leq Z \leq 100$) to lead-platinum ($74 \leq Z \leq 87$) abundance ratio compared to solar system and r-process abundances. The experimental data points are from our HEAO experiment, the Ariel-6 experiment, early balloon experiments [22], and the skylab experiment [23]. The predicted values [25] are solid lines without FIP fractionation and dashed lines with FIP fractionation, for sources with composition of the solar system at its time of formation or for a source with the composition of freshly synthesized r-process material. The dot-dash solar system line substitutes present-day abundances for the (solid line) abundances at the time of the solar system formation.

6. CONCLUSIONS

In conclusion we find that the cosmic ray abundances are in generally good agreement with solar system abundances with a step-FIP fractionation model applied. Two important exceptions to this are Ge and Pb which appear to be underabundant in the cosmic radiation. However both are volatile and if recent solar photospheric measurements are used for the solar system abundances they are brought into agreement with our data. For these cases it appears that the present precision in cosmic ray measurements is better than our current knowledge of the correct standard for comparison. Although the platinum/lead ratio and the actinides are consistent with some r-process enhancement, we have seen that the cosmic ray source is not dominated by the r-process up through the 50's as evidenced by the $_{38}$Sr to $_{37}$Rb ratio and by the abundance of $_{50}$Sn and $_{56}$Ba. In addition the actinides are not greatly enhanced, ruling out freshly synthesized r-process production as the primary source of the heavy cosmic rays.

ACKNOWLEDGEMENTS

I wish to acknowledge the HEAO C-3 collaboration which is responsible for the experiment described in this paper. In particular I want to acknowledge the work of B. J. Newport who is mainly responsible for the improved resolution data set described herein. This work has been supported by NASA contract NAG 8-498.

REFERENCES

1. Anders, E., and Ebihara, M., 1982, *Geochim. et Cosmochim.*, **46,** 2363.
2. Meyer, J. P., 1979, *Proc. 16th Intl. Cosmic Ray Conf.* **3**, 115.
3. Meyer, J. P., 1985, *Ap. J. Suppl.* **57**, 151.
4. Cameron, A. G. W., 1982, in *Essays in Nuclear Astrophysics*, Ed. Barnes, C. A., Clayton, D. D., and Schramm, D. N., Cambridge University Press, p.23.
5. Cameron, A. G. W., 1982, *Astrophys. and Space Sci.* **82**, pp. 123-131.
6. Letaw, J. R., Silberberg, R., and Tsao, C. H., 1984, *Ap. J.*, **279**, 144.
7. Newport, B. J., 1986 Thesis, California Institute of Technology.
8. Binns, W. R., Brewster, N. R., Fixsen, D. J., Garrard, T. L., Israel, M. H., Klarmann, J., Newport, B. J., Stone, E. C., and Waddington, C. J., 1985, *Ap. J.*, **297**, 111.
9. Israel, M. H., 1983, in *Composition and Origin of Cosmic Rays*, Ed. Shapiro, M. M., (Dordrecht: D. Reidel), p. 291.
10. Meyer, J. P., 1981, *Proc. 17th Intl. Cos. Ray Conf.*, **2**, 281.
11. Meyer, J. P., 1985, *Ap. J. Sup.*, **57**, 173.
12. Meyer, J. P., 1985, *Ap. J. Sup.*, **57**, 151.
13. Meyer, J. P., 1985, *Proc. 19th Intl. Cos. Ray Conf.*, **9**, 141.
14. Grevesse, N., and Meyer, J. P., 1985, *Proc. 19th Intl. Cos. Ray Conf.*, **3**, 5.

15. Binns, W. R., Grossman, D. P., Israel, M. H., Jones, M. D., Klarmann, J., Garrard, T. L., Stone, E. C., Fickle, R. K., and Waddington, C. J., 1983, *Proc. 18th Intl. Cos. Ray Conf.*, 9, 106.
16. Stone, E. C., Garrard, T. L., Krombel, K. E., Binns, W. R., Israel, M. H., Klarmann, J., Brewster, N. R., Fickle, R. K., and Waddington, C. J., 1983, *Proc. 18th Intl. Cos. Ray Conf.*, 9, 115.
17. Fowler, P. H., Masheder, M. R. W., Moses, R. T., Walker, R. N. F., Worley, A., and Gay, A. M., 1985, *Proc. 19th Intl. Cosmic Ray Conf.*, **2**, 115.
18. Fowler, P.H., Masheder, M. R. W., Moses, R. T., Walker, R. N. F., Worley, A., and Gay, A. M., 1985, *Proc. 19th Intl. Cos. Ray Conf.*, **2**, 119.
19. Brewster, N. R., Freier, P. S., and Waddington, C. J., 1985, *Ap. J.*, **294**, 419.
20. Howard, W. M., Mathews, G. J., Takahashi, K., and Ward, R. A., 1987, *Ap. J.*, Accepted for publication.
21. Waddington, C. J., Binns, W. R., Brewster, N. R., Fixsen, D. J., Garrard, T. L., Israel, M. H., Klarmann, J., Newport, B. J., and Stone, E. C., 1985, *Proc. 19th Intl. Cos. Ray Conf.*, **9**, 527.
22. Fowler, P. H., Alexandre, C., Clapham, V. M., Henshaw, D. L., O'Sullivan, D., and Gay, A. M., 1977, *Proc. 15th Intl. Cos. Ray Conf.*, **11**, 165.
23. Shirk, E. K., and Price, P. B., 1978, *Ap. J.*, **220**, 719.
24. O'Sullivan, D., 1985, *Irish Astron. J.*, **17**, 40.
25. Blake, J. B., Hainebach, K. L., Schramm, D. N., and Anglin, J. D., 1978, *Ap. J.*, **221**, 694.

GALACTIC COSMIC RAY HYDROGEN AND HELIUM

J. J. Beatty
Department of Physics
Boston University
590 Commonwealth Avenue
Boston, Massachusetts 02215
U.S.A.

The quartet of isotopes of hydrogen(H) and helium(He) are the four most abundant species in the arriving cosmic radiation. In this paper, a brief summary of the spectra and propagation of these species is given. The reader is referred to Beatty (1986) for a more detailed discussion.

Table 1 (Simpson 1983) summarizes the high energy H and He elemental spectra, as obtained from power law fits to the world data set. Both species are well described by power laws with index ~ -2.8 at energies above several GeV per nucleon. The high energy spectra result from the propagation of accelerated cosmic rays in the galactic magnetic field. In the leaky box model, the spectrum of a pure primary species is given by the simple relation $J = \Lambda Q$, where Q is the source spectrum and $\Lambda = (\lambda_{escape}^{-1} + \lambda_{spallation}^{-1})^{-1}$ is the mean pathlength for losses. Λ is expected to depend on particle rigidity at high energies because of the decreased scattering of more rigid cosmic rays by irregularities in the galactic magnetic field (see Cesarsky (1980) for a review of cosmic ray confinement). Based on the secondary/primary ratio B/C, the rigidity power law index of Λ is -0.6 ± 0.1. Thus, the power law index of the highy energy H and He source spectra is ~ -2.2. Furthermore, by correcting the H/He ratio for the rigidity dependence of Λ (which affects H and He differently at constant energy per nucleon due to the difference in A/Z) and for spallation, the source ratio is found to be 12.9 ± 2.5, which is remarkably consistent with the solar system value. Engelmann et al. (1985) have shown that H/He at the source is constant from 3 to 60 GeV per nucleon, that the source spectra are power laws in momentum with index ~ -2.2, and that the spectra are steeper than those of species with Z>2. The momentum power law spectrum $(dJ/dE) \sim p^{\gamma}, \gamma < -2$ is expected from models of stochastic acceleration, for which $(dJ/dE) \propto (dN/dP)$ is the relevant quantity, since the acceleration process can be represented as a diffusion process in momentum space.

At low energies, study of the H and He spectra is more complicated due to the effect of solar modulation. Because of the large difference in A/Z, the ratio H/He at earth is strongly dependent on the solar modulation level. The effect of a given level of adiabatic deceleration on the observed ratio at constant energy per nucleon E depends on the

M. M. Shapiro and J. P. Wefel (eds.), Genesis and Propagation of Cosmic Rays, 91–96.

TABLE 1

H	$(4.0\pm0.4)\times10^{-4}$	$(m^2 sr\ s\ MeV\ n^{-1})^{-1}$
He	$(1.9\pm0.2)\times10^{-5}$	$(m^2 sr\ s\ MeV\ n^{-1})^{-1}$
H/He	21 ± 3	

Simpson 1983; fluxes and ratio at 50 GeV per nucleon based on power law fits to the world data set; fitted spectral index is -2.75 ± 0.05.

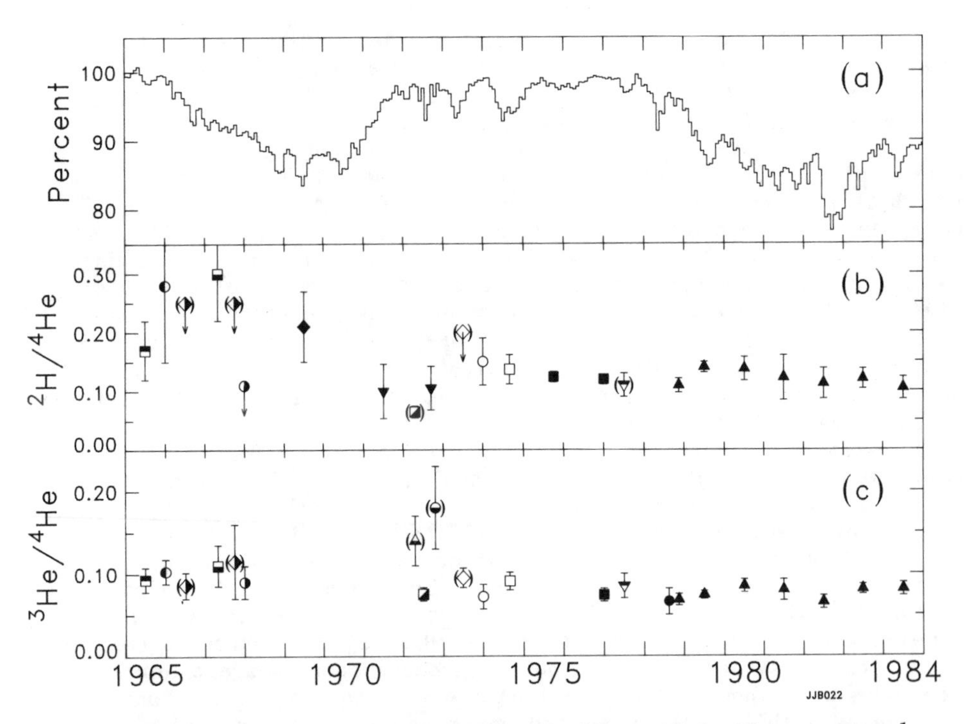

Figure 1-- (a)-The Climax neutron monitor count rate expressed as a percentage of the rate at the 1954 solar minimum. (b)- The ratio $^2H/^4He$ measured by various space and balloon experiments. (c)- The ratio $^3He/^4He$. See Beatty 1986 for symbol definitions and references to individual data points.

interstellar ratio at energies near $(Z\phi/A)+E$. Thus, the observed ratio depends on both the interstellar H/He ratio at constant energy and on the spectral shape.

In addition to the difficulties in unraveling modulation effects, one must contend with the fact that ^{4}He exhibits an anomalous component at low energy, which is believed to be due to acceleration in the outer heliosphere. This component must be corrected for in studies of the galactic component. Discrepancies between conclusions of workers on the quartet can often be traced to their choice of anomalous ^{4}He correction.

With the exception of the ^{3}He/^{4}He measurement of Jordan and Meyer (1984; see also Jordan 1985), no isotopic data for H and He exist at energies greater than 200 MeV per nucleon. This is because most high resolution techniques require that the particles stop in the detector, and H and He are more penetrating at a given energy per nucleon than heavier species. Thus, until higher energy experiments are performed, one must rely on the low energy isotopic data and make corrections for solar modulation effects.

Figure 1 shows the Climax neutron monitor counting rate (panel a) and the ratios ^{2}H/^{4}He and ^{3}He/^{4}He (panels b and c) as a function of time. (see Beatty (1986) for symbol definitions and references.) Most measurements are near 50 MeV per nucleon for ^{2}H; this is the end of range for a typical dE/dX vs. E telescope. Measurements enclosed in parentheses are, for various reasons, not suitable for direct comparison with the others. Because of the large backgrounds associated with measurements of singly-charged particles, the early measurements exhibit a large scatter. Measurements made since the early 1970's by a number of workers are in good agreement. The ratio varies by less than 30% over the solar cycle. This leads to the conclusion that the local interstellar ratio is approximately independent of energy, since A/Z=2 for both species.

In the case of ^{3}He/^{4}He the ratio is also remarkably independent of modulation. This is surprising, since the difference in A/Z would cause the ratio to decrease at solar maximum if the local interstellar ratio were energy-independent.

The local interstellar spectrum can be calculated from a model of interstellar propagation, and the result subjected to a model of solar modulation and compared with the observations. In the case of H and He interstellar propagation is particularly interesting because of the small number of species involved. Only H, He, and CNO contribute in appreciable amounts to the secondaries ^{2}H and ^{3}He. The cross sections can be deduced to within ~10%, although this involves relying on a minimal set of critical measurements and is thus subject to possible large systematic errors (Beatty, in preparation). Figure 2 shows the contributions to the local interstellar ^{2}H spectrum from the primary species ^{1}H, ^{4}He, and CNO. Spallation of ^{4}He is dominant at all energies. CNO makes an important contribution above several hundred MeV per nucleon (Ramaty and Lingenfelter 1969). The reaction $p+p \Rightarrow d+\pi^+$, which proceeds via the Δ^{++} nucleon resonance, is important at low energies. Its importance depends on the spectral shape, since the peak cross section is for incident protons at ~600 MeV, resulting in ~200 MeV per

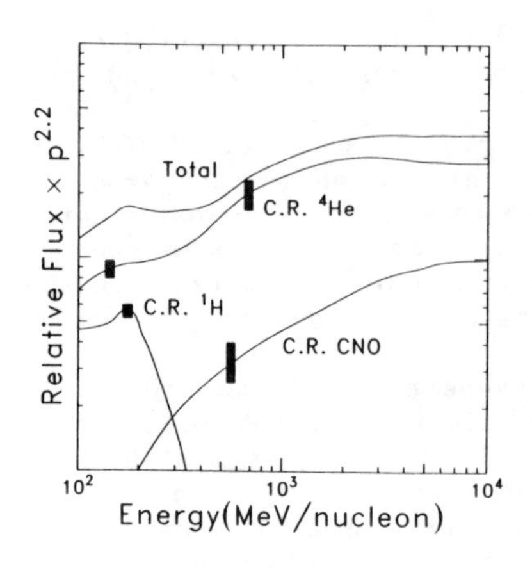

Figure 2--The local interstellar ^{2}H spectrum, normalized to the assumed source spectrum, is plotted as a function of energy. The individual contributions of ^{1}H, ^{4}He, and CNO are shown, with the black bars indicating the estimated uncertainty due to the cross sections.

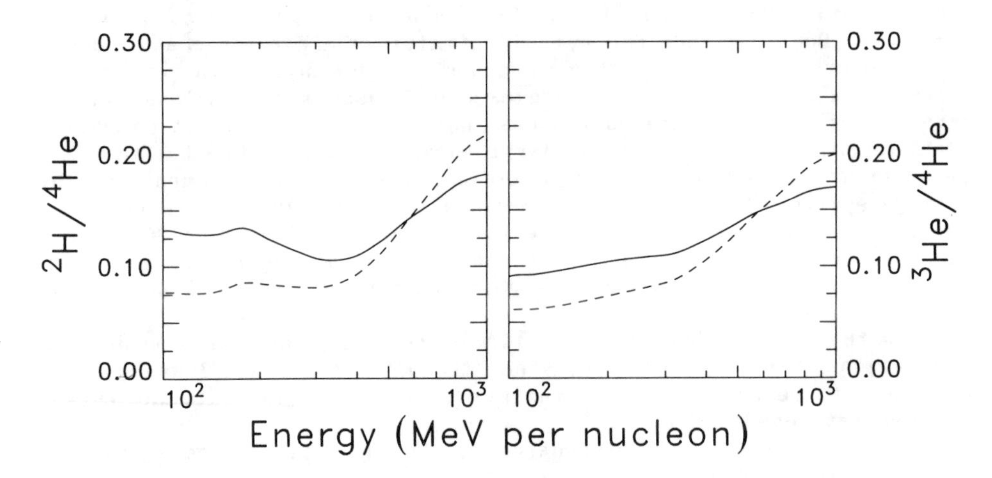

Figure 3--The local interstelar ^{2}H/^{4}He and ^{3}He/^{4}He ratios calculated for an energy-independent mean pathlength (solid line) and for the energy-dependent mean pathlength of Garcia-Munoz et al. 1981 (dashed line).

nucleon ^{2}H due to the kinematics of the reaction.

Two models for propagation have been chosen for study in the following discussion. The first, indicated by the solid curves, has an energy independent mean pathlength of 7.2 g cm^{-2}. The second uses the energy-dependent mean pathlength deduced from B/C by Garcia-Munoz et al. (1981), and is represented by dashed curves. In the latter case, the pathlength increases with increasing energy below ~1 GeV per nucleon. In both cases, momentum power law spectra with index -2.2 have been used. Figure 3 shows the computed interstellar ratios for the two models. The relevant energy range for low energy data over the solar cycle is 300-650 MeV per nucleon for ^{2}H, and 350-850 MeV per nucleon for ^{3}He. Taking solar modulation into account, and plotting the ratio below 100 MeV per nucleon versus the modulation parameter ϕ we see that in the energy independent case both ratios are nearly independent of energy, while the ratios for the energy dependent model exhibit a strong modulation dependence (Figure 4). The data are better fit by the energy independent model, particularly near solar minimum. The data are of insufficient accuracy to rule out some energy dependence in the mean pathlength at low energy, especially if the energy dependence were somewhat weaker than the best fit to B/C due to errors in experimental data or cross sections. (See Webber (1985) for possible changes in C(p,X)B in this direction.)

The low energy ^{1}H/^{4}He source ratio determined in this proceedure is 12.2$\pm$1.5 at constant energy per nucleon. The momentum power law source spectrum thus provides a consistent picture of H and He over a wide energy range. Both H and He are depleted by about a factor of 4 relative to species such as O and Ne with similar first ionization potential. This must be accounted for in the injection/acceleration process.

Areas requiring future study include further unraveling of the low energy spectrum and propagation. This is particularly enticing because of the data of Webber and Yushak (1983), which show the ^{2}H and ^{3}He fluxes increasing up to ~200 MeV per nucleon. Measurements of the isotopic ratios at intermediate energies are needed to further probe this phenomenon. The high value of ^{3}He/^{4}He at 6 GeV per nucleon reported by Jordan and Meyer (1984; see also Jordan 1985) was obtained using the geomagnetic technique, and is thus sensitive to the spectral index at the time of the measurement. Recent studies of the spectral index lead to a value of ^{3}He/^{4}He in better agreement with the propagation models. (See Meyer (1985) for further discussion.) A planned reflight of this experiment at two different sites with low and high geomagnetic cutoff at the same modulation level would be beneficial, since the spectral index folded together with the instrument response can be measured, leading to a more reliable ^{3}He/^{4}He measurement.

This work was supported in part by NASA grant NGL 14-001-006, NASA Contract NAG 5-706, and the Arthur H. Compton Fund at the University of Chicago, and by NASA grant NAGW 1032 at Boston University.

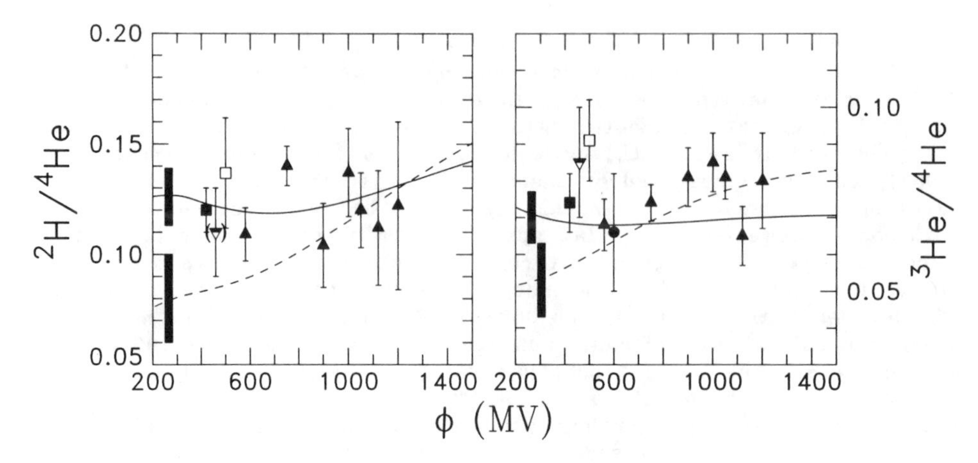

Figure 4--The modulated ratios $^2H/^4He$ and $^3He/^4He$ are plotted as a function of the adiabatic deceleration parameter. The solid line is calculated using an energy-independent mean pathlength, and the dashed line is the result for the energy-dependent mean pathlength of Garcia-Munoz et al. 1981. The data are plotted at the modulation level appropriate to the time of measurement. The black bars show combined cross section and mean pathlength uncertainties.

References:

Beatty, J. J. 1986, Ap. J., 311, 425.

Cesarsky, C. J. 1980, Ann. Rev. Astron. Astrophys., 18, 289.

Engelmann et al. 1985, Astron. Astrophys., 148, 12.

Garcia-Munoz et al. 1981, Proc. 17th Internat. Cosmic Ray Conf. (Paris), 9, 195.

Jordan, S. P. 1985, Ap. J., 291, 207.

Jordan, S. P. and Meyer, P. 1984, Phys. Rev. Lett., 53, 505.

Meyer, J. P. 1985, Proc. 19th Internat. Cosmic Ray Conf. (La Jolla), 9, 141.

Ramaty, R. and Lingenfelter, R. E. 1969, Ap. J., 155, 587.

Simpson, J. A. 1983, Ann. Rev. Nucl. Part. Sci., 33, 323.

Webber, W. R. and Yushak, S. M. 1983, Ap. J., 275, 391.

Webber, W.R 1985, in Proc. Workshop on Cosmic Ray and High Energy Gamma Ray Experiments for the Space Station Era, ed. W. V. Jones and J. P. Wefel(Baton Rouge:Louisiana State University), p. 283.

COSMIC RAYS OF THE HIGHEST ENERGIES

Jacek Szabelski
Institute of Nuclear Studies
ul. Uniwersytecka 5
90-950 Lodz 1, box 447, Poland

ABSTRACT. An analysis is made of the current status of measurements on the energy spectrum and anisotropy of the cosmic rays of the highest energy, above 10^{18} eV. Measurements from Haverah Park (northern hemisphere) and Sydney (southern hemisphere) are consistent in energy spectrum and anisotropy in the common sky area. Events with energy above $5x10^{19}$ eV are largely of extragalactic origin. Several particles arrive preferentialy from the general direction of the local supercluster. There is also an excess of particles coming from the Galactic South Pole direction, which is not understood.

1. INTRODUCTION

The problem of origin of the highest energy cosmic rays has been studied by a number of workers during the last few years. The idea of the present paper is to compare data from Haverah Park - the experiment seeing the northern hemisphere - with data from Sydney - the experiment seeing the southern hemisphere. Very recently the final data from the Sydney experiment have appeared (Winn et al., 1986a,b). In this analysis we have used Haverah Park data published in the Catalogue of the Highest Energy Cosmic Rays no. 1 by Cunningham et al., 1980.

We examine two major topics: the form of the energy spectrum above 10^{19} eV to the highest energies detectable, trying to understand differences between published spectra, and the nature of the anisotropies. Taking both together we conclude that there is a different origin for particles having energy below a few times 10^{19} eV, being mostly Galactic, from particles above a few times 10^{19} eV which seem to be of extragalactic origin.

2. THE ENERGY SPECTRUM ABOVE 10^{18} eV

The energy spectra from Haverah Park (crosses; Cunningham et al., 1983) and Sydney (circles; Winn et al., 1986a) are presented on figure 1. It can be seen that the Sydney spectrum is 20-30% higher than that from Haverah Park, even for the energy range 10^{18} - $4x10^{18}$ eV where there are

M. M. Shapiro and J. P. Wefel (eds.), Genesis and Propagation of Cosmic Rays, 97–104.

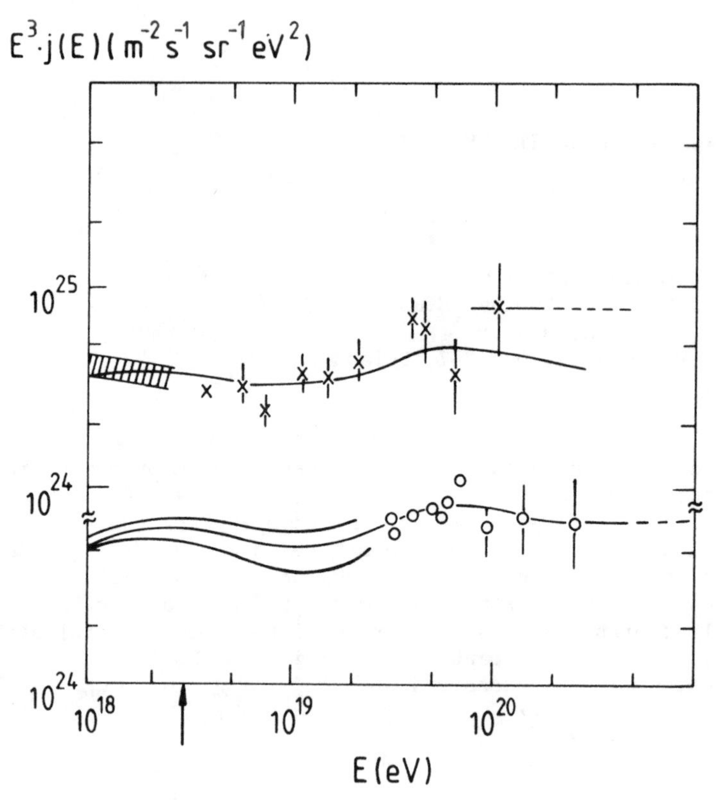

Figure 1. Energy spectra from Haverah Park (crosses) and Sydney (circles). The decrease of Sydney energy by 15% moves the bottom line to the upper line to normalize both spectra at Haverah Park energy 3×10^{18} eV (arrow).

good statistics and it is known that the particles are isotropic. For that reason we have applied a normalization of the Sydney spectrum to the Haverah Park results at Haverah Park energy 3×10^{18} eV. The normalization is by scaling the Sydney energy (not intensity) by 15%. (More precisely: we have scaled the energy evaluated by the Sydney group using the particular model of high energy interactions - Hillas "Model E"; Hillas et al., 1971). This is a reasonable value in view of the uncertainties in the high energy physics involved in extensive air showers (EAS). The scaling moves the best hand made curve through the Sydney points (circles) to the curve in the upper part of figure 1, which is in good agreement with Haverah Park data.

From the comparison of Haverah Park data with scaled Sydney result on the energy spectrum we see that the spectrum turns up above 10^{19} eV. This indicates that the 3°K cut-off in the highest energy end of the cosmic ray spectrum is not becoming significant below 10^{20} eV, and the opposite efect may suggest a different origin of cosmic rays above 10^{19} eV.

3. COMPARISON BETWEEN HAVERAH PARK AND SYDNEY OF ANISOTROPY FROM THE SAME SKY AREA

The Haverah Park experiment is located at 54° North and accepts EAS arrivivg with the zenith angle smaller than 60°. This gives an opportunity of seeing the sky from the North Pole to the celestial equator. The Sydney experiment is located at 30.5^{o} South and accepts EAS coming with zenith angles smaller than 72°. Therefore it was possible to see the sky from the South Pole to the declination $+40^{o}$ during the daily Earth rotation. (The sky coverage is not uniform and this will be discussed later). We see that the declination area between the equator and $+40^{o}$ can be monitored by both experiments.

As we are interested in the highest energy particles, we restrict ourself to Haverah Park events with energy greater than $4.7\text{x}10^{19}$ eV and to Sydney events with corresponding "Hillas E model" energy above $5.53\text{x}10^{19}$ eV. In this common sky area (declination between 0° and $+40^{o}$) Haverah Park has 9 events and Sydney has 9 events as well, all above the same energy limits. Because of such poor statistics we divide the common sky area into four parts in right ascesion (in hours: 3-9, 9-15, 15-21, 21-3). Numbers of events in each bin are presented in table I. There is a positive correlation between experiments located in opposite hemispheres in arrival directions of the highest energy cosmic rays from the common sky area.

Since both experiments have the same number (9 events) from this sky area and there is a good correlation in anisotropy we made a sum of the numbers of events in each bin to improve the statistics in the anisotropy examination. Results are presented in figure 2. The large excess between declination 0° and $+40^{o}$ and right ascesion between 9 and 15 hours may be due to the fact that the Virgo cluster of galaxies is in that part of the sky (declination $12^{o}40'$, R.A. 12h28').

Table I

Number of Haverah Park and Sydney events with energy above HP energy $4.7\text{x}10^{19}$ eV in the common sky area (declination between 0° and $+40^{o}$).

R.A. (hours)	HP	Sydney
3 - 9	3	1
9 - 15	5	6
15 - 21	0	1
21 - 3	1	1

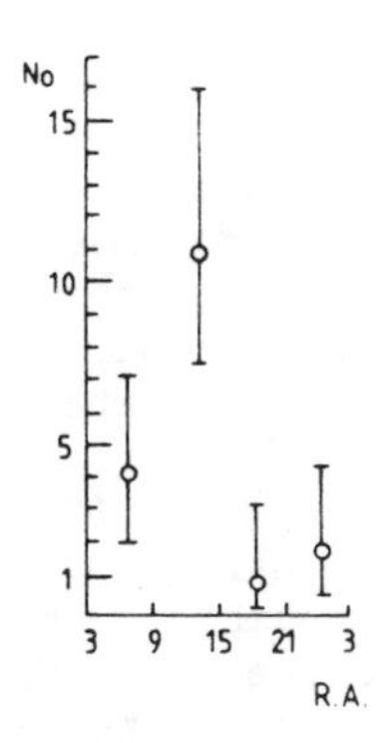

Figure 2. Distribution in right ascesion of events from HP and S arriving from the common sky area (declination between 0° and $+40^{o}$) with HP energy above $4.7\text{x}10^{19}$ eV.

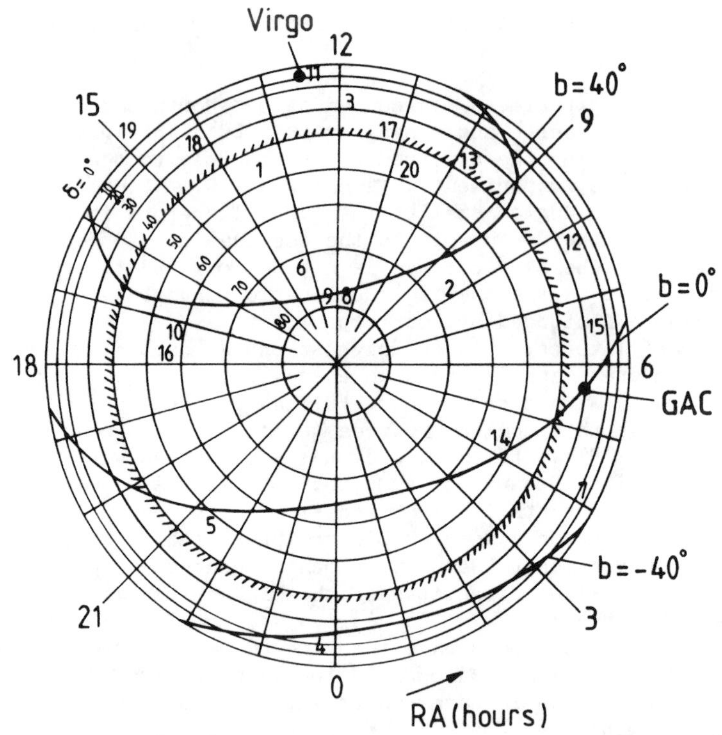

Figure 3. Haverah Park equal exposure map with the 20 most energetic events (above 4.7×10^{19} eV) ranked with decreasing energy. The border of the common area with Sydney has been shaded.

4. EQUAL EXPOSURE MAPS

We have indicated arrival directions of the most energetic EAS on equal exposure maps ranking the events in order of decreasing energy for both the Haverah Park and Sydney experiments. On the equal exposure map an area is proportional to the probability of observing a particle for an isotropic distribution. Some astronomical objects and some Galactic coordinates are indicated. Figure 3 presents the 20 most energetic ($E > 4.7 \times 10^{19}$ eV) events observed in Haverah Park and in figure 4 there are 38 most energetic ($E_{Hillas-E} > 5.53 \times 10^{19}$ eV) events observed by the Sydney experiment. In both experiments the most energetic events (indicated by number 1) arrived from directions near to the Virgo cluster or to the North Galactic Pole (NGP).

On the Sydney experiment map it is interesting to note a concentration of the very energetic events (numbers 2, 3, 6 and others) in the southern Galactic hemisphere. We have not found any candidate for a source in this part of the sky.

Another concentration of events (numbers 4, 7, 8, 9, 10 and others) near to the Galactic equator points to the direction of the Galactic spiral arm.

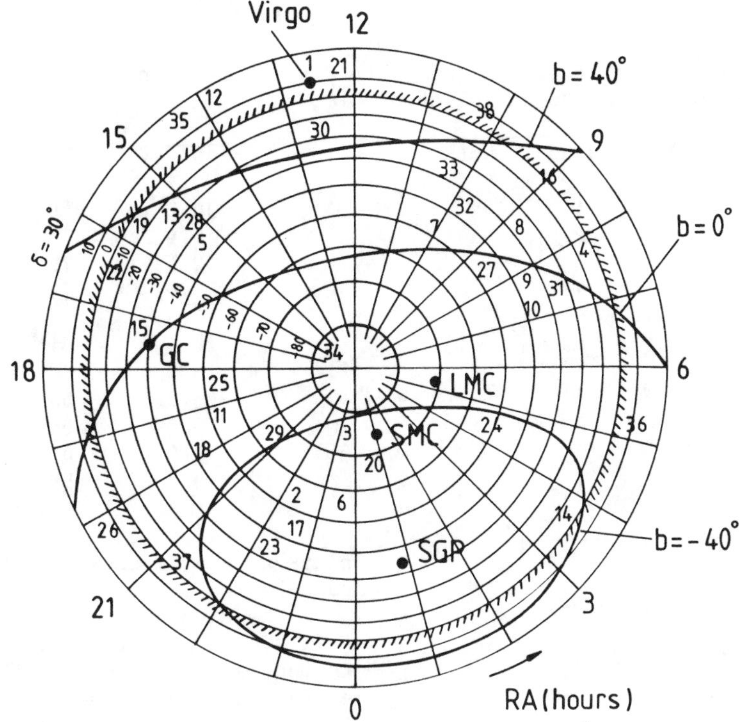

Figure 4. Sydney equal exposure map with the 38 most energetic events (above 5.53×10^{19} eV - equivalent to HP energy limit from figure 3) ranked with decreasing energy. The border of the common area with Haverah Park has been shaded.

5. THE PLAN VIEW OF THE NORTH GALACTIC POLE REGION

The North Galactic Pole region is a very interesting part of the sky since the Virgo cluster is near and the most energetic events come from that direction. It happens that combining the view from the northern hemisphere (Haverah Park) and from the southern hemisphere (Sydney) on the NGP area we achieve an almost equal exposure map of that region from NGP down to a Galactic latitude of 50°. To plot events from both experiments on the same figure we have to renumber our energy ranking for events near to NGP in the way presented in table II. Those events are plotted on figure 5. On that figure we have indicated the centre of energy of the plotted events. This point does not coincide with the Virgo cluster position (V).

Therefore another model has been considered, namely the possibility of our local group of galaxies being responsible for the highest energy events. We have plotted positions of the nearest galaxies and evaluated the direction of the mean of galactic mass divided by the distance squared. This direction is indicated on figure 5 by a circle. It is important to say that the contribution from the Virgo cluster still

Table II

New ranking of events from both Haverah Park and Sydney in the region of the North Galactic Pole.

new No	No(HP)	No(S)	E(10^{19}eV)	b(deg)
1		1	16.8	84
2	1		15.9	73
3	3		12.6	78
4		12	7.0	79
5	11		6.3	74
6	13		6.1	51
7		21	5.8	78
8	17		5.2	66
9	18		5.0	66
10		35	4.9	67
11	20		4.7	59

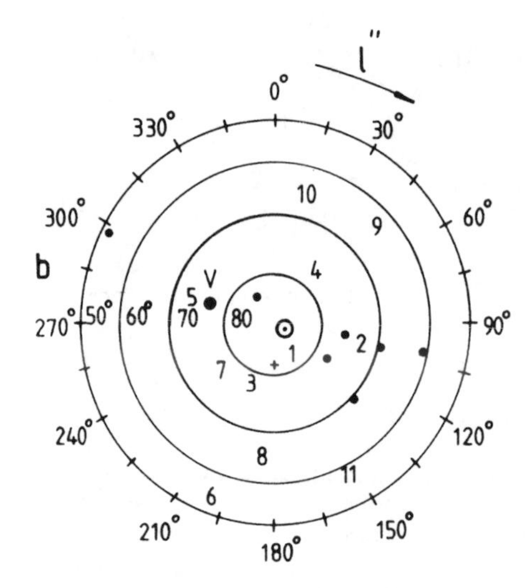

Figure 5. Almost equal exposure map of the North Galactic Pole region. Events from HP and S are ranked with decreasing energy (see table II). The nearest galaxies are indicated by dots.

V - indicates the centre of Virgo cluster,

+ - the centre of energy of plotted events,

circle - the direction of the mean of galactic mass devided by distance squared.

plays a very important role in the calculations.

The direction of the weighted galactic mass distribution is near to the centre of energy of the most energetic events observed as well as to NGP, so the NGP coordinates are very useful to see whether or not there is a correlation between particle momentum and the inclination of its arrival direction from the "centre of the source". Such correlation would indicate the magnetic pole of the galactic halo being responsible for the particle trajectory curvatures. We have calculated the average energy for four bands of inclination from the NGP direction. This is plotted as a stepped line on figure 6. Each air shower is indicated by its new number from Table II. The curve represents the ideal situation in which there is a point source in the direction of the NGP, the galactic halo is extended 10 kpc above the galactic plane, the effective magnetic field is equal to 2 micro-gauss and the particles are protons. There is a coincidence present on figure 6 although we do

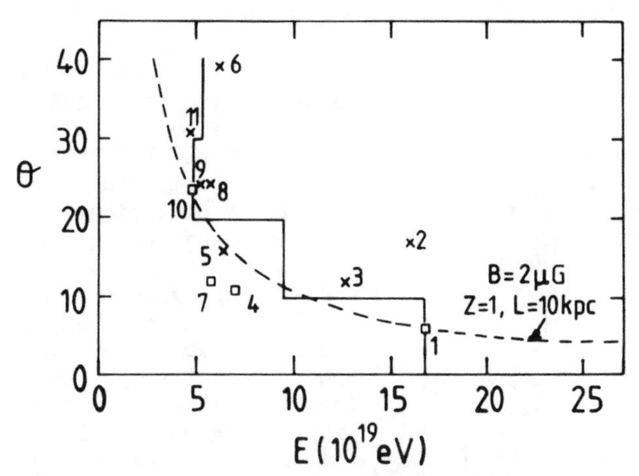

Figure 6. The correlation of particle arrival inclination from NGP direction with particle energy. See table II for ranking details. Stepped line represents average energy observed in inclination bands. Broken line - prediction on simple asssmption shown in figure.

not understand it, especially what is the magnetic field configuration needed to produce this correlation and the angular distribution seen in figure 5.

6. THE GALACTIC PLANE ENHANCEMENT PARAMETER f_E

The Galactic plane enhancement parameter f_E has been defined by the following expression (Wdowczyk and Wolfendale, 1984, Wdowczyk, 1986):

$$I(b) = I_0 \left((1-f_E) + f_E \exp(-b^2) \right)$$

where b is Galactic latitude in radians, I(b) is a ratio of observed number of particles to the number expected on isotropy in a Galactic latitude band and I_0 is a normalization factor. f_E is evaluated to get the best chi-square value varying both I_0 and f_E. For isotropy $f_E = 0$, for Galactic plane enhancement f_E is positive and it is negative for Galactic plane deficiency. We do not expect f_E to be unity for the case of all particles beeing of Galactic origin since the form of the Galactic latitude distribution, $\exp(-b^2)$, is very approximate, although f_E should reflect the general tendency.

Figure 7 presents f_E as a function of energy and comes basicaly from the paper by Wdowczyk and Wolfendale (1984) with a few new points calculated in this work for the highest energy cosmic rays. It can be seen that the magnitude of the Galactic plane enhancement is increasing up to an energy near 2×10^{19} eV but for the highest energy events there is evidence for Galactic plane avoidance. The big errors of f_E in this area can be due to the inadequate formula of Galactic latitude dependence.

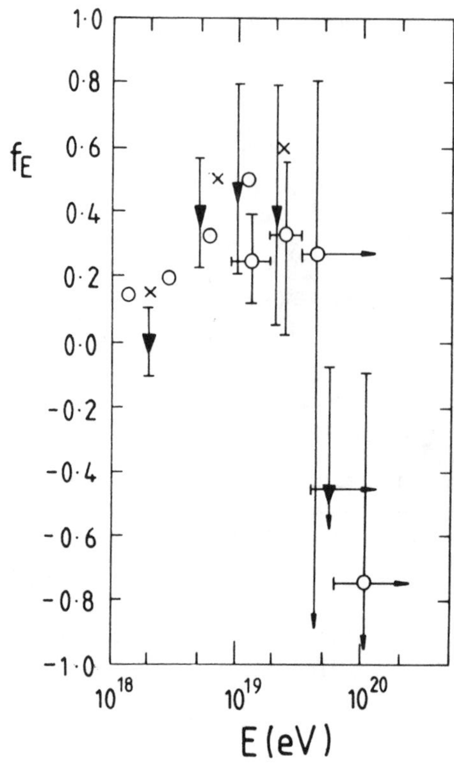

Figure 7. Galactic enhancement factor f_E dependence upon energy. X - Yakutck, ▼ - Haverah Park, O - Sydney. Part of figure after Wdowczyk and Wolfendale (1984).

7. CONCLUSIONS

Examining the most energetic events in the Haverah Park and Sydney experiments, we found that the change of spectrum power index near 10^{19} eV coincides with the change in anisotropy of arrival directions of the particles, being largely of Galactic origin below that energy and of extra-galactic origin above it. The change of slope may reflect the fact that the production spectrum is flatter than observed locally inside the Galactic propagation region. Both experiments indicate the Virgo cluster region or North Galactic Pole region as the source of the highest energy cosmic rays. The local group of galaxies may be responsible for that. The Sydney experiment has seen a concentration of arrival directions of the highest energy particles in the galactic spiral arm direction and near to the South Galactic Pole; neither is understood.

8. ACKNOWLEDGMENTS

I wish to thank the Directors of the Advanced Study Institute, Professors M. M. Shapiro and J. P. Wefel for providing me opportunity to attend the School, and for kind hospitality during the Course.

I would like to thank Professors J. Wdowczyk and A. W. Wolfendale, who made a great contribution to this paper.

REFERENCES

Cunningham, G. et al., 1980, the Catalogue of the Highest Energy Cosmic Rays no. **1**.

Hillas, A. M., et al., 1971, Proc. 12th Int. Cosmic Ray Conf., **3**, 1007.

Wdowczyk, J., 1986, 'Cosmic Radiation in Contemporary Astrophysics', Ed. M. M. Shapiro (Dordrecht : Reidel, **C 162**), 149.

Wdowczyk, J, and Wolfendale, A. W., 1984, Journal of Physics G., **10**, 1453

Winn, M. M. et al., 1986a, Journal of Physics G., **12**, 653.

Winn, M. M. et al., 1986b, Journal of Physics G., **12**, 675.

STARS AND COSMIC RAYS

I. COOL STARS

Thierry Montmerle

Service d'Astrophysique
Centre d'Etudes Nucléaires de Saclay
91191 Gif sur Yvette Cedex, France

ABSTRACT

The analysis of the composition of galactic cosmic rays gives fundamental clues to solve the problem of their genesis. The present conclusion that the bulk of the cosmic-ray nuclei originate in F to M stars appears to be independently supported by studies of stellar outer atmospheres and activity. We review our current knowledge on "solar-like activity" and its relation with stellar parameters at various locations of the Hertzsprung-Russell diagram. This "solar-stellar connection" emphasizes the role of solar studies in helping to understand the pre-acceleration states of the history of galactic cosmic rays.

1. STARS AND COSMIC RAYS: WHY ?

1.1 The Cosmic Ray "Sources"

In the astronomical sense of the term, there is no such thing as a "cosmic-ray source". Indeed, a particle having a rigidity $R = pc/Ze$ (in standard notations) has a Larmor radius $r_L = R/(cB)$, which is numerically, for an interstellar magnetic field B of typical strength 3 μG, $r_L \simeq 10^{12}$ R(GV)cm, i.e., much smaller than the thickness of the galactic disk.

Therefore, we cannot "see" a cosmic-ray source, except at the very highest energies (when the fluxes are so small, anyway), and this means that we have to rely only on indirect evidence to understand the origin of cosmic rays.

However, experimental and theoretical advances have been quite substantial in recent years, and we feel we can use the present data on composition and spectra to separate, at least in an approximate way, the various phases of cosmic ray generation:

M. M. Shapiro and J. P. Wefel (eds.), Genesis and Propagation of Cosmic Rays, 105–130.

(a) existence of a specific particle reservoir,

(b) injection and/or acceleration,

(c) propagation in the interstellar medium.

It turns out that the propagation can be adequately described by steady-state models (e.g., "leaky box") as a useful first approximation. More accurate models feature some reacceleration of the particles on their way throughout the Galaxy (Silberberg et al., 1983 ; Wandel et al., 1987 ; Shapiro, this volume), but it is small enough that such concepts as "source composition" or "source spectrum" remain meaningful. But even this does not guarantee that a "source" of cosmic rays actually exists in terms of an astronomical object: for instance the interstellar medium (ISM), as a whole, traversed by supernova (SN) shock waves, could be this "source" - and this has indeed been widely believed in the past.

The present conclusion that cosmic-ray nuclei originate neither in the ISM nor in SN ejecta comes from recent detailed analyses of the composition of cosmic rays, which give almost direct evidence for a well-defined composition of the cosmic rays "at the source". In turn, the composition, elemental as well as isotopic, gives strong constraints on the nature of the cosmic-ray sources (see e.g. Cassé 1983).

In the present lecture, which updates and completes the review by Montmerle (1984), we will make the case that the bulk of cosmic rays likely originate in stars, either directly through manifestations of surface activity, or indirectly through their interactions with the surrounding medium. As will be seen, the cool stars can be associated with the great majority of the cosmic-ray elements, while hot stars seem associated with more "exotic" cosmic-ray species: selected isotopes, γ-rays, antiprotons, etc... Hence the convenience to break down the lecture in two parts: I. Cool stars, II. Hot stars.

1.2 The Stellar Connection: Direct Evidence

The elemental source composition of cosmic rays ("Galactic Cosmic Ray Source" composition, or GCRS) is known for most elements between Z = 1 and Z = 60. The error bars, however, vary widely from one element to the next, depending on its relative abundance. In addition, the isotopic composition is known only for a limited number of elements (see below), when the relative abundance is not too small and the charge resolution of the detector is sufficient.

The elemental composition does not show large deviations with respect to the "local galactic" (LG) abundances (Meyer 1985a), hence reflects only the atomic processes in which the seed nuclei are involved. A first very important clue is obtained when one plots the cosmic ray composition at the source, normalized to the local galactic abundances of the same elements, as a function of the first ionization potential ("FIP"). As Fig. 1 shows, for the comparatively more abundant elements having $Z < 30$, for which the highest FIP is ~ 25 eV (helium), the relative abundance GCRS/LG, which is fairly well defined, can be represented by two plateaus: GCRS/LG = 1 for FIP $\leqslant 8$ eV,

GCRS/LG $\simeq 0.2$ for FIP > 10 eV (p and α particles, however, are one order of magnitude underabundant). In other words, with respect to "ordinary" matter, the elements become less easily cosmic rays if they have a high FIP, above ~ 9 eV (Cassé and Goret 1978, Meyer 1985a,b). This bias remains the same for the less abundant elements with $30 < Z \leqslant 60$ (Meyer, 1986 ; Binns, this volume). The second key clue is that the same FIP composition bias holds for Solar Energetic Particles (low-energy cosmic rays from the Sun), for the solar corona and for the solar wind, thus confirming the existence of an "ionization filter" at the chromospheric level, acting around 9 eV (see Fig. 1).

These two clues have prompted Meyer (1986, and refs. therein) to suggest that most of the cosmic-ray heavy nuclei, which have therefore essentially a "solar-mix" composition, belong originally to a plasma with temperature $\lesssim 8000$ K, corresponding to the chromospheres of F to M stars. We note, in passing, that the "normal" abundance of refractory elements rules out the interstellar medium, where they are locked into grains. Likewise, nuclei selectively synthesized in SN explosions (r-process) are not in excess in cosmic rays, ruling also out a large contribution of SN ejecta. On the other hand, we note also that, with respect to the "FIP-bias" shown on Fig. 1, the important elements C and O tend to be overabundant (see below).

Isotopic composition, on the other hand, is the signature of nucleosynthesis ; at this point, in view of the results on elemental composition, we can expect only deviations from the general "normal" trend, i.e., signatures of quantitatively minor (if qualitatively important) contributors. The main isotopic ratios known to date are $^{22}Ne/^{20}Ne$, $^{25,26}Mg/^{24}Mg$ and $^{29,30}Si/^{28}Si$. Only the first of these ratios appears to be in strong excess (a factor ~ 4) over the corresponding GCRS ratio (Meyer 1985b, 1986). These isotopes, along with the C and O nuclei corresponding to the excesses mentioned above, are efficiently produced during the helium-burning phase of massive stars, the so-called Wolf-Rayet phase (see Part II). This means that a fraction of the cosmic rays comes from diluted ($\sim 1/50$, see Cassé and Paul, 1982 ; Meyer 1986) material originally processed in Wolf-Rayet stars, which are able, in addition, to shed this enriched material in the interstellar medium via their huge stellar winds (Part II).

1.3 Indirect Evidence

In addition to the above connection between stars and cosmic rays based on composition, one must add other signatures of cosmic-ray production in terms of photon emission.

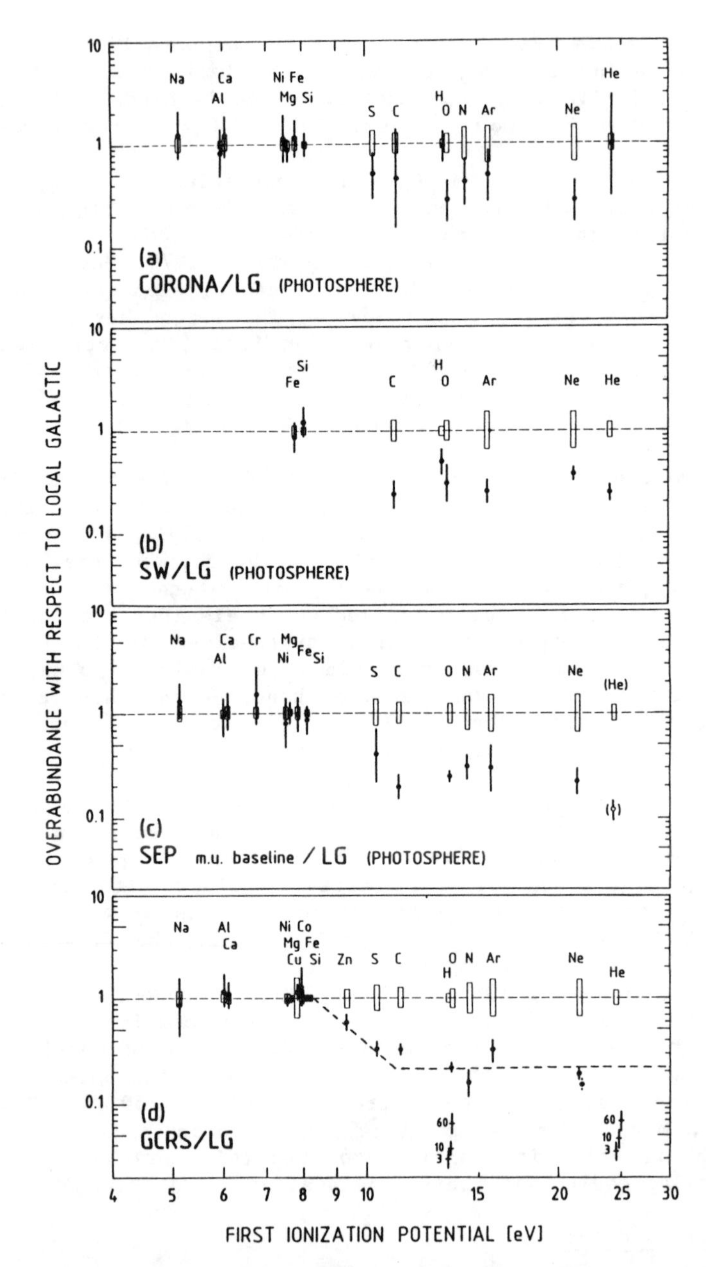

Fig. 1. Elemental abundance ratios with respect to Local Galactic abundances. (a) solar corona; (b) solar wind; (c) solar energetic particles; (d) galactic cosmic-ray sources (Meyer 1986). In this last panel, the added dotted line emphasizes the FIP bias, which is also apparent in the other panels.

First, it is now well established that the diffuse galactic γ-ray emission detected by the SAS-2 and COS-B satellites (i.e., from ~ 30 MeV to a few GeV) comes almost exclusively from the interaction of cosmic-ray protons (and α's) and electrons with the interstellar gas, via the π°-decay and bremsstrahlung mechanisms, respectively. The diffuse γ-ray flux observed may be entirely explained quantitatively using the amount of interstellar gas known from CO millimeter wave and HI 21 cm observations, as a target to a cosmic-ray flux equal to that prevailing in the vicinity of the Sun (e.g., Bloemen et al. 1986). Superimposed on this flux, a number of γ-ray sources have been found, mainly by COS-B (e.g. Bignami and Hermsen 1983). However, it is only very recently (owing to the completion of the galactic CO survey in the southern hemisphere by the Columbia/Harvard-Smithsonian group, Dame et al. 1987) that the "calculated" diffuse γ-ray flux has been substracted from the observed one for the whole galactic disk: in the process, "passive" sources (i.e., appearing as sources simply because of a localized enhancement of matter along the line of sight) are naturally removed, leaving only the "active" sources (i.e., showing a γ-ray excess). These sources may all be compact objects, analogous to the Vela and Crab pulsars already known as γ-ray sources, but it turns out that an important fraction of them appears to be associated with regions of star formation. If the observed γ-ray excess from these regions is explained in terms of cosmic-ray - matter interactions, this implies in situ cosmic-ray flux densities significantly higher than in the solar vicinity. We shall come back to this point in Part II.

For completeness, one should also mention the galactic non-thermal radio emission in the ~ 100 MHz to ~ GHz range. This emission, due to the synchrotron mechanism, is present up to very high galactic latitudes, and testifies to the presence of relativistic electrons throughout the galactic plane and in the halo (e.g., Kanbach 1983). Here also, "hot spots" are present, associated with supernova remnants and several star-forming regions. Finally, the stars themselves may be radio emitters, with good evidence for a non-thermal mechanism in a number of cases (see § 3).

Given this multi-faceted evidence for links between cosmic rays and all sorts of stars, it is therefore of interest to review what is known to date about various forms of stellar activity (keeping in mind the possible influence of the stellar environment), to look for independent evidence that stars can indeed be associated with the genesis of cosmic rays.

2. ACTIVITY PHENOMENA IN THE SUN AND STARS

2.1 Activity on the Sun

This is, of course, a huge topic in itself, which we only sketch here for comparison with the stars.

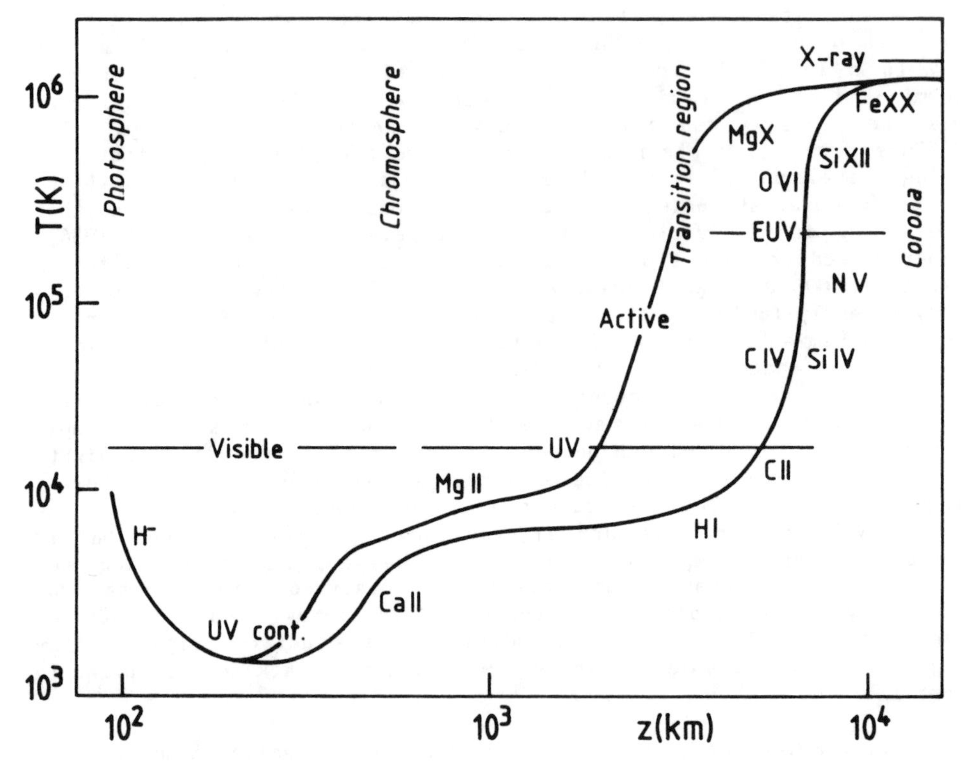

Fig. 2. Sketch of the outer layers of late-type stars, normal or active. The various regions, and the corresponding spectral ranges and tracers, are indicated (after Basri 1987b).

That the Sun is "active" has been known for a long time, with the early observations of transient or violent phenomena: flares, spots, plages, etc... Another form of activity, more recently discovered, is the existence of the solar wind (see, e.g., Axford 1985). Its structure is complex and shows a non-uniform time variable expansion velocity pattern ; the mass-loss rate is itself variable. Orders of magnitude are: $V_{wind} \sim 400$ km.s^{-1}, $\dot{M} \sim 10^{-14}$ $M_\odot$ yr^{-1}.

The role of the magnetic field has been first demonstrated in 1908 by Hale, who was able to detect the Zeeman effect in sunspot spectra where $B \lesssim 10^3$ G. But its global importance was recognized only after the "Skylab" results (1973) (see, e.g., Vaiana and Rosner 1978). The X-ray (~ keV) images, in particular, show that the magnetic field actually controls all the activity phenomena, not only on localized regions like spots, but also on a large scale like the corona. The wind itself appears channelled by the field lines when they are open, and originates in the "coronal holes".

Locally on the Sun, but globally on the stars, one can rely on a number of activity tracers, associated with different regions above the surface (Fig. 2). The most widely used tracers are: Hα (photosphere, $T \lesssim 10^4$ K, but its interpretation is often difficult, especially when strong mass loss is present), H and K lines of Ca II in the optical range, h and k lines of Mg II in the UV range (chromosphere), and $\lesssim$ keV X-rays (corona, flares). On the Sun, the Ca II and X-ray fluxes from localized active region have been shown observationally to be directly proportional to the magnetic flux across the corresponding area (Skumanich et al. 1975).

Activity phenomena and magnetic fields are thus intimately linked. The question of magnetic field generation is therefore crucial to the understanding of these phenomena. Although the problem is far from solved today, it is widely believed that magnetic fields are generated by the dynamo effect, resulting from movements of the solar plasma in its outer layers, on various scales (e.g., Belvedere 1985). The theory is extremely difficult, but recent results, notably numerical simulations, indicate that scales up to the "supergranulation" ($\ell \sim 10^4$ km), or even the recently discovered "giant rolls" ($\ell \sim 2\ 10^5$ km, Ribes et al. 1985) are involved. These movements, in turn, are of course the result of the existence of an extended outer convection zone (depth ~ $2\ 10^5$ km).

The solar evidence therefore suggests a close link between surface activity and outer convection zones.

2.2 Stellar Activity

Stellar activity in the form of flares visible in the optical range has been discovered by Hertzsprung in 1924, although the name "flare stars" was coined only in the fifties, following the recognition that a certain class of red dwarfs (dMe stars) were selectively undergoing rapid and irregular variations reaching a few magnitudes in a few minutes (Gurzadyan 1980). Other kinds of stars, like RS CVn or BY Dra binary systems, (pre-main sequence) T Tauri stars, and other young stars, are also known to undergo rapid optical variations.

But like for the Sun, the field of study came of age only with the acquisition of X-ray data. It has been one of the major discoveries made by the Einstein satellite, through its imaging capabilities, that X-ray emission is widespread among stars of all spectral types (see, e.g., Mewe 1984 ; Rosner et al. 1985) (Fig. 3).

X-ray flares have been observed in several members of the above classes, in the form of a steeply rising flux (seconds to minutes), followed by a slow decay, with an e-folding timescale of hours (Haisch 1983). The X-ray emission is due to optically thin thermal bremsstrahlung at $T \sim 10^6$ to 10^7 K. The phenomenon seems to take place only in late-type stars ; although variabilities have been observed in the X-ray flux of early-type stars, no evidence exists that they correspond to flares. Table 1 summarizes the available data on X-ray

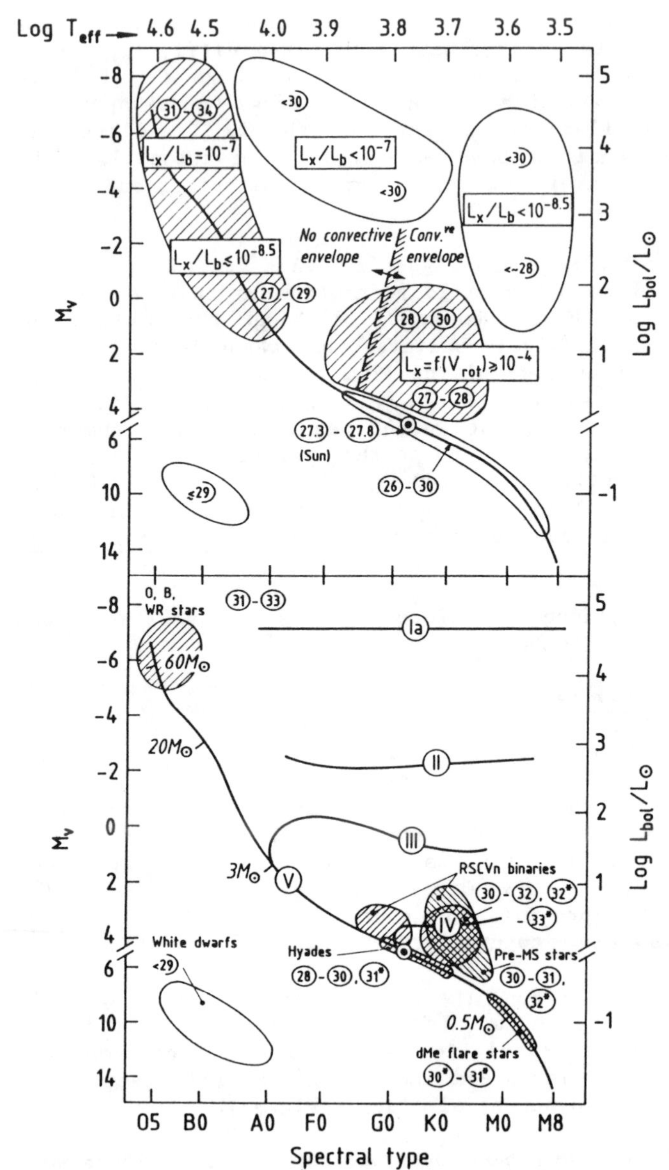

Fig. 3. *The H-R diagram of X-ray stars. The upper panel indicates the areas searched by the Einstein satellite. (i) Hatched areas correspond to positive detections, seen to lie in the vicinity of the main sequence; (ii) the hatched dotted line separates the regions where outer convection does, or does not, exist. The lower panel specifies the main classes of X-ray-detected stars (hatched areas). In both panels, the circled numbers give* $\log L_x$*; asterisks indicate flare values. (After Mewe 1984.)*

TABLE 1. Stellar X-ray flares

Star	$L_x(erg.s^{-1})$	T(K)	E(erg)	l(cm)	$n_e(cm^{-3})$
Sun	$\sim 10^{26}-10^{27}$	$1-2\ 10^7$	$\sim 10^{29}-10^{31}$	$\sim 10^8-10^{10}$	$10^{12}-10^{10}$
dMe	$\sim 2-16\ 10^{30}$	$1-3\ 10^7$	$>5\ 10^{32}$	$\sim 10^{10}$	10^{11}
T Tauri	$\sim 5-100\ 10^{30}$	$1-2\ 10^7$	$\sim 10^{34}-10^{36}$	$\sim 10^{10}-10^{12}$	10^{10}
OBA	-	-	-	-	-

flares, including a comparison with the Sun. It can be seen that stellar X-ray flares are much stronger than solar flares (factors 10^3 to 10^5), but this appears to be due mainly to the larger volumes of hot gas, since the densities and temperatures are quite comparable to the solar case.

In short, the available X-ray evidence shows that late-type stars undergo an enhanced solar-type activity, whereas this kind of activity seems to be absent in early-type stars. We will come back in more detail on this point in subsequent sections.

There is also evidence for quiescent "activity" in stars, i.e. release of energy other than luminous, but in a more or less continuous fashion (e.g. Cassinelli and McGregor 1986).

First, quiescent X-ray emission is observed throughout the main sequence or in its vicinity, contrary to pre-Einstein expectations. X-ray luminosities range from $L_x \sim 10^{27}$ erg.s^{-1} (typical solar value) to $\sim 10^{34}$ erg.s^{-1} (early-type stars), with L_x/L_{bol} varying from 10^{-7} to 10^{-3}. As shown in Fig. 3, it is important to notice that the low L_x/L_{bol} ratio applies to early-type stars, with little dependence on spectral type or other stellar parameters, whereas the high ratio applies to cool stars, and appears correlated with rotation (see below). There is another important difference between these two categories of stars: for late-type stars, the X-ray spectrum appears essentially thermal and suggests the existence of coronae ; for early-type stars, this spectrum is more complex, and cannot be explained in terms of a corona in the solar sense (e.g., see the discussion in Lucy 1982, and Rosner et al. 1985).

TABLE 2. Stellar winds

Star	$M(M_\odot.yr^{-1})$	v_∞ $(km.s^{-1})$	Remarks
A-B	$\gtrsim 10^{-9}$	$\gtrsim 1000$	winds show
O	$\gtrsim 10^{-6}$	2-3000	short-scale
Of	$\sim 10^{-5}$	2-3000	variabilities
WR	$> 10^{-5}$	3-4000	$\Rightarrow$ instabilities
< A	$< 10^{-10}$	?	not detectable
Sun	10^{-12}-10^{-14}	400-800	-
T Tauri	$< 10^{-9}$-10^{-8}	$\lesssim 300$	see also jets, bipolar flows, etc.

A second form of "activity" is the existence of mass loss. Here also, there is a clear distinction, in the vicinity of the main sequence, between early-type and late-type stars (e.g., Cassinelli 1979 ; Dupree 1986 ; De Jager et al. 1986): winds from early-type stars are both strong and fast, $\dot{M}$ up to 4×10^{-5} $M_\odot$ yr^{-1} and velocities up to 4000 $km.s^{-1}$ in the case of Wolf-Rayet stars (see Schmutz and Hamann 1986), whereas winds from late-type stars are slow (~ 200-500 km^{-1}), with smaller mass loss rates (down to 10^{-14} $M_\odot$ yr^{-1} for the Sun, as mentioned above). Note, however, that, for observational reasons, no stellar wind less strong than ~ 10^{-10} $M_\odot$ yr^{-1} is yet detectable. Table 2 summarizes the main characteristics of known stellar winds.

The activity of hot stars therefore appears to be dominated by mass loss. The mass loss, in turn, is widely thought to be the result of radiation pressure on heavy ions (Abbott and Lucy 1985, Kudritzki et al., 1987), even though some problems remain for a quantitative interpretation, especially of the highest values (WR stars), for which L_w $(= \frac{1}{2} \dot{M} v_w^2)/L_{bol}$ may reach ~ 10%. At the risk of over-simplifying reality, we will consider in what follows that early-type stars are "radiation-dominated", as opposed to the "magnetic field-dominated" late-type stars (Noyes 1985). Hot stars will be discussed in Part II.

As a final remark, nothing can be said of stars away from the main sequence: only upper limits exist (see fig. 3) for red giants and supergiants, as well as for isolated white dwarfs (e.g., Fontaine et al., 1982), so that these stars will not be discussed further here (see Linsky 1985).

3. SOLAR-TYPE ACTIVITY IN THE HR DIAGRAM

3.1 Along the Main Sequence

As indicated above, and mainly on the basis of X-ray observations, solar-type activity in the form of flares or standard coronae appears restricted to late F-M5 stars, with a very weak X-ray emission associated with early F stars (Wolff et al., 1986), essentially upper limits for A stars (Schmitt et al., 1985), and a strong decline after M5 (Rosner et al., 1985).

But other evidence exists, which supports the idea that, in late-type stars, there are a lot of similarities with respect to the Sun, making up what is now widely referred to as the "solar-stellar connection" (e.g., Pallavicini 1985): starspots, stellar cycles, correlations between activity tracers.

Starspots are "seen" in a variety of stars. In reality, the observational evidence comes only from periodic fluctuations of an activity indicator (like the H and K lines of Ca II, if the star is bright enough) or in multiband photometry, although in this case the interpretation is sometimes ambiguous (see examples below, § 3.2). We note that, in order to be detected, the starspots must have a distribution deviating strongly from circular symmetry: for instance, numerous starspots, equally spaced along the equator, would not produce periodic luminosity variations.

If starspots are well visible, it is possible to monitor the corresponding tracers over long periods of time. The field has been pioneered by O. Wilson (see Baliunas and Vaughan 1985), who discovered the existence of activity cycles by monitoring the Ca II (H + K) flux of 91 F5 to M2 stars over several decades. For such stars, the known stellar cycles span a range of $\sim$ 3 to $\sim$ 20 (?) years, i.e., are roughly comparable to the 11-year solar cycle. Photometry on archival plates allow to look for the existence of much longer cycles, found to be $\gtrsim$ 60 yrs in selected RS CVn and BY Dra binaries.

Finally, a good argument supporting the solar nature of the activity in late-type stars relies on the rather tight mutual correlations one finds between the intensity of different activity tracers. For instance, Fig. 4 shows the correlation of the X-ray flux F_x with the excess Ca II (H + K) flux ΔF_{H+K} (Mewe 1984) or with the excess Mg II flux ΔF_{h+k} (Rutten and Schrijrer 1987), all directly related to the magnetic field (§ 2.1) ; note that such correlations span 4 to 5 orders of magnitude in F_x and 2 to 3 in ΔF_{H+K} or ΔF_{h+k}. In this context, dMe

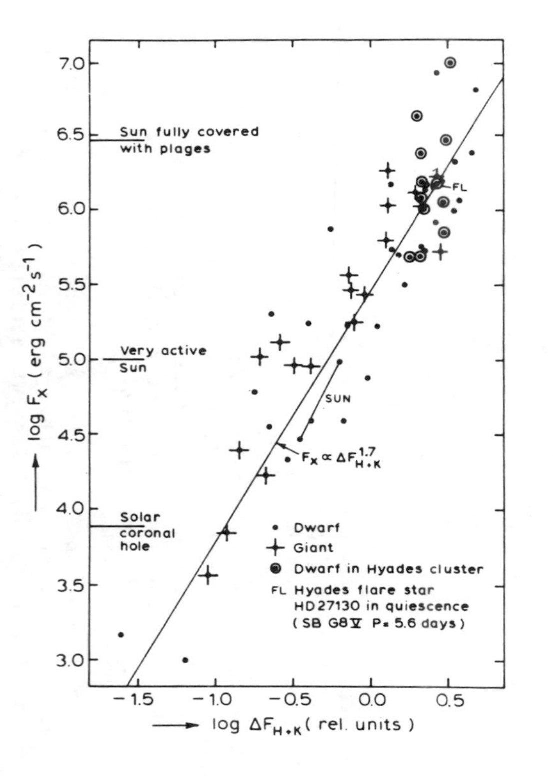

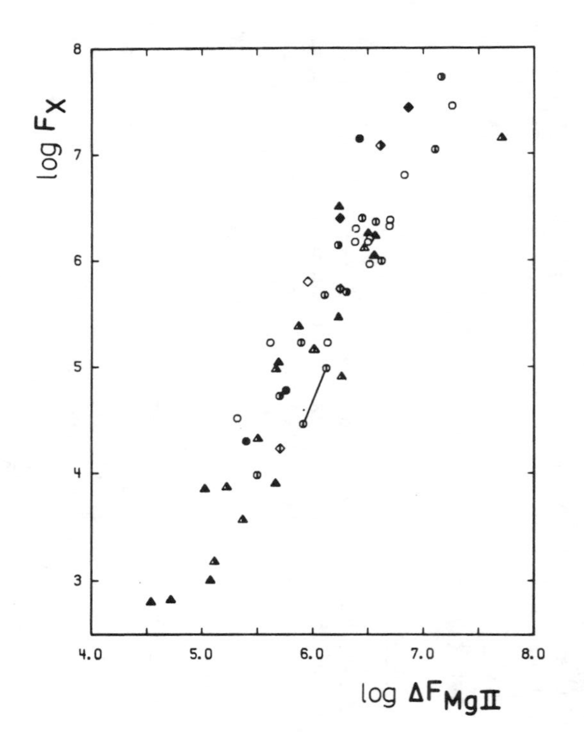

Fig. 4. Correlations between activity tracers: X-rays vs. CaII(H+K), and MgII(h+k). Left panel: X-ray flux vs. excess flux ΔF(H+K) (over an observational lower limit F(H+K)), for a sample of dwarfs and giants, including some binaries and young dwarfs in the Hyades (Mewe 1984). Right panel: id., but for the excess flux ΔF(h+k); the sample comprises F, G, and K stars of various luminosity classes (fluxes in erg.$cm^{-2}.s^{-1}$). For details, see Rutten and Schrijver (1987).

("flare") stars appear to have a strongly enhanced activity, but do not appear otherwise very different from ordinary dwarfs.

3.2 Before the Main Sequence

The most commonly known pre-main sequence stars are T Tauri stars (e.g., Bertout 1984, Cohen 1984), discovered by A. Joy in 1945 though their emission-line properties, later refined by G. Herbig. Their young age is attested mainly by their proximity to molecular ("dark") clouds, and the presence of lithium absorption. (Lithium, presumably of Big-Bang origin, is destroyed by convection in the course of evolution: the Sun, a former T Tauri star, contains only 1% of its initial amount of lithium.) Evolutionary tracks, although still highly uncertain, give an age of $\lesssim 10^6$ yrs, and a mass range ~ 0.5 to $\sim 3\ M_\odot$; spectral types run from ~ G to late M.

It is therefore clear that T Tauri stars represent only a relatively evolved subset of pre-main sequence stars, seen mainly at the periphery of the molecular clouds in which, presumably, they were born. Observational data on more deeply embedded objects is now growing, leading to the general - if admittedly vague - concept of "young stellar objects" (YSO) of various masses. Those are obtained by observing at wavelengths less sensitive to absorption, mostly keV X-rays, infrared, and radio.

The most clearcut evidence for solar activity in YSOs again comes from X-rays (see Feigelson 1984, 1987). Two separate sets of observations are available to date: on the one hand, X-ray surveys of cataloged T Tauri stars, on the other, X-ray surveys of star-forming regions, without reference of a specific class of YSO.

The second approach, making full use of the wide field of view (1°x1°) of the Imaging Proportional Counter (IPC) aboard *Einstein*, proved extremely fruitful. The best documented case concerns the nearby ρ Ophiuchi dark cloud, ~ 160 pc away from the Sun (Montmerle et al. 1983). Indeed, in a single image pointed towards the densest parts of the cloud, more than a dozen X-ray sources could be detected, including a few T Tauri stars. Because of the known variability of these stars in the optical, additional images were obtained at different epochs, showing that, indeed, variability by large factors is aslo a characteristic feature of the X-ray emission of T Tauri stars and other young stellar objects (fig. 5a). The T Tauri star surveys, on the other hand, yielded a low detection rate (1/3 out of 40 stars investigated), a result fully compatible with the repeated ρ Oph survey, given that most of these T Tauri stars were observed only once.

On this (admittedly still limited) basis, the X-ray emission from YSOs in the ρ Oph cloud was found to be due to thermal bremsstrahlung, mostly from flares, very similar to solar flares, as suggested by their power-law intensity distribution, and by the fact that the temperature

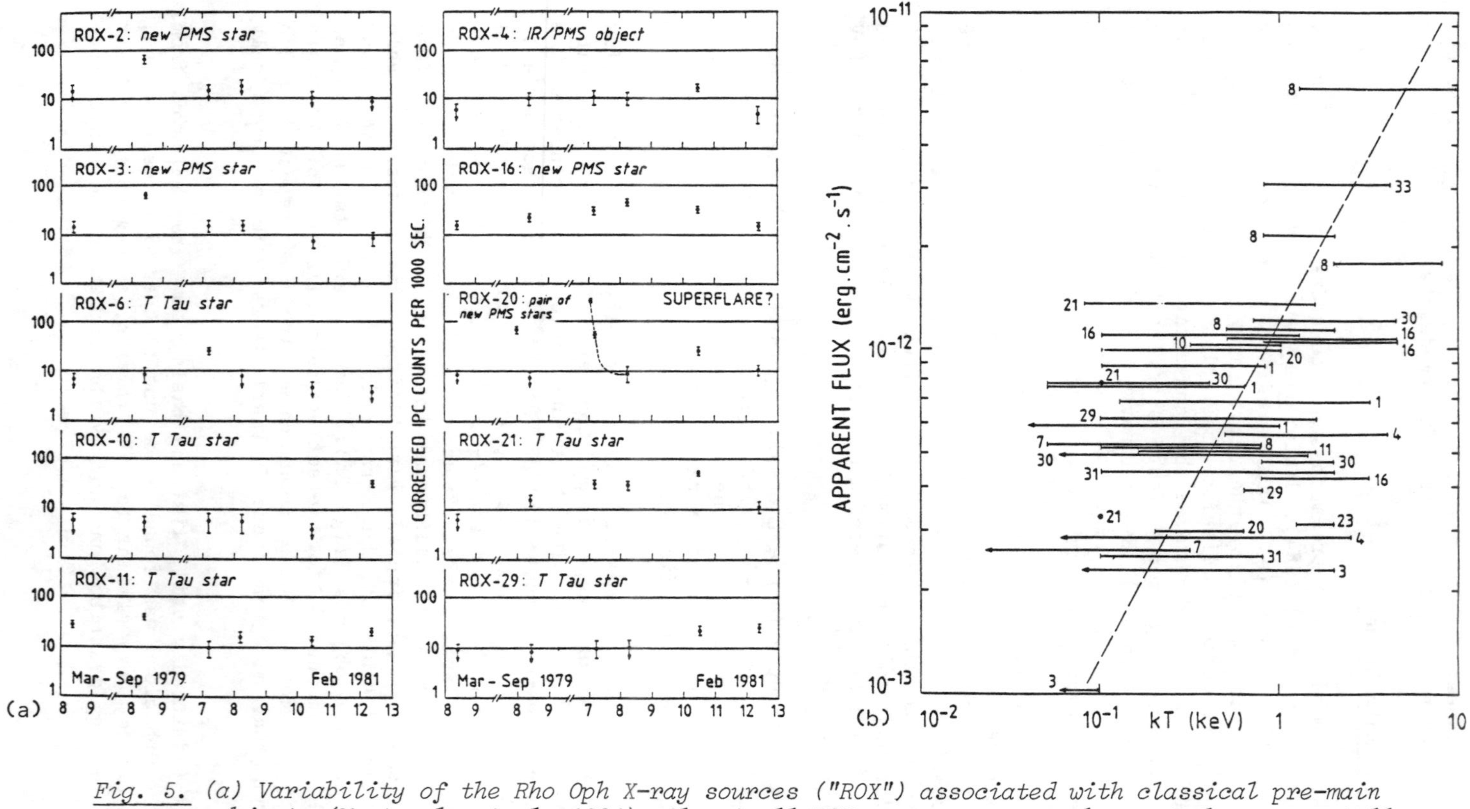

Fig. 5. (a) Variability of the Rho Oph X-ray sources ("ROX") associated with classical pre-main sequence objects (Montmerle et al. 1984). Almost all ROX sources are now known to be young stellar objects, mostly faint T Tauri-like stars, moderately embedded in the outer layers of the cloud ($A_v < 5$). (b) X-ray flux distribution, as a function of the temperature T. The error bars on T result from the uncertainties on the *Einstein* IPC gain. The dotted line is merely a guide to the eye, and shows the trend for the X-ray flux to increase with T, in a fashion reminiscent of solar flares. Numbers identify the ROX sources (Montmerle et al. 1984).

of order ~ 1 keV, tends to increase with increasing intensity (Fig. 5b). (For a more extended discussion, see Montmerle et al., 1983, 1984.) This result has been confirmed for a few stars in the radio range by interferometry with the NRAO Very Large Array, yielding in particular the discovery of the first radio flare in a pre-main sequence star (DoAr 21, Feigelson and Montmerle, 1985).

Another strong indication of the solar nature of the activity in these stars is provided by optical and near-IR photometry. Recent observations (Bouvier et al., 1986a, 1986b) have shown, in several counterparts to the X-ray sources in ρ Oph, as well as in classical T Tauri stars, periodic variations in the UBVRI bands (fig. 6). These variations may be interpreted, most of the time, in terms of a rotating single large region cooler than the stellar surface. This region can be a large starspot, but also a tight group of smaller ones. Ca II photometry has also been used, with comparable results (Hartmann et al., 1987).

But, for YSOs, the situation is more complex than for main-sequence stars of the same spectral type, owing to the existence of strong mass loss, sometimes in exotic forms (jets, see e.g., Mundt et al., 1984 ; CO bipolar flows, e.g., Lada 1985 ; see also Peimbert and Jugaku, 1987). The winds are usually seen in the optical (for instance, from P Cygni line profiles), giving velocities of a few 100 $km.s^{-1}$ (e.g., Sâ et al., 1986), with evidence for rotation in some cases (see the example of the Ae/Be star AB Aur, Catala et al., 1986).

Radio observations, when interpreted in terms of emission from an ionized wind, can give only the mass-loss rate. These have shown that cases of isotropic mass loss with $\dot{M} \gtrsim 10^{-8} \dot{M}_\odot$ do exist, but this is now thought to be the exception rather than the rule, upper limits to the mass loss rate on the order of a few 10^{-9} $M_\odot yr^{-1}$ being common (see, e.g., the VLA survey of the ρ Oph cloud by André et al., 1987 ; see also André 1987).

It thus appears that YSOs have a mixed type of activity, the dominating solar component being supplemented by a significant non-solar mass-loss component. We note right away that this other component cannot be of radiative origin either, luminosities and temperatures of YSOs being too small in most cases. The mechanism is not clear, but is probably related to the stellar magnetic field (hydromagnetic waves, see e.g., Lago 1984 and refs. therein). Other forms of non-solar activity also exist, for instance linked with the possible presence of accretion disks (e.g., Bertout 1987, Montmerle 1987).

3.3 In Close Binaries

Binary systems in which activity phenomena have been well studied are mainly the RS CVn systems, made of a late-type main sequence star with a post-main sequence companion, or two evolved stars, orbiting

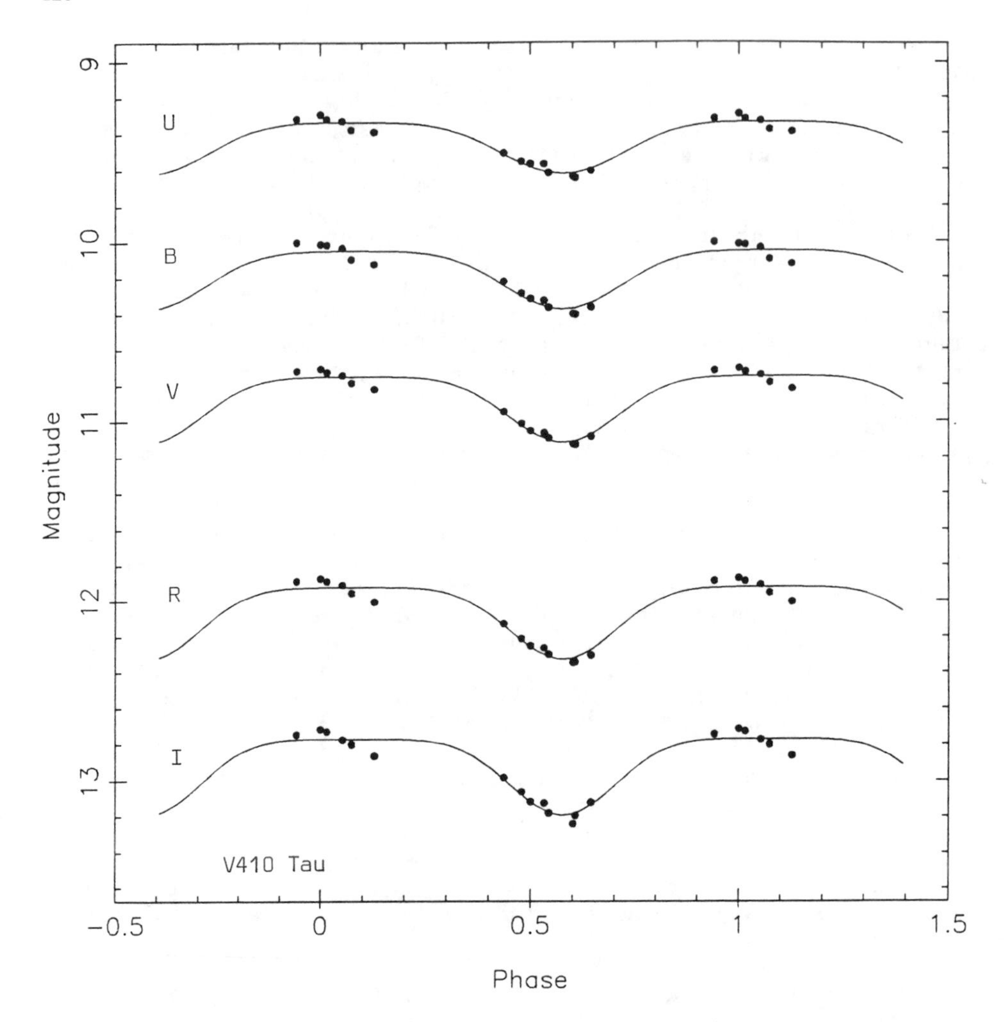

Fig. 6. UBVRI photometry of the active T Tauri star V410 Tau, in Jan. 1986. The data are interpreted in terms of a single starspot, covering 8% of the star's surface, at a latitude of 44° above its equator, and cooler than T_{eff} by 1315 K. Owing to the presence of this starspot, a rotation period can be accurately determined; here, P_{rot} = 1.88 day (from Bouvier 1987).

within one or two stellar radii of each other. Detailed X-ray observations, performed all along the orbital period of a few days, have allowed to decompose the detected flux into a flux from flares close to the stellar surface(s) and an extended corona, the size of which can be determined in the case of an eclipsing system like AR Lac (Walter et al., 1983). Another sign of activity is seen in the radio (Mutel and Lestrade 1985) where the (variable) GHz flux is likely due to inhomogeneous gyrosynchrotron radiation as in HR 1099 (Klein and Chiuderi-Drago 1987, and refs. therein). There are in fact a lot of similarities between RS CVn and T Tauri (or similar) stars, an important remark as regards the origin of their activity (e.g., Rosner et al., 1985 ; see § 4). However, they do not show the non-solar activity component displayed by T Tauri stars in the form of mass loss.

Another well-known class of close binaries are the BY Dra systems. Their activity has been known for a long time through the observations of very large starspots (e.g., Baliunas and Vaughan 1985), but they have not been systematically studied in X-rays. As a result, they are not frequently used for quantitative analyses of stellar activity (§ 4).

4. NATURE OF SOLAR ACTIVITY: LINKS WITH STELLAR PARAMETERS

4.1 Empirical Correlations

The above discussion has shown that "solar-type activity" is a meaningful term which, through ubiquitous tracers like Mg II, Ca II, X-rays, etc..., may be really measured. Furthermore, this activity appears to be confined in a clearcut manner to cool, late-type stars (late F to M). To understand the nature of the activity, which is clearly linked to magnetic fields, one first looks for correlations with stellar parameters, knowing two facts:

(i) on the Sun, the "α-ω" dynamo, which combines convection and differential rotation, is thought to be the main mechanism for generating magnetic fields (Fig. 7) (e.g., Gilman 1983 ; Belvedere 1985). At the cost of a few "free" (i.e. unknown from first principles) parameters, numerical calculations have been able to reproduce even the migration of sunspots towards the equator during a cycle ("butterfly diagram") (e.g., Belvedere et al., 1980).

(ii) on the HR diagram, solar-type activity is confined to stars having outer convection zones (Fig. 1). In this respect, it is highly significant that the stars with shallow convection zones (F stars, see Wolff et al., 1986), or stars which have outer envelopes "forced" to be quiet by an external agent (like a fixed, large scale magnetic field), and that reveal themselves through abundance anomalies requiring slow element diffusion to appear (the Ap-Bp stars, for instance ; see e.g., Montmerle and Michaud 1976 ; Michaud 1980 ; Vauclair and Vauclair 1982) have little, if any, X-ray flux.

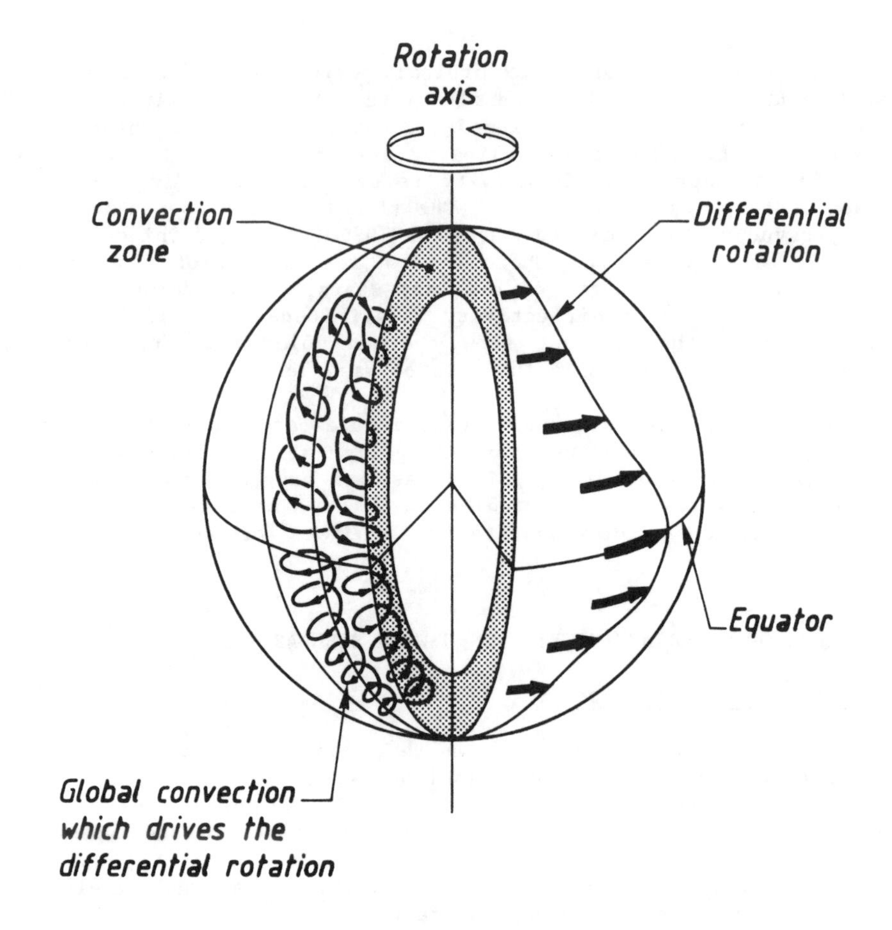

Fig. 7. *Structure of a $1M_{\odot}$ star. The combination of outer convection and differential rotation gives rise to magnetic field generation via the dynamo mechanism (after Gilman 1983).*

Clearly, however, convection is not enough. For instance, RS CVn stars are much more active than isolated stars in the same region of the HR diagram, and T Tauri stars are much more active than main sequence stars of the same spectral type. In the dynamo picture of stellar activity, an obvious idea is to try and correlate activity tracers with rotation tracers (v_{rot}, period), or with a "mixed" tracer of both rotation and convection like the "Rossby number" R_o = Period/τ_c (Mangeney and Praderie 1984, Noyes et al., 1984), where τ_c is the theoretical convective turnaround time (e.g., Gilliland 1985, 1986). Looking at recent results, though (Bouvier 1987, Fig. 8) one has the feeling that these tracers are perhaps not yet the physically relevant ones, since the correlations are not too good. In some cases, the correlations look better without including τ_c (see discussions in Rutten and Schrijver 1987 ; Basri 1987a,b). Part of the scatter is perhaps attributable to the fact that a number of late-type main sequence stars appearing in the diagrams may be in reality binaries, as has been recently discovered in an X-ray-selected sample of such stars (Silva et al., 1987).

Nevertheless, trends certainly emerge, and rotation does appear to be important. In this framework, one explains for instance, at least qualitatively, why RS CVn stars and T Tauri stars are more active than their main-sequence counterparts: although evolved, RS CVn stars do not lose angular momentum like the Sun, because their rotation is forced through the almost synchronous rotation between the two components (e.g., Uchida and Sakurai 1983). In an analogous fashion, T Tauri stars rotate faster than the Sun because they are young. This last conclusion is substantiated by several studies of the X-ray activity of the Hyades and Pleiades, which have shown a significant decline of stellar activity with age. The exact functional form of the decline, however, is still debated, from $t^{-\frac{1}{2}}$ to $e^{-\alpha t}$ (e.g., Caillault and Helfand 1985).

Conversely, the fact that main-sequence stars of spectral types later than M5 show very little activity in X-rays, although they are fully convective, may be explained by their very slow rotation: T Tauri stars of very late spectral type are active, and rotate significantly faster.

4.2 Stellar Magnetic Fields

Although it is probably safe to say that it is today reasonably well understood, the stellar-solar connection is not very well established on a quantitative basis. The reason for this may lie in the fact that we have a poor knowledge of the magnetic fields in stars. Not only are the measurements difficult, being to date limited to the strongest values, but also the interpretation of the results is often delicate especially as regards the actual structure of the magnetic field. For instance, the Sun has a global, essentially bipolar, magnetic field which averages ~ 1 G at the surface, but this value may reach several 10^2 G in magnetic loops (this is essentially the

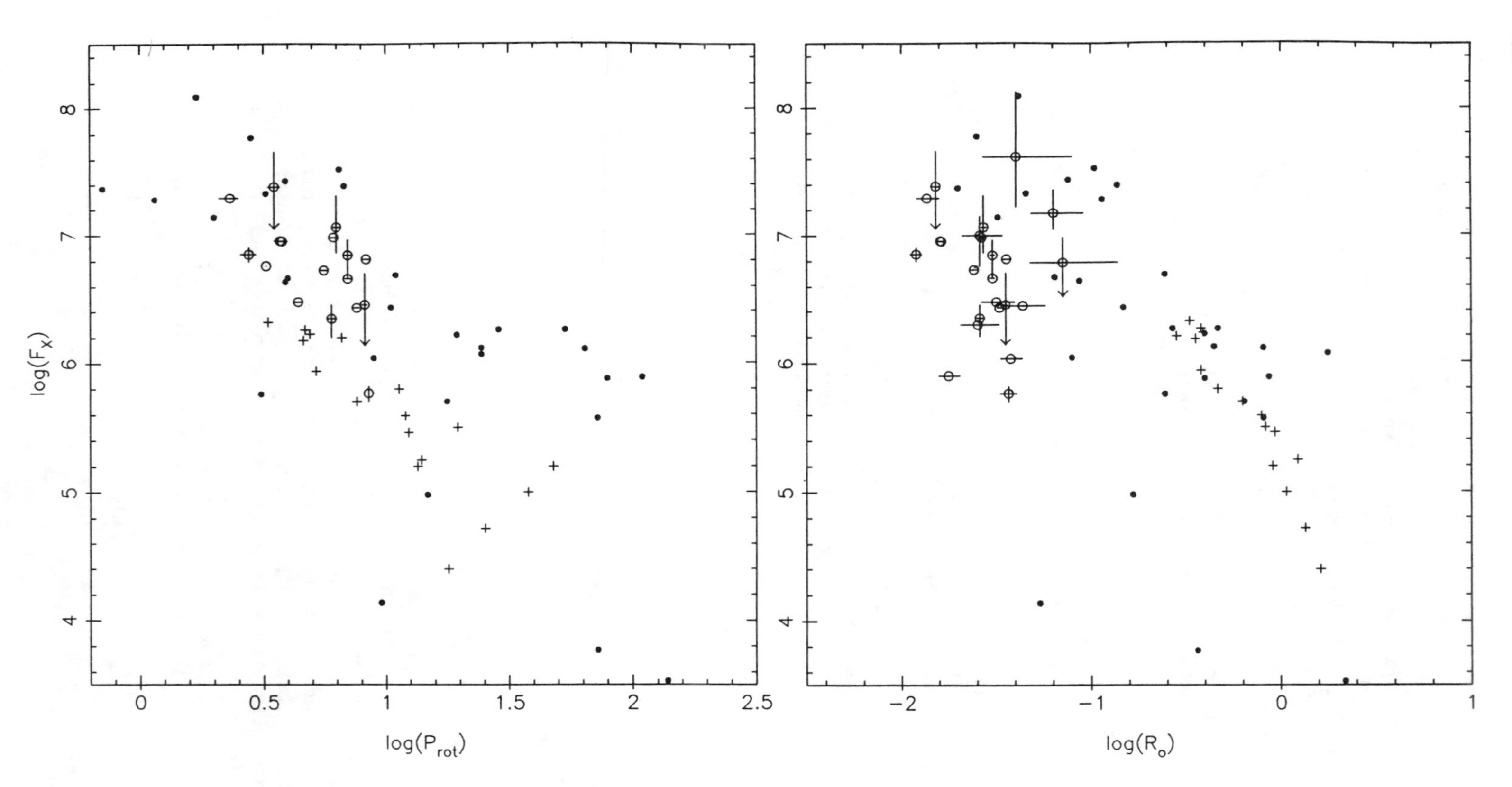

Fig. 8. Plot of the X-ray flux F_x (in erg.cm^{-2}.s^{-1}) vs. the period P_{rot} (days; left), and vs. the Rossby number (R_o = period/convective turnover time; right). Open circles: T Tauri stars, including some ROX sources; crosses: dwarfs; filled circles: RS CVn binaries. These plots show a trend for an increase of activity (= X-ray flux, here) with rotation, but there is apparently no significant advantage of using R_o (i.e., rotation + convection) over P_{rot} (i.e., rotation alone). (Bouvier 1987)

equipartition value if the loop confines an X-ray-emitting plasma), or even several 10^3 G in localized active regions like spots. In fact, the surface magnetic field of the nearest star, α Cen, almost a twin to the Sun, cannot be measured with present methods.

Measurements of stellar magnetic fields can be done in several ways (Borra et al., 1982 ; Marcy 1983). They all rely on the Zeeman effect ; the main obstacle is the broadening of the spectral lines by the star's rotation.

Through Zeeman polarization measurements, one can determine the component B_L of the magnetic field along the line of sight. For "magnetic" Ap and Bp stars, B_L ranges from several 10^2 G to $\sim 2 \times 10^4$ G. For the strongest measured fields, and using the star's rotation when not too rapid, the structure can be approximately modelled ; this structure turns out to be significantly more complex than a pure dipole but is certainly a global field, of typical size on the order of a stellar radius or larger.

For cooler stars, the "Robinson method", which uses selected spectral lines with a different sensitivity to magnetic fields (i.e., different Landé factors), has yielded positive results for about 20 main-sequence G and K stars. The detected magnetic fields are on the order of 500-3000 G, hence are likely to correspond to active regions rather that to the star as a whole. When it can be determined, the filling factor can reach high values (~ 0.5), suggesting that the corresponding stars are perhaps at the maximum of an activity cycle. The Robinson method is now being refined, especially though a more careful treatment of line transfer, yielding a better sensitivity: the first results are becoming available (Saar et al., 1986), giving a field of 2500 ± 300 G for the flaring BY Dra-type binary system EQ Vir.

We note for completeness that, for O and non-magnetic B stars, one has only fairly high upper limits ($B_L < 2000$ G), because of observational difficulties (large broadening of spectral lines formed in the wind).

5. COOL STARS AND COSMIC RAYS

Even though all the physical apsects of the solar-stellar connection are not yet well understood, the observational ones appear to be better established: reversing the argument, the Sun does appear in this connection also like an average star. In § 1.2, we have seen that the elemental source composition of cosmic-ray heavy nuclei give a strong constraint (the FIP bias) on the medium out of which they are initially extracted, in view of its similarity with that of the solar corona, the solar wind, and solar energetic particles, suggesting that most cosmic - ray heavy nuclei indeed come initially from the chromospheres of F-M stars.

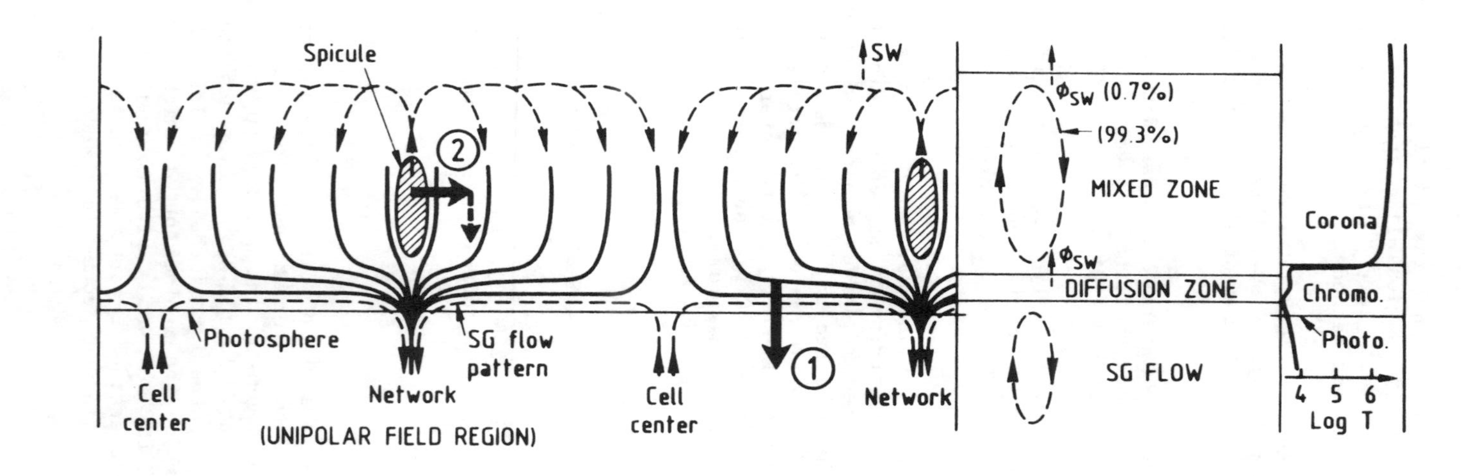

Fig. 9. The framework for two proposed scenarios to explain the underabundance of heavy nuclei of FIP 9 eV in the solar corona. The material brought into the base of the corona (see at right) in the form of spicules cycles by following the high-altitude magnetic field lines, parallel to the photosphere, before coming down; a small fraction $_{SW}$ is lost in form of the solar wind. In the outer layers of the photosphere, the convective flow cycles in the supergranulation pattern (SG). In Model 1 (Vauclair and Meyer 1985), the neutral elements (i.e., of high FIP) of the chromosphere diffuse downward by gavitational settling across the field lines, whereas ionized elements (low FIP) cannot. In Model 2 (Geiss and Boschler 1986), the neutral elements diffuse out of spicules, and eventually fall also down. (Meyer 1986)

The good news is that the stellar evidence, on the other hand, independently gives three important arguments, all going in the same direction:

(i) it is reasonable to assume that all stars displaying solar-type activity should also be producing stellar energetic particles by the same mechanism by which solar energetic particles are produced ;

(ii) chromospheres exist only in the cool star spectral domain (they correspond to a temperature range in which hydrogen is mostly neutral, hence is able to act as a thermostat) ;

(iii) the temperature of the chromospheric plateau (Fig. 2) is essentially uniform (~ 6500 K), irrespective of the spectral type in the F-M domain, hence the corresponding stellar energetic particles associated with activity should have the same composition - i.e., that of the solar energetic particles.

One may add that, because the stellar mass distribution goes as $M^{-1.4}$, cool stars are much more numerous than hot stars, giving even more weight to the above arguments, and therefore perhaps explaining qualitatively the relative sharpness of the FIP transition around 9 eV.

The bad news, however, is that there remains the problem of actually demonstrating how to transform stellar chromospheric particles into cosmic rays. To date, the best understood stage is certainly the final one, i.e., acceleration to relativistic energies, likely by supernova shocks (e.g., Drury 1983 ; Völk, this volume), in our scenario boosting particles having leaked from stellospheres of cool stars out to the interstellar medium. Prior to this, however, one must transfer them from the chromosphere, where they acquire their ionization state (FIP bias), to the corona, or accelerate them into energetic particles, while freezing this ionization state. Clearly, the transport from the chromosphere to the corona is crucial to the whole scenario.

In fact, this very first stage is not even understood on the Sun. Recent attempts feature the diffusion of neutral particles across the surface magnetic field lines (Fig. 9): out of spicules (Geiss and Boschler 1986), or downwards from the chromosphere (Vauclair and Meyer 1985). The problem is obviously very complex, as one must consider the various possibilities of diffusion (ion-ion, ion-neutral, parallel ou perpendicular to the magnetic field, etc.) ; the relevant diffusion coefficients themselves are not always known.

The above "stellar connection" also fails to explain the proton and α deficiencies in cosmic rays with respect to ordinary, photospheric material. A somewhat embarrassing problem, as these two components make up 99% of the cosmic-ray particles... But here again, the key may well lie with the Sun itself. Indeed, the status of H and He in solar energetic particles is reminiscent of the general cosmic ray situation, in that these nuclei are observed to behave very differently from the other nuclei:

(i) in relation with solar flare events, variations in particle flux are recorded, but they are well ordered in the case of heavy nuclei (hence the possibility to define a "mass-unbiased baseline", Meyer 1985a), whereas variations of He, and especially of H, are very erratic ;

(ii) averages may be defined for He (see Breneman and Stone 1985) - but not for H -, and they tend to show that He is already underabundant in solar energetic particles.

Understanding particle acceleration on the Sun thus appears today as a necessary (but not sufficient !) condition to understand the origin of galactic cosmic rays.

REFERENCES

Abbott, D.C., Lucy, L.B.: 1985, Ap. J. 288, 679.
André, P.: 1987, in Montmerle and Bertout (1987).
André, P., Montmerle, T., Feigelson, E.D.: 1987, Astr. J., 92, 163.
Axford, W.I.: 1985, Sol. Phys., 100, 575.
Baliunas, S.L., Vaughan, A.H.: 1985, Ann. Rev. Astr. Ap., 23, 379.
Basri, G.: 1987a, Ap. J., 316, 377.
Basri, G.: 1987b, in Montmerle and Bertout (1987).
Belvedere, G.: 1985, Sol. Phys., 100, 363.
Belvedere, G., Paterno, L., Stix, M.: 1980, Astr. Ap., 86, 40.
Bertout, C.: 1984, Rep. Progr. Phys., 47, 111.
Bertout, C.: 1987, in Circumstellar Matter, eds. I. Appenzeller and C. Jordan (Dordreicht: Reidel), in press.
Bignami, G.F., Hermsen, W.: 1983, Ann. Rev. Astr. Ap., 21, 67.
Bloemen, J.B.G.M., Strong, A.W., Blitz, L., Cohen, R.S., Dame, T.M., Grabelsky, D.A., Hermsen, W., Lebrun, F., Mayer-Hasselwander, H.A., Thaddeus, P.: 1986, Astr. Ap. 154, 25.
Borra, E.F., Landstreet, J.D., Mestel, L.: 1982, Ann. Rev. Astr. Ap., 20, 191.
Bouvier, J.: 1987, Ph. D. Thesis, University of Paris.
Bouvier, J., Bertout, C., Benz, W., Mayor, M.: 1986, Astr. Ap., 165, 110.
Bouvier, J., Bertout, C., Bouchet, P.: 1986, Astr. Ap., 158, 149.
Breneman, H.H., Stone, E.C.: 1985, Ap. J. (Letters), 299, L57.
Byrne, P.B., Rodono, M. (Eds): 1983, Activity in Red Dwarf Stars (Dordrecht: Reidel).
Caillault, J.P., Helfand, D.J.: 1985, Ap. J., 289, 279.
Cassé, M.: 1983, in Composition and Origin of Cosmic Rays, ed. M.M. Shapiro (Dordrecht: Reidel), p. 193.
Cassé, M., Goret, P.: 1978, Ap. J., 221, 703.
Cassé, M., Paul, J.A.: 1982, Ap. J., 258, 860.

Cassinelli, J.P.: 1979, Ann. Rev. Astr. Ap., 17, 275.
Cassinelli, J.P., Mac Gregor, K.B.: 1986, in Physics of the Sun, Vol. III, ed. P.A. Sturrock (Dordrecht: Reidel), p. 47.
Catala, C., Felenbok, P., Czarny, P., Talavera, A., Boesgaard, A.M.: 1986, Ap. J., 308, 791.
Cohen, M.: 1984, Phys. Rep., 116, 173.
Dame, T.M., Ungerechts, H., Cohen, R.S., De Geus, E., Grenier, I., May, J., Murphy, D.C., Nyman, L.-A., Thaddeus, P.: 1987, Ap. J., in press.
De Jager, C., Nieuwenhuijzen, H., Van der Hucht, K.A.: 1986, in Luminous Stars and Associations in Galaxies, ed. C. de Loore, A.J. Willis, and P. Laskarides (Dordrecht: Reidel), p. 109.
Drury, L. O'C.: 1983, Rep. Progr. Phys., 46, 973.
Dupree, A.K.: 1986, Ann. Rev. Astr. Ap., 24, 377.
Feigelson, E.D.: 1984, in Cool Stars, Stellar Systems, and the Sun, eds. S. Baliunas and L. Hartmann (Berlin: Springer-Verlag), p. 27.
Feigelson, E.D.: 1987, in Montmerle and Bertout (1987).
Feigelson, E.D., Montmerle, T.: 1985, Ap. J. (Letters), 289, L19.
Fontaine, G., Montmerle, T., Michaud, G.: 1982, Ap. J., 257, 695.
Geiss, J., Bochsler, P.: 1986, in The Sun and the Heliosphere in Three Dimensions, ed. R.G. Marsden (Dordrecht: Reidel), p. 173.
Gilliland, R.L.: 1985, Ap. J., 299, 286.
Gilliland, R.L.: 1986, Ap. J., 300, 339.
Gilman D.A.: 1983, in Stenflo (1983), p. 247.
Gurzadyan, G.A.: 1980, Flare Stars (Oxford: Pergamon).
Haisch, B.M.: 1983, in Byrne and Rodono (1983), p. 255.
Hartmann, L.W., Soberblom, D.R., Stauffer, J.R.: 1987, Astr. J., 93, 907.
Kanbach, G.: 1983, Sp. Sci. Rev., 36, 273.
Klein, K.L., Chiuderi-Drago, F.: 1987, Astr. Ap., 175, 179.
Koch-Miramond, L., Montmerle, T. (Eds.): 1984, Very Hot Astrophysics Plasmas, Phys. Scr. T7.
Kudritzki, R.P., Pauldrach, A., Puls, J.: 1987, Astr. Ap., 173, 293.
Lada, C.J.: 1985, Ann. Rev. Astr. Ap., 23, 267.
Lago, M.T.V.T.: 1984, M.N.R.A.S. 210, 323.
Linsky, J.L.: 1985, Sol. Phys., 100, 333.
Lucy, L.B.: 1982, Ap. J., 255, 286.
Mangeney, A., Praderie, F.: 1984, Astr. Ap., 130, 143.
Marcy, G.W.: 1983, in Stenflo (1983), p. 3.
Mewe, R.: 1984, in Koch-Miramond and Montmerle (1984), p. 1.
Meyer, J.P.: 1985a, Ap. J. Suppl., 57, 173.
Meyer, J.P.: 1985b, Rapporteur Paper, 19th Int. Cosmic Ray Conf., (La Jolla), 9, 141.
Meyer, J.P.: 1986, in Vangioni-Flam et al. (1986), p. 393.
Michaud, G.: 1980, Astr. J., 85, 589.
Montmerle, T.: 1984, Adv. Space Res., 4, n°2-3, 357.
Montmerle, T.: 1987, in Solar and Stellar Physics, eds. E.H. Schröter and M. Schüssler (Berlin: Springer-Verlag), in press.
Montmerle, T., Bertout, C. (Eds.): 1987, Protostars and Molecular Clouds (Saclay: CEA/Doc), in press.
Montmerle, T., Koch-Miramond, L., Falgarone, E., Grindlay, J.E.: 1983, Ap. J., 269, 182.

Montmerle, T., Koch-Miramond, L., Falgarone, E., Grindlay, J.E.: 1984, in Koch-Miramond and Montmerle (1984), p. 59.
Montmerle, T., Michaud, G.: 1976, Ap.J. Suppl., 31, 489.
Mundt, R., Bührke, T., Fried, J.W., Neckel, T., Sarcander, M., Stocke, J.: 1984, Astr. Ap., 140, 17.
Mutel, R.L., Lestrade, J.F.: 1985, Astr. J., 90, 493.
Noyes, R.W.: 1985, Sol. Phys., 100, 385.
Noyes, R.W., Hartmann, L.W., Baliunas, S.L., Duncan, D.K., Vaughan, A.H.: 1984, Ap. J., 279, 763.
Pallavicini, R.: 1985 in Radio Stars, Eds. R.M. Hjellming and D.M. Gibson, (Dordrecht: Reidel), p. 197.
Peimbert, M., Jugaku, J. (Eds.): 1987, Star Forming Regions (Dordrecht: Reidel).
Ribes E., Mein, P., Mangeney, A.: 1985, Nature, 318, 170.
Rosner, R., Golub, L., Vaiana, G.S.: 1985, Ann. Rev. Astr. Ap., 23, 413.
Rutten, R.G.M., Schrijver, C.J.: 1987, Astr. Ap., 177, 155.
Sà, C., Penston, M.V., Lago, M.T.V.T.: 1986, M.N.R.A.S., 222, 213.
Saar, S.H., Linsky, J.L., Beckers, J.M.: 1986, Ap. J., 302, 777.
Schmitt, J.H.M.M., Golub, L., Harnden, F.R., Jr., Maxson, C.W., Rosner, R., Vaiana, G.: 1985, Ap. J., 290, 307.
Schmutz, W., Hamann, W.R.: 1986, Astr. Ap., 166, L11.
Silberberg, R., Tsao, C.H., Letaw, J.R., Shapiro, M.M.: 1983, Phys. Rev. Letters, 51, 1217.
Silva, D.R., Gioia, I.M., Maccacaro, T., Mereghetti, S., Stocke, J.T.: 1987, Astr. J., 93, 869.
Skumanich, A., Smythe, C., Frazier, E.N.: 1975, Ap. J., 200, 747.
Stenflo, J.O. (Ed.): 1983, Solar and Stellar Magnetic Fields: Origin and Coronal effects (Dordrecht: Reidel).
Uchida, Y., Sakurai, T.: 1983, in Byrne and Rodono (1983), p. 629.
Vaiana, G., Rosner, R.: 1978, Ann. Rev. Astr. Ap., 16, 393.
Vangioni-Flam, E., Audouze, J., Cassé, M., Chièze, J.P., Trân Thanh Vân J. (Eds.): 1986, Advances in Nuclear Astrophysics (Gif-sur-Yvette: Editions Frontières).
Vauclair, S., Meyer, J.P.: 1985, 19th Int. Cosmic Ray Conf. (La Jolla) 4, 232.
Vauclair, S., Vauclair, G.: 1982, Ann. Rev. Astr. Ap., 20, 37.
Walter, F.M., Gibson, D.M., Basri, G.S.: 1983, Ap. J., 267, 665.
Wandel, A., Eichler, D., Letaw, J.R., Silberberg, R., Tsao, C.H.: 1987 preprint.
Wolff, S.C., Boesgaard, A.M., Simon, T.: 1986, Ap. J., 310, 360.

STARS AND COSMIC RAYS

II. HOT STARS

Thierry Montmerle

Service d'Astrophysique
Centre d'Etudes Nucléaires de Saclay
91191 Gif sur Yvette Cedex, France

ABSTRACT

Because they are short-lived and subject to an intense mass loss, hot stars interact strongly throughout their lifetime with their parent molecular clouds: they excite giant HII regions, inside which they create large wind cavities. The possibility of particle acceleration by wind shocks then allows to introduce the concept of thick cosmic-ray sources. The high-energy particles generated by such sources are very different from those associated with cool stars: protons, antiprotons, γ-rays, neutrinos, but also X-rays (> keV) and nuclear γ-ray lines. In spite of these differences, common features between hot and cool stars as regards their connection with cosmic rays are singled out.

1. MASSIVE STARS: ACTIVITY, EVOLUTION, ENVIRONMENT

1.1 Activity and Evolution

The internal structure of stars of spectral type earlier than A is very different from that of later types (Fig. 1). O and B stars ($M >$ 20 $M_\odot$) have a convective core surrounded by a thick outer radiative zone. They are characterized by a strong stellar wind, from $\sim 10^{-9}$ $M_\odot$ yr^{-1} for intermediate B stars to $\sim 10^{-6}$ $M_\odot$ yr^{-1} for O giants, at velocities $\sim$ 2 to 3000 $km.s^{-1}$. Their luminosity is very high (L_{bol} goes roughly as $\sim M^3$ in this mass range), and drives the wind by radiation pressure on the lines of heavy ions, although the theory is able to reproduce only the main features of the phenomenon (see Part I). The mass loss rate is high enough to significantly affect the evolution of the stars, which takes place over typical time scales of a few 10^6 yrs (e.g., Chiosi and Stalio 1981 ; Chiosi and Maeder 1986). As mentioned in Part I, magnetic activity is weak or non-existent ; the "radiative activity" in the form of a stellar wind dominates by far.

If the initial mass of an O star is larger than ~ 35 $M_\odot$, the star

M. M. Shapiro and J. P. Wefel (eds.), Genesis and Propagation of Cosmic Rays, 131–151.

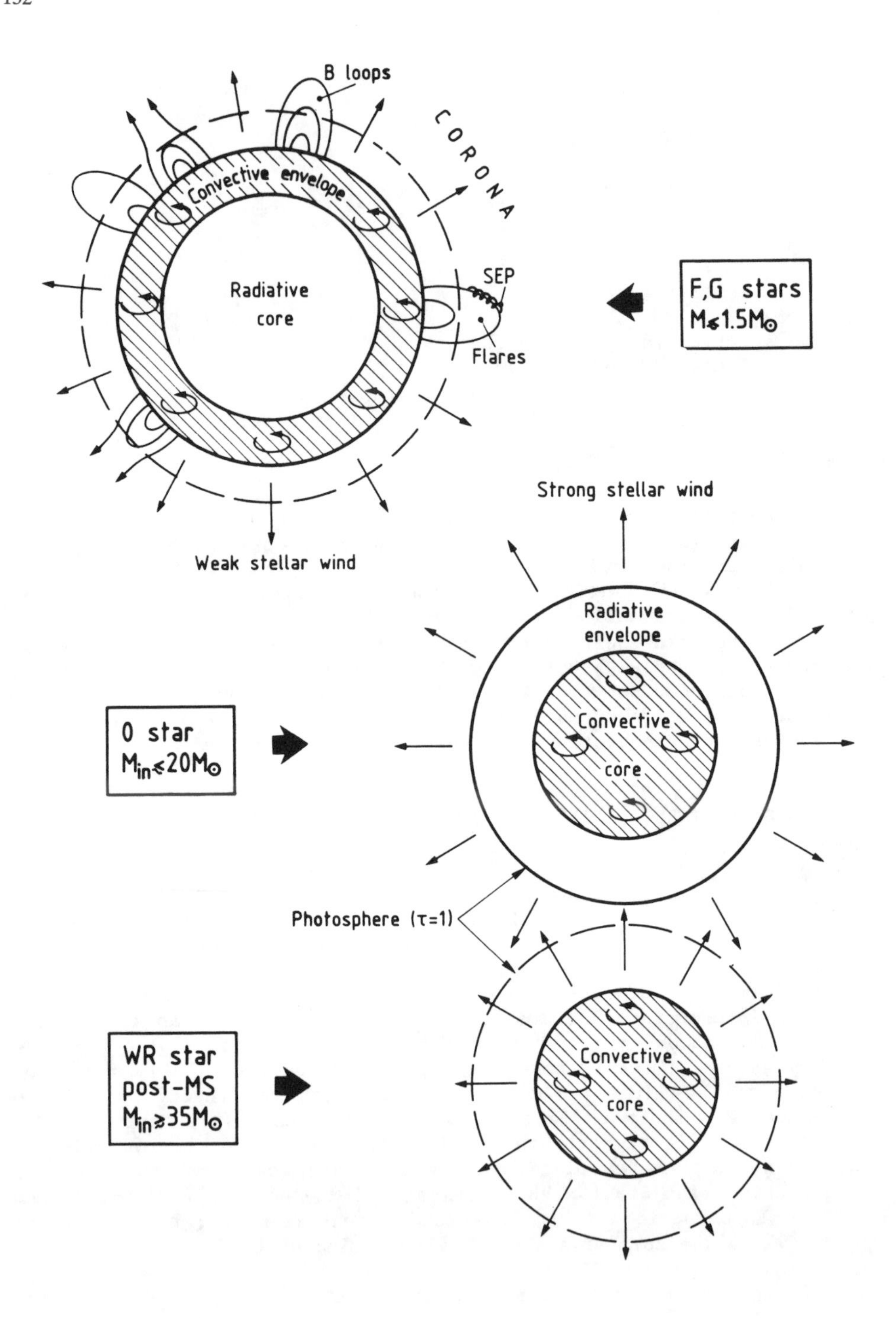
B loops
CORONA
Convective envelope
Radiative
core
SEP
Flares
F,G stars
M≤1.5M☉
Weak stellar wind
Strong stellar wind
Radiative
envelope
Convective
core
O star
Min≤20M☉
Photosphere (τ=1)
WR star
post-MS
Min≥35M☉
Convective
core

evolves into the so-called "Of" and "Wolf-Rayet" stages, characterized by a further increase in mass loss over timescales $< 10^5$ yrs, to reach $\lesssim 4 \times 10^{-5}\ M_\odot\ yr^{-1}$ at $\sim 4000\ km.s^{-1}$. Although the exact nature of the objects defined spectroscopically as Wolf-Rayet stars is somewhat a matter of definition (essentially because radiative transfer in such massive winds has not been solved accurately enough), these stars essentially correspond to the post-main sequence, helium-burning state (Fig. 2). During this phase, the mass loss is so large that the photosphere (i.e., the surface of the star in the conventional sense) is moved into the wind: the star itself is a "bare core", and the wind is able to transport directly outwards the freshly synthesized material (see Fig. 1, and De Loore and Willis 1982 ; Chiosi and Maeder 1986). This fact will have a lot of consequences, which will be the main topic of this review.

Table 1 summarizes the relevant basic data on hot stars.

1.2 Environment

Most massive stars are not isolated, but belong to "OB associations", comprising often up to several tens of luminous O and B stars (Humphreys 1978). These OB associations are surrounded by, or buried inside, giant molecular clouds, in which they were born. An important

TABLE 1. Basic data on hot stars

parameter	late O B0-O7	early O O7-O3	Of	WR
$M(M_\odot)$	20-35	35-120+	x0.8	x0.7
$L_{bol}/L_\odot$	$\lesssim 10^5$	$\lesssim 10^6$	$\sim 2\times10^6$	$\sim 2\times10^6$
$\dot{M}(M_\odot.yr^{-1})$	$\sim 10^{-7}$	$\sim 10^{-6}$	10^{-6}x	10^{-5}x
v_∞ (km.s^{-1})	2-3000	2-3000	2-3000	3-4000
$L_w=\frac{1}{2}\dot{M}v_\infty^2$(erg.s^{-1})	10^{36}x	10^{37}x	10^{37}x	10^{38}x
L_w/L_{bol}	$\lesssim 1\%$	$\sim 1\%$	$\sim 1\%$	$\gtrsim 10\%$
lifetime(10^6 yr)	~10	<4-6	~2-4	~0.5

Fig. 1. (Facing page) Hot stars vs. cool stars: internal structure. Low-mass, cool F-M stars have an extended convective envelope (see part I). On the contrary, high-mass, hot OB stars have an extended convective core, surrounded by a radiative envelope. Because of the large radiation pressure, a strong wind is established. For Wolf-Rayet stars (into which O stars evolve after the main sequence if their initial mass M_{in} is larger than $\sim 35\ M_\odot$), the mass loss is even larger, and only a bare core remains, surrounded by a dense wind.

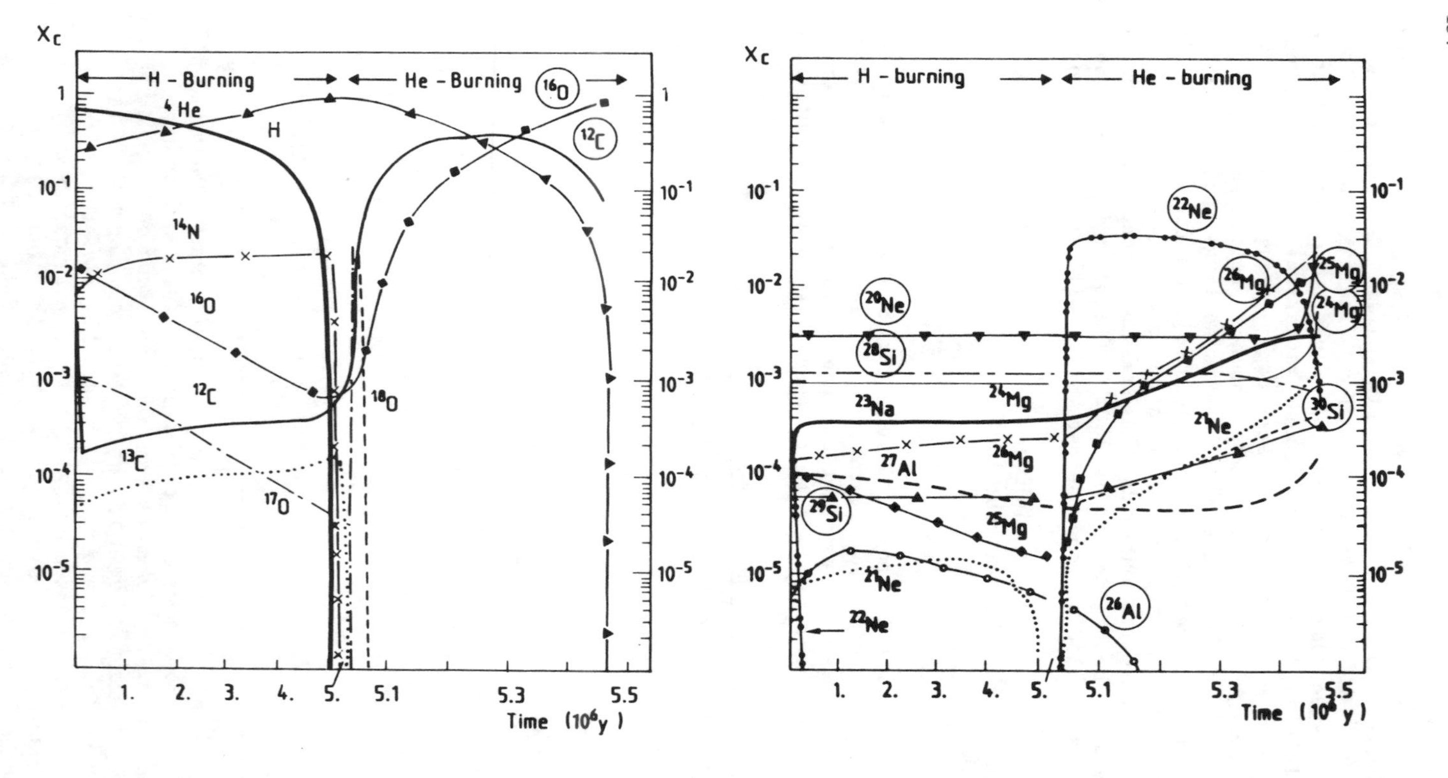

Fig. 2. Chemical evolution of the convective core of a star with $M_{in} = 60\ M_{\odot}$, for A<20 (left panel), and for A≥20 (right panel)(Prantzos et al. 1986). The He-burning phase (note the different timescale) corresponds spectroscopically to a Wolf-Rayet star. Isotopes of particular cosmic-ray interest are circled: ^{12}C and ^{16}O because they are overabundant; Ne, Mg, and Si because their isotopic ratios have been measured (see Part I). ^{26}Al has also been included, as it emits a characteristic 1.8 MeV line, which has been recently detected in the galactic plane (see §5.2).

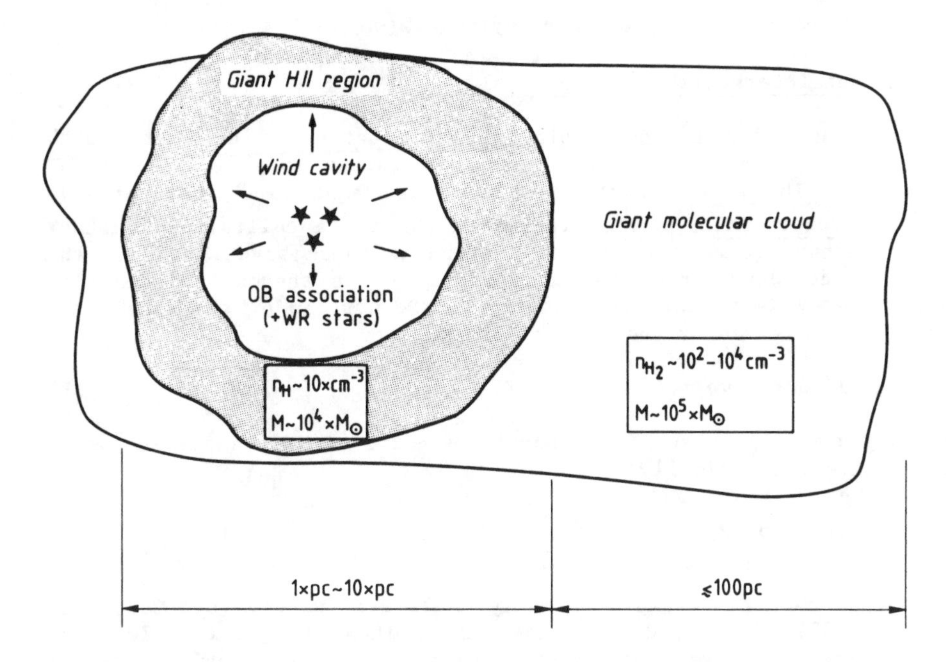

Fig. 3. The environment of an OB association. The OB and WR stars dig a large, hollow HII region into their parent molecular cloud. Particles accelerated by the exciting star winds therefore must traverse a dense material, leading to the concept of a thick cosmic-ray sources (grammages ~10-100 g.cm^{-2}). Typical sizes, densities, and masses of the ionized and neutral media are shown.

feature of OB associations is that their members, because of their comparatively short lifetime, will spend essentially all their lives at the same location. The observations show that at least 60% of the O stars, and at least 40% of the WR stars, belong to OB associations (Lundström and Stenholm 1984).

The O, B, and WR stars interact with their surroundings in essentially two ways: their intense flux of Lyman continuum photons creates a huge volume of ionized gas, up to several tens of pc in size ("giant HII regions") ; their strong winds create a large inner cavity. Since typical HII region densities are ~ several 10 cm^{-3}, and typical molecular cloud H_2 densities of order ~ 10^2 - 10^4 cm^{-3} (cf. Fig. 3), any material created by the stars of the association has to interact with a surrounding dense material. (By contrast, the lower-mass stars, also born in molecular clouds, live much longer and "diffuse" out in the low-density interstellar medium.) In the next section, we discuss in greater detail the structure of HII regions, since we will argue they may be the target of cosmic rays after acceleration has taken place inside them.

2. STRUCTURE OF HII REGIONS WITH STELLAR WINDS

2.1 The Hot Interstellar Bubble Model

The standard model describing the interaction of an O star with the interstellar medium is the "hot interstellar model" of Weaver et al. (1977). The model starts from the fact that winds from massive stars are supersonic, hence, presumably through a collisionless shock (like supernova remnants), create a standing shock wave, far from the star. At the shock, the wind of velocity v_w is thermalized, and its kinetic energy is transformed into heat. Downstream of the shock, the temperature is given by the standard formula:

$$T = (3/16\ k) \langle m \rangle v_w^2 \tag{1}$$

($\langle m \rangle$ is the average mass of the particles $\sim 1.2\ m_p$, for a normal composition), i.e., numerically:

$$T \simeq 2.7 \times 10^7\ K \times \left(\frac{v_w}{1000\ km.s^{-1}} \right)^2 \tag{2}$$

A plasma at this temperature normally cools very slowly (see Gaetz and Salpeter 1983): if its density is 1 cm^{-3}, it will cool in $\sim 10^7$ yrs, i.e., in a time of the same order or longer that the lifetime of an OB star. The hot downstream gas will therefore expand into a large interstellar "bubble", compressing the HII region excited by the O star into a thin ionized shell. This shell itself plows into the cooler interstellar medium to generate an outer HI shell. The structure of the bubble, sketched in Fig. 4a, thus consists mainly in a thick X-ray emitting bubble, surrounded by a thin HII shell. The expansion of the bubble is essentially adiabatic, but a small fraction is lost into evaporating HII region at its interface with the bubble, via electron conduction. This is only energy loss mechanism, given the inefficiency of the radiative losses at X-ray temperatures.

2.2 Evolution with Stellar Evolution and Dissipation

However, in many cases, the above structure is not supported by observation:

(i) giant HII regions, excited by early O stars and WR stars, are hollow and thick, the size of the HII shell being of the order of the size of the cavity (see, e.g., the well-known Rosette nebula, NGC2244);

(ii) the kinetic energy of the outer shell, when measured, is small with respect the calculated fraction (20%) of the wind energy of the exciting star it should have (Chu 1983) ;

(iii) in bubbles around WR stars, overabundant heavy ions (He,N), presumably transported by the stellar wind, have been observed in the optical, implying cool temperatures downstream of the shock (Kwitter 1981).

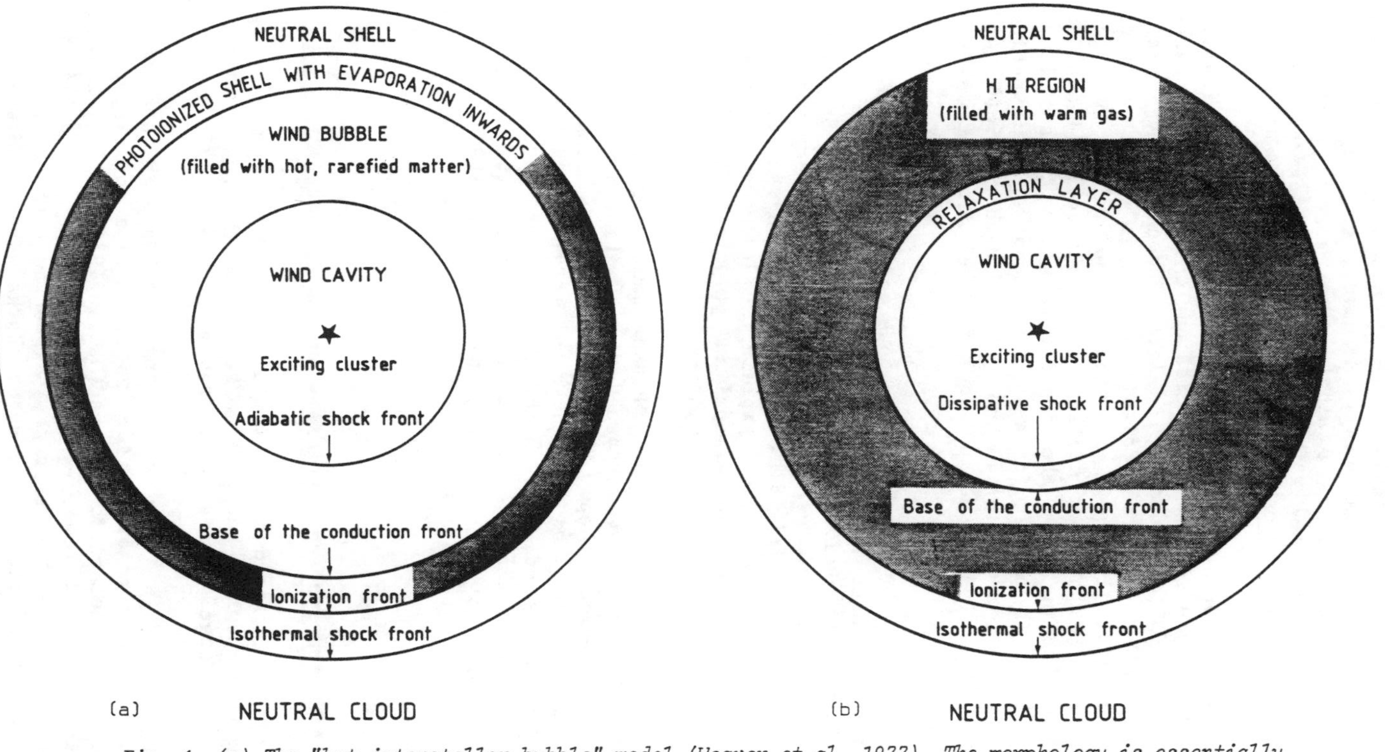

Fig. 4. (a) The "hot interstellar bubble" model (Weaver et al. 1977). The morphology is essentially that of a thin HII region, surrounding a large, hot cavity, energized at the supersonic stellar wind shock generated by the central stars. (b) On the contrary, a hollow, thick morphology is often observed in giant HII regions. This structure can be explained by an efficient dissipation of the stellar winds downstream of the shock, which is then separated from the HII region only by a thin, hot layer (Dorland et al. 1986).

All these observations strongly suggest that most of the wind energy has to be dissipated efficiently. Moreover, the standard "hot interstellar bubble" model assumes a single, non-evolving exciting O star, whereas in reality, we deal with associations of several O and B stars, each evolving separately on comparatively short timescales.

This has prompted Dorland, Montmerle and Doom (1986) to revisit the problem. First, they assumed a total dissipation of the wind energy downstream of the shock, thus turning the large "shocked wind region" of Weaver et al. (1977) into a thin "dissipation layer". Second, they described more realistically the excitation of the bubble by using actual OB associations and refined stellar evolution models (featuring mass loss and overshooting from the convective cores ; Doom 1985, Prantzos et al., 1986). (Arguments for a smaller amount of overshooting have been recently put forward by Mermilliod and Maeder (1986), and Maeder and Meynet (1987), but should not affect significantly their conclusions.)

In this way, the agreement with the observed structure of real nebulae is found to be quite satisfactory: the HII region, basically, is thick <u>at all times</u>. It evolves like an ordinary, windless Strömgren sphere with stellar evolution, inside which a wind cavity with momentum conserving evolution is included (Fig. 4b). Hence, wind energy dissipation does appear as a key factor to explain the disagreement between the classical hot interstellar bubble model and observations.

A few proposals exist in the literature to explain the energy dissipation (e.g., McKee et al., 1984 ; Hanami and Sakashita, 1987) ; Dorland and Montmerle (1987) have proposed that it is linked with the fact that, in presence of steep temperature gradients such as are already present in the standard hot interstellar bubble model, <u>non-linear conduction</u> must be used, whereas the usual linear conduction law was used by Weaver et al. (1977). The details not directly relevant to the present review may be found in Dorland and Montmerle (1987), but we shall come back to this mechanism in § 4.1.

3. THICK COSMIC-RAY SOURCES

3.1 Wind Shock Acceleration and Particle Trapping

The medium just described may be highly relevant to cosmic rays: strong shocks, huge mechanical energy from stellar winds (up to several times 10^{38} erg.s^{-1}), large volumes of ionized material.

Indeed, as mentioned in § 1.1, winds from OB or WR stars are ionized, and variable, or even chaotic. For instance, in a rotating wind, one expects corotating interaction regions (Mullan 1986) ; also, instabilities are likely to exist (Owocki and Rybicki 1985). Variabilities are in fact observed in the wind spectra, suggesting that the wind may be turbulent (e.g., Gry et al., 1984). Hence shocks within the wind are probably present, and via the now classical diffusive shock acceleration mechanism, particles may be accelerated at either or both locations: (i) within the wind (Cesarsky and Montmerle 1983, White 1985) ; (ii) at the edge of the wind cavity (e.g., Webb et al., 1985). The recent evidence for non-thermal radiation from O stars (Abbott et al., 1986 ; Persi et al., 1985), and from WR stars (Pollock 1987) would tend to support these views, at least for electrons. It is not yet completely clear that nuclei energies $>$ GeV/n may be attained in these conditions, but this is likely if we add the shocks from successive explosions of supernovae from O, B and WR stars.

In any case, the accelerated particles will be trapped in the surrounding thick, dense ionized material, for instance by resonant scattering on Alfvén waves (Cesarsky 1980, Montmerle and Cesarsky 1981). The fast particles will then essentially stream outwards with at most a few times the Alfvén velocity $v_A = B \times (8\pi\, n\, m_p)^{-\frac{1}{2}}$, i.e., on the order of a few $km.s^{-1}$ for typical values $n = 10\ cm^{-3}$ and $B = 3 \times 10^{-6}$ G. The thickness of the ionized region being typically L = a few 10 pc, the particles will thus traverse source grammages $X_s = n\, m_p\, Lc/v_A \sim 10\text{-}100$ $g.cm^{-2}$, i.e., of the same order as the collision mean free path of most cosmic-ray nuclei.

Giant HII regions therefore appear as thick cosmic-ray sources, out of which only the lightest particles (p, α, and secondary p, γ, neutrinos) can escape. As a consequence, they cannot be sources of galactic cosmic-ray nuclei, an additional reason not to see the signature of hot stars in the FIP bias (Part I). In particular, the $^{22}Ne/^{30}Ne$, ^{12}C and ^{16}O excesses attributed to WR stars in general (Part I ; fig. 2), cannot come from such regions ; the relevant WR stars must belong to the fraction ($\lesssim$ 60%) that are isolated. The Ne isotopes could be accelerated either in the WR wind itself, or in the terminal shock with the ISM, or yet in the shock associated with the final explosion of the WR star as a supernova.

3.2 Gamma-ray Sources

The $\gtrsim$ GeV protons and $\gtrsim$ 0.1 GeV electrons, which are plausibly produced by stellar wind or supernova shocks in the cavity, will interact with the HII region to produce γ-rays of energies $\gtrsim$ 50 MeV, by a combination of reactions (e.g., Stecker 1971):

$$p + H \rightarrow \pi^\circ \rightarrow 2\gamma \ (\pi^\circ \text{ decay})$$

$$\rightarrow \pi^\pm \rightarrow \mu^\pm \rightarrow e^\pm + \nu$$

$$e^\pm + H \rightarrow \gamma \ (\text{bremsstrahlung})$$

(This last reaction takes place between secondary positrons and electrons, as well as primary electrons, with gas protons.) These reactions are the same as those which dominate the production of diffuse galactic γ-rays (e.g. Pollock et al., 1985). Here also, the a priori possible Compton effect of energetic electrons with stellar or IR photons from dust can be neglected (Montmerle and Cesarsky 1980).

In a region of massive star formation, there will therefore be two co-existing sources of γ-rays, both generated by the above mechanisms:

(i) $S_{\gamma,I}$: interaction of ambient galactic cosmic rays with the molecular, neutral, and ionized material of that region;

(ii) $S_{\gamma,II}$: interaction of cosmic rays produced by the massive stars (winds, supernova explosions) with the ionized gas (see Montmerle 1979,1981).

If $S_{\gamma,I}$ dominates over $S_{\gamma,II}$, since the amount of ionized matter is always much smaller than the total mass, one detects only a "passive" molecular cloud. These clouds are generally not seen individually, but make up most of the diffuse galactic γ-ray flux.

But if $S_{\gamma,II}$ dominates over $S_{\gamma,I}$, one may have a γ-ray <u>source</u>, associated with a star-forming region, provided the locally produced cosmic-ray flux is sufficiently high. (This source will appear extended if close enough.) It is thus of interest to look whether such sources are present in the COS-B data.

Out of the first COS-B catalog of γ-ray sources, Montmerle (1979) noted that about 1/3 to 1/2 were in the line-of-sight of OB associations linked with supernova remnants ("SNOBs"). This catalog has since been updated (second catalog, see Bignani and Hermsen 1982), and, thanks to the completion of the galactic CO survey (see § 1.2), a number of γ-ray emitting regions are now known to be associated with molecular clouds - hence "false sources" of type $S_{\gamma,I}$ above, leaving 20 "true" sources. Among these sources, two are the well-known Crab and Vela pulsars, and four others have been found to be variable over time-scales of months or years ; they are still unidentified but their variability suggest they are small objects. In fact, WR stars have been proposed (Pollock 1987) ; they might be also compact objects. There remains therefore 14 non-variable sources along the galactic plane. Generalizing the initial concept of "SNOBs" to OB associations linked with giant HII regions, some known to include energetic agents like early O, Of, and/or WR stars, one finds a list of 9 regions of massive star formation being candidate γ-ray sources (Table 2, Montmerle 1985 and references therein) of type $S_{\gamma,II}$ above.

TABLE 2. Non-variable, active γ-ray sources in the galactic plane (*)

name (2CG)	proposed id.	ref.	d (kpc)	HII region	opt. diam.	exc. class	most active star	SNR
006-00	M8	(1,2)	1.5	G6.0-1.2	90'	f	WR	W28
013+00	W33 complex	(*)	4.2	G12.8-0.2		m	n.a.	-
			5.8	G13.2+0.0		f	n.a.	-
			4.0	G14.6+0.1		f	n.a.	-
075+00	-	(*)	5.7	G75.8+0.4		f	n.a.	-
078+01	DR3+DR4	(1,2)	5.0	G78.5+2.1 (+)		f	n.a.	DR3+DR4
121+04	-	-	-	-		-	-	-
135+01	IC1805	(1,3)	2.3	G134.8+1.0	150'	m	O4If	-
218-00	-	-	-	-		-	-	-
235-01	-	-	-	-		-	-	-
284-00	RCW49	(*)	4.7	G284.3-0.3	90'	b	n.a.	-
288-00	Carina complex	(4)	2.6	G287.9-0.8 ...	180'	b	WR	-
311-01	SGMC (++)	(*)	15.5	?		?	?	G311.5-0.3
333+01	RCW106 complex	(1)	4.2	G333.6-0.1 ...	35'	b	WR	MSH16-51
342-02	-	-	-	-		-	-	-
359-00	W24? (§)	(*)	10.0	G0.5-0.0 ...		m	n.a.	-

NOTES. (*) Not including already known identifications : 2CG363-02 = Vela, 2CG184-05 = Crab, 2CG353+16 = ρ Oph (see Montmerle 1985), 2CG195 +04 = Geminga. Restricted to $|b| < 5°$.

(+) Smith et al. (1978)

(++) "Supergiant" molecular cloud (Cohen et al. 1985). This cloud is by far the most massive cloud in the Carina arm ($M = 7.8 \times 10^6 M_\odot$) Position of SNR G311.5-0.3 highly uncetain (see Clark and Caswell 1976). Size of cloud ~ 1 sq. deg.

(§) Identification doubtful. From Smith et al.(1978), Astr.Ap. 66, 65

(*) For references, see Montmerle (1985)

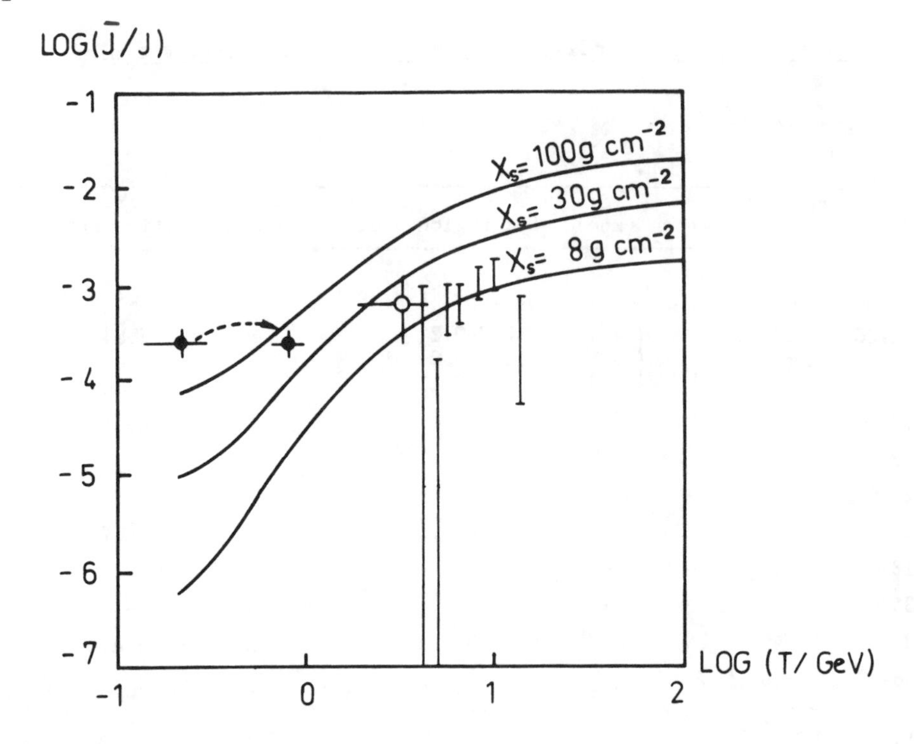

Fig. 5. Antiproton to proton flux ratio, after traversal of a thick cosmic-ray source material with grammages X_s = 8, 30, and 100 g.cm^{-2}, assuming a proton spectrum proportional to $p^{-2.1}$. The open circle and error bars are the experimental data which the models can fit (see Lagage and Cesarsky 1985); the filled circles are the data of Buffington et al. (1981), with and without taking the solar modulation into account (arrow).

Taking as representative values a typical γ-ray flux ~ 10^{-6} ph. cm^{-2}s^{-1} above 100 MeV, a distance ~ 2.5 kpc, and an ionized hydrogen mass of ~ 10^4 $M_\odot$, one finds that the value of the cosmic-ray density ε must be ~ 10 times its value in the solar vicinity $\varepsilon_\odot$ (with a standard bremsstrahlung + π° source function, see discussion in Bloemen 1987). Note that this high value of $\varepsilon/\varepsilon_\odot$ cannot affect the structure of the HII regions, where the energy density of the gas is ≳ 10 eV.cm^{-3}. Pending confirmation, these active HII regions, associated with γ-ray sources, can be therefore considered as true cosmic-ray sources (or cosmic-ray "bubbles") even if they cannot be the sources of galactic cosmic rays, owing to their "optical thickness" to nuclei.

3.3 Antiproton and Neutrino Production

Another by-product of putative proton-proton interactions in giant HII regions is the production of secondary antiprotons, with a threshold of ~ 10 GeV.

Comparing the calculated $\bar{p}$ spectrum in thick sources of various grammages, Lagage and Cesarsky (1985 ; Fig. 5) have found that a value of $X_s \sim 10$ g.cm^{-2} may explain the observed $\bar{p}$ spectrum above several GeV (produced by p of energies above several 10 GeV). (The still controversial high flux at lower energies measured by Buffington et al. 1981 cannot be explained in this model.) This value of X_s is towards the low end of the values associated with HII regions, and tends to overproduce the total γ-ray flux in the Galaxy by factors 2-3, if a "normal" proton spectrum α (momentum)$^{-2.1}$ is used. A harder proton spectrum could however reconcile the $\bar{p}$ data at high energies without conflicting with the total galactic γ-ray flux. We shall come back to this point later (§ 5.3).

Also as a result of proton-proton interactions, we have seen (§ 3.2) that $\pi°$-decay γ-rays are produced. At the same time, charged pions are created ; these decay ultimately into secondary electrons, but also in muon and electron neutrinos and antineutrinos, in numbers approximately equal to that of the $\pi°$ γ-rays. Hence thick cosmic-ray sources are also good <u>sources of neutrinos</u>, which bear unambiguously the signature of p-p interactions (Lagage and Cesarsky 1985). No such source has yet been detected ; above 1 TeV, neutrino fluxes in the range of ~ 10^{-10} ν cm^{-2}s^{-1} are expected, if the corresponding γ-ray sources have a spectral index ~ -2.1. Harder spectra of locally produced protons would lead to a higher flux. In the future, such low fluxes may be detected by the DUMAND experiment, now in the course of preliminary tests in the Pacific Ocean (see, e.g., Grieder 1986).

4. MONITORING STELLAR WINDS: DIFFUSE KeV X-RAY PRODUCTION

4.1 Structure of Dissipative Wind Shocks in Giant HII Regions

In the presence of steep temperature gradients ($T/\nabla T$ < heat-conducting electron mean free path), experimental, as well as theoretical work on laser- heated fusion plasmas (Luciani et al., 1985) has shown that heat conduction is non-linear and strongly modified. This is because the electron distribution function deviates strongly from a Maxwellian: roughly speaking, a non-thermal high-energy tail is present (corresponding in our context to the shock-heated electrons), in addition to a low-energy, approximately Maxwellian "warm" component (corresponding to the electrons of the ambient HII gas heated via Coulomb collisions by the hot component). (For details, see Dorland and Montmerle 1987.) Thanks to their high mass, the ions have a very nearly Maxwellian distribution at the warm temperature. Downstream of the shock, schematically, the successive dominating electron populations are:

(i) the hot electrons from the wind thermalization (the "shocked wind region" in the standard picture but here their distribution is not thermal ; energy in the keV range) ;

(ii) the warm electrons, which have a temperature corresponding to a maximum radiative loss efficiency, i.e., $\gtrsim 10^5$ K, blending progressively as it cools with:

(iii) the ambient HII region electrons at $\lesssim 10^4$ K.

In many ways, this structure resembles an "inverted" transition region, like in the case of the Sun: the analogy is profound, since there is increasing evidence that the structure of the solar transition region is governed by non-linear conduction (Shoub 1983 ; Smith 1986 ; Owocki and Canfield 1986).

In the Dorland-Montmerle model, there is an energy balance between the thermalized wind energy and the energy radiated at the warm temperature: a fraction of the wind energy is radiated in X-rays, but the bulk will be radiated in the UV range. (For an outside observer, though, it is likely that this energy will be mostly re-radiated in the infrared after traversing the HII region.) The post-shock dissipation layer is found to have a thickness:

$$\lambda_D \simeq 5.2 \text{ pc } (v_w/3000 \text{ km.s}^{-1})^2 (n/10 \text{ cm}^{-3})^{-1}. \quad (3)$$

For typical values of the stellar wind velocity v_w and of the HII region density n, we see that the dissipation layer is indeed thin with respect to the thickness of the HII region (see Fig. 4b).

4.2 X-ray Production

Given the terminal velocity of the stellar winds we consider, we expect some X-ray emission in the keV range, i.e., an energy domain in which a wealth of data is available. In the standard situation, the temperature would be determined by eq. (2), but the result is not applicable here because of the specific features of the dissipation layer described above.

To compute how much of the energy radiated goes into X-rays, we need to know how much energy goes into heat conduction, i.e., how much energy is dumped by Coulomb collisions in the surrounding medium. The amount of heat conduction q is conventionally defined in terms of the "free-streaming value" q_{FS}, given by:

$$q_{FS} = n_e \, m_e \, v_e{}^3 = n_e \, m_e (kT_e/m_e)^{3/2}. \quad (4)$$

The actual value of q is by definition:

$$q = \zeta \, q_{FS}, \quad (5)$$

where we have introduced the so-called "flux-limit factor" ζ, which is therefore a measure of the departure from standard conduction.

The value of ζ depends on the physical conditions surrounding the heat source: for laser-heated fusion plasmas, the experimental value ζ_{exp} is $0.03 < \zeta_{exp} < 0.10$, whereas using a collimated electron beam leads to a maximum possible value $\zeta_{max} = 3.2$ (see, e.g., Luciani et al., 1985).

In eq. (4) above, T_e is, strictly speaking, the temperature of the conduction electrons, which, in our case, are the hot electrons with a non-Maxwellian distribution. We approximate this distribution by a characteristic energy kT_h ; further, we approximate their X-ray emission by bremsstrahlung at T_h from collisions with the cool ions, far more numerous than the hot ones. With these approximations, T_h and the X-ray luminosities in the dissipative layer depend on q, i.e., on ζ. In principle, plasma physics should allow to calculate ζ, but in practice, this is not yet possible, the calculated values of ζ being still larger than the measured ones (see discussion in Dorland and Montmerle, 1987).

It is therefore preferable to leave ζ be a free parameter ; Figs. 6a and 6b show L_x and kT_h as a function of both v_w and ζ. Then, ζ may be adjusted to astronomical X-ray observations. This has been done for the Carina nebula, one of the nebulae best studied in X-rays (e.g., Chlebowski et al., 1984), with the result $\zeta_{obs} \simeq 0.085$, i.e., well within the range of laboratory values ζ_{exp}. For typical early-O and WR stellar wind velocities, one finds $kT_h \sim 7.5$-12 keV, i.e., significantly smaller than the standard temperatures given by eq. (2). One also finds a typical ratio L_x/(wind kinetic energy rate) $\sim 2 \times 10^{-4}$.

Based on these numbers, the > keV X-ray flux and temperature of several giant HII regions (like Orion) can be computed and compared with (sometimes old !) otherwise still unexplained observations, and found to be essentially in agreement - although all the relevant parameters are not always known independently from observation (Dorland and Montmerle 1987 ; Montmerle, in preparation).

5. STELLAR WINDS AND COSMIC RAYS

5.1 The Galactic X-ray Ridge: Stellar Winds ?

While keV X-ray emission appears to be present in several well-studied individual HII regions, recent results from the japanese "Tenma" and european EXOSAT satellites (Koyama 1986, Warwick et al., 1985) have shown that this emission extends all over the galactic plane, being stronger in its inner regions ($|\ell| \lesssim 60°$). The thermal nature of the emission of this so-called "galactic ridge" is demonstrated by a ubiquitous 6.7 keV line, due to helium-like iron, present in all observed lines-of-sight. The "Tenma" satellite, with a comparati-

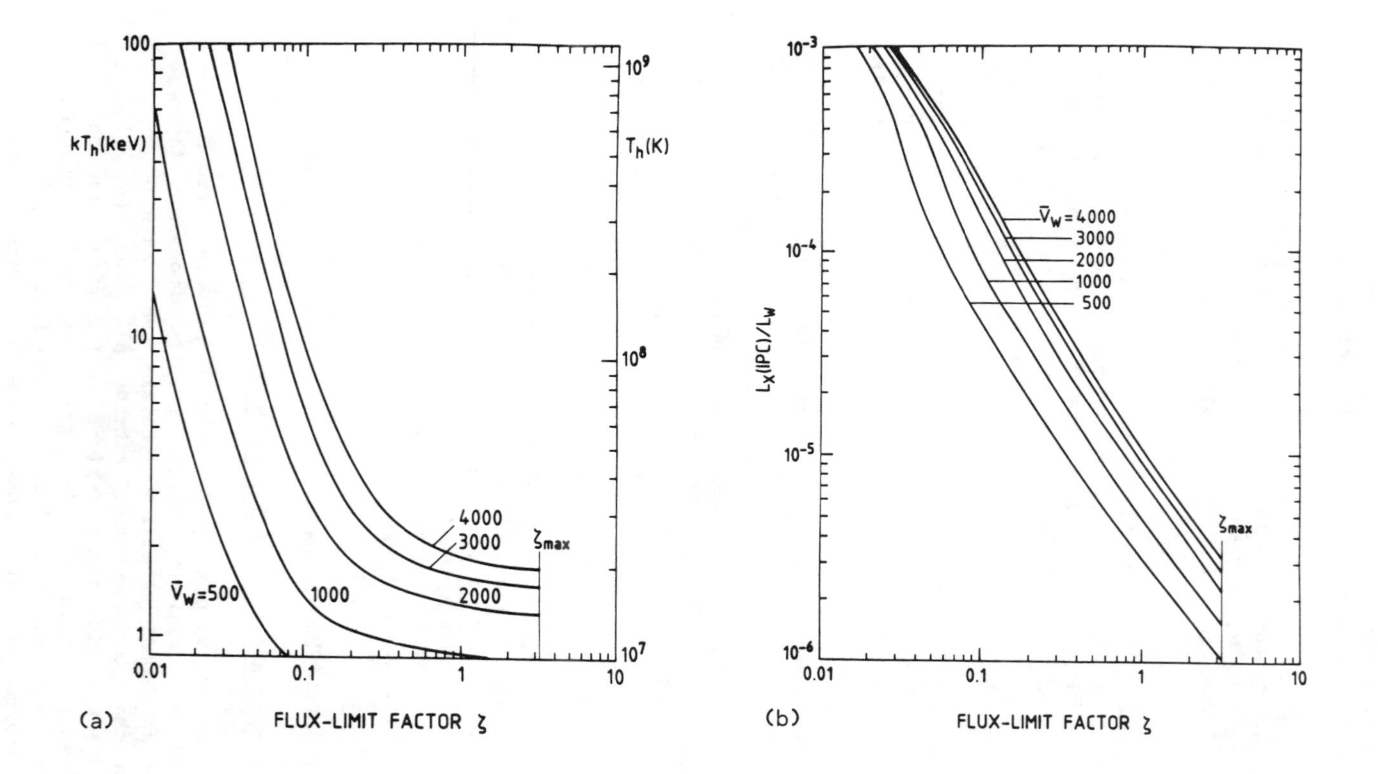

Fig. 6. (a) The flux-limit factor ζ gives a measure of the non-linearity of heat conduction in the presence of steep temperature gradients. The "astrophysical" value, normalized to X-ray observations, is $\zeta_o = 0.085$, and ζ_{max} is the theoretical maximum value in plasma physics. T_h is the temperature of the hot electron component present in the dissipation layer. For $\zeta = \zeta_o$, T_h is in the keV range for typical values of stellar wind velocities V_w in the 1000x km.s^{-1} range. (b) Same as (a), for the corresponding diffuse luminosity L_x, normalized to the wind "luminosity" L_w. Typical values of L_x/L_w are 1-2 10^{-4}. (Dorland and Montmerle 1987)

vely narrow field of view (3.1° FWHM), and good spectral resolution, has allowed to detect this high-temperature gas in particular in directions in which no supernova remnant or X-ray source stronger than ~ 2 mCrab is known ; the temperatures range from ~ 2 to ~ 16 keV. It is possible to interpret this "X-ray ridge" in terms of very young, otherwise undetected supernova remnants, but excesses of emission are seen in the direction of several star-forming regions in which no SNR is known (e.g., Perseus), as was already the case in Orion (~ 20° below the galactic plane). In addition, the SN rate required to explain the total luminosity of $L_x \sim 10^{38}$ erg.s^{-1} in the 2-6 keV band is uncomfortably high, about one every 10 years (Koyama et al., 1986).

In our view, these observations are better explained in terms of stellar winds. If a value $\zeta = \zeta_{obs}$ is assumed to be valid throughout the Galaxy (meant here as a superposition of regions of star formation associated with HII regions - hence mainly the galactic arms), this corresponds to a total wind "luminosity" $L_w \simeq 5 \times 10^{41}$ erg.s^{-1}, assuming a characteristic wind velocity of ~ 3000 km.s^{-1}. This figure is important, since if we take a SN rate of one every 30 years, each releasing 10^{51} erg of mechanical energy, the total "luminosity" they dump into the interstellar medium is $L_{SN} \simeq 10^{42}$ erg.s^{-1}, i.e., of the same order of magnitude.

5.2 Links with Massive Stars

What kind of stars may shed such a huge amount of energy ? Obviously, the very powerful WR and Of stars are the best candidates. Indeed, each WR star releases ~ 10^{38} erg.s^{-1} ; each early O or Of star releases > 10^{37} erg.s^{-1}. The exact figure may be higher if one takes into account the observed increase in metallicity towards the inner Galaxy, which implies an increase in the wind terminal velocity (see Part I). One needs therefore $\lesssim$ 5000 WR stars in the whole Galaxy (Montmerle, 1986). This figure is probably somewhat too high, but finds some support if, as argued by Prantzos and Cassé (1986) in one of their models, the γ-ray line at 1.8 MeV of ^{26}Al, recently detected in the inner galactic plane (see Ballmoos et al., 1987, and refs. therein) is explained in terms of freshly synthesized material, released by WR stellar winds in the course of their evolution (see above ; § 1.1). The interpretation of the galactic X-ray ridge in terms of a by-product of winds from WR and early O stars is therefore at least tantalizing.

5.3 Acceleration of Cosmic Rays

We have seen that, irrespective of the actual dissipation mechanism, stellar wind shocks in HII regions must be strongly dissipative. This leads to high compression ratios, r = v (upstream)/v (downstream) ~ 100 or more.

If particle acceleration takes place at such shocks, the resulting power-law spectrum α $E^{-\Gamma}$ will not have the canonical $\Gamma \simeq 2$ spectral index. Indeed, in the linear theory, the spectral index Γ is Γ =

TABLE 3. Stars and Cosmic Rays: an Overview

Topic	COOL STARS (F-M)	HOT STARS (O,B,WR)
Dominant activity	Magnetic (solar-like)	Radiative (mass loss)
Origin	Outer convection	UV photons
Characteristic "filter"	Chromosphere (thin filter: FIP bias, H mostly neutral)	HII region (thick filter: high grammages, H mostly ionized)
Pros	Right composition: injection of stellar energetic particles (subsequent shock acceleration by SN in the ISM ?)	Lots of energy, + SN explosions (i) in thick sources: p, $\bar{p}$, γ-ray sources, neutrinos X-rays, nuclear γ-ray lines (ii) not in thick sources: ^{12}C, ^{16}O, ^{22}Ne excesses
Cons	Adiabatic energy losses in stellosphere (reacceleration ?) p and α deficiencies (solar problem ?)	no heavy GCR
Theoretical problems to solve	Transport out to the ISM: chromosphere $\longrightarrow$ corona (diffusion ?) stellosphere $\longrightarrow$ ISM	CR acceleration: (i) in winds (ii) in dissipative shocks (harder spectrum ?)
Observational tests	Particle acceleration on the Sun and in the heliosphere out to large distances; corotating interaction regions	Spectrum at high energies of γ-ray sources in star formation regions ^{26}Al γ-ray line

$(2+r)/(r-1)$ for a plane parallel shock, hence $\Gamma \to 1$ as $r \to \infty$. The linear theory is not strictly applicable (curvature, high r, etc.), but this suggests that the protons, if accelerated by this mechanism (they are essentially the only primary particles able to leak out of the HII regions), have a spectrum harder than usual. Given the huge total energy output likely coming from stellar winds (§ 5.2), this suggests that a significant fraction of galactic cosmic-ray protons might have harder spectra than the bulk of the cosmic-ray nuclei, especially in the vicinity of regions of massive star formation.

The calculations have not been done, but we know, from Lagage and Cesarsky (1985), that a harder p spectrum would tend to reduce the excess γ-rays produced in thick sources (§ 3.3). Also, such a harder p spectrum is perhaps already present in if one puts together the HEAO-B and JACEE data (see discussion in Engelmann et al., 1985). On the other hand, the γ-ray spectrum itself from such sources would be harder (at energies $\gg$ 100 MeV, the π°-decay γ-ray spectrum and the proton spectrum have the same slope). This could provide an observational test for a (future) experiment having an angular resolution good enough to isolate the γ-ray emission of the HII region from the strong γ-ray emission from surrounding regions (molecular clouds, in most cases).

6. STARS AND COSMIC RAYS: CONCLUDING REMARKS

Table 3 summarizes the many topics we have addressed in Part I and Part II of this lecture. The links between stars and cosmic rays are numerous and diversified ; they support the view that the field of "cosmic-ray astrophysics" not only exists, but is coming of age.

The crude reality, however, is that any major cosmic-ray experiment in space is at least a decade away. Progress can nevertheless be expected on the theoretical side, relying on the presently available data, possibly reprocessed or used along new avenues. Indeed, it must be realized that a number of important theoretical problems can be attacked without the need for new cosmic-ray data: for instance, transfer of solar chromospheric atoms upwards to the corona (see Part I), or acceleration of nuclei within stellar winds or dissipative shocks at their boundaries with HII regions (Part II).

Also, we point out that, in spite of the diversity displayed in Table 3, deep unifying features exist as regards the early stages of cosmic ray generation in stars, i.e., before their transfer into the interstellar medium at large. Indeed, in both cool stars and hot stars, the original pool is radiation-dominated (photosphere, wind), and a "filter" has to be traversed: chromospheres of cool stars constitute a thin filter (atomic: FIP bias), in which H is predominantly neutral ; hot stars do not possess chromospheres, but most are surrounded by HII regions, which constitute a thick filter (nuclear: heavy nuclei cannot leak out), in which H is predominantly ionized. In both cases, the magnetic field likely plays a central role: it regulates "atomic" diffusion of particles at the interface between the chromosphere and the corona, as well as "magnetic" diffusion (trapping by Alfvén waves) in HII regions. Furthermore, these two filters have fairly uniform properties over wide stellar spectral ranges (F-M stars on one hand, O-B on the other): they are both low-temperature thermostats, each regulated by the cooling properties of, mainly, hydrogen. For their lack of all the above properties, A stars appear as pivotal.

In addition to theoretical problems such as outlined above, we may therefore expect significant progress from solar and stellar physics. True enough, space astronomy is also in limbo for the moment, as major observatories will not be sent into orbit before a number of years. Contrary to cosmic rays (at least at low energies), however, observational studies of the Sun and stars can be conducted from the ground, and thus may help keeping cosmic-ray physicists busy.

ACKNOWLEDGEMENTS

It is a pleasure to thank John Wefel and Maurice Shapiro for their hospitality in Erice and for their patience during the preparation of this manuscript. I thank also Pierre-Olivier Lagage and Jean-Paul Meyer, as well as many students during the School, for useful discussions.

REFERENCES

Abbott, D.C., Bieging, J.H., Churchwell, E., Torres, A.V.: 1986, Ap. J., 303, 239.
Ballmoos, P.V., Diehl, R., Schönfelder, V.: 1987, Ap. J., in press.
Bloemen, J.B.G.M.: 1987, Ap. J. (Letters), in press.
Buffington, A., Schindler, S.M., Pennypacker, C.R.: 1981, Ap. J., 248, 1179.
Cesarsky, C.J.: 1980, Ann. Rev. Astr. Ap., 18, 289.
Cesarsky, C.J., Montmerle, T.: 1983, Sp. Sci. Rev., 36, 173.
Chiosi, C., Maeder, A.: 1986, Ann. Rev. Astr. Ap., 24, 329.
Chiosi, C., Stalio, R. (Eds.): 1981, Effects of Mass-Loss on Stellar Evolution (Dordrecht: Reidel).
Chlebowski, T., Seward, F.D., Swank, J., Szymkoviak, A.: 1984, Ap. J., 281, 665.
Chu, Y.H.: 1983, Ap. J., 269, 202.
De Loore, C., Willis, A.J. (Eds.): 1982, Wolf-Rayet Stars, Observations, Physics, Evolution,(Dordrecht: Reidel).
Doom, C.: 1985, Astr. Ap., 142, 143.
Dorland, H., Montmerle, T.: 1987, Astr. Ap., 177, 243.
Dorland, H., Montmerle, T., Doom, C.: 1986, Astr. Ap., 160, 1.
Engelmann, J.J., Goret, P., Juliusson, E., Koch-Miramond, L., Lund, N., Masse, P., Rasmussen, I.L., Soutoul, A.: 1985, Astr. Ap., 148, 12.
Gaetz, T.J., Salpeter, E.E.: 1983, Ap. J. Suppl., 52, 155.

Grieder, P.K.F.: 1986, in Neutrinos and the Present-Day Universe, eds. T. Montmerle and M. Spiro (Saclay: CEA/Doc), p. 171.
Gry, C., Lamers, H.G.J.L.M., Vidal-Madjar, A.: 1984, Astr. Ap., 137, 29.
Hanami, H., Sakashita, S.: 1987, Astr. Ap., in press.
Humphreys, R.M.: 1978, Ap. J. Suppl., 38, 309.
Koyama, K.: 1986, preprint ISAS 324.
Koyama, K., Makishima, K., Tanaka, Y., Tsunemi, H.: 1986, Pub. Astr. Soc. Japan, 38, 121.
Kwitter, K.B.: 1981, Ap. J., 245, 154.
Lagage, P.O., Cesarsky, C.: 1985, Astr. Ap., 147, 127.
Luciani, J.F., Mora, P., Pellat, R.: 1985, Phys. Fluids, 28, 835.
Lundström, I., Stenholm, B.: 1984, Astr. Ap. Suppl., 58, 163.
Maeder, A., Meynet, G.: 1987, Astr. Ap., in press.
Mermilliod, J.C., Maeder, A.: 1986, Astr. Ap., 158, 45.
McKee, C.F., Van Buren, D., Lazareff, B.: 1984, Ap. J. (Letters), 278, L115.
Montmerle, T.: 1979, Ap. J., 231, 95.
Montmerle, T.: 1981, Phil. Trans. Roy. Soc. London, A301, 505.
Montmerle, T.: 1985, 19th Int. Cosmic Ray Conf. (La Jolla), 1, 209.
Montmerle, T.: 1986, in Vangioni-Flam et al. (1986), p. 335.
Montmerle, T., Cesarsky, C.J.: 1980, in Non-Solar Gamma-Rays, eds. R. Cowsik and R.D. Wills (Oxford: Pergamon).
Montmerle, T., Cesarsky, C.J.: 1981, Proc. Int. School and Workshop on Plasma Astrophysics (Varenna) ESA SP-161, p. 319.
Mullan, D.J.: 1986, Astr. Ap., 165, 157.
Owocki, S.P., Canfield, R.C.: 1986, Ap. J., 300, 420.
Owocki, S.P., Rybicki, G.B.: 1985, Ap. J. 299, 265.
Persi, P., Ferrari-Toniolo, M., Tapia, M., Roth, M., Rodríguez, L.F.: 1985, Astr. Ap., 142, 263.
Pollock, A.M.T.: 1987, Astr. Ap., 171, 135.
Pollock, A.M.T., Bennett, K., Bignami, G.F., Bloemen, J.B.G.M., Buccheri, R., Caraveo, P.A., Hermsen, W., Kanbach, K., Lebrun, F., Mayer-Hasselwander, H.A., Strong, A.W.: 1985, Astr. Ap., 146, 352.
Prantzos, N., Cassé, M.: 1986, Ap. J., 307, 324.
Prantzos, N., Doom, C., Arnould, M., De Loore, C.: 1986, Ap. J., 304, 695.
Shoub, E.C.: 1983, Ap. J., 266, 339.
Smith, D.F.: 1986, Ap. J., 302, 836.
Stecker, F.W.: 1971, Cosmic Gamma Rays (Washington: NASA SP-249).
Vangioni-Flam, E., Audouze, J., Cassé, M., Chièze, J.P., Trân Thanh Vân J. (Eds.) 1986, Advances in Nuclear Astrophysics (Gif sur Yvette: Editions Frontières).
Warwick, R.S., Turner, M.J.L., Watson, M.G., Willingale, R.: 1985, Nature, 317, 218.
Weaver, R., McCray, R., Castor, J., Shapiro, P., Moore, R.: 1977, Ap. J., 218, 377.
Webb, G.M., Forman, M.A., Axford, W.I.: 1985, Ap. J. 298, 684.
White, R.L.: 1985, Ap. J., 289, 698.

ON THE POSSIBLE CONTRIBUTION OF WC STARS TO ISOTOPIC ANOMALIES IN COSMIC RAYS AND METEORITES

J. B. Blake
Space Sciences Laboratory
The Aerospace Corporation
P. O. Box 92957, Los Angeles, CA, 90009
and
D.S.P. Dearborn
Lawrence Livermore National Laboratory
P. O. Box 808, Livermore, CA, 94550

ABSTRACT. Potential contributions of WC stars to isotopic anomalies in cosmic rays and meteorites are considered. If the Cassé-Paul (1982) hypothesis concerning the origin of the excess of ^{22}Ne in the galactic cosmic rays (GCR) is correct, then other detectable anomalies in the GCR should occur. Wolf-Rayet stars make interesting amounts of ^{26}Al (Dearborn and Blake 1985) which may account, at least in part, for the gamma-ray observations of Mahoney et al. (1984). K-shell x-ray emission is expected from ^{59}Co created in the decay of ^{59}Ni made in the core helium burning s-process. In addition WC stars could be the source of meteoritic anomalies; the case of ^{107}Pd/^{107}Ag is especially interesting.

1. INTRODUCTION

The very strong stellar winds of Wolf-Rayet (WR) stars have attracted the attention of cosmic-ray physicists for two reasons -- the large energy and the large amount of nuclearly processed mass injected by these winds into the ISM. In a study of massive stars within 3 kpc of the sun, Abbott (1982) has found that the total energy input to the ISM from stellar winds is 2×10^{38} ergs/sec-kpc^2, and that one half of this energy is due to WR stars. He also estimates the total energy input from supernovae to the ISM is 1×10^{39} ergs/sec-kpc^2. Thus the WR wind energy input is ~ 10% of the total input from supernovae. Such a large energy input clearly is interesting in the study of cosmic ray acceleration (cf. Cassé 1984; Cassé and Cesarsky 1984). Furthermore, van der Hucht, Cassinelli, and Williams (1986) recently have pointed out that the radio measurements of mass loss have underestimated the mass-loss rate by a factor of ~ 2 because almost pure He atmospheres were assumed in the radio analyses, whereas substantial abundances of C and O are present in the winds of WC stars.

Abbott (1982) also estimated the mass flow from WR stars as 4.8×10^{-5} M_o/yr-kpc^2. Since WR stars show nuclearly processed material in their stellar winds, obviously this material is being supplied to the ISM. Cassé and Paul (1982) made the provocative suggestion that the

M. M. Shapiro and J. P. Wefel (eds.), Genesis and Propagation of Cosmic Rays, 153–162.

"excess" ^{22}Ne seen in the galactic cosmic rays (cf. review by Wiedenbeck 1984) could have originated in the stellar winds of Wolf-Rayet stars during the WC_2 phase when the products of core helium burning, including substantial ^{22}Ne, appear at the surface. Only of the order of 2% of the WC material needs to be mixed with solar-system abundances (Anders and Ebihara 1982) according to the recipe of Cassé and Paul 1982 (Figure 1).

Figure 1. This cartoon illustrates the essential features of the Cassé-Paul (1982) hypothesis concerning the origin of the ^{22}Ne anomaly in the GCR.

This approach also has been discussed by Maeder (1983). The Cassé-Paul hypothesis led to a search for other possible contributions to the cosmic rays from the winds of WR stars. Dearborn and Blake (1984, 1985) showed that WR stars produce significant amounts of ^{26}Al, and that of the order of 0.3 M_o of ^{26}Al now exists in the ISM from production in WR stars. This is approximately 10% of the amount of ^{26}Al which Mahoney et al. (1984) estimate to be present in the ISM based upon observations with the HEAO 3 gamma-ray spectrometer. Prantzos and Cassé (1986) carried out a parameterized study of the ^{26}Al production in WR stars. They came to the same conclusions as did Dearborn and Blake (1985) and, in addition, found that for an extreme parameter selection, WR stars could account for the entire amount of ^{26}Al suggested by the observations of Mahoney et al. (1984).

In the case of ^{26}Al, its presence is seen both in the cosmic (gamma) rays and in meteoritic anomalies (cf. Wasserberg 1985); it is unlikely that the nucleosynthesis of the ^{26}Al "seen" these two ways is totally unrelated. Evidence for production of isotopes detected in some meteoritic anomalies in WR stars adds additional insight to the present understanding of nucleosynthesis in WR stars, and the hypothesis of their significant contribution to certain isotopes in the cosmic-ray population.

Blake and Dearborn (1984), and Prantzos (1984a, 1984b) considered some of the possible contributions to the cosmic-ray population of the products of the core helium-burning (CHeB) s-process (Lamb et al. 1977) which will be present in the stellar winds of WC stars. Similar conclusions were reached in these studies. One of the most interesting for experimental verification is that the isotopic ratio of $^{58}Fe/^{56}Fe$ is expected to be significantly increased in the cosmic rays if the Cassé-Paul hypothesis is correct.

The CHeB s-process does not have sufficient neutrons to produce a large elemental yield beyond the neutron magic number at Zr (Wefel, Schramm and Blake 1977). However the isotopic ratios of nuclei up to Pb will be substantially modified by neutron captures, electron captures, and high temperature β decays. While the yield may not be high, if this material constitutes a substantial fraction of some observed reservoir (as Cassé and Paul proposed for the cosmic rays), WR stars could be responsible for isotopic anomalies. The dust observed around some WC stars must have isotopic anomalies characteristic of the CHeB s-process.

2. CALCULATIONS

Models have been evolved with X = 0.70, Z = 0.02, and masses of 50, 100, and 150 M_o. This mass range was chosen as representative of the initial mass of most single Wolf-Rayet stars; detailed nucleosynthesis calculations have been performed for the 100 M_o model.

An important feature of the evolution of Wolf-Rayet stars is the mass loss. Without ad hoc overshooting prescriptions and with no mass loss, the 100 M_o star develops a core of ~ 40 M_o. A mass-loss rate in the observed range (~ few × 10^{-5} M_o/yr) removes the hydrogen envelope, and leaves a helium core in the range ~ 28-35 M_o (depending on the mass-loss parameterization). The models pass through a stage that we associate with WN (hydrogen deficient, nitrogen rich), and WC (hydrogen depleted, helium-burning product rich) stars. The formula that we used was discussed by Dearborn and Blake (1979) and normalized for a zero-age main sequence (ZAMS) mass-loss rate of 1.1 × 10^{-5} M_o/yr for the 100 M_o star. The mass loss increased during the evolution, but was inadequate to prevent the star from becoming a red supergiant at the end of core hydrogen burning. The mass-loss rate was enhanced for a brief period to near 10^{-4} M_o/yr in order to prevent the model from becoming a red supergiant. This procedure was adopted because studies of luminous stars in nearby galaxies (Humphreys 1978) have shown an absence of very luminous red supergiants which would be expected to result from the post main-sequence evolution of the most luminous blue stars that are observed. One concludes that the most luminous stars do not become red supergiants. The average mass-loss rate of the model was 2.3 × 10^{-5} M_o/yr.

The time spent in the WN and WC stages depends sensitively on the mass-loss rate. An average mass-loss rate of 2.1 × 10^{-5} M_o/yr results in a WN model at the end of core helium burning, while with an average rate of 2.5 × 10^{-5} M_o/yr it passes quickly through the WN stage and becomes a WC star.

Even among those models that become WC-like, the mass-loss rate causes some variation in the material exposed. A very high mass-loss rate causes the convective helium core to gradually retreat, and the abundance of helium burning products to slowly build. This material is representative of the early stages of helium burning in which $C/O > 7$. A somewhat slower mass-loss rate allows the convective helium core to grow for a while and reach the WC stage at a later phase of core helium burning. In this case, semiconvection does moderate the composition profile, but the buildup of helium burning products is faster. The material also shows the results of more advanced helium burning ($C/O \sim 1$).

As noted above, we have not incorporated any assumptions regarding overshoot. We believe it is quite likely that convective elements overshoot beyond the point of convective stability, but we consider the subsequent mixing of those elements (as opposed to simply oscillating about that point transferring mechanical and radiative energy) to be very uncertain. Prantzos et al. (1986) have claimed that larger core sizes aid in understanding the spread of the upper main sequence of the HR diagram and suggested that overshooting may be the cause. It should be noted that there are many other incomplete features in the models of massive stars such as rotational mixing (Maeder 1982) and the buoyancy resulting from magnetic fields (Dearborn and Hubbard 1980). In any case, for our purpose, the problem is one of associating the mass of the Wolf-Rayet star with the mass of the ZAMS progenitor. If core sizes are enhanced by some process, the initial mass of a 30 M_o Wolf-Rayet star becomes ~ 80 M_o instead of 100 M_o.

The nucleosynthesis network used was that described by Dearborn and Blake (1985). The reaction rates used were from Caughlan et al. (1985) with the exception of a revised triple α rate (Fowler, private communication 1985). The neutrons produced by this nuclear network, as well as the structure of each model, were used with a new code that calculates the heavy isotope production. In addition to neutron captures, this code, derived from the n-process code of Blake and Schramm (1976), evaluates an arbitrary network of electron captures, β decay, and charged particle reactions. The neutron capture rates were taken from Holmes et al. (1976) and Woosley et al. (1978). The electron capture and beta decay rates were from the work of Takahashi and Yokoi (1985).

3. RESULTS

Figure 2 shows the path of the s-process in the iron region; it illustrates the situation as the s-process begins in the core. The isotopes ^{40}Ca, ^{50}Cr, ^{54}Fe, ^{58}Ni and ^{64}Ni will be burned up as they are not on the s-process path. The situation for these isotopes is analogous to the p-process only and r-process only isotopes in the region of the periodic table well above iron.

Examination of Figure 2 makes it clear why ^{58}Fe is greatly enhanced in the CHeB s-process. It is fed, with only two neutron captures, by ^{56}Fe which is ~ 300 times more abundant in solar system material. Note that the situation is different for ^{64}Ni, the heaviest isotope of nickel. Although its relative abundance is low, as is the case for ^{58}Fe, it is separated from the other nickel isotopes by 100 yr halflife ^{63}Ni. The neutron density is sufficiently low that the mean time between neutron captures is thousands of years. Thus, the Cassé-Paul hypothesis (Figure 1) leads to a clear prediction: ^{58}Fe will be enhanced in the cosmic rays but ^{64}Ni will not.

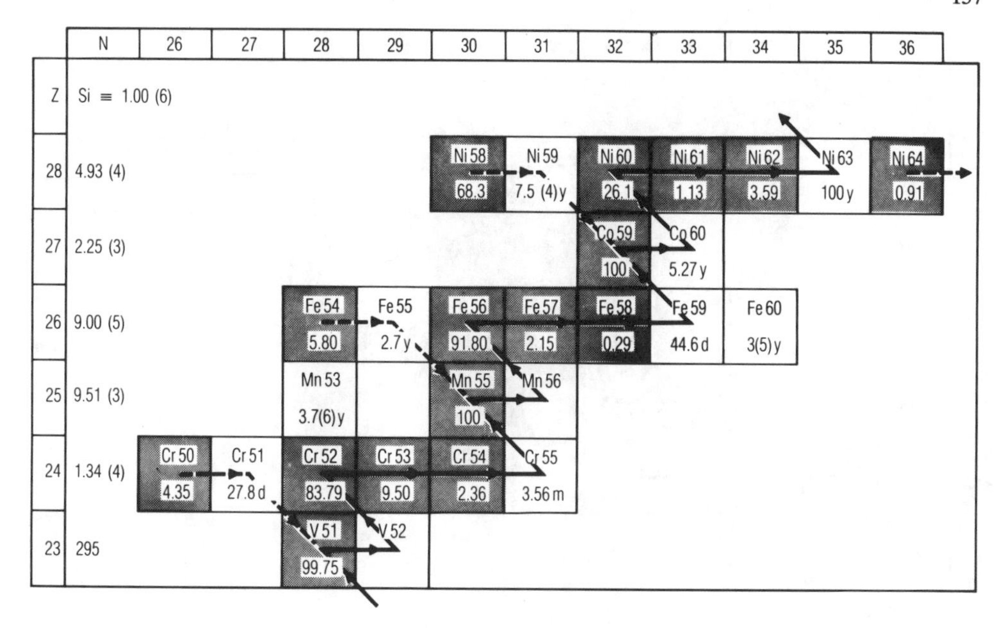

Figure 2. The path of the CHeB s-process is shown in the iron region.

Another isotope of potential interest is illustrated in Figure 2. The most abundant isotope of nickel, ^{58}Ni, feeds ^{59}Ni which has a half-life of 7.5×10^4 yrs. It decays to ^{59}Co by electron capture; usually this decay will be followed immediately by a K-shell x-ray of 6.9 keV. In our 100 M_o model, ~ 10^{-7} M_o of ^{59}Ni is lost in the stellar wind by the time core helium burning ends. Most of the ^{59}Ni is emitted in a time period of ~ 1/2 the halflife of ^{59}Ni. The x-ray intensity will depend upon where in the loss process one observes the WC star and its mass; one expects that some WC stars will be surrounded by a x-ray halo emitting ~ 10^{35} cobalt K-shell x-rays/sec. Observation of this x-ray emission would add credence to present theoretical studies, and could indicate the length of time the star had been in the WC phase.

Figure 3 shows the path of the s-process in the sulfur region. The heaviest sulfur isotope, ^{36}S, can be seen to have very low solar-system abundance and thus is a potential candidate for being enhanced in the cosmic rays. The 87.3 day halflife of ^{35}S prevents significant production of ^{36}S by n,γ reactions. However significant production arises from the n,p reaction on ^{36}Cl. The production of ^{36}Cl is from ^{35}Cl; the 3.01×10^5 yr halflife of ^{36}Cl makes it effectively stable over the time period of the CHeB s-process. Some additional production comes from the n,α reaction on ^{39}Ar but, because of its 269 yr halflife, ^{39}Ar does not make a large contribution to the production of ^{36}S. Some of the ^{36}S so produced in the helium core is lost through subsequent neutron capture, but a detectable anomaly in the cosmic rays is expected from the Cassé-Paul mechanism.

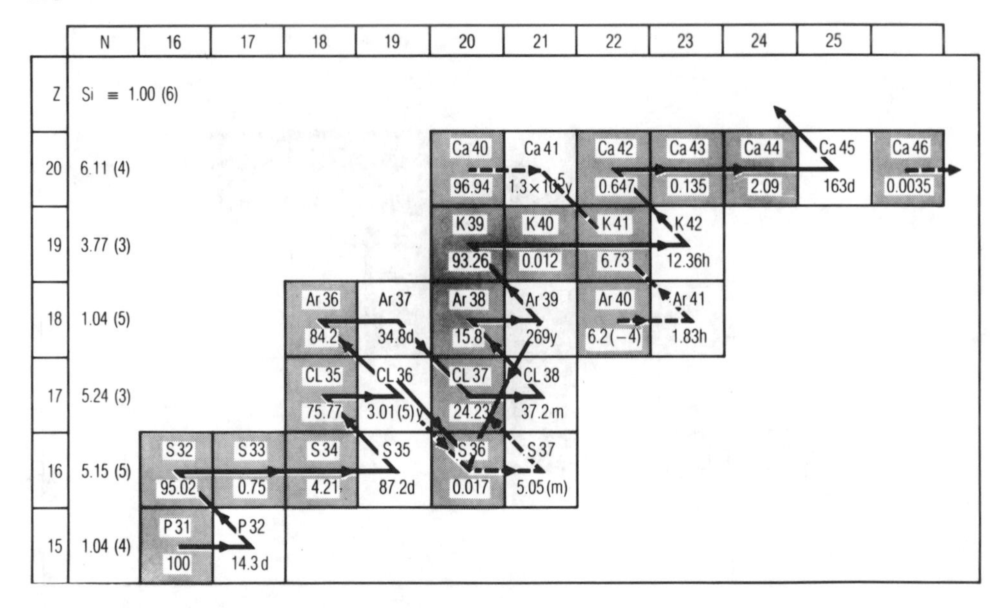

Figure 3. The path of the CHeB s-process is shown in the sulfur region.

A major difficulty in calculating the enrichment expected in the cosmic rays is in the estimation of the nature of the dilution process. The stellar wind of a WC star is highly enriched in ^{22}Ne. This enrichment is quite constant throughout most of the WC phase (cf. Prantzos 1984a), although the ^{22}Ne abundance does fall at the end of helium burning as the ^{22}Ne is destroyed thereby generating neutrons [^{22}Ne (α,n) ^{25}Mg] for the CHeB s-process. However, in contrast to the case of ^{22}Ne, the products of the s-process are much more abundant in the stellar wind near the end of helium burning than in the beginning. The zero-order process depicted in Figure 1 considers an average WC wind; substantial variation would be expected if the recipe calls for a selected portion of the WC wind material. One might argue that the material produced at the end of core helium burning could be most important for the GCR since it would be closest to the subsequent supernova (the putative GCR accelerator).

Wasserberg (1985) has given values for the abundances of certain presently extinct radioactivities that existed at the time of formation of the solar system, Table 1.

The 100 M_o model gives ^{26}Al/^{27}Al $\approx$ 6 × 10^{-2} in the material lost in the stellar wind. This value is approximately three orders of magnitude larger than the value given in Table 1; thus substantial dilution and free decay is allowed.

Table 1
(from Wasserberg 1985)

Isotope	Solar System Abundance At Formation		
^{26}Al	$^{26}Al/^{27}Al$	$\approx$	5×10^{-5}
^{41}Ca	$^{40}Ca/^{41}Ca$	$\leq$	$(8\pm4) \times 10^{-9}$
^{107}Pd	$^{107}Pd/^{108}Pd$	$\approx$	2.0×10^{-5}
^{129}I	$^{129}I/^{127}I$	$\approx$	1.0×10^{-4}

Examination of Figure 3 indicates that ^{41}Ca is made from a single neutron capture on ^{40}Ca. Significant quantities are made in the CHeB s-process; our model gives a ratio of $^{41}Ca/^{40}Ca$ of 3×10^{-4} in the ^{22}Ne rich ejecta. A free decay period of 2×10^6 yrs is required to reduce the $^{41}Ca/^{40}Ca$ ratio to the value given in Table 1.

The ratio of $^{129}I/^{127}I$ in the ^{22}Ne rich ejecta was calculated to be $\sim 6 \times 10^{-4}$. However, when the effect of the stellar environment upon the beta rates was included in the s-process calculations using the results of Takahashi and Yokoi (1985), the production of ^{129}I was zero. The halflife of ^{129}I is reduced in the helium burning core by several orders of magnitude. The winds of WC stars cannot be the source of the ^{129}I which existed at the time of formation of the Solar System.

The case of the ^{107}Pd - ^{107}Ag pair is promising. At the temperature and electron density existing in the WC helium core, stable ^{107}Ag can decay to ^{107}Pd as depicted in Figure 4. In fact at certain temperatures and electron densities found in the helium core the decay rate of ^{107}Ag is greater than that of ^{107}Pd. The situation is similar for the ^{205}Pb-^{205}Tl pair which is discussed in a related context by Yokoi, Takahashi and Arnould (1985). No anomaly due to ^{205}Pb has been detected in solar system material; it will not be discussed further here but should be present in WC winds.

The present calculations give a production ratio of $^{107}Pd/^{108}Pd$ $\approx 5 \times 10^{-3}$, a value very much larger than the value for the ratio given in Table 1. Thus substantial dilution and free decay can occur before the $^{107}Pd/^{108}Pd$ ratio drops below the observed value. It is interesting to note that the modified beta rates, and not neutron capture, are in large measure responsible for the large $^{107}Pd/^{108}Pd$ ratio. Calculations using a much finer grid of beta rates in T,ρ space (Takahashi, private communication 1986) are in process.

The question of dilution is especially complicated in the case of meteoritic anomalies. Grains are observed in the winds of some WC stars (Hackwell et al. 1976; Hackwell, Gehrz and Grasdalen 1979; Allen, Barton and Wallace 1981); thus anomalies might be preserved without undergoing further dilution if stellar grains could have been preserved in the formation of some meteorites (Clayton 1982; for another viewpoint see Wasserberg 1985). This situation is characterized in Figure 5. Study of the incorporation of isotopic anomalies in the grains formed in WC star winds appears to be well worth pursuing.

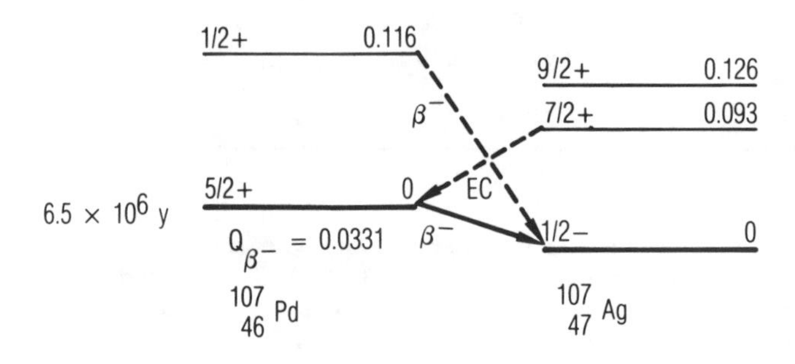

Figure 4. The weak interactions behavior of the ^{107}Pd-^{107}Ag isobaric pair in the WC helium core is schematically illustrated.

Figure 5. This cartoon illustrates the Cassé-Paul (1982) hypothesis modified for the case of stellar grains and meteoritic anomalies.

4. CONCLUSIONS

The Cassé-Paul hypothesis concerning the origin of the $^{22}Ne/^{20}Ne$ anomaly in the galactic cosmic rays leads to the prediction of other anomalies in the cosmic rays and perhaps in meteoritic materials.

The prediction of a significant enhancement in ^{58}Fe and not ^{64}Ni is straightforward. If this prediction is fulfilled it will be strong evidence that the Cassé-Paul hypothesis is correct. A ^{36}S enhancement relies upon the n,p and n,α reaction rates in radioactive isotopes; these rates are uncertain and are difficult to check by experiment.

Beta decay from an excited state of ^{107}Ag was found able to produce substantial amounts of ^{107}Pd at the core temperatures of WR stars during helium burning. Similarly the $^{205}Pb/^{205}Tl$ ratio can be significant. These anomalies should be present in the dust which forms around WC stars. The precise ratios expected depend on both the mass-loss profile of the star (how much helium burned material is exposed) and the amount of subsequent dilution. We are engaged in further studies to determine the isotopic ratios expected as a function of the ZAMS mass and mass-loss rates.

5. ACKNOWLEDGMENTS

This work was supported at The Aerospace Corporation by the company-sponsored research program, and at the Lawrence Livermore National Laboratory under the auspices of the U. S. Department of Energy.

6. REFERENCES

Abbott, D. C. 1982, Ap. J., 263, 723.
Allen, D. A., Barton, J. R. and Wallace, P. T., Mon. Not. R. Ast. Soc., 196, 797.
Anders, E. and Ebihara, M. 1982, Geochim. et Cosmo. Acta, 46, 2363.
Blake, J. B. and Schramm, D. N. 1976, Ap. J., 209, 846.
Blake, J. B. and Dearborn, D. S. P. 1984, Adv. Sp. Res., 4, 89.
Cassé, M. and Paul, J. A. 1982, Ap. J., 258, 860.
Cassé, M. and Cesarsky, C. J. 1984, in High Energy Astrophysics, Proceedings of the Nineteenth Rencontre de Moriond Astrophysics Meeting, ed. by J. Tran Thanh Van, editions Frontieres, Gif sur Yvette, France, p. 363.
Cassé, M. 1984, Adv. Sp. Res., 4, 411.
Clayton, D. D. 1982, Q. J. R. Astr. Soc., 23, 174.
Caughlan, G. R., Fowler, W. A., Harris, M. J., and Zimmerman, B. A. 1985, At. Nucl. Data Tables, 32, 197.
Dearborn, D. S. P. and Blake, J. B. 1979, Ap. J., 231, 193.
Dearborn, D. S. P. and Blake, J. B. 1984, Ap. J., 277, 783.
Dearborn, D. S. P. and Blake, J. B. 1985, Ap. J. Lett., 288, L21.
Hackwell, J. A., Gehrz, R. D., Smith, J. R. and Strecker, D. W. 1976, Ap. J., 210, 137.
Hackwell, J. A., Gehrz, R. D. and Grasdalen, G. L. 1979, Ap. J., 234, 133.
Holmes, J. A., Woosley, S. E., Fowler, W. A. and Zimmerman, B. A. 1976, At. Nucl. Data Tables, 18, 306.
Humphreys, R. M. 1978, Ap. J. Suppl., 38, 309.

Hubbard, E. N., and Dearborn, D. S. P. 1980, Apl. J., 239, 248.
Lamb, S. A., Howard, W. M., Truran, J. W. and Iben, I., Jr., Ap. J., 217, 213.
Maeder, A. 1982, Ast. Ap., 105, 149.
Maeder, A. 1983, Ast. Ap., 120, 130.
Mahoney, W. A., Ling, J. C., Wheaton, W. A. and Jacobson, A. S. 1984, Ap. J., 286, 578.
Prantzos, N. and Cassé, M. 1986, Ap. J., 307, 324
Prantzos, N. 1984a, in High Energy Astrophysics, Proceedings of the Nineteenth Rencontre de Moriond Astrophysics Meeting, ed. by J. Tran Thanh Van, éditions Frontières, Gif sur Yvette, France, p. 341.
Prantzos, N. 1984b, Adv. Sp. Res., 4, 109.
Prantzos, N., Doom, C., Arnould, M. and de Loore, C. 1986, Ap. J., 304, 695.
Takahashi, K. and Yokoi, K. 1985, Preprint UCRL-93799, prepared for Atomic Data and Nuclear Data Tables.
van der Hucht, K. A., Cassinelli, J. P. and Williams, P. M. 1986, Ap. J., in press.
Wasserberg, G. J. 1985, in Protostars and Planets II, ed. by D. C. Black and M. S. Matthews, Univ. of Arizona Press, Tucson, p. 703.
Wefel, J. P., Schramm, D. N. and Blake, J. B. 1977, Astro. Sp. Sci., 49, 47.
Wiedenbeck, M. E. 1984, Adv. Sp. Res., 4, 15.
Woosley, S. E., Fowler, W. A., Holmes, J. A. and Zimmerman, B. A. 1978, At. Nucl. Data Tables, 22, 371.
Yokoi, K., Takahashi, K. and Arnould, M. 1985, Ast. Ap., 145, 339.

GAMMA-RAY VIEWS ON THE GALACTIC COSMIC-RAY DISTRIBUTION

Hans Bloemen
Astronomy Department
University of California, Berkeley, CA 94720

ABSTRACT. Recent γ-ray studies of the distribution of cosmic rays throughout the Galaxy are discussed, based on the observations obtained with the *SAS*-2 and *COS-B* satellites. For reasons addressed in this paper, the picture arising from recent analyses of both data bases differs from the one generally advocated in the past.

(*i*) It seems inevitable to conclude at the moment that the mean density of GeV cosmic-ray nuclei in the galactic plane can vary only weakly — by a factor less than two — from the inner ($R \simeq 4$ kpc) to the outer ($R \simeq 15 - 20$ kpc) regions of the Galaxy ($R_{\odot} \equiv 10$ kpc). The allowed fall-off with increasing galacto-centric distance R is much smaller than the radial gradient in the distribution of the types of objects generally believed to have injected the particles and/or accelerated them, either directly or indirectly by producing shocks in the interstellar medium. It is unlikely that this is due to extensive diffusion of the particles into the outer Galaxy. It is suggested here that it might be due to an increase of the mean escape lifetime with increasing galacto-centric distance.

(*ii*) There seems no doubt that the interpretation of the *COS-B* γ-ray data requires an energy-dependent model, such that the γ-ray intensity spectrum towards the inner Galaxy is softer than for the remainder of the disk. This suggests a softer γ-ray emissivity spectrum in the inner regions, but it is still debatable whether this holds only for the molecular clouds (concentrated in the inner Galaxy) or whether it is a ubiquitous phenomenon (a gradual steepening from the outer towards the inner Galaxy), which could, for instance, be ascribed to a stronger galacto-centric gradient for the CR electrons (<1 GeV) than for the protons. A concentration of steep-spectrum γ-ray sources in the inner Galaxy can also not be excluded.

1. INTRODUCTION

Gamma rays in the energy range of ~50 MeV to several GeV, covered by the experiments flown on the *SAS*-2 and *COS-B* satellites, originate in interstellar space from inelastic collisions between cosmic-ray (CR) nuclei (those containing the bulk of the energy density, so mainly 1–10 GeV protons) and gas particles (via the immediate decay of the produced π°-mesons) and from the bremsstrahlung losses of electrons with energies $\lesssim 1$ GeV. Various studies have indicated that these CR-matter interactions probably dominate the observed diffuse γ-ray emission; an additional contribution (generally estimated to be $\lesssim 10\%$) results from the inverse-Compton losses of CR electrons (>1 GeV) in the

M. M. Shapiro and J. P. Wefel (eds.), Genesis and Propagation of Cosmic Rays, 163–174.

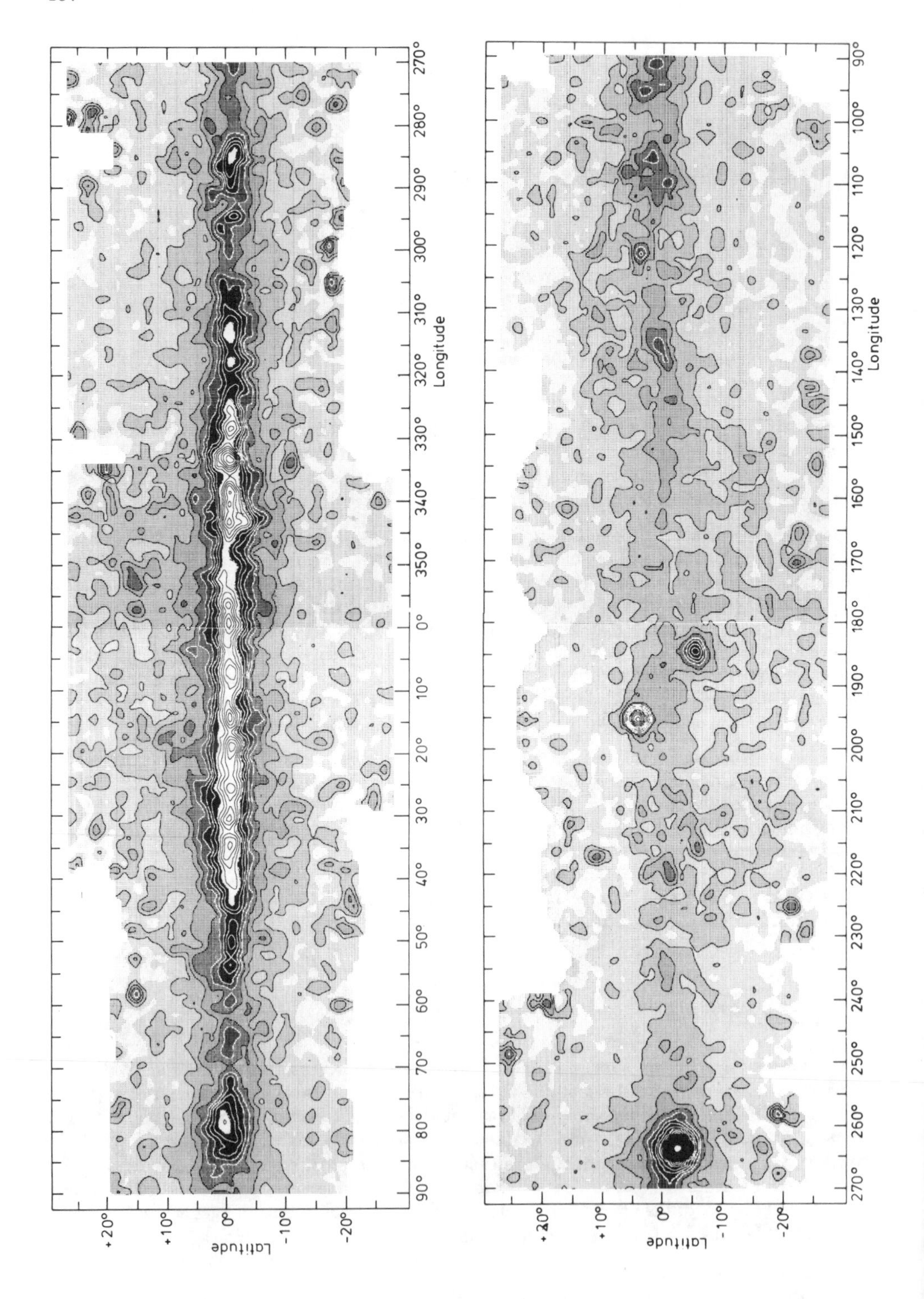
Longitude
Latitude
Longitude
Latitude

interstellar photon field, mainly in the optical and infrared range. Gamma-ray astronomy offers therefore an excellent means to study the distribution of CR particles in the Galaxy. If the galactic γ-ray emission is indeed mainly due to CR-matter interactions, then the observed intensities are basically a tracer of the product of the CR density and the interstellar gas density, integrated along the line of sight: $I_\gamma \propto \int \rho_{cr}\rho_{gas}dl$ (the Galaxy is transparent to γ-rays). In order to use γ-ray observations as a CR tracer throughout the Galaxy, it is essential to know the distribution of the target interstellar gas particles.

Given the limited angular resolution of the γ-ray experiments that have flown thus far (one degree or worse), it should be kept in mind that unresolved γ-ray emission from stellar-like objects might be hidden and would appear part of the diffuse emission to these experiments. At least the Crab and Vela pulsars have been identified as strong γ-ray sources and there are strong indications for several others (see Pollock *et al.* 1985ab), although some could be mimickedby a local concentration of cosmic rays. Nevertheless, the assumption that unresolved γ-ray sources do not contribute strongly to the observed emission in the studies described in this paper seems to be justified *a posteriori*; if their contribution is significant, it tends to strengthen the conclusions.

This paper is concentrated on recent γ-ray studies of the large-scale CR distribution in the Galaxy. There is some confusion in this field at the moment. This is *not* due to uncertainties in the point-source contribution or to uncertainties in the instrument sensitivities and instrumental background (although these should of course be treated carefully), as sometimes believed. It seems to have three major origins, discussed in further detail throughout this paper. (*i*) *A lack of large-scale CO surveys until recently.* Although the large-scale distribution of the atomic hydrogen (HI) component had been mapped using the characteristic 21-cm line emission, the numerous γ-ray studies of the galactic CR distribution performed in the past suffered severely from uncertainties in the galactic distribution of the other major component, molecular hydrogen (H_2). The best tracer of the large-scale distribution of H_2 is the CO molecule, but the sky coverage of the CO observations was poor and the relation between the measured CO intensities and H_2 column densities was uncertain. Large-scale CO surveys are available now, which enable more refined analyses as described in §2. (*ii*) *Recalibration of the SAS-2 data.* The situation seemed to be less confusing for the outer Galaxy, particularly because the H_2 contribution to the gas column densities appears to be small on average. Dodds *et al.* (1975), analyzing the *SAS*-2 γ-ray data, first suggested a decrease in the density of cosmic rays with increasing distance beyond the solar circle; other studies of the *SAS*-2 data confirmed this gradient and it has been adopted ever since as the main observational evidence against a universal origin of CR nuclei in the GeV range. However, this strong gradient barely shows up in recent studies of the (final) *SAS*-2 data ($E_\gamma > 100$ MeV), which turned out to be a direct consequence of an improvement in the calibration of the *SAS*-2 data around 1978 (§3). The final *SAS*-2 and *COS-B* data bases are in basic agreement. (*iii*) *Frequent careless treatment of statistical (and systematic) uncertainties.* It is evidently most important that the procedures of data analysis be sufficiently sophisticated to justify firm conclusions and to make the best possible use of the data, particularly for counting experiments. The low count rate of γ-ray experiments requires even additional care.

Figure 1: *COS-B* γ-ray intensity maps for the 150 MeV - 5 GeV range. Although a smoothing algorithm has suppressed fluctuations of angular scale smaller than the point-spread function, not all individual features in the maps are statistically significant, particularly at medium latitudes near the edges of the field of view. Contour values are indicated at multiples of 5×10^{-5} photon cm^{-2} s^{-1} sr^{-1}. The isotropic background (mainly instrumental) is subtracted.

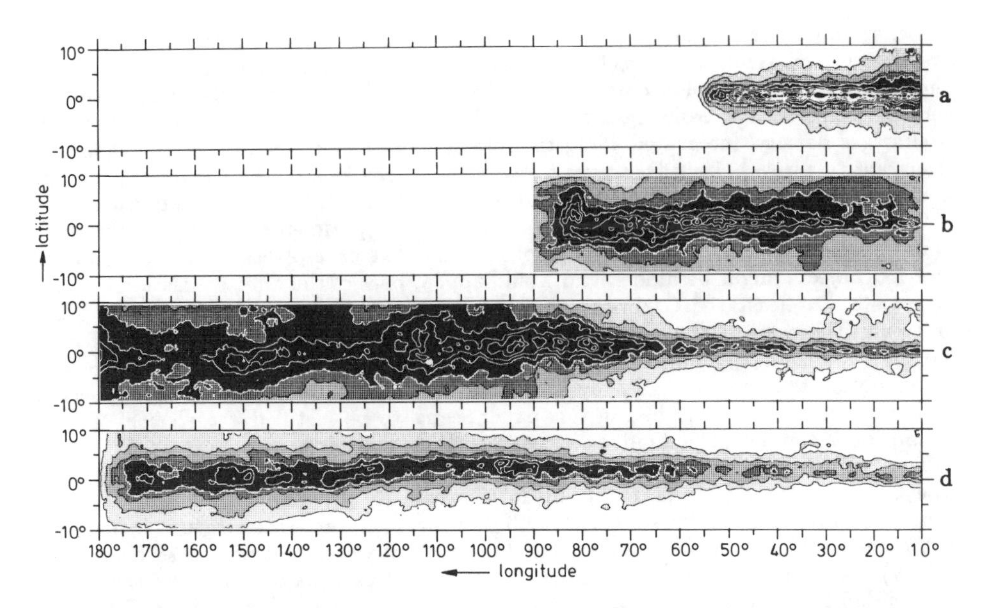

Figure 2: Atomic hydrogen column-density maps of the first and second galactic quadrants corresponding to four galacto-centric distance intervals: (*a*) 2 kpc $< R <$ 8 kpc, (*b*) 8 kpc $< R <$ 10 kpc, (*c*) 10 kpc $< R <$ 15 kpc, and (*d*) $R >$ 15 kpc. Contour values: (0.4, 1, 2, 3, 5, 7, ...)$\times 10^{21}$ atom cm^{-2}.

2. RESULTS FROM RECENT ANALYSES OF THE *COS-B* DATA

A large-scale mm-wave CO survey obtained with the 1.2m telescopes of Columbia University & Goddard Institute for Space Studies, covering the entire Milky Way up to $|b| \simeq 7° - 10°$, was completed recently; the large sky coverage and 0.5° angular resolution make it very suitable for correlation studies with the γ-ray observations. The *COS-B* observations have sufficient resolution and sensitivity to constrain the relation between the integrated CO line intensity W_{CO} and the H_2 column density $N(H_2)$ (Lebrun *et al.* 1983; Bloemen *et al.* 1984a). The Columbia CO survey was used by Bloemen *et al.* (1986) in a large-scale γ-ray study of the CR distribution as well as the H_2 distribution throughout the Galaxy. The good counting statistics of the *COS-B* observations (about 25 times higher than for the *SAS*-2 observations), are of crucial importance in this broad approach. They used the velocity information of the HI and CO line observations as a distance indicator to ascertain the spatial distribution of the interstellar gas. Using this distance information, the galacto-centric distribution of the γ-ray emissivity (the production rate per H atom) was determined for different energy ranges from a correlation study of the γ-ray intensity maps and the HI and CO gas-tracer maps for selected galacto-centric distance intervals, taking into account the expected inverse-Compton contribution. On the assumption that unresolved γ-ray point sources do not contribute significantly to the observed γ-ray emission, the γ-ray emissivity is proportional to the CR density. In addition, since the spectral shapes of the γ-rays produced by the CR nuclei and electrons are different (the π°-decay spectrum has a maximum at ~70 MeV, whereas the bremsstrahlung spectrum decreases monotonically with energy), the spatial distribution of the γ-ray emissivity spectrum provides information on the electrons and nuclei separately, if the shapes of the CR spectra

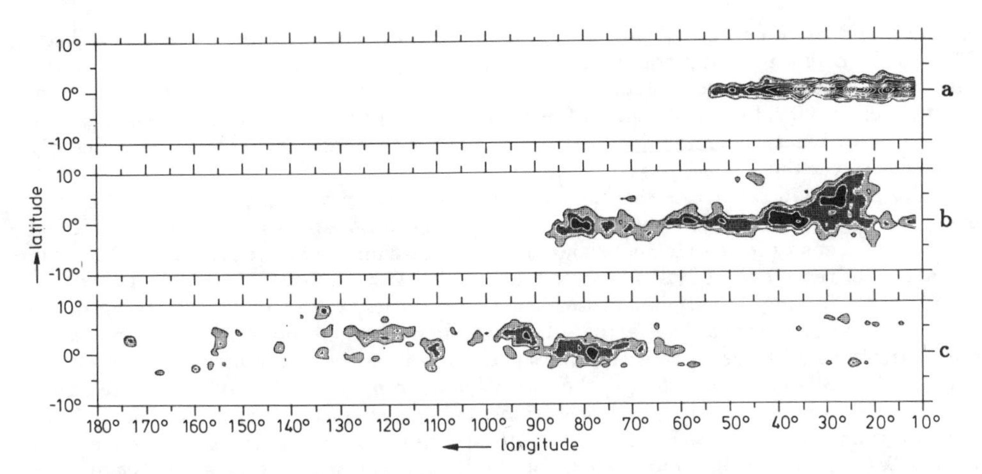

Figure 3: Carbon-monoxide intensity maps of the first and second galactic quadrants corresponding to three galacto-centric distance intervals: (*a*) 2 kpc $< R <$ 8 kpc, (*b*) 8 kpc $< R <$ 10 kpc, and (*c*) $R >$ 10 kpc. Contour values: 4, 8, 15, 25, 35, ... K km s^{-1}.

do not strongly vary throughout the Galaxy. At high energies ($E_\gamma \gtrsim$ 300 MeV) the diffuse γ-ray emission is dominated by π°-decay, whereas the emission at lower energies has a large electron bremsstrahlung contribution. In an internally consistent approach Bloemen *et al.* determined also the ratio $X \equiv \mathrm{N(H_2)/W_{CO}}$ and studied possible variations of this quantity throughout the Galaxy.

Skymaps of HI column densities, $\mathrm{N(HI)}_i$, and integrated CO line intensities, $\mathrm{W}_{\mathrm{CO},i}$, were constructed in four ($i = 1,2,3,4$) galacto-centric distance ranges (2 kpc $< R <$ 8 kpc, 8 kpc $< R <$ 10 kpc ($=R_\odot$), 10 kpc $< R <$ 15 kpc, and $R >$ 15 kpc) and convolved with the energy-dependent *COS-B* point-spread function. These four distance intervals were selected because the angular distributions of the gas in each interval show distinct differences (figures 2 and 3), which are needed to ascertain the contribution of the gas in each interval to the observed γ-ray intensities. Similarly, determining the conversion factor X requires distinct differences between the structures in the HI and CO maps. Due to the still limited CO coverage available at that time, the correlation analysis was restricted to the first and second galactic quandrants and the Carina region ($270° < \ell < 300°$), and $|b| \lesssim 5°$. Using a likelihood analysis for three different energy ranges (70–150 MeV, 150–300 MeV, 300 MeV – 5 GeV), it was investigated which combination of the free parameters (q_i, Y_i, and I_B) best describes the observed γ-ray intensity distributions I_γ for each energy range, assuming a relation of the form:

$$I_\gamma = \left\{ \sum_i \frac{q_i}{4\pi} [\mathrm{N(HI)}_i + 2Y_i \cdot \mathrm{W}_{\mathrm{CO},i}] \right\} + I_\mathrm{IC} + I_\mathrm{B}.$$

The term enclosed by braces represents the intensities that originate from CR collisions with atomic and molecular hydrogen (q_i is the γ-ray emissivity for annulus i), I_IC represents the modelled (small) inverse-Compton contribution (Bloemen 1985), and I_B is the isotropic background, including the (dominant) instrumental background. The parameter Y_i is related to the $\mathrm{N(H_2)/W_{CO}}$ ratio X_i for each annulus: $Y_i = \{q_i(H_2)/q_i\} \cdot X_i$. If CR particles are not excluded from, or concentrated in, molecular clouds, then Y_i equals

X_i, independent of energy. Y was first assumed to be constant throughout the Galaxy. Starting from this general model with six parameters for each energy range, it was tested (using the likelihood ratio) whether various simpler models with fewer parameters (i.e. constant emissivity distributions as a function of R, identical values of Y for each energy range and/or identical emissivity distributions for each energy range) give significantly worse fits to the data.

The likelihood-ratio tests showed that the hypothesis of an identical Y-value for each energy range gives an almost equally good fit to the data as the general model. If the average CR density inside the molecular clouds is the same as in the HI gas [(1)], then the Y-parameter represents X; its value was found to be $X = (2.75 \pm 0.35) \times 10^{20}$ mol. cm^{-2} K^{-1} km^{-1} s. Taking into account systematic uncertainties, a more realistic estimate of the uncertainty on X is probably a factor $\sim 2\times$ the formal error. To investigate possible large-scale variations of X throughout the Galaxy, the Y-values were determined separately for 2 kpc $< R <$ 8 kpc and $R >$ 8 kpc; they were found to be identical within uncertainties.

The following discussion is concentrated on the resulting emissivity distributions and the implications for the galactic CR distribution; for a detailed discussion on the $N(H_2)/W_{CO}$ ratio and the implications for the H_2 content of the Galaxy, see Bloemen *et al.* (1986).

Bloemen *et al.* (1984bc) applied this correlation method to the *COS-B* observations and HI observations of the second and third galactic quadrants alone (they showed that H_2 can be neglected in the outer Galaxy within the uncertainties of the analysis), and found that the γ-ray emissivity for the low-energy range decreases in the outer Galaxy, whereas surprisingly the high-energy emissivity shows barely any fall-off. The extension of this analysis, including the inner Galaxy as described above, resulted in a similar picture for the Galaxy as a whole. For the inner Galaxy, however, the interpretation of the low-energy emission is debatable. The problem is the following. The likelihood tests mentioned above showed very convincingly that the interpretation of the *COS-B* data requires an energy-dependent model, such that the intensity spectrum towards the inner Galaxy is softer than for the remainder of the disk. In the modelling considered here, this has to be ascribed to either a higher Y-value for low energies than for high energies (note that most of the H_2 is located inside the solar circle, contrary to the HI; compare figures 2 and 3) or to a steeper γ-ray emissivity gradient for low energies. In principle, these two effects are distinguishable in the correlation analysis because of the differences between the angular distributions of HI and H_2. In practice, this method applies less satisfactorily to the observations at low energies than at high energies because the angular resolution of *COS-B* is degrading with decreasing energy. In other words, the γ-ray data cannot tell very distinctively whether the relatively steep γ-ray emissivity spectrum in the inner Galaxy holds only for the γ-ray emission originating from the molecular gas or whether it is a ubiquitous phenomenon, holding for both the HI and H_2 gas. The hypothesis tests favoured the latter and we concentrate in the following on this case; a correlation analysis for the whole Galaxy is in progress and will give further insight.

On the assumption that the shapes of the CR spectra do not vary strongly throughout the Galaxy, the energy dependence of the emissivity distribution was interpreted as a stronger galacto-centric gradient for the electrons than for the nuclei. Bloemen *et al.*

---

(1) There is at least no indication that cosmic rays with energies $\gtrsim$ 100 MeV/nucleon are excluded from molecular clouds. The theoretical work of Skilling and Strong (1976) and Cesarsky and Völk (1978) indicates that only very low-energy cosmic rays ($<$ 50 MeV/nucleon) may not penetrate a dense cloud completely. Furthermore, the γ-ray observations of the Orion molecular complex show conclusively that cosmic rays do penetrate at least the local giant molecular clouds (Bloemen *et al.* 1984a).

(1986) showed that the data are consistent with radial CR density distributions of an exponential form for $R \gtrsim 4$ kpc: $\rho_{cr}(R) \propto \exp(-R/L)$. There are some small-scale discrepancies, which might be due to genuine point sources or local enhancements of the CR density (or H_2/CO ratio), but all deviations are fairly small. For the CR nuclei, the radial scale length L was found to be > 18 kpc (and the data are consistent with a constant density throughout the entire galactic plane); for CR electrons, $L = 5 - 11$ kpc. When the radial distributions of the electrons and nuclei are forced to be the same, in which case the energy dependence is attributed to Y, $L = 15 \pm 4$ kpc. These CR gradients are upper limits if a population of unresolved galactic γ-ray sources exists with a latitude distribution similar to that of the gas but with a stronger concentration towards the inner parts of the Galaxy. Obviously, if such a population would have a steeper γ-ray spectrum than the diffuse emission, it might be responsible for the softer γ-ray emission towards the inner Galaxy.

For $R \lesssim 4$ kpc, there is no conclusive evidence from the *COS-B* data for a significant increase or decrease of the CR density. The γ-ray flux (>300 MeV) from the central few hundred parsecs of the Galaxy is however an order of magnitude smaller than the value expected from the large H_2 masses generally estimated to be present near the centre and from the average γ-ray emissivity measured in the disk (Blitz *et al.* 1985). If the cosmic rays do penetrate the clouds (see footnote 1) this deficit of γ-rays suggests that either the density of GeV CR nuclei is anomalously small near the galactic centre relative to the disk or that H_2 is nearly an order of magnitude less abundant than estimates made from CO observations.

3. PREVIOUS FINDINGS & RECENT RESULTS FROM THE *SAS*-2 DATA

For the inner Galaxy, discrepancies between previous works and the findings from analyses of the *COS-B* data described above can largely be ascribed to the poor sky coverage of the CO observations and to the large uncertainties in the relation between W_{CO} and $N(H_2)$ in the past. Usually the sparse CO observations and the 21-cm HI observations were converted into (spiral arm) models of the gas distribution. By assuming that the CR density on a large scale is proportional to the gas density, some studies showed that the observed and predicted γ-ray intensities agreed (e.g. Bignami *et al.* 1975; Kniffen *et al.* 1977; Fichtel and Kniffen 1984) while others required modifications, such as abundance gradients or a partial exclusion of cosmic rays from molecular clouds (e.g. Cesarsky *et al.* 1977). Without *a priori* assumptions on the proportionality between CR and gas density, several studies indicated that the CR density in the inner Galaxy is higher than in the solar vicinity (e.g. Stecker *et al.* 1975; Higdon 1979; Issa *et al.* 1981; Harding and Stecker 1985). With the Columbia CO survey and the large *COS-B* counting statistics, which enable a more sophisticated analysis, some of the difficulties that confronted earlier studies of the inner Galaxy have been overcome.

For the outer Galaxy, the discrepancy between previous works using *SAS*-2 data (e.g. Dodds *et al.* 1975; Cesarsky *et al.* 1977; Higdon 1979; Issa *et al.* 1981) and the results summarized in §2 cannot be ascribed to uncertainties in the H_2 contribution to the γ-ray intensities, which is negligible. The pioneering analysis of the γ-ray data from the *SAS*-2 satellite by Dodds *et al.* (1975) first suggested a strong gradient in the density of CR nuclei in the outer Galaxy. The disturbing disagreement with the findings described in §2, which indicate a near constancy of the density of CR nuclei throughout the Galaxy, seems, after all, to have a straightforward explanation: the γ-ray intensities (>100 MeV) measured by the *SAS*-2 satellite, presented by Fichtel *et al.* (1975), and used by Dodds *et al.*, are about a factor of two lower than the final outer-Galaxy intensities released by the *SAS*-2

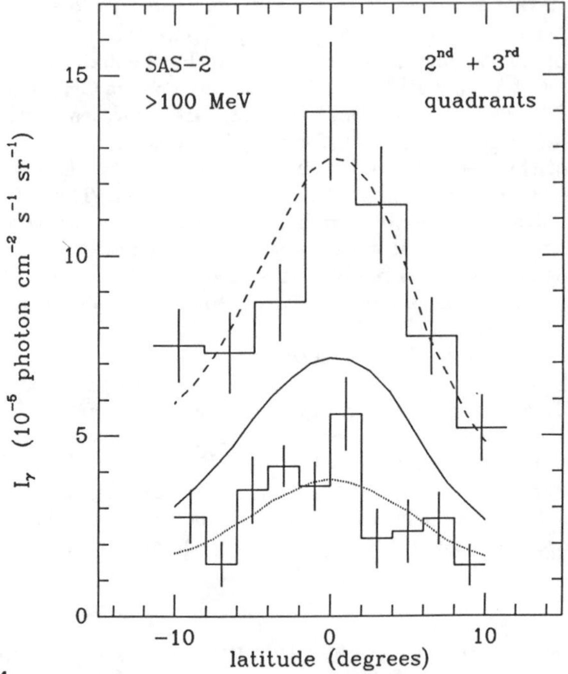

Figure 4:

histograms	:	measured γ-ray intensities by *SAS*-2 lower – as used by Dodds *et al.* (1975) upper – from final *SAS*-2 data (from Hartman *et al.* 1979)
dotted line	:	prediction from "galactic" model of Dodds *et al.* (i.e. decrease in CR density towards outer Galaxy)
full line	:	prediction from "extragalactic" model of Dodds *et al.* (i.e. constant CR density)
dashed line	:	as full line, but 70% higher γ-ray emissivity to be consistent with more recent estimates of the local emissivity

Note: these are only coarse reproductions of figures in the original papers and the longitude ranges for the two *SAS*-2 histograms are not exactly the same.

group (Fichtel *et al.* 1978). This intensity difference may be ascribed to the improvement in calibration of *SAS*-2 after the first presentation of the data in 1975 (see Thompson *et al.* 1977). Using the *final SAS*-2 data, it can be seen that the "extragalactic" model of Dodds *et al.* (i.e. a constant density of CR nuclei) predicts γ-ray intensities that are even too low for the outer Galaxy, but there is good agreement if recent estimates of the local γ-ray emissivity for energies above 100 MeV are used ($\sim 1.7 \times 10^{-26}$ photon H atom^{-1} s^{-1} sr^{-1}), which are $\sim$70% higher than the value used by Dodds *et al.* The situation is clarified in figure 4.

Other works (e.g. Cesarsky *et al.* 1977; Higdon 1979) essentially confirmed the findings of Dodds *et al.*, because they relied on the same *SAS*-2 data base. Using the final *SAS*-2 data, Strong *et al.* (1978), Arnaud *et al.* (1982), and Bloemen *et al.* (1984c) found no indication of a strong emissivity gradient (>100 MeV). Although Strong *et al.* and Arnaud *et al.* ascribed this result to a weaker gradient in the second quadrant (which they studied) compared to the third quadrant, the disappearance of the strong gradient found by Dodds *et al.* is really due to the use of the recalibrated *SAS*-2 data. Issa *et al.* (1981), analysing the final *SAS*-2 data and *COS-B* data, found a steep emissivity gradient outside the solar circle, independent of γ-ray energy, but they used a radial-unfolding technique that gives uncertain results at the solar circle and is essentially inapplicable to regions beyond the solar circle. A recent paper by Bhat *et al.* (1985) is largely based on this analysis by Issa *et al.* (1981).

4. COMPARISON WITH SYNCHROTRON EMISSION

The observed diffuse radio-continuum emission of the Galaxy at low frequencies is primarily synchrotron emission from CR electrons spiraling around the magnetic-field lines. Observations around 100 MHz are most appropriate for a large-scale study, because they do not suffer from free-free absorption and they have only a small thermal contribution. In a typical interstellar field of a few μG, 100 MHz observations mainly trace electrons with energies of a few GeV, which is approximately an order of magnitude larger than that of the electrons traced by the γ-ray observations around 100 MeV, but the galactocentric distributions can be expected to be similar. For CR electrons with a power-law spectrum $kE^{-\alpha}$, the synchrotron emissivity is proportional to $kB_{\perp}^{(1+\alpha)/2}$, where $B_{\perp}$ is the magnetic-field component perpendicular to the line of sight. For the compilation of the local interstellar electron spectrum given by Webber (1983a), $(1+\alpha)/2 \simeq 1.8$. Phillipps *et al.* (1981) and Beuermann *et al.* (1985) both found, from different unfolding techniques applied to the 408 MHz survey of Haslam *et al.* (1981ab), that the synchrotron volume emissivity in the galactic plane for 5 kpc $\lesssim R \lesssim$ 15 kpc can be represented by an exponential distribution $\propto \exp(-R/L_{\rm syn})$ with a radial scale length L_{syn} of 3.9 kpc. This result appears to depend not strongly on the large-scale geometry and fluctuations of the galactic magnetic-field. The synchrotron emissivity distribution is steeper than the CR electron distribution derived from the γ-ray observations, as expected if the magnetic-field strength has a (weak) galacto-centric gradient; representing the magnetic-field strength also by an exponential distribution, this comparison shows that the field strength has a radial scale length $\gtrsim$ 10 kpc, even if the electron distribution would have a radial scale length as large as 15 kpc.

5. IMPLICATIONS FOR CR ORIGIN/CONFINEMENT

The exponential scale length of the CR nuclei (and electrons?) is much larger than for the type of objects generally considered as candidates for CR sources or injectors; e.g. supernova remnants, pulsars, early-type stars, and disk stars in general, all have radial scale lengths of typically ~5 kpc for $R \gtrsim 4$ kpc (e.g. Kodaira 1974; Lyne *et al.* 1985; Mathis *et al.* 1983 and references therein). It seems that the distribution of at least the CR protons in the Galaxy does not reflect the distribution of the objects that have probably injected the particles and/or accelerated them, either directly or indirectly by producing shocks in the ISM. In the case of diffusive shock acceleration (Axford *et al.* 1977; Bell 1978; Blandford and Ostriker 1978), however, the impact of the ambient medium needs further investigation in view of the probably widely different structure of the ISM in the inner and outer regions of the Galaxy (see e.g. Heiles 1986). If the production rate of cosmic rays indeed increases towards the inner Galaxy with a radial scale length of ~5 kpc, it is unlikely that extensive diffusion of the particles into the outer Galaxy can reproduce the near constancy. This is an inevitable consequence of the rapid increase of the volume to be filled as the radius increases (although this problem would be alleviated if the particles propagated predominantly in the disk instead of 3-dimensionally). Moreover, it was pointed out by Ormes and Protheroe (1983) that the steep rigidity dependence of the escape length of 1–10 GeV/nucleon CR nuclei found recently ($\lambda_e \propto \Re^{-0.7}$), together with the absence of structure in the proton spectrum up to $\sim 10^5$ GeV, implies that these particles can probably not diffuse more than a few hundred parsecs from their souces during their $(1-2) \times 10^7$ yr lifetime (at least in the solar vicinity). Reacceleration of cosmic rays in the interstellar medium by diffusive shock acceleration in successive interstellar shocks (Blandford and Ostriker 1980) may affect this argument. It is not obvious, however, that this mechanism can significantly widen the galacto-centric CR distribution, because the number density of supernovae (which induce the shock waves) decreases strongly with increasing galacto-centric distance. Moreover, continuous stochastic acceleration in general throughout the interstellar medium may be of only minor importance because it does not lead simultaneously to the observed decrease in the CR secondary-to-primary ratio with increasing energy and the power-law shape of the observed CR spectra (Eichler 1980; Cowsik 1980); a suggestion by Lerche and Schlickeiser (1985) may alter this situation, but this is not clear yet (Cowsik 1986). There remains an interesting alternative that has not been considered hitherto, namely that the mean escape lifetime for cosmic rays is smaller in the inner Galaxy than in the outer Galaxy. This interpretation is not unlikely, but needs to be studied in detail, if the confinement of cosmic rays is due to scattering by the hydromagnetic waves they excite (Kulsrud and Pearce 1969; Wentzel 1969; Skilling 1971; Holmes 1975). The near constancy of the density of the CR nuclei throughout the Galaxy also does not exclude an extragalactic origin, with the cosmic rays filling up for instance the volume of a (super) cluster (Brecher and Burbidge 1972; Burbidge 1974, 1983); the gradient of the CR nuclei is so weak, if it exists, that it can no longer be regarded as evidence against a universal origin of CR nuclei in the GeV range.

ACKNOWLEDGEMENTS

I gratefully acknowledge receipt of a Miller Fellowship and thank my colleagues and co-authors of the *COS-B* gamma-ray work summarized in this paper for many helpful discussions.

REFERENCES

Arnaud, K., Li Ti pei, Riley, P.A., Wolfendale, A.W., Dame, T.M., Brock, J.E., and Thaddeus, P. 1982, *M.N.R.A.S.*, **201**, 745.

Axford, W.I., Leer, E., and Skandron, K.G. 1977, *Proc. 15th Int. Cosmic Ray Conf.*, **1**, 132.

Bell, A.R. 1978, *M.N.R.A.S.*, **182**, 147.

Beuermann, K., Kanbach, G., and Berkhuijsen, E.M. 1985, *Astr. Ap.*, **153**, 17.

Bhat, C.L., Issa, M.R., Houston, B.P., Mayer, C.J., and Wolfendale, A.W. 1985, *Nature*, **314**, 511.

Bignami, G.F., Fichtel, C.E., Kniffen, D.A., and Thompson, D.J. 1975, *Ap. J.*, **199**, 54.

Blandford, R.D., and Ostriker, J.P. 1978, *Ap. J. Lett.*, **221**, L29.

Blandford, R.D., and Ostriker, J.P. 1980, *Ap. J.*, **237**, 793.

Blitz, L., Bloemen, J.B.G.M., Hermsen, W., and Bania, T.M. 1985, *Astr. Ap.*, **143**, 267.

Bloemen, J.B.G.M., Caraveo, P.A., Hermsen, W., Lebrun, F., Maddalena, R.J., Strong, A.W., and Thaddeus, P. 1984a, *Astr. Ap.*, **139**, 37.
Bloemen, J.B.G.M., Blitz, L., and Hermsen, W. 1984b, *Ap. J.*, **279**, 136.
Bloemen, J.B.G.M., Bennett, K., Bignami, G.F., Blitz, L., Caraveo, P.A., Gottwald, M., Hermsen, W., Lebrun, F., Mayer-Hasselwander, H.A., and Strong, A.W. 1984c, *Astr. Ap.*, **135**, 12.
Bloemen, J.B.G.M. 1985, *Astr. Astr.*, **145**, 391.
Bloemen, J.B.G.M., Strong, A.W., Blitz, L., Cohen, R.S., Dame, T.M., Grabelsky, D.A., Hermsen, W., Lebrun, F., Mayer-Hasselwander, H.A., and Thaddeus, P. 1986, *Astr. Ap.*, **154**, 25.
Brecher, K., and Burbidge, G.R. 1972, *Ap. J.*, **174**, 253.
Burbidge, G.R. 1974, *Phil. Trans. Roy. Soc. London*, **A 277**, 481.
Burbidge, G.R. 1983, in *Composition and Origin of Cosmic Rays*, ed. M.M. Shapiro, Reidel, Dordrecht, p. 245.
Cesarsky, C.J., Cassé, M., and Paul, J.A. 1977, *Astr. Ap.*, **60**, 139.
Cesarsky, C.J., and Völk, H.J. 1978, *Astr. Ap.*, **70**, 367.
Cowsik, R. 1980, *Ap. J.*, **241**, 1195.
Cowsik, R. 1986, *Astr. Ap.*, **155**, 344.
Dodds, D., Strong, A.W., and Wolfendale, A.W. 1975, *M.N.R.A.S.*, **171**, 569.
Fichtel, C.E., Hartman, R.C., Kniffen, D.A., Thompson, D.J., Bignami, G.F., Ögelman, H.B., Özel, M.E., and Tümer, T. 1975, *Ap. J.*, **198**, 163.
Eichler, D. 1980, *Ap. J.*, **237**, 809.
Fichtel, C.E., Hartman, R.C., Kniffen, D.A., Thompson, D.J., Ögelman, H.B., Tümer, T., and Özel, M.E. 1978, *NASA Technical Memorandum* 79650.
Fichtel, C.E., and Kniffen, D.A. 1984, *Astr. Ap.*, **134**, 13.
Harding, A.K., and Stecker, F.W. 1985, *Ap. J.*, **291**, 471.
Haslam, C.G.T., Klein, U., Salter, C.J., Stoffel, H., Wilson, W.E., Cleary, M.N., Cooke, D.J., and Thomasson, P. 1981a, *Astr. Ap.*, **100**, 209.
Haslam, C.G.T., Salter, C.J., Stoffel, H., and Wilson, W.E. 1981b, *Astr. Ap. Suppl.*, **47**, 1.
Heiles, C. 1986, *Ap. J.*, in press.
Higdon, J.C. 1979, *Ap. J.*, **232**, 113.
Holmes, J.A. 1974, *M.N.R.A.S.*, **166**, 155.
Holmes, J.A. 1975, *M.N.R.A.S.*, **170**, 251.
Issa, M.R., Riley, P.A., Strong, A.W., and Wolfendale, A.W. 1981, *J. Phys.*, **G7**, 973.
Kniffen, D.A., Fichtel, C.E., and Thompson, D.J. 1977, *Ap. J.*, **215**, 765.
Kodaira, K. 1974, *Publ. Astr. Soc. Japan*, **26**, 255.
Kulsrud, R.M., and Pearce, W.P. 1969, *Ap. J.*, **156**, 445.
Lebrun, F., Bennett, K., Bignami, G.F., Bloemen, J.B.G.M., Buccheri, R., Caraveo, P.A., Gottwald, M., Hermsen, W., Kanbach, G., Mayer-Hasselwander, H.A., Montmerle, T., Paul, J.A., Sacco, B., Strong, A.W., Wills, R.D., and Dame, T.M., Cohen, R.A., and Thaddeus, P. 1983, *Ap. J.*, **274**, 231.
Lerche, I., and Schlickeiser, R. 1985, *Astr. Ap.*, **151**, 408.
Lyne, A.G., Manchester, R.N., and Taylor, J.H. 1985, *M.N.R.A.S.*, **213**, 613.
Mathis, J.S., Mezger, P.G., and Panagia, N. 1983, *Astr. Ap.*, **128**, 212.
McKee, C.F., and Ostriker, J.P. 1977, *Ap. J.*, **218**, 148.
Ormes, J.F., and Protheroe, R.J. 1983, *Ap. J.*, **272**, 756.
Phillipps, S., Kearsey, S., Osborne, J.L, Haslam, C.G.T., Stoffel, H. 1981, *Astr. Ap.*, **98**, 286.

Pollock, A.M.T., Bennett, K., Bignami, G.F., Bloemen, J.B.G.M., Caraveo, P.A., Hermsen, W., Kanbach, G., Lebrun, F., Mayer-Hasselwander, H.A., and Strong, A.W. 1985a, *Astr. Ap.*, **146**, 352.

Pollock, A.M.T., Bennett, K., Bignami, G.F., Bloemen, J.B.G.M., Caraveo, P.A., Hermsen, W., Kanbach, G., Lebrun, F., Mayer-Hasselwander, H.A., and Strong, A.W. 1985b, *Proc. 19th Int. Cosmic Ray Conf.*, **1**, 338.

Skilling, J. 1971, *Ap. J.*, **170**, 265.

Skilling, J., and Strong, A.W. 1976, *Astr. Ap.*, **53**, 253.

Stecker, F.W., Solomon, P.M., Scoville, N.Z., and Ryter, C.E. 1975, *Ap. J.*, **201**, 90.

Strong, A.W., Wolfendale, A.W., Bennett, K., Wills, R.D. 1978, *M.N.R.A.S.*, **182**, 751.

Thompson, D.J., Fichtel, C.E., Kniffen, D.A., and Ögelman, H.B. 1977, *Ap. J.*, **214**, L17.

Webber, W.R. 1983a, in *Composition and Origin of Cosmic Rays*, ed. M.M. Shapiro, Reidel, Dordrecht, p. 83.

Webber, W.R. 1983b, in *Composition and Origin of Cosmic Rays*, ed. M.M. Shapiro, Reidel, Dordrecht, p. 25.

Wentzel, D.G. 1969, *Ap. J.*, **156**, 303.

RADIO ASTRONOMY AND COSMIC RAYS

Kurt W. Weiler
E.O. Hulburt Center for Space Research
Naval Research Laboratory, Code 4131
Washington, DC 20375-5000
U.S.A.

ABSTRACT. It is generally accepted that the electronic component of cosmic rays must be of galactic origin, is the source of the galactic radio background, and possibly, or even probably, originates in supernovae and/or supernova remnants. Therefore in consideration of this electronic component, we review the general history and observed properties of supernovae with concentration on the more recent results on radio supernovae, the observed properties of supernova remnants of both the centrally driven and shock driven subclasses, and the general properties of the galactic non-thermal radio background. These results are then compared with the measured properties of the e^- component of cosmic rays in the solar neighborhood. It is found that even though this component of the cosmic rays is a rather minor one in terms of either number or energetics, it is the component which appears best explained by known phenomena at the present time.

1. INTRODUCTION

I am quite fortunate here since I plan to discuss only the electronic, and in fact only the negative electronic (e^-), component of cosmic rays which makes up only about 2% of the cosmic rays which are measured in the vicinity of the Earth. However, I have been given a much larger fraction of the available time and space than 2% in order to discuss it. Therefore, I will try to cover in some detail an area which impinges on the study of cosmic rays in general, supernovae. However, I will still restrict myself mainly to the electronic component since the nuclear component will be more capably discussed by other authors. Also, even for the electronic component, I will concentrate on only the negative electrons and generally ignore the positrons (e^+) [as well as a fraction (~10%) of the e^-] which are thought to arise from the interaction of cosmic ray nuclei (principally protons) with the interstellar gas producing pions which decay into electrons and positrons. Also, even though radio measurements show that extragalactic radio sources and, in particular, the strong and active radio galaxies and quasars are copious generators of relativistic (and thus cosmic ray) electrons, I will also

M. M. Shapiro and J. P. Wefel (eds.), Genesis and Propagation of Cosmic Rays, 175–203.

ignore them since it is generally accepted that all of the cosmic rays which we measure in the Earth's vicinity must be of galactic origin.

However, in spite of these limitations, cosmic ray electrons are still a very important component of the cosmic radiation since they provide us with a great deal of information on the propagation and confinement of cosmic rays, since they are the only component of the cosmic rays which can be observed at a distance, and since they are thought to be responsible for much of the x- and γ-ray distributed emission in the Galaxy as well as for the well known non-thermal galactic background radio emission. Also, through observations of cosmic ray e^-, we can investigate one of the likely origins of all types of cosmic rays -- supernovae and supernova remnants. Two references which I can highly recommend for much more information on the cosmic ray aspects of relativistic electrons and on which I have relied heavily in preparing the present lectures are Webber (1983) and Longair (1981). This latter is a particularly lucid and interesting discussion of the cosmic ray-astrophysics connection in general.

To approach this subject then, I will review the known types and properties of supernovae, include some of the most recent results on the radio emission from supernovae, continue to the properties and types of supernova remnants, and then expand to the observations of the non-thermal galactic radio background. Finally, I will attempt to connect all of these to the e^- component of cosmic rays which we can measure in the Earth's vicinity.

2. SUPERNOVAE

2.1. Background

The modern study of supernovae started slightly more than 100 years ago when Hartwig at the Dorpat Observatory in the U.S.S.R. found a new star near the nucleus of M31 with $m \sim 6.5$ where previously there had been nothing brighter than $m \sim 15$. When in 1919 a distance estimate became available for the Andromeda Nebula of ~200 kpc, this implied an absolute magnitude for Hartwig's star of $M \sim -15$, much, much brighter than ordinary novae which have $M \sim -6$ to -7. Thus, there was a problem. For example, this extraordinarily bright "nova" was used by Shapley as an argument against the "island universe" theory in his debates with Curtis in the 1920s.

However, when the existence of extragalactic nebulae became accepted, it was necessary to explain these occasional, extremely powerful "novae," since between 1885 and 1920 about 10 were observed by chance in nearby galaxies. Also, since our Galaxy is, in principle, no different from many of the others, people realized that these unusual "novae" should also occur locally and Lundmark found in the early years of the century that ancient astronomical chronicles indeed record the occasional appearance of bright and varying stars. One of these, observed in the constellation of Taurus by the Chinese in 1054, was connected in 1928 by none other than Edwin Hubble to the stellar explosion which created M1, the Crab Nebula. Finally, in 1934 Fritz

Zwicky and Walter Baade, who might be considered the true founders of this field of study, coined the name "supernova" for the phenomenon.

Now, even though the objects had a name and were recognized as an unusual astrophysical phenomenon, almost nothing was known about them. The last galactic supernova which had been observed was in 1604 and was studied by Johannes Kepler. For the more modern extragalactic examples, only a few rough light curves existed. Therefore, Zwicky began to assemble data on the statistics and properties of the faint extragalactic supernovae, first in 1934 with a 3.25 inch refracting telescope on the rooftop of the Robinson building at Caltech in Pasadena. Zwicky did not have much luck in the beginning, and was able to find no new examples for the first two years of his work. However, in 1936 he switched his efforts to the newly completed 18 inch Schmidt telescope on Mt. Palomar and found his first supernova in March 1937. During the period up to 1939, he discovered 12 supernovae (oddly enough all of which were the so-called Type I) and firmly established supernova research as a respectable area of astrophysics.

This period of research in the late '30s and early '40s was probably the most productive period of supernova research which has occurred until the end of the '60s when new instruments and new interest in active astronomical phenomena sparked the application of the most modern instruments to the problem. During the early period, the basic types of supernovae were established (Zwicky preferred 5 types labelled, simply, Types I, II, III, IV, and V), their differences in nature from normal novae were determined, their rough absolute magnitudes were estimated, their rough occurrence rates were calculated, their connection to a number of flaring stars recorded in ancient historical chronicles was found, and even the best run of spectra with time on a Type I supernova to be recorded until 1973 was taken on a bright supernova in 1937.

2.2. Light Curves

Now, what do we know about the gross properties of supernovae (SNe)? Modern usage considers only two basic types, simply enough Type I and Type II, and lumps the last 10% of the observed supernovae which do not fit well into these two basic classes as "other." I must warn that life for the student of supernovae is getting more complicated these days, but let me give you the simpler picture and allude to the complications as we go along. So, how are these two classes identified? The two ways, quite naturally enough, are through spectroscopy and photometry. To discuss photometry first, reference to Figure 1 shows that, in the simplest case, the Type I supernova has a sharp decline of about 3 magnitudes during the first 30 days immediately after maximum light and then a relatively sharp exponential decline after that at a rate of ~0.016 mag day^{-1} (Barbon, 1978), while Type II supernovae have a slower decline immediately after maximum light, a pause or "knee" in the decline, and then, again a relatively slower exponential decrease with time (Ciatti, 1978). Let us keep in mind, however, that real life is not quite this simple and is rapidly getting more complicated. Barbon (1978) already points out that the Type I SNe can be divided into two light curve classes of "fast" and "slow" and, recently, a new type of subluminous Type I supernova called

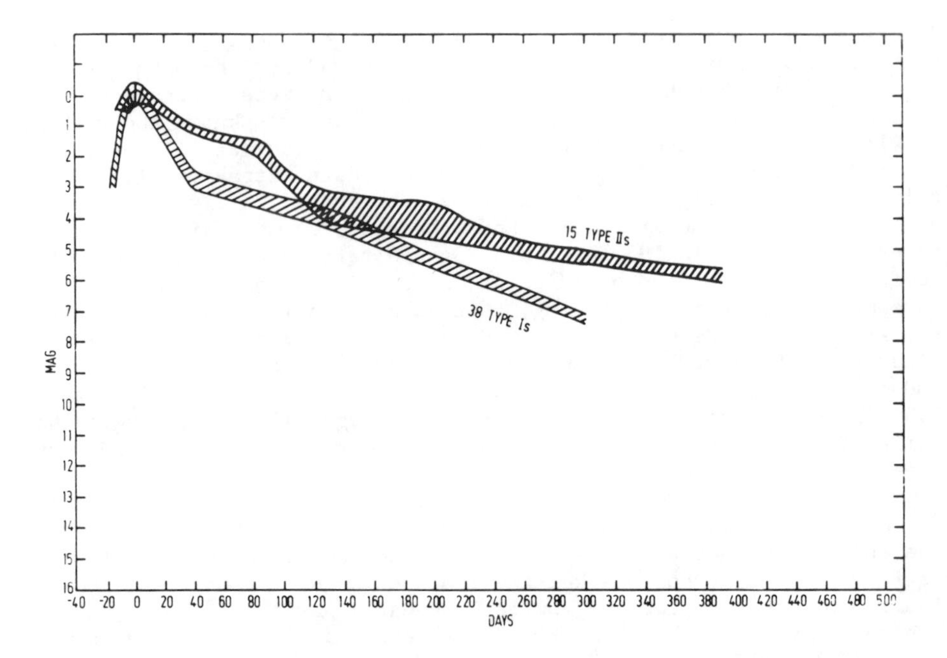

Figure 1: Schematic light curves for the "classical" Types I and II supernovae. The modern situation is far more complicated, however.

Type Ib (the original Type Is now being known as Type Ia) has been identified [see e.g., Panagia, Sramek, and Weiler (1986) and references therein]. Similarly for the Type IIs, Ciatti divides them into the two classes of Type IIP and IIL, where "P" are the "normal" plateau light curves and "L" (linear) are a class of Type II supernovae without the characteristic plateau, having an almost pure exponential decline after the initial sharp drop. This "L" class also has the disconcerting property that its light curves strongly resemble those of Type I supernovae. Thus, with only photometric data such as exists for a number of the historical supernovae, there can be endless arguments as to which Type they likely were.

2.3. Spectra

Because of these problems of identifying supernovae from their light curves, most modern identifications depend on spectrographic information to define the two general classes. It is generally accepted that the two classes of supernovae are separated by the presence or lack of hydrogen lines in their spectra. Type I supernovae show no hydrogen emission while Type II do. As I have already mentioned, for both of the main classes several subclasses also exist.

2.4. Statistics

The question then arises as to what are the statistical properties that we know about supernovae? These are summarized in Table I.

Table I: Supernova Statistics[a]

Average in SNU (1 SNU = 1 SN $(10^{10} L_{\odot})^{-1}$ $(10^2 \text{ yr})^{-1}$

Galaxy	SNe	Type	
Type	(All)	I	II
E	0.22	0.22	0
S0	0.12	0.12	0
S0a, Sa	0.28	0.28	0
Sab, Sb	0.69	0.37	0.32
Sbc, Sc, Scd, Sd	1.38	0.77	0.61
Sdm, Sm, Im	1.02	0.83	0.19
I0	---------- Undetermined ----------		

[a]Tammann (1982)

Several nearby galaxies appear to be rapid producers of supernovae with NGC6946 having had 6, NGC4321 (M100) having had 5 and NGC5236 (M83) having had 5 supernovae detected within the last century. In the particular case of our own Galaxy the Milky Way, if it is something like an Sb, Sbc, or Sc, it is expected to have a supernova rate of ~1.04 SNU which works out to be about 1 supernova every 25 years with roughly half being Type I and half being Type II.

From these extragalactic studies, the average luminosity of supernovae of the two types has been estimated. The best current value for Type I is $M_B = -19.69 \pm 0.13$ for $H_o = 50$ km s^{-1} Mpc^{-1} (Cadonau, Sandage, and Tammann, 1985). In other words, the Type I supernovae are extremely bright objects and appear to be rather good standard candles. In fact, Sandage has recently claimed that they appear to be one of the best cosmological distance indicators known at the present time (Cadonau, Sandage, and Tammann, 1985). Discovery of the previously mentioned subluminous Type Ib supernovae has somewhat complicated this simple picture, but the present evidence is that, while these Type Ib supernovae are ~1.5 to 2 mag fainter than the Type Ia, they still have quite constant absolute magnitudes at maximum light (Panagia, Sramek, and Weiler, 1986). Type II supernovae are a weaker and more diverse class, having absolute magnitudes about 1 magnitude fainter than the Type I with considerably more variation in their absolute magnitudes ($M_B \sim -18.5 \pm 1.5$ for $H_o = 50$ km s^{-1} Mpc^{-1}; Ciatti, 1978 corrected to $H_o = 50$ km s^{-1} Mpc^{-1}).

2.5. Explosion

It is fairly well accepted, since the Type I supernovae are observed in E galaxies (as well as in other types of galaxies) where it is thought that little or no star formation is going on at the present time, that they are the result of the explosion of a low mass star $\leq 1.5\ M_\odot$ which has reached the end of its evolutionary path. This is then consistent with the fact that Type I supernovae show no hydrogen lines in their spectra and, thus, presumably lack it in their outer envelopes. (As mentioned earlier, this simple picture is complicated somewhat by the fact that Type Ib supernovae have so far been found only in spiral galaxies, sometimes near dusty regions, and may be the result of the explosion of very massive stars of 10 to 20 $M_\odot$; see e.g. Wheeler and Levreault, 1985.) On the other hand, since Type II supernovae are found only in late type galaxies and often in association with HII and star forming regions, they are thought to be the result of the explosion of massive stars ($\geq 8\ M_\odot$) which are young and have evolved very quickly. This is also consistent with the detection of hydrogen in their spectra.

The most popular models at the present time for the origin of the observed Type I supernova light curves explain the initial peak and rapid, first 30 day decline as a diffusive release of shock energy in an extended stellar envelope. The exponential decay after that time is thought to be due to the radioactive decay chain

$$Ni^{56}\ (6\ \text{day half-life}) \rightarrow Co^{56}\ (77\ \text{day half-life}) \rightarrow Fe^{56}$$

where the light curve observed is modified by the changing optical depth of the supernova shell to give the measured ~60 day exponential decline. The amount of Ni^{56} needed to accomplish this and to provide the ~10^{51} ergs of energy observed in Type I supernovae is quite large, being ~1 $M_\odot$ and most models predict total stellar disruption. However, even though the models for the light curve are rather successful, there is a theoretical problem to make a star with a mass as low as 1.5 $M_\odot$ explode. Although an explosion can be obtained by taking a much more massive star of ~6 to 8 $M_\odot$ and letting it lose its hydrogen envelope to become a ~2 $M_\odot$ He star before exploding, the most popular models at the present time involve an old white dwarf star in a binary system. When the companion evolves into a giant star and begins to lose mass rapidly, the white dwarf accretes enough of this mass to collapse and explode as a supernova (see, e.g., Iben and Tutukov, 1984).

The light curves for Type II supernovae are less well understood theoretically, although they may also involve the radioactive decay of Ni^{56} into Fe^{56} in the more complicated physical environment in the shells of these much more massive stars. However, the explosion itself is thought to be well explicable in terms of the last stages of evolution of a massive star ($>8\ M_\odot$) in which the stellar core has exhausted its reserves of nuclear fuel. As it begins to contract, the core becomes unstable. The protons convert into neutrons through electron capture which decreases the density of electrons and lessens the degeneracy pressure which was the main support of the stellar core. Also, the breakup of iron into α particles is endothermic and further decreases the thermal energy and pressure in the core. Finally, the neutrinos created by these reactions carry additional energy away from the core, cooling it and leading to an accelerating collapse. The collapse then becomes so rapid that it overshoots and rebounds, causing an outgoing shock wave which, under the proper conditions, can transfer enough energy to the outer layers of the star to accelerate them beyond the escape velocity and blow them off, giving us a supernova. I won't go into the difficulties associated with actually getting this to work on a computer, but the problem has been well investigated by a number of researchers (see, e.g., Brown, 1982; Woosley and Weaver, 1982).

Now, rather than going into the vast amount of information which is available on the observation of SNe in all wavelength ranges, let me go immediately to that which is the newest, (to me) the most interesting, and of the most direct consequence to our goal of studying the e^- component of cosmic rays: that is, the study of the radio emission from supernovae.

2.6. Radio Supernovae (RSNe)

This subject has recently been discussed quite extensively by Weiler, Sramek, and Panagia (1986) and by Weiler et al. (1986), but let me summarize it here briefly.

Even though several tens of the extragalactic supernovae which have been recorded during the last century were examined with the largest and best radio telescopes available (de Bruyn, 1973; Brown and Marscher, 1978; Ulmer et al., 1980; Weiler et al., 1981; Cowan and Branch, 1982) no emission was found before 1980 except for a few weak detections of the supernova SN1970g shortly after maximum light (Gottesman et al., 1972; see also, Allen et al. 1976). (It should be noted that the designations of supernovae are by year of discovery followed by alphabetical lettering of the order of discovery.) In April 1979, however, a bright supernova, designated SN1979c, was discovered in the nearby galaxy M100 (NGC4321) and one year later in April 1980 the VLA detected 6 cm radio emission from it (Weiler and Sramek, 1980; Weiler et al., 1981). The detection of this bright radio emission, subsequent measurements at a number of wavelengths over several years, and the detection and study of other new examples of radio emission from supernovae have established the study of supernovae as an active area of research for radio astronomy. The results give direct evidence for the acceleration of relativistic electrons early in the supernova explosion process. There are now a number of supernovae which have been detected in the radio range, and the study of most of them is still continuing. For convenience, a list of the most recent examples is given in Table II and a detailed discussion of them is given in Weiler et al. 1986.

For brevity, we shall discuss only 2 RSNe here, a Type II, SN1979c already mentioned, and a Type Ib, SN1983n. So far, there has never been any detection of radio emission from a "normal" Type Ia supernova. The radio light curves for these two supernovae have been measured in considerable detail at both 6 and 20 cm wavelengths with the VLA and these are shown in Figures 2 and 3.

Table II: Radio Supernovae

Name	Opt. Type	Optical Max. Date (YY/MM/DD)	Brightness m_B[a]	Brightness M_B[b]	Obs. Radio Max. at 6 cm Age (yrs)	Peak (mJy)	Spectral luminosity (erg s^{-1} Hz^{-1})
SN1950b	II?	50/03	≤14.5	<-15.2	30	0.5	~$3x10^{25}$
SN1957d	II?	57/12	≤15.0	<-15.7	23	1.9	~$1x10^{26}$
SN1970g	II	70/08/01	11.7	-18.2	1.4	~2.5	~$1x10^{26}$
SN1979c	IIL	79/04/19	11.6	-20.0	1.2	8.3	~$2x10^{27}$
SN1980k	IIL	80/10/30	11.6	-18.9	0.4	2.6	~$1x10^{26}$
SN1981k	II?	~81/08/15	<16	<-14	0.5	~2	~$1x10^{26}$
SN1983n	Ib	83/07/17	11.8	-18.5	0.08	18.5	~$1x10^{27}$
SN1984l	Ib	84/08/30	13.9	-18.5	0.14	0.7	~$4x10^{26}$

[a]The apparent optical magnitude in blue light at the time of maximum observed brightness.

[b]The estimated maximum absolute optical blue magnitude after correction for distance and extinction.

In each of these cases, the light curves have been fitted by the relation

$$S(mJy) = K_1 \left(\frac{\nu}{5\ GHz}\right)^{\alpha} \left(\frac{t-t_o}{1\ day}\right)^{\beta} e^{-\tau} \quad (1)$$

where

$$\tau = K_2 \left(\frac{\nu}{5\ GHz}\right)^{-2.1} \left(\frac{t-t_o}{1\ day}\right)^{\delta} \quad (2)$$

This formulation assumes that the change of flux density S and of optical depth τ with time after the explosion date t_o are described by power law functions of the supernova age ($t-t_o$) with powers β and δ; that the change of τ with frequency ν is due to pure, external thermal absorption in an ionized medium with frequency dependence $\nu^{-2.1}$; and that the intrinsic radio emission is produced by the nonthermal synchrotron mechanism with an optically thin spectral index α. K_1 and K_2 are two scaling parameters for the units of choice: millijanskys, gigahertz, and days. The best fit values for the parameters are as follows (Weiler *et al.* 1986). SN1979c: $K_1 = 9.3 \times 10^2$, $\alpha = -0.72$, $\beta = -0.71$, $K_2 = 5.1 \times 10^7$, $\delta = -3.01$; SN1983n: $K_1 = 4.4 \times 10^3$, $\alpha = -1.03$, $\beta = -1.59$, $K_2 = 5.3 \times 10^2$, $\delta = -2.44$. These curves are also shown as the solid lines in Figures 2 and 3.

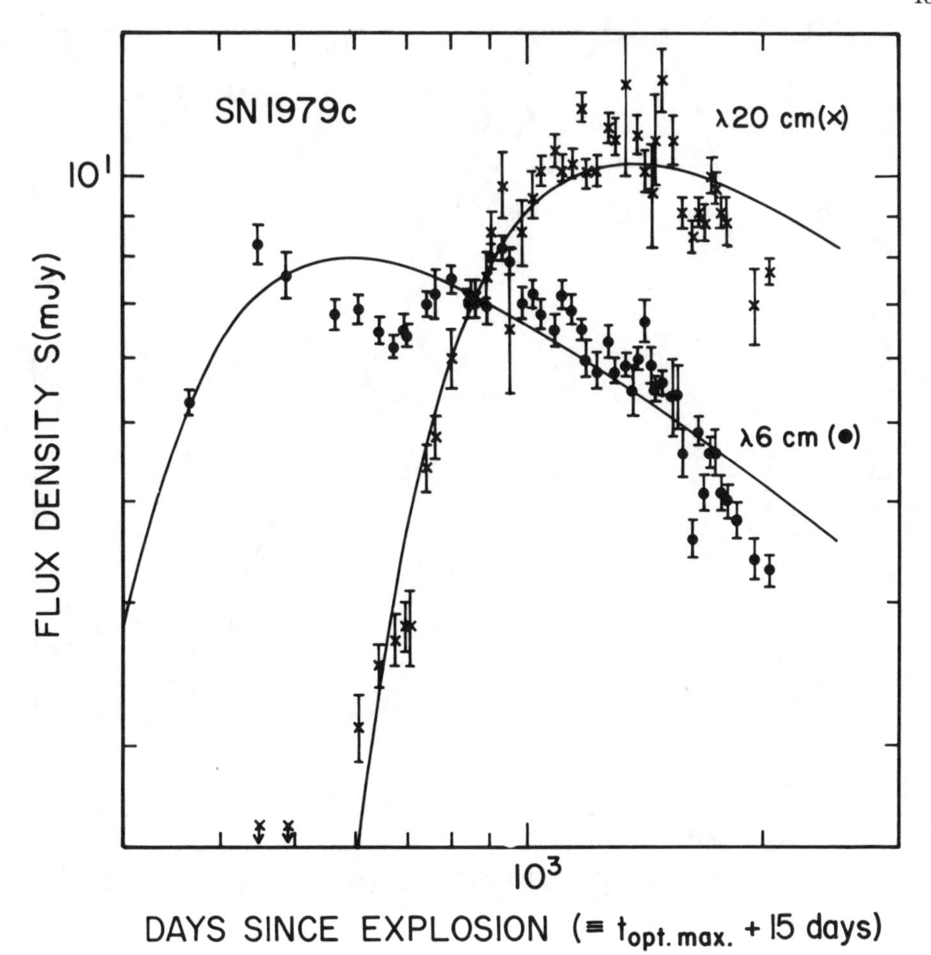

Figure 2: Radio light curves for the Type II supernova SN1979c in NGC4321 (M100). The age of the supernova is measured in days from the estimated date of explosion on 4 April 1979, 15 days before the date of maximum optical light.

The rising, optically thick part of the light curves is almost certainly due to absorption by a thermal ionized gas, which probably originated as mass lost from the presupernova system in a relatively high density, low velocity stellar wind, external to the radio emitting regions. By adopting a reasonable velocity (10 km s^{-1}) and temperature (10^4 K) for this wind and assuming that it is of normal chemical abundance and is fully ionized, it is possible to calculate the time

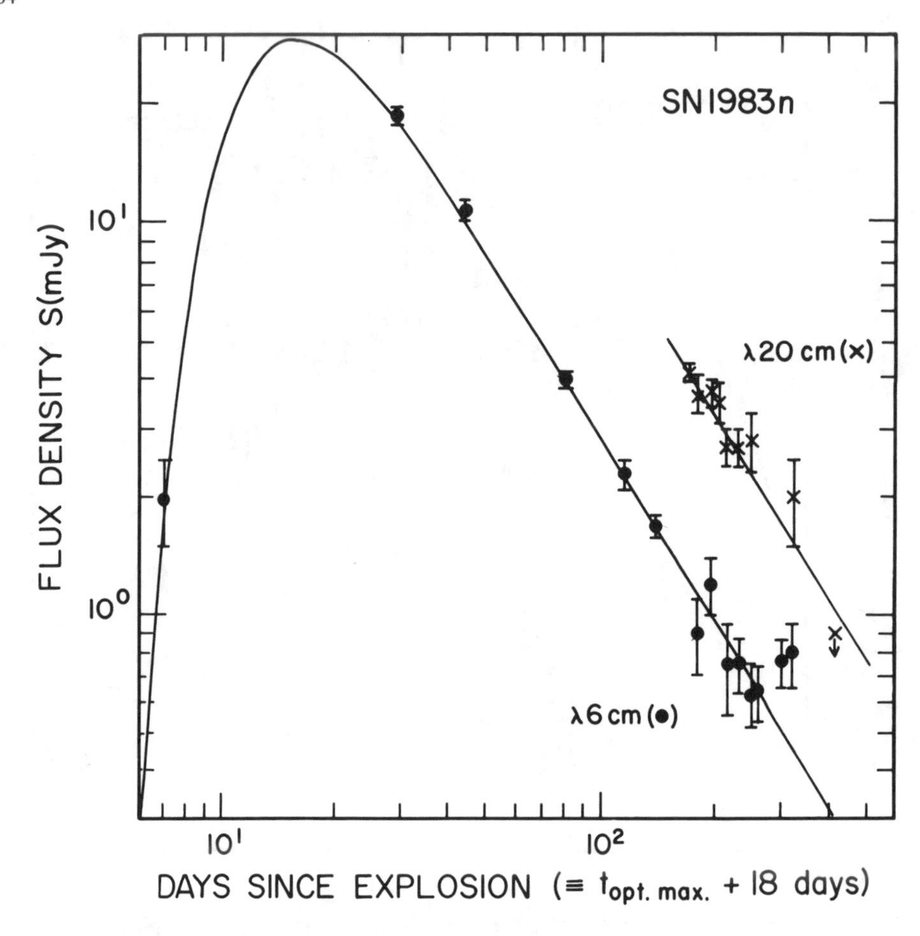

Figure 3: Radio light curves for the Type Ib supernova SN1983n in NGC5236 (M83). The age of the supernova is measured in days from the estimated date of explosion on 29 June 1983, 18 days before the date of maximum optical light.

dependence of the optical depth effects for comparison with the fitting parameter δ in Equation 2. Further, the absolute radius of the supernova ejecta can be calculated from the velocity measured for the optical emission lines during the early phase of the explosion and its evolution with time. All of this information can then be combined to estimate the rate of mass loss from the pre-supernova stellar system. The values thus obtained are all quite high ranging from about 2×10^{-6} $M_{\odot}$ yr^{-1} for SN1983n to about 5×10^{-5} $M_{\odot}$ yr^{-1} for SN1979c.

The optically thin, non-thermal radio emission has two suggested mechanisms for accelerating the needed relativistic electrons: (i) shock acceleration in a region external to the supernova photosphere and (ii) pulsar acceleration by the supernova stellar remnant. For simplicity, let me call these: (i) mini-shell, from the resemblance of the shock model to that seen in shell-type supernova remnants (SNR) such as Tycho's supernova (SN1572, 3C10) and (ii) mini-plerion, from the resemblance of the pulsar acceleration model to that found in plerionic supernova remnants such as the Crab Nebula (SN1054). (These two types of SNR will be discussed later.)

The mini-shell model involves the external generation of relativistic particles and magnetic field by the shock wave of the supernova explosion interacting with a high-density gas envelope surrounding the supernova system. This high-density material is presumably the same as that which provided the initially observed absorption; that is, the mass lost from the stellar system in a wind preceding the supernova explosion. This model has mainly been explored by Chevalier (1981a, b, 1984), who predicts an optically thin ($\tau = 0$) flux density dependence of

$$S \alpha \nu^{\alpha} t^{\beta} = \nu^{(1-\gamma)/2} t^{-(\gamma+5-6m)/2} \quad \text{(mini-shell)} \qquad (3)$$

where γ is the power law dependence of the relativistic electron injection spectrum and m is a model parameter ($0 \leq m \leq 1$) related to the time dependence of the radius R of the supernova shock wave ($R \alpha t^{m}$) and dependent upon the amount of deceleration experienced.

The mini-plerion model involves a central source of generation of the relativistic electrons and magnetic fields, presumably by the remnant of the supernova progenitor star which has become something like a rapidly spinning pulsar or a black hole. This model has been explored by several investigators (Pacini and Salvati, 1973, 1981; Shklovskii, 1981; Bandiera, Pacini, and Salvati, 1983, 1984) with Pacini and Salvati (1973, 1981) predicting

$$S \alpha \nu^{\alpha} t^{\beta} = \nu^{(1-\gamma)/2} t^{\alpha+3(1-\alpha)(1-m)/2} \qquad (4)$$

$$\text{(mini-plerion; } \nu < \nu_b\text{)}$$

where ν_b is a discontinuity or critical break frequency at which adiabatic losses and synchrotron losses become equal.

Although I will not go into details here (see, e.g., Weiler _et al._, 1986), the observational evidence presently available shows a preference for the mini-shell models for describing this early phase of the development of the radio emission from both Types I and II supernovae. However, it must be emphasized that the centrally driven mini-plerion models can give, with the right choice of parameters, exactly the same predictions so that they cannot yet be ruled out. In particular, several "middle aged" supernovae (ages >10 and <100 years) have now been discovered (Cowan and Branch, 1985) and these are conceivably driven by the mini-plerion mechanism. Thus, although not yet proven or even supported by very strong evidence, it is possible that the radio supernovae generate their initial radio emission through shock interaction with the surrounding circumstellar material in the first few (<10 years) of their lives, and then some (small?) fraction of them which contain active stellar remnants like pulsars can either continue to be or can again become radio sources at ages (>10 and <100 years) where little radio emission from supernovae has so far been found.

To summarize, it appears that at the time of or shortly after the initial explosion, the shock wave from the supernova interacts with the high density "cocoon" of matter lost by the stellar system during the last stages of evolution before the explosion and gives rise to intense radio emission which appears to fade within a few years. This decline can be attributed to the supernova shock running out of the high density stellar wind region. A central pulsar, if it exists, may play a role in this, but that is less certain. So, at an age of ≥10 years, no more detectable radio emission is seen from the supernova (with the few exceptions mentioned above) until the interaction of the supernova shock wave with the interstellar medium begins to become important and we enter the range of supernova remnants at age >100 years. The plerions and mini-plerion RSNe may provide some exceptions to this simplified picture, but the evidence is still accumulating.

Although the statistics on RSN are too poor to draw any conclusions on their relative types or numbers at the present time, we shall arbitrarily assume that the estimated rate of one SN per ~25 years, with half being Type I and half being Type II, is roughly correct and that normal Type I (Type Ia) supernovae do not give rise to significant acceleration of relativistic electrons (radio emission) while the subclass Type Ib supernovae do. Type II SNe may always produce significant radio emission.

3. SUPERNOVA REMNANTS

For the supernova remnants, as for the supernovae, I will mainly discuss their radio properties since this is where we directly see the relativistic electrons which are our main concern for the e^- component of cosmic rays.

As was alluded to in the above discussion, supernova remnants (SNR) can be divided into at least two classes (Weiler and Seielstad, 1971; Weiler and Panagia, 1978): the shell-type remnants which show an obvious shell-like structure with no sign of a central energy source [e.g., the remnant of Tycho's supernova (SN1572), 3C10; the remnant of Kepler's supernova (SN1604), 3C358; Cassiopeia A (SN~1670), 3C461; etc.] and the plerionic supernova remnants which often show no sign of a shell and have strong evidence for the existence of a central energy source with their distinct filled center, blob-like form [e.g., the Crab Nebula, SN1054; 3C58, SN1181; etc.]. It should be mentioned that the true picture is not quite this simple since Weiler and Panagia (1980) pointed out that there also exists a class of "composite" supernova remnants which combine some of the properties of both the plerionic and shell types. However, this is a complication which we will ignore here for simplicity (for a recent review, see Weiler, 1985). Let us first briefly discuss what is known about the shell-type SNR.

3.1. Shell-type Supernova Remnants

After age ~100 years, the interaction of the supernova shock wave sweeping up the interstellar medium begins to become important and it is possible for the supernova to again become a prominent radio emitter as a shell-type supernova remnant. This theory has been well developed by Gull (1973a, b) and other workers referenced therein. The evolution of the shell-type supernova remnants are generally described by four phases (Woltjer, 1972):

1. The sweeping phase where the swept up interstellar mass is much less than the ejected mass in the supernova envelope.

2. The adiabatic phase where the swept up matter now dominates but the radiative losses are negligible and energy is conserved. This phase is also known as the Sedov (1959) phase.

3. The isothermal phase where the radiative losses have now become important and the shock cools quickly.

4. The dispersal phase where the shock velocity has dropped to that of the random motions of the interstellar medium (~10 km s^{-1}) and the remnant begins to break up.

Although there has been some disagreement, it is generally accepted that most of the supernova remnants which we can study are in the Sedov phase of evolution with the young historical remnants like Kepler (SN1604), Tycho (SN1572) and Cassiopeia A (SN~1670) in between Phases 1 and 2 and the oldest remnants like the Cygnus Loop in Phase 3. There is also discussion as to whether the large, faint galactic background features such as the North Polar Spur might be very old and near supernova remnants in between Phases 3 and 4 (Berkhuijsen, Haslam, and Salter, 1971) and it has been shown that such spurs contribute a significant fraction of the galactic background emission (Berkhuijsen, 1971).

An example of a well formed Phase 1 - 2 shell-type remnant (3C10) is shown in Figure 4 and of a fading, distorted Phase 3 shell-type remnant (Cygnus Loop) in Figure 5.

3.2. Plerionic Supernova Remnants

The plerionic SNR appear to be primarily driven by a central energy source like a pulsar formed from the central core of the supernova stellar remnant rather than by the shock energy from the supernova explosion interacting with the interstellar medium as was the case with the shell-type remnants. The prototype of the plerions is, of course, the Crab Nebula (SN1054) where a central driving pulsar is definitely identified and which shows the characteristic blob shape at all

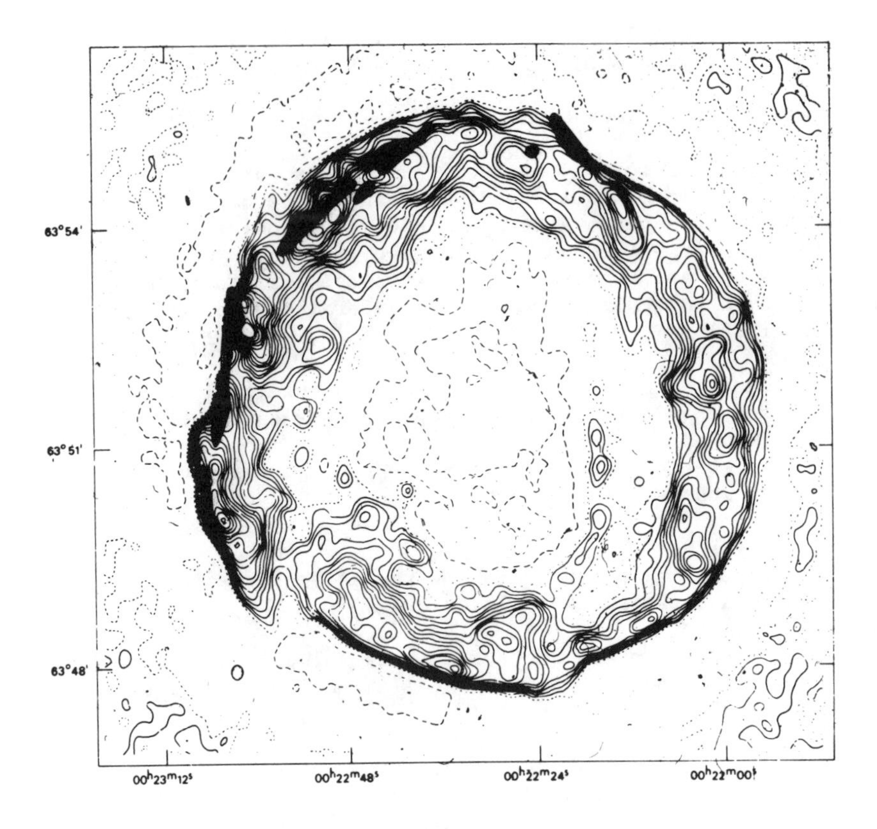

Figure 4: Contour map of the total intensity distribution of 3C10, the remnant of Tycho's supernova (SN1572) at 21 cm wavelength (Strom and Duin, 1973).

wavelengths. Another plerion which has been well studied is 3C58 (SN1181), illustrated in Figure 6. It has the characteristic centrally filled radio form, a flat radio spectrum (also a characteristic of the class), and, although it has not been proven to contain a pulsar, contains centrally condensed x-ray emission which indicates a continuing acceleration process.

The evolution of plerions, as might be expected, takes place differently than that of the shell-type SNR. In general, however, it is expected that plerions have a complex evolution and are visible in the radio range for a shorter period of time (~30,000 years) than the shell-type remnants (~100,000 years) (see, e.g., Weiler and Panagia, 1980; Reynolds and Chevalier, 1984).

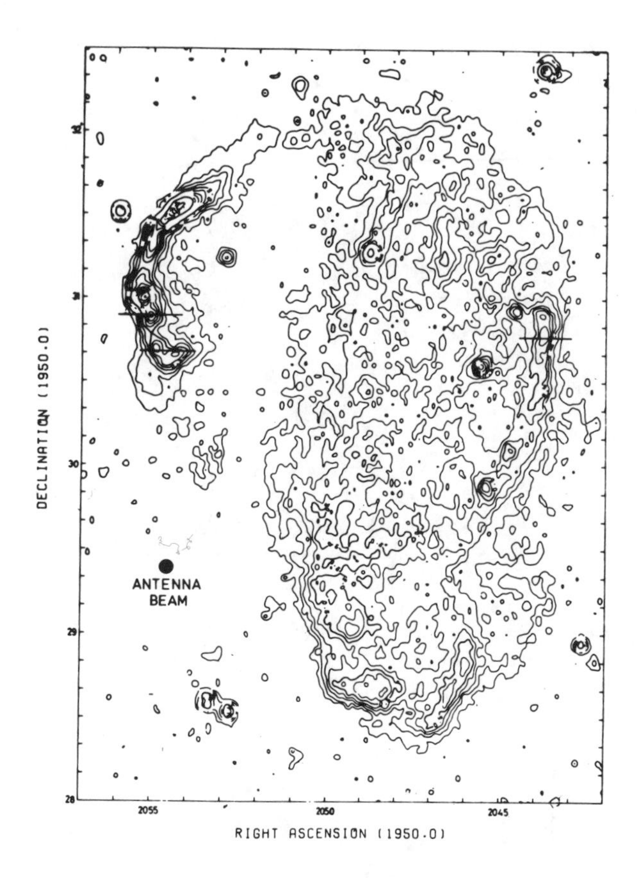

Figure 5: Contour map of the total intensity distribution of the Cygnus Loop at 11 cm wavelength (Keen et al., 1973)

3.3. Composite Supernova Remnants

Finally, as was alluded to above, there exists a class of supernova remnants which appear to exhibit the properties of both the shell-type and plerionic SNR, the composite SNR. I will not dwell on these since they appear, in first approximation, to be describable as a simple combination of the above two classes. However, as an example, Figure 7 shows the radio map of G327.4+0.4 where the central condensation appears to have the flat radio spectrum and blob-like form characteristic of the plerions while the surrounding emission has the steep radio spectrum and partial shell-like form characteristic of shell-type SNR.

The broad conclusions of relevance to the study of cosmic rays is that we see in the supernova remnants not only a transition phase from the supernovae themselves to the broad non-thermal galactic background

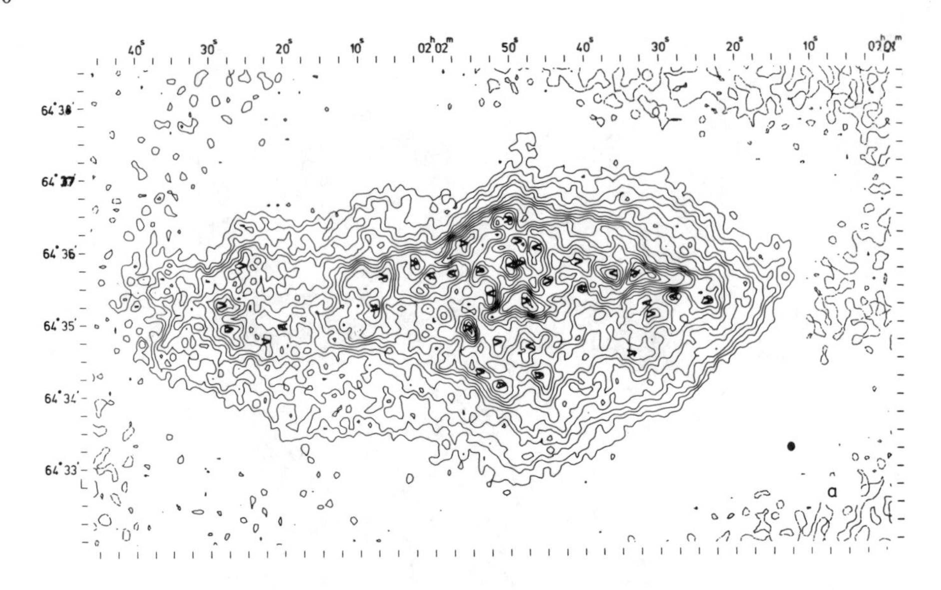

Figure 6: Contour map of the total intensity distribution of 3C58 at 6 cm wavelength (Wilson and Weiler, 1976).

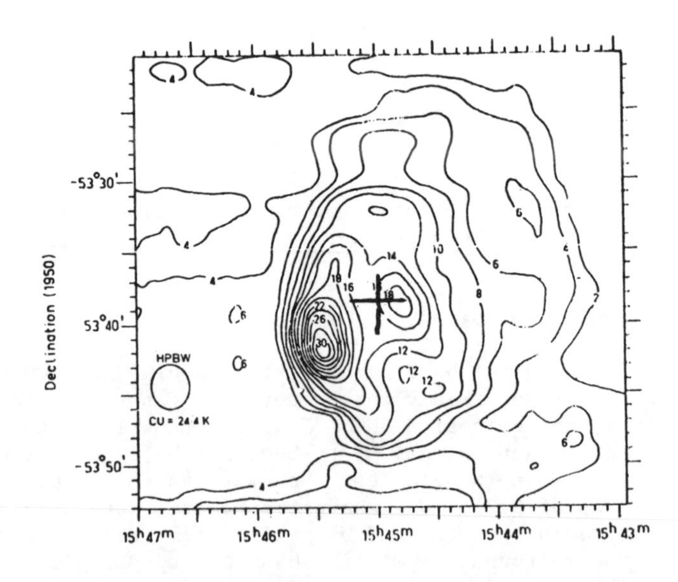

Figure 7: Contour map of the total intensity distribution of G327.4+0.4 at 75 cm wavelength (Caswell, Clark, and Crawford, 1975)

radio emission but also we see clear evidence of two possible acceleration mechanisms for relativistic electrons, pulsar acceleration and shock acceleration. These are, not by chance, the same two mechanisms which are suspected as being the origin of the relativistic electrons in the radio supernovae. However, in the remnants we have direct evidence that both mechanisms work well.

One additional point should be discussed before going on to the analysis of the galactic background radiation, the statistics of the relative numbers of the two types of SNR. This is a very difficult problem since a great deal of conjecture is needed to arrive at an answer. First, there are "not enough" SNR of all types to match the expected supernova rate for our Galaxy of 1 SN per ~25 years. The best numbers for SNR indicate 1 SN leaving an identifiable remnant every ~150 years (Clark and Caswell, 1976). Second, there appear to be too few plerionic SNR, including those found in the composite SNR. One supernova resulting in a plerion every ~240 years appears to be sufficient to produce the number found (Srinivasan, Battacharya, and Dwarakanath, 1984). Then, arguments rage as to whether plerions are indeed shorter lived than normal SNR (Panagia and Weiler, 1980), or are simply not being found because of selection effects. (With their flat spectrum and filled center form, they tend to look like HII regions.) We cannot solve these problems here, but for estimation purposes for cosmic rays I will simply postulate that the estimated supernova rate for our Galaxy of 1 per ~25 years is approximately correct and that roughly half of these form shell-type remnants and half form plerions.

4. NON-THERMAL GALACTIC BACKGROUND RADIATION

If the results of Berkhuijsen (1971) discussed above are correct, and there is no reason to doubt them, then the old shell-type SNR gradually become larger and fainter until they are distinguishable only as slight perturbations on the galactic background. So on something like $\sim 10^5$ years they fade completely into that background, and, of course, contribute to its existence. The ultimate fate of plerions is less well understood. It is very likely that, with their intrinsic blob-like morphology and their probably shorter visible lifetimes, they are even more rapidly absorbed into the general non-thermal galactic background.

The broad, non-thermal galactic background radio radiation was, of course, the first discovery and the reason for the existence of a field called radio astronomy. Karl Jansky detected the most intense part of this emission, that from the galactic center, during his search for the sources of communications static at a wavelength of 14.6 meters in 1932 (Jansky 1932, 1933a, b, c). Also, until the post-WWII period when better instrumentation built for the war became available, when more interest by professional astronomers developed, and when a better economic climate than the Depression arrived, study of the galactic background radiation constituted one of the few areas of work in radio astronomy. The most significant results of this period were, in fact, obtained by Grote Reber, a radio amateur who built, with his own funds, a 10 meter parabolic reflector in his back yard in Wheaton, Illinois in 1937 and produced an accurate map of the distributed galactic radio emission at 2 meters wavelength (see, e.g., Sullivan, 1984).

Although the main efforts of radio astronomy are these days concentrated at cm-mm wavelengths and on high resolution studies of individual radio sources, mapping of the galactic background is still carried out by a few workers. Such surveys remain important, of course, both for finding discrete sources and for studying the distribution and properties of cosmic ray electrons (see, e.g., Haslam _et al._, 1974). A recent map of the galactic background radio radiation is shown in Figure 8.

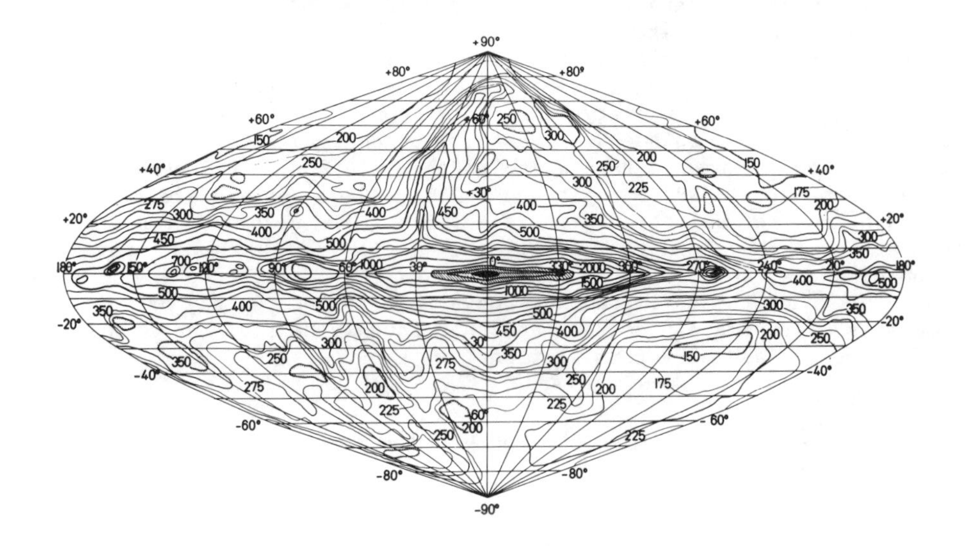

Figure 8: Radio emission in the Galaxy at 150 MHz plotted in galactic coordinates (From Landecker and Wielebinski, 1970).

The study of this radiation and its relevance for cosmic ray e^- work is well discussed by Longair (1981). He points out that it has a relatively flat radio spectrum at low frequencies (<200 MHz) of $\alpha \sim 0.4$ ($S \propto \nu^{-\alpha}$), has a gradual "break" in the spectrum between ~200 to ~400 MHz, and steepens to a spectral index of $\alpha \sim 0.8$ to 0.9 at higher frequencies (see Figure 9).

The study of the galactic radio background not only provides us with direct proof that the e^- which we can measure at the top of the Earth's atmosphere are a general phenomenon, but it also allows us to investigate the large scale distribution of cosmic ray e^- and magnetic fields throughout the Milky Way. It has also led to a long running argument as to whether our Galaxy possesses a "halo" or not. I will not enter into this fray except to express the view that the best present evidence does indicate the presence of a "thick disk," at least for the relativistic

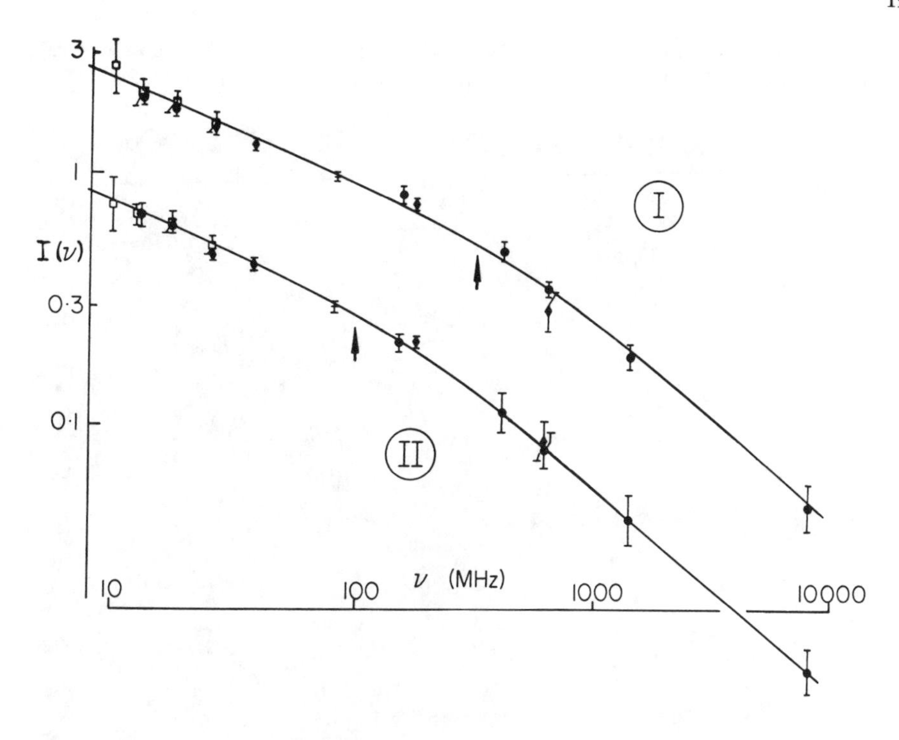

Figure 9: The radio spectrum of the galactic, non-thermal background emission in two regions of the Galaxy: (I) the anticenter region at high galactic latitudes and (II) the inter-arm regions (from Webster, 1971; see also, Longair, 1981).

electrons and magnetic fields responsible for the galactic background emission, with rough dimensions of ~9 kpc radius in the galactic plane and thickness ~1 kpc perpendicular to the plane (see, e.g. Baldwin 1967, 1976), further imbedded in a large spheroidal halo with semi-major axis ~10 kpc in the plane and semi-minor axis ~3 to 4 kpc perpendicular to the plane (Longair, 1981). Also, there is good evidence that a further concentration of particles and fields occurs in the galactic spiral arms. The best way of viewing this is, of course, by looking at other galaxies similar to our own where we are not so confused by sitting inside of what we are trying to study. For example, the spiral galaxy NGC891 (Figure 10) shows just such a radio halo and the galaxy M51 (Figure 11) shows enhanced, non-thermal emission from its spiral arms. We will see that the existence of a large scale halo for the Milky Way is of great assistance in providing confinement of the galactic e^- cosmic rays.

5. THE ELECTRONIC COMPONENT OF COSMIC RAYS

Finally, we reach our goal of discussing the actual electronic component of cosmic rays which I mentioned in the beginning was our aim. The

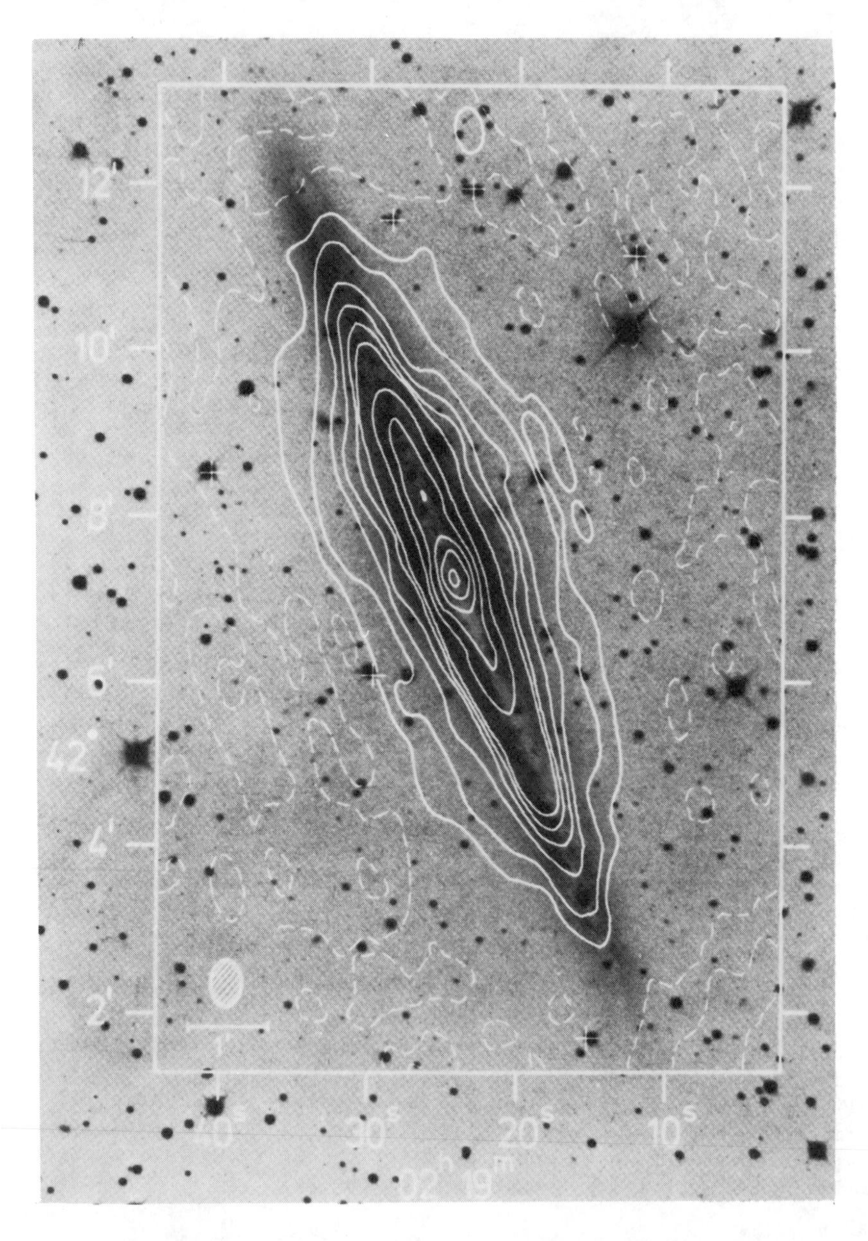

Figure 10: The distribution of non-thermal radio emission from the edge-on spiral galaxy NGC891 showing its distinct halo (from Allen, Baldwin, and Sancisi, 1978; see also Longair, 1981).

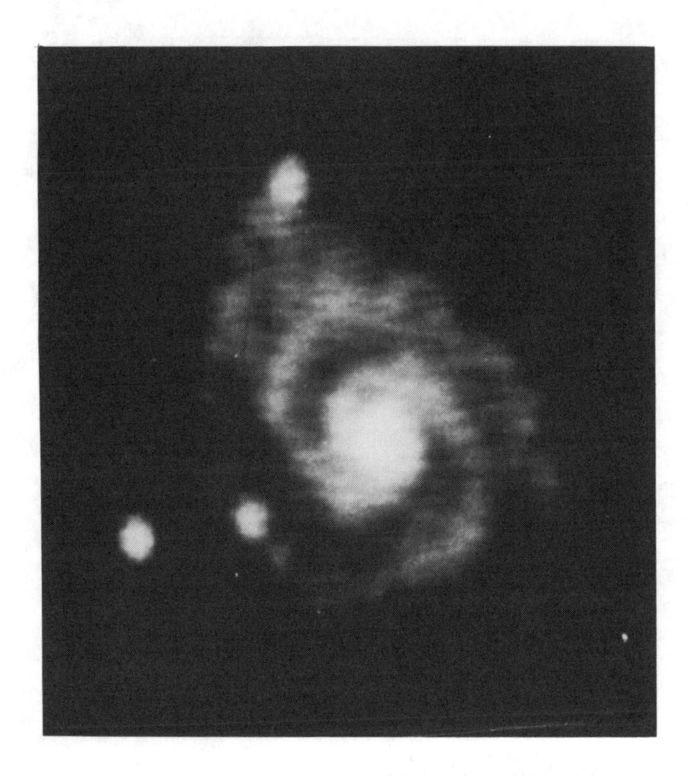

Figure 11: The radio emission distribution at 21 cm wavelength from the almost face-on spiral galaxy M51 (from Mathewson, van der Kruit, and Brouw, 1972; see also Longair, 1981).

energy spectrum of this e^- component measured in the vicinity of the Earth is shown in Figure 12. This is what cosmic ray physicists actually measure, and now we have to see if we can connect it back to the various astrophysical processes which we have been discussing at such length.

5.1. Connection to the Galactic Background Radio Emission

The start at least is promising and relatively concrete. We will become more inexact and speculative as we proceed, however. The energy spectrum of the locally measured cosmic ray e^- between about 4 and 40 GeV is well determined and has been found to be $\gamma = 2.66$. If these are the same electrons which are giving us the observed non-thermal galactic background radio emission discussed above, then synchrotron theory, which I will not go into in detail, shows us that the observed radio emission should have a spectrum of (see, e.g., Longair, 1981)

$$I = 1.7 \times 10^{-21} \; a(\gamma) V \omega B^{(\gamma+1)/2} \left(\frac{6.26 \times 10^{18}}{\nu}\right)^{(\gamma-1)/2} \tag{5}$$

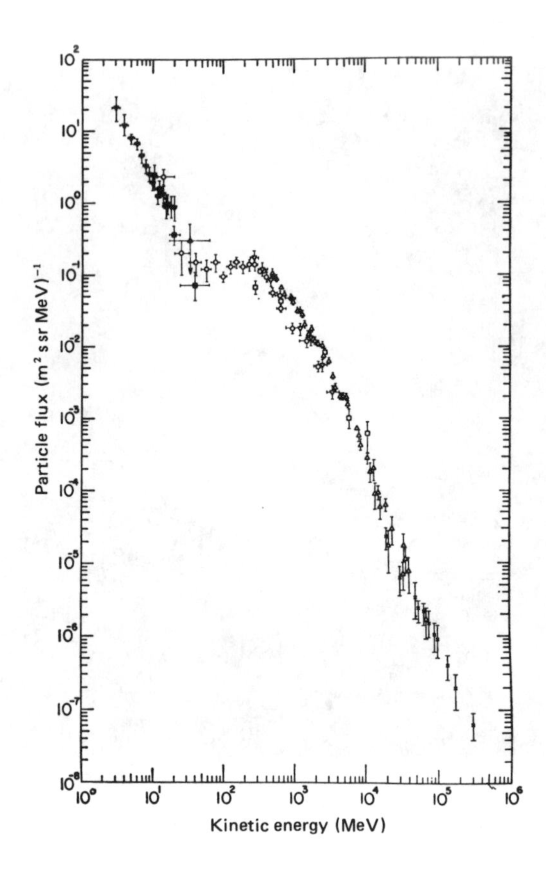

Figure 12: The energy spectrum of cosmic ray electrons measured at the top of the atmosphere (from Meyer, 1969; see also, Longair, 1981).

where I is the source intensity in ergs s^{-1} Hz^{-1}, $a(\gamma)$ is a known and tabulated function, V is the volume of the source region, B is the magnetic field strength in Gauss, γ is the power law index of the electron energy spectrum, and ω is the proportionality constant of the electron density power law distribution (i.e., $N(E)dE = \omega E^{-\gamma}dE$, where N(E)dE is the number of particles per unit volume). Keeping in mind that the radio spectral index α which we measure for radio sources has the form

$$S \propto \nu^{-\alpha} \tag{6}$$

we find from comparison of Equation 6 with Equation 5 that

$$\alpha = (\gamma-1)/2 \ . \tag{7}$$

Since, at least at these energies, the cosmic ray e^- spectrum has γ = 2.66, this implies that the spectral index of the galactic background radio emission should be $\alpha = 0.83$. This, as has been discussed above, is in excellent agreement with what is measured in the appropriate frequency range (see Figure 9). Thus, we can feel confident that, at the higher energies where solar modulation of the cosmic ray flux is felt to be unimportant, the relativistic electron spectrum which we observe at the top of the atmosphere is representative of the spectrum of the electrons which fill up interstellar space and provide the non-thermal galactic background radio emission. As Longair (1981) points out, it would be nice if we could use this correspondence and Equation 5 to measure the strength of the galactic magnetic field, but all we can solve for is $\omega B^{1+\alpha}$, and we do not know the value of the constant ω independently. However, Figure 9 also shows that there is a "break" in the radio spectrum of the galactic background at ~300 MHz, and Figure 12 shows a "break" in the cosmic ray electron spectrum at ~3 GeV. If these two breaks can be compared, a galactic magnetic field strength of $\sim 2 \times 10^{-6}$ Gauss can be estimated, which is thought to be about the right magnitude. However, it is also known that the effect of solar modulation of the cosmic ray flux is becoming important for the e^- component at just about these energies (1 to 3 GeV) so that we cannot really trust the result. In fact, the problem is often reversed with the interstellar magnetic field strength estimated from other sources and used to estimate the effects of solar modulation on the cosmic ray flux.

Nevertheless, let us be content. At least we have been able to show that high energy cosmic ray e^- are almost certainly the same ones which we observe in the radio. Therefore, we can study the distribution, energy spectrum, and density of high energy cosmic ray e^- in our own and other galaxies with the relative ease of radio astronomy techniques.

5.2. Origin (General)

Before we proceed to the more speculative area of trying to specify the sites of cosmic ray e^- acceleration, let me address the more general question of whether we need search the entire universe, or only our own Galaxy. Fortunately, for the case of the e^- the answer is almost certainly that we need look no further than the Milky Way. Longair (1981) points out that due to inverse Compton scattering on the omnipresent photons of the universal 3 K background radiation, a relativistic electron cannot survive more than about

$$\tau = \frac{2.8 \times 10^{12}}{1.1\xi} \text{ years} \tag{8}$$

where ξ is the relativistic Lorentz factor $[\xi = 1/\surd(1-v^2/c^2)]$. Thus, a 100 GeV ($\xi \sim 2 \times 10^5$) cosmic ray electron, even if it could travel in a straight line (which is unlikely since an intergalactic magnetic field almost certainly exists at some level), cannot propagate more than a few (~4) Mpc to reach us from extragalactic sources. A galactic cosmic ray e^- component due to the strong extragalactic radio sources is therefore unlikely. Also, even the local supercluster is too large for an extragalactic source to contribute significantly to the measured cosmic ray e^- flux. Thus, if we wish to look for sources of cosmic ray electrons, at least we need search no further than the Milky Way.

5.3. Lifetime and Injection Spectrum

Since the galactic magnetic field is known to be very inhomogeneous, it is reasonable to assume that cosmic ray e^- diffuse (See also Ginzburg and Syrovatskii, 1964, for a more detailed treatment.) through the interstellar medium to us rather than in straight lines or even in regular arcs through the galactic magnetic field. This is also consistent since diffusion can provide the high degree of isotropy observed for cosmic rays with no obvious sources indicated. Longair (1981) discusses this in much more detail and points out that the relevant loss processes can be expressed as an additive series with a term logarithmic in the energy for ionization losses, a term linear in the energy for the bremsstrahlung and adiabatic losses, and a term varying as the square of the energy for the synchrotron and inverse Compton losses. This means that the observed electron energy spectrum will differ from the true injection spectrum by:

1. one power of E flatter if ionization losses dominate,
2. unchanged if bremsstrahlung or adiabatic losses dominate, and
3. one power of E steeper if inverse Compton or synchrotron losses dominate.

What we want, of course, is to obtain the true injection spectrum of the cosmic ray e^- in order to better determine their origin. It appears that a definite conclusion is difficult at present (see, e.g., Longair, 1981), but if our local measurements of the energy spectrum between about 4 and 40 GeV are indicative of the injection spectrum, then the injection spectrum has $\gamma \sim 2.7$. This is consistent with other estimates of the cosmic ray injection spectrum.

Cosmic ray e^- measurements at higher energies (>40 GeV) show a detectable steepening of the spectral index which implies the presence of inverse Compton losses to the 3° K background radiation and synchrotron losses. An age estimate can thus be obtained of $\sim 10^7$ years (see, e.g., Prince, 1979; Tang, 1984).

5.4. Origin (Specific, Old)

Like most things in the universe which we cannot explain, and as you might guess from the way I have approached the problem, the best loved origin for cosmic rays of all types, and of cosmic ray e- in particular, is supernovae and their products, supernova remnants, pulsars, x-ray binaries, and, if you believe in them, black holes.

Let us first see if there is even enough energy available in supernovae to produce the cosmic rays which are detected in the vicinity of the Earth. Looking at a young supernova remnant such as Cassiopeia A, we need to estimate how much energy it has available in the form of relativistic particles and fields. They are tightly coupled and to produce the observed synchrotron radiation, we can either have many relativistic electrons in weak fields or fewer e^- in strong fields. However, even though we cannot separate the two cases directly, there is a minimum <u>total</u> energy requirement for a remnant if we have approximately the same energy in particles as in fields. This is the equipartition assumption which is always used. Not only does it allow us to obtain a solution, but it seems logical that systems will generally seek the state of lowest total energy. Longair (1981) shows that the minimum total energy for a supernova remnant can be expressed as

$$W_{min} \approx 6 \times 10^{41} \nu_8^{2/7} L_\nu^{4/7} R^{9/7} \eta^{4/7} \text{ ergs} \tag{9}$$

where ν_8 is the lowest frequency at which the source is observed in units of 10^8 Hz, L_ν is the power radiated at ν_8 Hz in W Hz^{-1} sr^{-1}, R is the source radius in kiloparsecs, and $\eta = (1 + \beta)$ where β is the ratio of the energy of cosmic ray protons to cosmic ray electrons in the source (often taken to be 100). Putting in the numbers for Cas A, we arrive at $W_{min} \sim 10^{49}$ ergs, if we only count the electrons we actually see ($\beta = 0$, $\eta = 1$), and $\sim 10^{50}$ ergs if we take $\beta = 100$. This is a lot of energy and, since the kinetic energy of a supernova is generally estimated to be 1 to 2 x 10^{51} ergs, shows that SNe are very efficient accelerators of relativistic cosmic rays.

Now, are there enough supernova remnants to give us all of the observed cosmic rays? If there are no losses of the cosmic ray content of the Galaxy except escape, then the energy density of cosmic rays in the Galaxy in terms of a SNR origin can be written (Longair, 1981) as

$$\epsilon_{CR} = \frac{t_c}{t_{SN}} \frac{E_0}{V} \tag{10}$$

where t_{SN} is the average time interval between galactic supernovae, V is the volume of the Galaxy within which cosmic rays are confined, t_c is the characteristic escape time of the cosmic rays from the confining volume V, and E_0 is the average energy release in cosmic rays per supernova. If we adopt a confinement volume of a disk with radius 10 kpc and thickness 700 pc, a supernova rate of 1/25 years, an energy per supernova of $\sim 10^{50}$ ergs, and a cosmic ray age of $t_c \sim 10^7$ years, we would obtain an average energy density of cosmic rays from supernova remnants in the Galaxy of ~5 eV cm^{-3}. Since this greatly exceeds the observed local cosmic ray energy density of ~1 eV cm^{-3}, the energy in relativistic particles which we can observe directly and postulate in supernova remnants <u>could</u> be the source of all cosmic rays. This, of course does not necessarily prove that it is.

At any rate, we have a possible source of cosmic ray e^- and perhaps all cosmic rays. Note that in this estimation we have basically lumped in all types of supernova remnants and have not distinguished between shell-types with their shock acceleration and plerions with their pulsar acceleration. In any complete model, such differentiation should be made. It should also be noted that we have not mentioned that the radio spectral indices for SNR are on the average flatter (for shells $\alpha \sim 0.45$; for plerions $\alpha \sim 0.2$), implying power law injection indices which are much flatter (for shells $\gamma \sim 1.9$; for plerions $\gamma \sim 1.4$) than those actually observed for cosmic rays ($\gamma_{CR} \sim 2.7$). Only Cas A, which has an unusually steep radio spectrum ($\alpha \sim 0.75$), comes close to giving us the desired result ($\gamma \sim 2.5$). So, we have some good suggestions, but, as yet, no proof of the origin of cosmic rays.

Things could be even worse! Although we appear to have more than enough energy in SNR to make the required number of cosmic rays, the question remains as to whether we can get those particles out of the SNR and into the general galactic volume without severe losses. Our supernova remnants are expanding with time and therefore "cooling" due to adiabatic energy losses ($E \propto r^{-1}$), which could quench their cosmic ray content before it can be released into the interstellar medium. For example, an SNR with initial total energy W_0 at radius r_0 will only have energy $W = (r_0/r)W_0$ by radius r as it expands. Thus, if r_0 is small the remnant will quickly run out of steam. This is a bit of a red herring, however, since observations show that shell-type remnants must have continuing particle acceleration and magnetic field amplification to maintain their observed intensities, and plerions obviously have a continuing energy supply from the conversion of pulsar rotational energy into fields and relativistic particles. Thus, there appear to be very reasonable ways in which an SNR can defeat the adiabatic energy loss problem.

5.5. Origin (Specific, New)

Above I have given what might be called the "classical" arguments for the acceleration of cosmic rays in supernovae where, because supernovae were "known" to not be strong radio emitters, all of the acceleration was assumed to take place in the visible radio remnant. It more than works, so it can't be all bad! However, we now know that some types of SNe can also be strong radio emitters, from 10 to 250 times more luminous in the radio than Cas A. Therefore, they can obviously accelerate relativistic electrons and presumably other types of particles. Also, in general the radio supernovae (RSN) have steeper spectral indices ($\alpha \sim 0.5$ to 1.0) [and thus electron injection spectra ($\gamma \sim 2.0$ to 3.0)] than most SNR, so they should be better at producing the observed cosmic ray e^- energy spectrum of $\gamma \sim 2.7$.

Let us first see if we have enough energy in the RSNe. Since above we used Cas A, which is atypically bright for an SNR, let us use for a comparison SN1979c which, based on our very limited sample, is also atypically bright. SN1979c is 250 times more intense than Cas A at 6 cm wavelength, but has essentially the same spectral index, so its total radio luminosity is ~250 times greater. On the other hand, while Cas A is ~4.1 pc in diameter, SN1979c, when it was at its brightest at age ~1.5 years, was only ~0.03 pc in diameter. Putting these differences into Equation 9, yields the result that SN1979c, in spite of its much greater luminosity, has only ~2% of the minimum total energy of Cas A because of its small radius. Also, since r_0 is so small for these young and compact RSNe, the adiabatic loss problem will be much more severe and even though the models for both the mini-shell and mini-plerion RSNe discussed above involve continuing acceleration of relativistic particles, it does not seem, from what we presently know about RSNe, that they are likely sources of a major component of the observed cosmic rays. The major direct contribution of supernovae to cosmic ray e^- thus appears to come when they age and form large SNR.

6. SUMMARY

I have tried to give here, with concentration on the astronomical results and observations, the thread which leads from supernovae to supernova remnants to the non-thermal galactic radio background and, finally, to the observed cosmic ray e^- component in the Earth's vicinity. To do this, I have followed the route from observations of supernovae to observations of supernova remnants to observations of the galactic background to local measurements of cosmic rays. Then I have worked my way back through the same chain trying to point out the possible relations which connect them at each stage. As you have seen, some of the links are well established, while others are weak and based on much hand waving. However, at least some plausibility exists in each case, and it is the most reasonable connection which we have at the present time. Obviously, new measurements and new theoretical results will have to be obtained to refine, or alter, this picture and there remains much to be done in all areas. You will note that I have not even mentioned the problems of how the particles actually are accelerated by the shocks and pulsars which obviously do so very effectively. That is an area which I will not delve into and have left to other speakers and to some excellent texts [Longair (1981) for a short overview or Ginzburg and Syrovatskii (1964) and Hayakawa (1969) for more detail]. The uncertainty and changeability of all of this is, of course, part of the thrill of doing astrophysics. We can be almost assured that at the next and following schools, we will already have new results to fit into our understanding of cosmic rays and their origins.

REFERENCES

Allen, R.J., Baldwin, J.E., and Sancisi, R. 1978, A & A **62**, 398.

Allen, R.J., Goss, W.M., Ekers, R.D., and de Bruyn, A.G. 1976, A & A **48**, 253.

Baldwin, J.E. 1967, in Radio Astronomy and the Galactic System, ed. H. van Woerden (London, Academic Press), p.337.

Baldwin, J.E. 1976, The Structure and Content of the Galaxy and Galactic Gamma Rays, ed. C.E. Fichtel (GSFC Publication), p. 206.

Bandiera, R., Pacini, F., and Salvati, M. 1983, A & A **126**, 7.

Bandiera, R., Pacini, F., and Salvati, M. 1984, A & A **285**, 134.

Barbon, R. 1978, Mem. S.A. It. **49**, 331.

Berkhuijsen, E.M. 1971, A & A **14**, 359.

Berkhuijsen, E.M., Haslam, C.G.T., and Salter, C.J. 1971, A & A **14**, 252.

Brown, G.E. 1982, in Supernovae: A Survey of Current Research, eds. M.J. Rees and R.J. Stoneham (Dordrecht, Reidel), p.13.

Brown, R.L. and Marscher, A.P. 1978, Ap. J. **220**, 467.

Cadonau, R., Sandage, A., and Tammann, G.A. 1985, in Supernovae as Distance Indicators, ed. N. Bartel (Berlin, Springer-Verlag), p. 151.
Caswell, J.L., Clark, D.H., and Crawford, D.F. 1975, Aust. J. Phys. **37**, 39.
Chevalier, R.A. 1981a, Ap. J. **246**, 267.
Chevalier, R.A. 1981b, Ap. J. **251**, 259.
Chevalier, R.A. 1984, Ap. J. Lett. **285**, L63.
Ciatti, F. 1978, Mem. S. A. It. **49**, 343.
Clark, D.H. and Caswell, J.L. 1976, M.N.R.A.S. **174**, 267.
Cowan, J.J. and Branch, D. 1982, Ap. J. **258**, 31.
Cowan, J.J. and Branch, D. 1985, Ap. J. **294**, 400.
de Bruyn, A.G. 1973, A & A **26**, 105.
Ginzburg, V.L. and Syrovatskii, S.I. 1964, The Origin of Cosmic Rays, (Oxford, Pergamon Press).
Gottesman, S.T., Broderick, J.J., Brown, R.L., Balick, B., and Palmer, P. 1972, Ap. J. **174**, 383.
Gull, S.F. 1973a, M.N.R.A.S. **161**, 47.
Gull, S.F. 1973b, M.N.R.A.S. **162**, 135.
Haslam, C.G.T., Wilson, W.E., Graham, D.A., and Hunt, G.C. 1974, A & A Suppl. **13**, 359.
Hayakawa, S. 1969, Cosmic Ray Physics, (New York, John Wiley & Sons).
Iben, I.I. Jr. and Tutukov, A.V. 1984, Ap. J. Suppl. **54**, 335.
Jansky, K.G. 1932, Proc. Inst. Radio Eng. **20**, 1920.
Jansky, K.G. 1933a, Nature **132**, 66.
Jansky, K.G. 1933b, Proc. Inst. Radio Eng. **21**, 1387.
Jansky, K.G. 1933c, Pop. Astron. **41**, 548.
Keen, N.J., Wilson, W.E., Haslam, C.G.T., Graham, D.A., and Thomasson, P. 1973, A & A **28**, 197,
Landecker, T.L. and Wielebinski, R. 1970, Aust. J. Phys. Ap. Suppl. **16**, 31.
Longair, M.S. 1981, High Energy Astrophysics, (Cambridge, Cambridge University Press).
Matthewson, D.S., van der Kruit, P.C. and Brouw, W.N. 1972, A & A **17**, 473.
Meyer, P. 1969, Ann. Rev. Astron. and Ap. **7**, 18.
Pacini, F. and Salvati, M. 1973, Ap. J. **186**, 249.
Pacini, F. and Salvati, M. 1981, Ap. J. Lett. **245**, L107.
Panagia, N., Sramek. R.A., and Weiler, K.W. 1986, Ap. J. Lett. **300**, L55.
Prince, T.A. 1979, Ap. J. **227**, 676.
Reynolds, S.P. and Chevalier, R.A. 1984, Ap. J. **278**, 630.
Sedov, L.I. 1959, in Similarity and Dimensional Methods in Mechanics, (New York, Academic Press).
Shklovskii, I.S. 1981, Soviet Astron. Lett. **7 (No. 4)**, 263.
Srinivasan, G., Battacharya, D., and Dwarakanath, K.S. 1984, J. of A & A **5**, 403.
Strom, R.G. and Duin, R.M. 1973, A & A **25**, 351.
Sullivan, W.T. III 1984, in The Early Years of Radio Astronomy, (Cambridge, Cambridge University Press), p. 43.
Tammann, G.A. 1982, in Supernovae: A Survey of Current Research, eds. M.J. Rees and R.J. Stoneham (Dordrecht, Reidel), p. 371.

Tang, K.-K. 1984, Ap. J. **278**, 881.
Ulmer. M.P., Crane, P.C., Brown, R.L., and van der Hulst, J.M. 1980, Nature **285**, 151.
Webber, W.R. 1983, in The Composition and Origin of Cosmic Rays, ed. M.M. Shapiro (Dordrecht, Reidel), p.83.
Webster, A.S. 1971, Ph.D. Dissertation, University of Cambridge.
Weiler, K.W. 1985, in The Crab Nebula and Related Supernova Remnants, eds. M.C. Kafatos and R.B.C. Henry (Cambridge, Cambridge University Press), p. 227.
Weiler, K.W. and Seielstad, G.A. 1971, Ap. J. **163**, 455.
Weiler, K.W. and Panagia, N. 1978, A & A **70**, 419.
Weiler, K.W. and Panagia, N. 1980, A & A **90**, 269.
Weiler, K.W. and Sramek, R.A. 1980, I.A.U. Circ. No. 3485.
Weiler, K.W., van der Hulst, J.M., Sramek, R.A., and Panagia, N. 1981, Ap. J. Lett. **243**, L151.
Weiler, K.W., Sramek, R.A., and Panagia, N. 1986, Science **231**, 1251.
Weiler, K.W., Sramek, R.A., Panagia, N., van der Hulst, J.M., and Salvati, M. 1986, Ap. J. **301**, 790.
Wheeler, J.C. and Levreault, R. 1985, Ap. J. Lett. **294**, L17.
Wilson, A.S. and Weiler, K.W. 1976, A & A **49**, 357.
Woltjer, L. 1972, Ann. Rev. A & A **10**, 129.
Woosley, S.E. and Weaver, T.A. 1982, in Supernovae: A Survey of Current Research, eds. M.J. Rees and R.J. Stoneham (Dordrecht, Reidel), p. 79.

PARTICLE ACCELERATION IN GALACTIC SUPERNOVA REMNANTS

D.A. Green,
Mullard Radio Astronomy Observatory,
Cavendish Laboratory,
Cambridge CB3 OHE,
United Kingdom.

ABSTRACT. This paper discusses the radio spectral indices of Galactic SNRs in relation to the particle acceleration mechanisms responsible for their radio emission. Young, bright shell remnants generally have spectral indices α larger than the 0.5 expected from basic shock acceleration theory. Fainter, older shell remnants have a wide range of α. Evidence on whether or not particle acceleration is continuing in SNRs is also discussed.

1. INTRODUCTION

Statistical studies of Galactic SNRs based on the Σ–D relation have revealed a wide spread of properties (Green 1984a; cf. Mills et al. 1984 for remnants in the Magellanic Clouds), implying that current catalogues of Galactic SNRs are incomplete both for faint objects and also for young but distant remnants. The realization of this incompleteness and the wide spread of properties raise doubts about many previous statistical studies of Galactic remnants which depend on the Σ–D relation. Studies of the distance-independent properties of Galactic SNRs do not, however, suffer in this way, and here I discuss the radio spectral indices of Galactic SNRs in relation to the mechanisms of particle acceleration (specifically for the electrons) and evidence for continuing acceleration in Galactic remnants.

2. GALACTIC REMNANTS

The great majority of the objects identified as Galactic SNRs (e.g. van den Bergh 1983; Green 1984a) are 'shell' remnants, recognised by their limb-brightened radio structures. Two processes have been put forward to explain the radio emission from shell remnants, either the enhancement of existing cosmic ray electrons and magnetic field by compression through the SNR shock (e.g. van der Laan 1962), or Fermi acceleration of relativistic particles in the shock (e.g. Bell 1978a,b, also see Völk 1984). There are a few Galactic remnants with centre-brightened radio structure,

M. M. Shapiro and J. P. Wefel (eds.), Genesis and Propagation of Cosmic Rays, 205–213.

of which the Crab Nebula is the best known. Identification of such 'filled-centre' remnants is not easy as their amorphous structures like those of some HII regions. Also their radio spectra are flat, similar to the thermal spectra of HII regions but without a low-frequency turnover. In the case of the Crab Nebula, it is clear that its non-thermal synchrotron emission and centre-brightened structure are due to energetic particles produced by a central source, the Crab's pulsar. There are a few SNRs (e.g. Weiler 1983) which share some of the characteristics of both shell and filled-centre remnants — these are 'composite' SNRs: it is not yet clear what are the criteria for remnants of this type, although a flat-spectrum core within a steeper spectrum shell is seen in several. These remnants may represent the later stages of evolution of filled-centre SNRs, when the influence of the central power source has diminished, and the shock wave produced by the energy input into the interstellar medium from both the parent SN explosion and the central source becomes apparent. Finally, there some peculiar objects, W50 and CTB80 included, which are generally regarded as SNRs, but for which classification by type is inappropriate, as each has unique features.

There are two observational selection effects which apply to all remnants and make current catalogues incomplete. Firstly, almost all remnants have originally been identified by their radio emission which, for the fainter objects, is difficult to distinguish from background Galactic emission. Secondly, many young but distant remnants will not have been resolved by radio surveys. Neither of these selection affects has any strong bias towards objects with spectral indices at the extremes of the observed range.

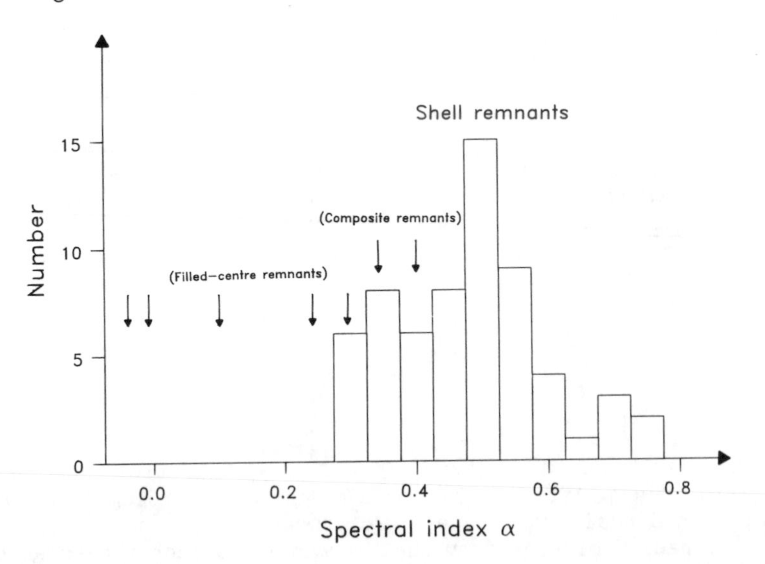

Figure 1. Histogram of the spectral indices α of 62 shell remnants with $\Delta\alpha\leq 0.12$, together with the spectral indices of some filled-centre and composite remnants.

3. THE DATA

Of the 150 or so objects catalogued as Galactic SNRs, there are about 80 definitely identified as shell remnants. I have derived radio spectral indices α (with uncertainties $\Delta\alpha$) for these objects from the available data, and Fig.1 shows a histogram of spectral indices for the 62 with $\Delta\alpha \leq 0.12$. The majority have α between ~0.3 and ~0.6, while few have larger α's up to ~0.8. These steeper spectrum SNRs, as we shall see later, are young, bright remnants. Filled-centre remnants have spectra that are flatter than those of shell remnants, with spectral indices between -0.04 and 0.3 (e.g. G20.2-0.2, G21.5-0.9, G74.9+1.2, 3C58 and the Crab Nebula), while two of the better studied composite remnants (G6.4-0.1 and G320.4-1.2) have spectral indices of 0.34 and 0.40, lower than those found for most shell remnants.

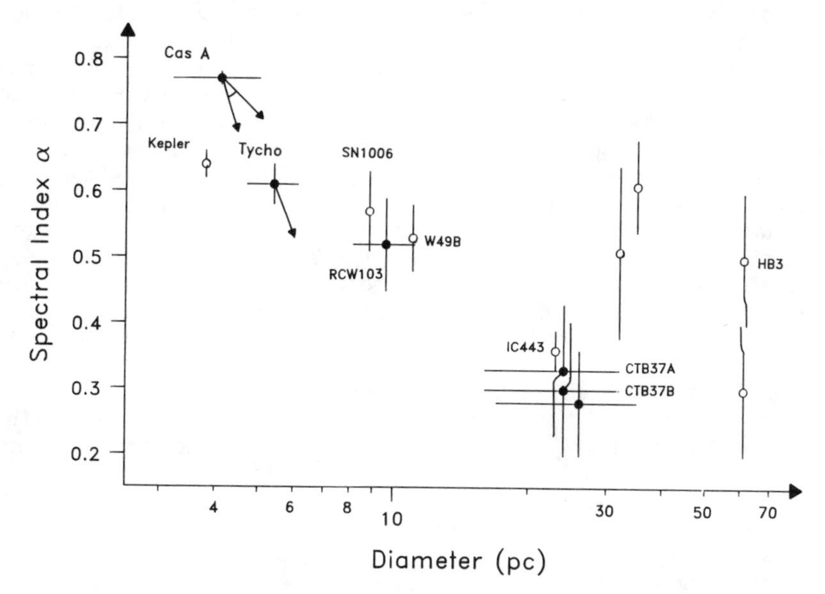

Figure 2. Spectral index versus linear diameter for 14 shell remnants with distance measurements (filled circles) or estimates (open circles).

3.1 A dependence on age/size

Fig.2 shows spectral indices of 14 shell remnants for which measurements or good estimates of their distances are available. (The distances are taken from Green 1984a, with the addition of IC443 at 2 kpc, as its association with S249 has recently been strengthened, Fesen 1984.)

The trend in Fig.2, that the spectra are flatter for the larger remnants, is due to the relatively steep spectra of young, small remnants (with diameters less than ~10 pc). These have spectral indices greater than the 0.5 expected from simple shock acceleration theories, whereas older, larger remnants have a wide spread of α. In the case of Cas A and Tycho, two of the young remnants with steeper spectra, secular flattening of their spectra has been observed (e.g. Stankevich 1979). The values of $(\Delta\alpha/\Delta t)$ are -1.25×10^{-3} and -10^{-3} yr^{-1} approximately for Cas A and Tycho respectively which, when coupled with radio expansion ages of 300–750 and ~880 yr (see below), give the evolutionary paths in the α–D plane shown in Fig.2.

Although it is tempting from Fig.2 to suggest that the spectra of shell remnants continue to flatten throughout their evolution, there are several problems about this interpretation of the data. Firstly, many of the older shell SNRs may have developed from filled-centre remnants rather than from young shells, and in this case 'seeding' of the relativistic particle distribution from that of the filled-centre remnant may have an important influence on the subsequent radio spectrum of the remnant (cf. some composites have spectral indices that are between those of shell and filled-centre remnants). Secondly, because of the α-dependence of the adiabatic fading of remnants (see Section 4), remnants with spectra steeper than average will fade faster. Nevertheless, the fact remains that young, small remnants, whether having a shell or filled-centre structure, have spectra that are steeper or flatter respectively than larger shell remnants. This suggests that either the mechanisms at work in young shell and filled-centre remnants produce progressively flatter or steeper spectra respectively as the remnant evolves, or that another mechanism (presumably van der Laan compression) takes over in large-diameter remnants. The fact that young remnants like Cas A, Tycho and Kepler have steeper than average spectra was noted by Clark & Caswell (1976), but because there were other young (i.e. high surface brightness) remnants with flatter than average spectra Clark & Caswell concluded that there were no correlations between spectral index and diameter, since they failed to make the distinction between shell and filled-centre remnants.

Only a little of the spectral index information for Galactic SNRs has been used so far, because reliable distance measurements or estimates (and hence diameters) are only available for a small fraction of the remnants. To investigate the evolutionary trend of spectral index further, the surface brightness of each remnant can be used as a crude measure of its age, since Σ–D studies show that larger, older remnants are <u>on average</u> fainter. Fig.3 shows the distribution of surface brightness against spectral index for the 62 shell remnants have $\Delta\alpha\leq0.12$. The trend that the young remnants, which are on average the brighter ones, have steeper than average spectra is also apparent in Fig.3, but less clearly than in Fig.2. At low surface brightnesses where remnants have a wide range of spectral indices, the historical or suspected historical shell remnants generally have steeper spectra. This supports the idea that the radio spectra of shell remnants, or at least those less than a few thousand years old, flatten as the remnants expand.

3.2 Other observational approaches

There are other observational properties of an SNR which may help in distinguishing which acceleration mechanism is at work within it.

1) The structure of a shell remnant. A remnant whose radio emission is solely or largely due to van der Laan compression are likely to have a radio structure different from that produced by a shock acceleration

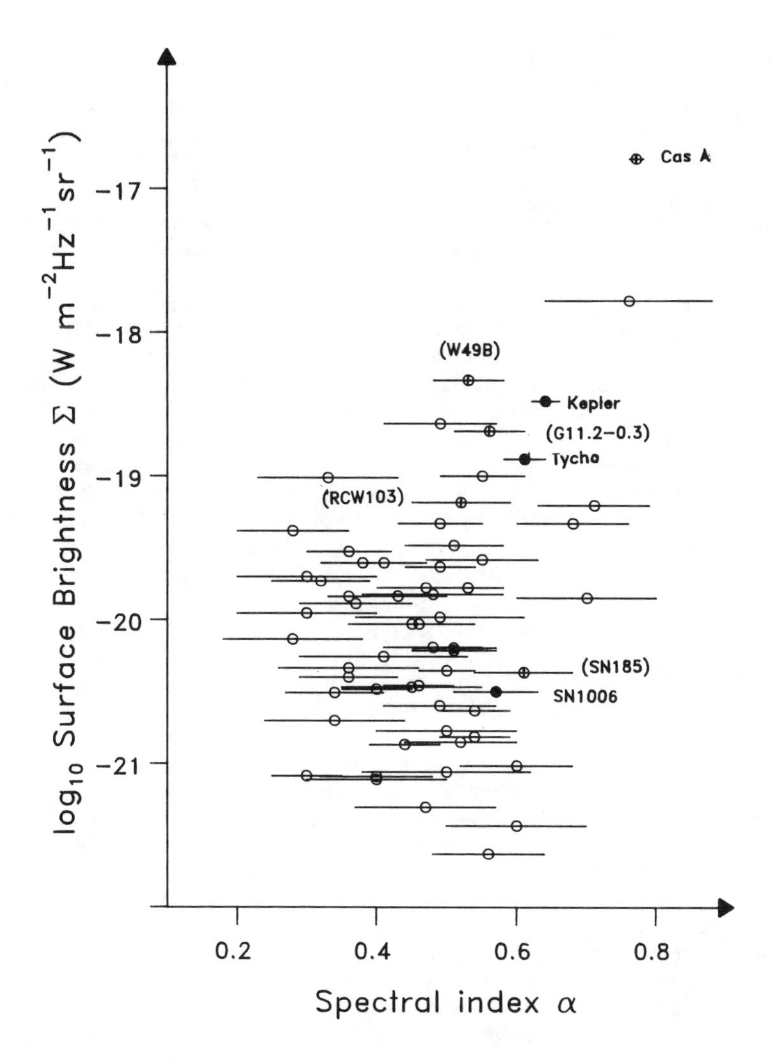

Figure 3. Surface brightness Σ versus spectral index α for 62 shell remnants with $\Delta\alpha \leq 0.12$.

mechanism. If the magnetic field in the interstellar medium into which the remnant is expanding is ordered on a scale comparable with or larger than the size of the remnant, then the SNR shock would produce a toroid of compressed, enhanced magnetic field, aligned with the initial direction of the magnetic field. From most viewing angles, this would produce a limb-brightened double-arc radio structure, although the shell would appear more complete for viewing directions close to the initial magnetic field direction. A shock acceleration mechanism, on the other hand, if it applies in a particular SNR, might work in all regions of the shock, giving a complete radio shell structure. These expected structures are clearly ideals: for the van der Laan compression, the interstellar magnetic field may not be ordered on the scale of the remnant, and in both cases there could be important, but unknown factors that dictate whether or not the mechanism works efficiently. Many Galactic remnants do indeed show 'complete' shells or 'double' shells, although many others are best described as having a one-sided or 'horseshoe' shaped radio structure, presumably because of interaction with regions of the surrounding ISM that differ on the scale of the remnant shock.

2) Spectral variations and breaks. Scheuer (1984) has noted that, although shell remnants have a wide range of spectral indices, there is little evidence of spectral variations within individual remnants (except for those now classified as composites). This is surprising if the emission is due to a shock acceleration mechanism, since the spectral index will depend on the local acceleration conditions and these presumably vary over remnants which have evident asymmetries. The lack of variations implies that the spectrum of the radio emission from each remnant depends on properties of the surrounding ISM (presumably the cosmic ray electron spectrum if van der Laan's mechanism is at work) which change on a large scale throughout the Galaxy, but not on the scales comparable with the sizes of individual remnants. The observational evidence on spectral variations is limited, however. Tycho's SNR, a relatively bright remnant, has a spectral index (between 610 MHz and 10.7 GHz) constant to within 0.05 over the face of the remnant (Klein et al. 1979), but the limits for most other, fainter remnants are usually much larger. (One of the best limits on the spectral variation across a remnant is that claimed by Erickson & Mahoney (1985) for IC443, namely $\Delta\alpha<0.02$ between a few tens of megahertz and a few gigahertz, although it is not clear to what portion of the remnant this limit applies. Green 1986, on the other hand, does find spectral variations across this source.) If the van der Laan compression mechanism were responsible for the radio emission from older, fainter remnants, then any break in the spectrum of the Galactic background emission would be shifted to higher frequencies, and might be apparent in the integrated spectrum of the remnant. Such breaks have been reported for a few faint remnants, but further confirmatory observations are required. Variations in spectral index across remnants, or breaks in their integrated spectra each imply that a simple one parameter (i.e. α) description of the radio emission is an over simplification, and that detailed studies of the emission from individual remnants are likely to be more useful than comparisons of many objects.

3) Magnetic field orientations. Tycho's SNR has a radial magnetic field (Duin & Strom 1975), whereas older remnants have predominantly tangential fields (e.g. Dickel & Milne 1976), presumably because cooling has become important, and highly compressed shells develop. This difference in magnetic field structure may reflect the transition between dominance by shock acceleration and van der Laan compression.

4) Optical studies. These can be used to decide whether or not a remnant is in a radiative phase, where cooling is important and highly compressed shells are expected.

5) Radio 'filaments'. If there are density structures within the ISM that are aligned with the large-scale interstellar magnetic field, distinct radio filaments may be produced by a shock with a high compression ratio, as has been noted in a few cases by Green (1984b).

4. IS THERE CONTINUING ACCELERATION IN SNRs

It is also appropriate to ask whether observations can tell us if particle acceleration is currently at work (at least for electrons) within SNRs. For expansion with no continuing acceleration and a frozen-in magnetic field, Shklovskii (1960) showed that the surface brightness of a remnant is expected to evolve with

$$\Sigma \propto D^{-4(1+\alpha)}.$$

The observed properties of Galactic SNRs (and those in the Magellanic Clouds) have generally been fitted (e.g. Clark & Caswell 1976; Milne 1979) with a Σ–D relation significantly flatter than the $\Sigma \propto D^{-6}$ that would be expected for an 'average' spectral index of 0.5, and this has been taken as an indication of continuing acceleration in SNRs (e.g. Drury 1983). Backward extrapolation at $\Sigma \propto D^{-6}$ of young Galactic remnants predicts luminosities for even younger remnants that are in excess of the observed values both for extragalactic radio SN a few years old and the possible remnants of SN that occurred a few decades or more ago. The explanation may be either continuing acceleration and/or that the radio emission 'switches-on' after a hundred years or so (e.g. Cowsik & Sarkar 1984).

Although the observed properties of Galactic SNRs show that the brightest remnants for a given diameter are related by $\Sigma \propto D^{-3}$ this does <u>not</u> necessarily mean that acceleration is required throughout the observable lifetime of Galactic remnants. Firstly, <u>if</u> there are rare objects that produce exceptionally bright SNRs ('super-Cas' remnants), they are more likely to be evident as large rather than small remnants, on simple considerations of the time spent in each state. Secondly, the assumption that field and particles are conserved is unlikely to be applicable throughout the lifetime of a remnant. At early stages of evolution, the production or decay of magnetic field within a remnant is likely to have a large influence on the radio emission and, at later stages, van der Laan compression of existing ISM cosmic ray electrons and magnetic field is likely to grow in importance.

Another approach is to study the evolution of individual remnants, although this is naturally restricted to the young ones. A SNR evolving with $\Sigma \propto D^{\beta}$ has

$$\beta = -2 + (1/S)(\Delta S/\Delta t)\tau,$$

where S is the flux density of the remnant, ΔS the flux density change in a time Δt, and τ the current expansion timescale of the remnant (i.e. its size divided by its expansion velocity).

Cas A has a spectral index of 0.77±0.01, so Shklovskii's prediction for no acceleration is $\Sigma \propto D^{-7.08 \pm 0.04}$. The change of the flux density of

Cas A is about −1 per cent yr^{-1} (Stankevich 1979, although it varies systematically with frequency, leading to the spectral flattening noted above). It is, however, not clear how to apply Shklovskii's simple, scaled expansion model to the complex shell of radio emission from Cas A, and neither of the expansion timescales in the literature is appropriate. The expansion age of ~300 yr (van den Bergh & Dodd 1970), which has been used by some previous authors (e.g. Clark & Caswell 1976), is derived from the motion of apparently undecelerated optical knots. The radio expansion age of ~950 yr that has more recently become available (Tuffs 1986) refers to the motions of compact features in the shell which are responsible for only a small proportion of the total radio emission. Cas A's radio structure, a complex ring of emission and a larger, fainter 'plateau' of emission (e.g. Bell, Gull & Kenderdine 1975; Tuffs 1986) has been taken (Gull 1973, 1975) to indicate that the remnant is in transition between the free-expansion and Sedov phases where Rayleigh-Taylor instabilities are important. If this is the case, and the radio-emitting region expands in a simple scaled way, then the appropriate timescale is between 1 and 2.5 times the age of the remnant, i.e. 300–750 yr (assuming the optical knots are in free expansion and their expansion timescale gives the age of the remnant). A timescale of 500 yr, or more, would imply that Cas A is evolving with $\Sigma \propto D^{-7}$ or steeper so that, with Shklovskii's model, no particle acceleration is needed at present.

On the other hand, Tycho's SNR has a sharply bounded limb-brightened shell, shown by studies of its expansion (Strom, Goss & Shaver 1982; Tan & Gull 1985) to be in the Sedov phase. Provided that the radio emission region expands in a self-similar way, the expansion age of ~880 yr derived by Tan & Gull is appropriate for comparison with Shklovskii's model which implies $\Sigma \propto D^{-6.44 \pm 0.12}$ for Tycho's SNR (which has $\alpha = 0.61 \pm 0.03$), and a flux density change of only −0.5 per cent yr^{-1}. The flux density change reported for Tycho's SNR (Stankevich 1979) is −0.8 per cent yr^{-1}, which implies that acceleration is not needed at present in Tycho's SNR (Tan & Gull reach a similar conclusion from comparisons of brightness). Unfortunately, the significance of this result is uncertain, and a small systematic error in the flux density change would be sufficient to imply acceleration.

For both Cas A and Tycho there are indications that for Shklovskii's simple model with frozen-in magnetic field continuing acceleration is not needed to explain their radio emission. Alternatively, there could be appreciable acceleration provided that there are decreasing magnetic fields within the remnants.

ACKNOWLEDGMENTS

I thank Churchill College, Cambridge for a Junior Research Fellowship, and Prof. H. Völk and Dr P.A.G. Scheuer for stimulating (theoretical) discussions.

REFERENCES

Bell, A.R., 1978a. Mon. Not. R. astr. Soc., **182**, 147.
Bell, A.R., 1978b. Mon. Not. R. astr. Soc., **182**, 443.
Bell, A.R., Gull, S.F. & Kenderdine, S., 1975. Nature, **257**, 463.
Clark, D.H. & Caswell, J.L., 1976. Mon. Not. R. astr. Soc., **174**, 274.
Cowsik, R. & Sarkar, S., 1984. Mon. Not. R. astr. Soc., **207**, 745.
Dickel, J.R. & Milne, D.K., 1976. Aust. J. Phys., **29**, 435.
Drury, L.O'C., 1983. Space Sci. Rev., **36**, 57.
Duin, R.M. & Strom, R.G., 1975. Astr. Astrophys., **39**, 33.
Erickson, W.C. & Mahoney, M.J., 1985. Astrophys. J., **290**, 596.
Fesen, R., 1984. Astrophys. J., **281**, 658.
Green, D.A., 1984a, Mon. Not. R. astr. Soc., **209**, 449.
Green, D.A., 1984b. Mon. Not. R. astr. Soc., **211**, 433.
Green, D.A., 1986. Mon. Not. R. astr. Soc., in press.
Gull, S.F., 1973. Mon. Not. R. astr. Soc., **162**, 135.
Gull, S.F., 1975. Mon. Not. R. astr. Soc., **171**, 263.
Klein, U., Emerson, D.T., Haslam, C.G.T. & Salter, C.J., 1979. Astr. Astrophys., **76**, 120.
Mills, B.Y., Turtle, A.J., Little, A.G., & Durdin, J.M., 1984. Aust. J. Phys., **37**, 321.
Milne, D.K., 1979. Aust. J. Phys., **32**, 83.
Scheuer, P.A.G., 1984. Adv. Space Res., **4**, 337.
Shklovskii, I.S., 1960. Soviet Astr., **4**, 243.
Stankevich, K.S., 1979. Aust. J. Phys., **32**, 95.
Strom, R.G., Goss, W.M. & Shaver, P.A., 1982. Mon. Not. R. astr. Soc., **200**, 473.
Tan, S.M. & Gull, S.F., 1985. Mon. Not. R. astr. Soc., **216**, 949.
Tuffs, R.J., 1986. Mon. Not. R. astr. Soc., **219**, 13.
van den Bergh, S., 1983. In Supernova Remnants and their X-ray Emission, Eds Danziger, I.J. & Gorenstein, P., (Reidel, Dordrecht), p597.
van den Bergh, S. & Dodd, W.W., 1970. Astrophys. J., **162**, 485.
van der Laan, H., 1962. Mon. Not. R. astr. Soc., **124**, 179.
Völk, H.J., 1984. In High Energy Astrophysics, ed J. Tran Tranh Van, (Editions Frontieres, Gif Sur Yvette), p281.
Weiler, K.W., 1983. Observatory, **103**, 85.

PULSARS AS COSMIC RAY PARTICLE ACCELERATORS - NEW RESULTS ON THE DYNAMICS OF PROTONS IN VACUUM FIELDS

K.O. Thielheim
Institut für Reine und Angewandte Kernphysik
Olshausenstr. 40
2300 Kiel
Federal Republic of Germany

ABSTRACT. Particle acceleration in pulsar magnetospheres appears to be a very complicated process. Therefore we have started a careful analysis of a very special aspect of the possible acceleration mechanism. This investigation is made on the basis of the vacuum field of an orthogonal rotator. Effects of special relativity and classical radiation theory are taken into account. Results are discussed in terms of the "critical surface" and "acceleration boundary".

HISTORICAL BACKGROUND

Shortly after the discovery of pulsars[1] and their identification as rapidly rotating, strongly magnetized neutron stars[2], the existence of which had already been predicted by theory much earlier[3], people have tried with their help to answer one of the greatest questions modern astrophysics is confronting its reasearch workers with since primary cosmic particle radiation has been observed many decades ago[4,5].

The latter is known to be constituted essentially of nuclei, mostly protons, with energies reaching up to 10^{21}eV. Of course it is of great interest to understand in which cosmic objects and by which physical mechanisms nature is able to accelerate particles up to energies so high. Also one can perhaps hope in this way to find new concepts of technical accelerating machines even more powerful than those which are already in operation and which have led to so important and fundamental discoveries in high energy elementary particle physics.

Among the many papers which since then have been published to explain quantitatively the physical processes involved in the supposed accelerating mechanism I just mention the pioneer work by J. P. Ostriker and J. E. Gunn[6] and by Goldreich and W. H. Julian[7]. Gunn and Ostriker have investigated the acceleration of electrically charged particles in the

M. M. Shapiro and J. P. Wefel (eds.), Genesis and Propagation of Cosmic Rays, 215–225.

vacuum field of a magnetic point dipole rotating with its axis inclined to the dipole vector under the pure wave field approximation. But in the meantime it has been found[8] that the dynamics of those particles which attain very high energies is governed by the near field contributions which are neglected in the aforementioned theory. The work of Julian and Goldreich is largely devoted to the structure of the magnetosphere of a magnetized star rotating with its axis parallel to the dipole vector. Such a configuration, obviously, is not able to produce "lighthouse" signals so typical for pulsar radiation and therefore certainly cannot be considered to be very realistic.

We are still far away from a comprehensive self-consistent theory of pulsar dynamics including the structure of the pulsar magnetosphere and the acceleration of electrically charged particles. The object one has to deal with appears so complicated that probably it is not wise to try to build such a comprehensive, self-consistent theory in just one big throw. It is more likely that one has to study carefully and in great detail the various different aspects of the mechanisms involved before one can hope to produce a satisfying physical theory of pulsars as cosmic ray particle accelerators[9,10].

STANDARD MODEL

This is the attitude which I have taken in the work presented here. As in the early theory of Gunn and Ostriker I have started with the vacuum field of a magnetic point dipole rotating at a constant angular velocity vector $\underline{\omega}$ inclined to the dipole vector $\underline{\mu}$. It seemed to be appropriate for a first systematic survey to restrict the present work to just one specified angle of inclination. The one that has been chosen to obtain the present results is equal to $\pi/2$. This object is sometimes called an orthogonal rotator.

Effects of special relativity and of classical radiation theory have been taken into account. The equations of motion have veen integrated numerically applying among others, a CRAY-I computer. Therefore the present results can be considered as being correct within the limits of special relativity and classical electrodynamics and, of course, within the limits of accuracy of the computer generated data. In the unfortunately rather laborious development of an appropriate software a group of then undergraduate students took part. The bulk of these data produced will be published elsewhere.

It should be made clear that these premises are different in many respects from the basic assumptions of Gunn-Ostriker theory, which neglects radiation reaction together with the terms governing azimuthal motion and those describing the latitudinal dependence of the field strength. Also this theory makes use of the constant phase approximation.

In performing the work which I am going to talk about here it appeared reasonable for a first survey to restrict the computations to just one set of parameters ω and μ with the intention to study the variation of results as functions of these parameters later. The parameter values chosen for the standard model are of no special distinction. But they lie well within the range of observed parameter values. The absolute value of the angular velocity is $\omega = 20\pi\ s^{-1}$ corresponding to a frequency of rotation $\nu = 10\ s^{-1}$ with a resulting light radius $r_L = 4775$ km. The absolute value of the magnetic dipole moment has been chosen as $\mu = 10^{30}$G cm^3. The equivalent magnetic field strength at a distance $r_p = 10$ km from the point dipole is about $B_{pole} = 2 \times 10^{12}$G in the polar region.

The basic idea underlying this work is to perform systematic orbit calculations for protons (as well as for electrons) within the field configuration specified above and to see what happens to these particles as far as a topography of trajectories and the development of energy is concerned.

Initial positions therefore are distributed systematically with respect to the radial coordinate r as well as with respect to the two angular coordinates, the longitudinal (azimuthal) angle ϕ (as measured against the dipole vector in the corotating system for zero retardation) and the latitudinal angle θ (as defined against the axis of rotation). It turned out to be sufficient to restrict the present survey to the orbits of particles starting with zero velocity.

PROTON ORBITS

Figure 1 and figure 2 are showing the same sample of proton trajectories as seen from two different directions of observation respectively. These particles start at a distance from the pulsar which corresponds to one unit of length, i.e. one light radius r_L. Their initial latitude is 70°. The values of initial longitude are distributed at an equal distance of 5°. These orbits exhibit some typical features: there is a certain range of initial longitude from which protons are drawn to the pulsar surface. As one looks more closely, they actually turn out to be focused to one of the magnetic poles of the pulsar. Protons from the remaining range of initial longitude are finally propagated outward into the interstellar space. These may be candidates for primary cosmic ray particles. Still one can also see that to

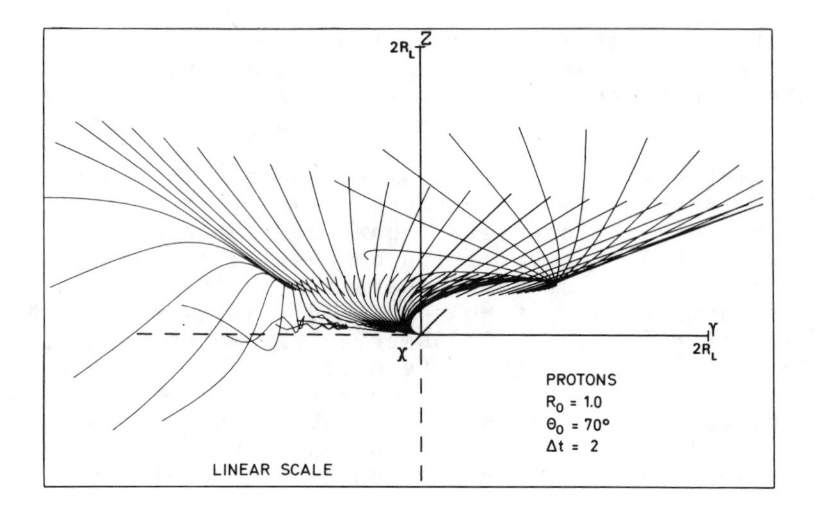

Fig. 1 Proton orbits calculated on the basis of the standard model

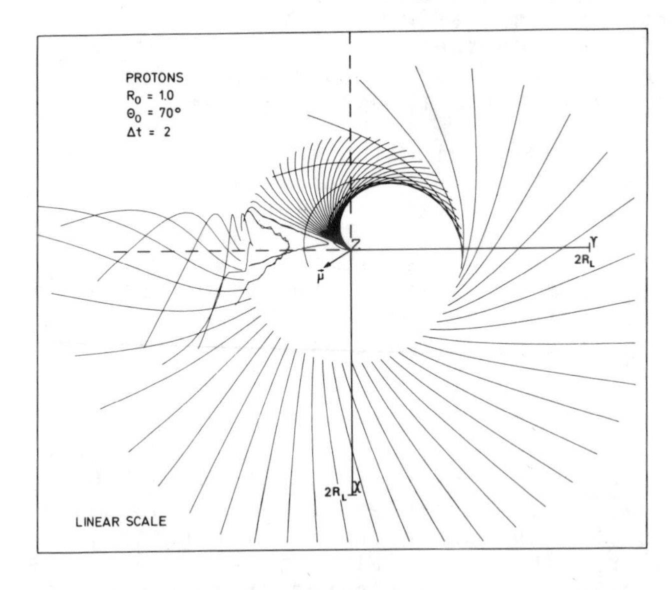

Fig. 2 The same orbits as in fig. 1 but projected onto the equatorial plane of rotation

a large extent the high energy protons do not follow the direction of the magnetic field lines.

The trajectories of protons starting from positions further out look quite different as is illustrated by figure 3 and figure 4. These pictures pertain to an initial radial distance from the pulsar, which corresponds to 10 units of length, i.e. 10 light radii r_L. The initial latitude is 25°

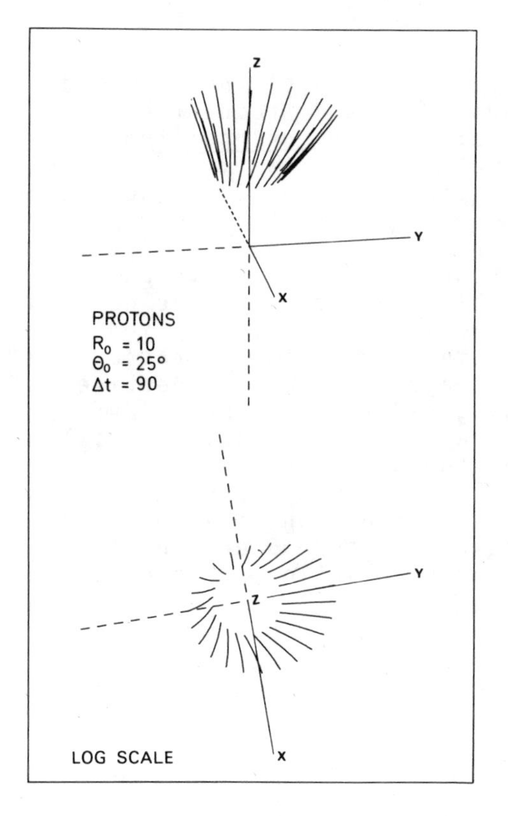

Fig. 3

Proton orbits starting well outside the critical surface, calculated on the basis of the standard model

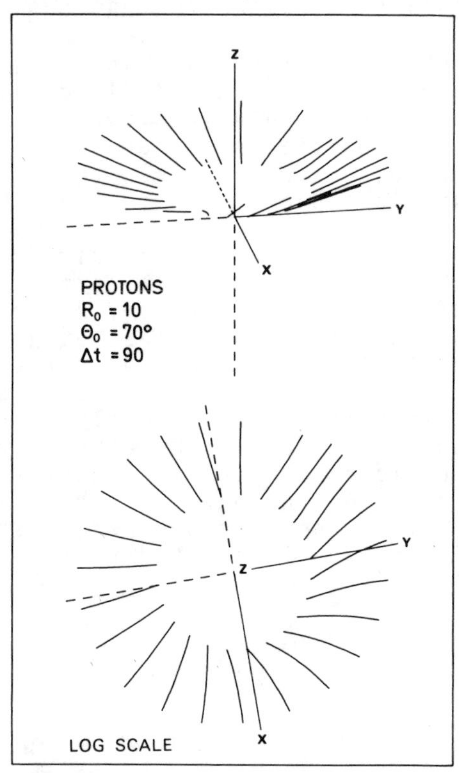

Fig. 4

Proton orbits starting at the same initial radial coordinate as in fig. 3, but at different initial latitude

in the case of figure 3, corresponding to starting positions comparatively near to the axis of rotation, and 70° in the case of figure 4, corresponding to initial positions near to the equator of rotation. In each of the two figures the orbits are shown from two different positions of observation. (A logarithmic scale, suitably defined for this purpose, has been used for the presentation of the radial coordinate r.) In any case for all practical purposes these protons move outward straight in radial direction. None of the particles is attracted onto the pulsar surface. Obviously in these regions, comparatively far away from the rotating dipole the pulsar is unable to act as a vacuum cleaner for protons. Instead it is functioning more like a snowplough for protons.

CRITICAL SURFACE FOR PROTONS

These findings give rise to the definition of the "critical surface for protons". It divides the space around the spinning dipole into two regions: an interior one, from which particles are drawn to the pulsar surface, and an exterior one, from which they are propelled out into the interstellar space.

The evaluation of the critical surface turns out to be somewhat laborious. I will not in this place go into the details of the numerical specifications which have been applied to produce these pictures but just mention that also those particles which on the initial stages of their respective trajectories exhibit a decreasing radial coordinate, but then turn to continuously move outward, with their initial positions have to be attributed to the exterior to the critical surface consequently. The one that has been obtained on the basis of the standard model which is made use of throughout this paper is shown in figure 5 and in figure 6 for two different directions of observation, respectively.

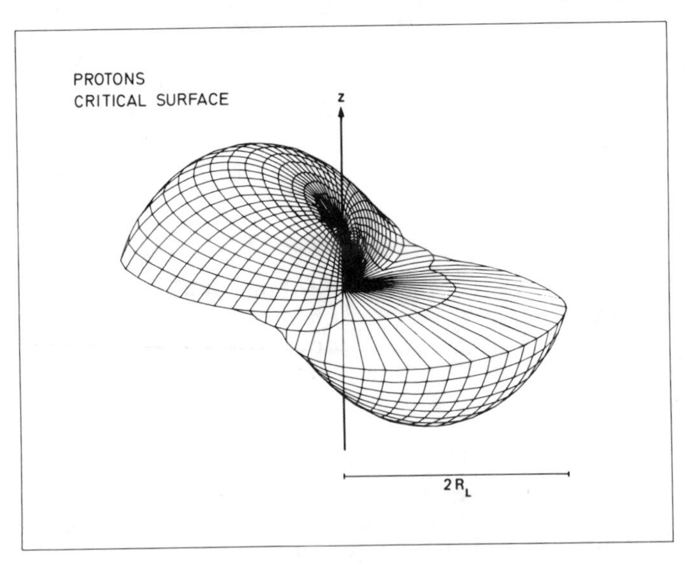

Fig. 5 Critical surface for protons calculated on the basis of the standard model

The perspective view given in figure 5 is the one seen by an observer at $\theta_o = 70^o$ latitude and $\phi_o = 30^o$ longitude. Figure 6 shows the projection onto the equatorial plane of rotation.

As a consequence of symmetries inherent in the equations of

motion the latter turns out to be point symmetric with respect to the position of the spinning dipole. Somehow the critical surface for protons resembles two half spheres shifted against each other along the direction of the dipole vector (for zero retardation) on each of the two sides of the equatorial plane of rotation. But actually its shape is more complicated than that. A close inspection of its projection onto the equatorial plane shows that there is no mirror symmetry with respect to the plane spanned by the

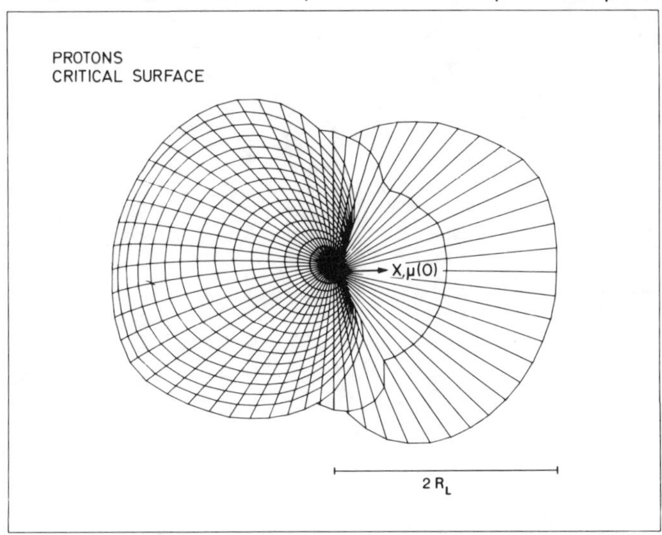

Fig. 6 The same critical surface as in fig. 5, but projected onto the equatorial plane of rotation

vectors of angular velocity $\underline{\omega}$ and dipole moment $\underline{\mu}$ (for zero retardation), which of course could hardly have been expected to exist.

In certain regions which are near to the magnetic poles the critical surface extends down very near to the point dipole suggesting the existence of initial positions of protons near to the pulsar surface from where they may be accelerated outward to become very high energy cosmic ray particles.

Under the premises of the standard model the largest radial extension of the critical surface for protons corresponds to about two units of length, i.e. two light radii r_L.

ENERGY DEVELOPMENT OF PROTONS

It is not possible to present and describe the many details

of the energy development of protons in the field configuration under consideration. Therefore I will just make a few comments on the maximum energy achieved by protons starting at a comparatively large radial distance from the spinning dipole well outside the critical surface as shown in figure 7. This picture shows on a decadic, logarithmic

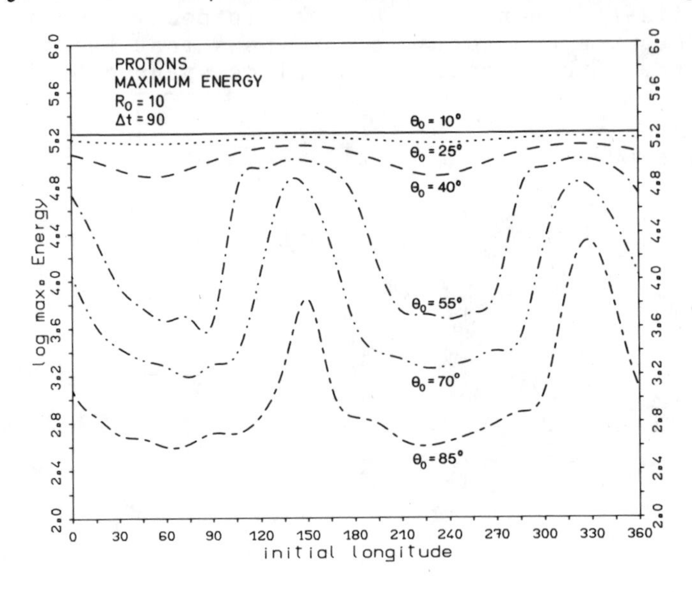

Fig. 7 Proton energy (for specified values of initial radial coordinate and of coordinate time interval) as a function of initial longitude. Different curves are for different values of initial latitude.

scale the energy (in units of particle rest energy) attained by protons after the elapse of 10 units of time (corresponding to 1.6 full rotations of the magnetic dipole) as a function of initial longitude. Different curves are for different values of initial latitude.

One of the remarkable features is the monotonous increase of maximum energy (which in this case is equal to the final energy as defined above) with decreasing initial latitude (which as has been said before is defined against the axis of rotation). Protons starting from positions near to one of the poles of rotation attain the highest energy values. Some of these final energy values are higher by more than a factor 100 as compared with the corresponding values for particles starting from inside the equatorial zone of rotation.

The final energy of particles originating from places near to the axis of rotation is practically independent of the initial longitude. Under the specification of the standard model and for the initial radius adopted here the maximum value of final energy in this sense is about 10^{15}eV. (Not much, probably, for those who wish to learn about particles of 10^{21} eV. But in this context it should be noted that much higher values of final energies can be achieved by protons starting from certain different initial positions and even more so by protons being accelerated under certain parameter values different from the ones underlying the standard model which I have adopted here).

The final energy of protons exhibits an increasing variability as far as its dependence on initial longitude is concerned if the initial position as measured in latitude approaches the equatorial region of rotation. The outstanding feature of this variability is the appearance of two maxima of final energy. Their respective positions correspond to places where the electric field strength attains its maximum value at the time when the proton starts to move from its initial position. The appearance of the two maxima also is observed in the development of electron energy and is supposed to have a bearing on the time structure of pulsar signals.

ACCELERATION BOUNDARY FOR PROTONS

It is obvious that the ability of the given field configuration to accelerate protons (or other types of charged particles) to very high energies and to large distances must break down beyond a certain range of distance from the rotating magnetic dipole. The limits of this range of influence define the "acceleration boundary for protons".

The existence of the latter has been predicted theoretically referring to the phase development of particles within the wave zone of a spinning magnetic dipole[13] and is clearly confirmed by the results of numerical integration methods similar (though different in some technical details) to the ones described before[14]. This is illustrated by figure 8 showing (on decadic, logarithmic scales) the maximum Lorentz factor of protons (the initial position of which in this case correspond to the phase of maximum field strength) as a function of initial radius in units of light radius r_L. (Here the maximum Lorentz factor has been defined as the Lorentz factor attained by the particle at a radial distance 8 times the initial radial distance). As one can clearly see the accelerating mechanism breaks down beyond a radial distance of slightly more than 10^4 light radii.

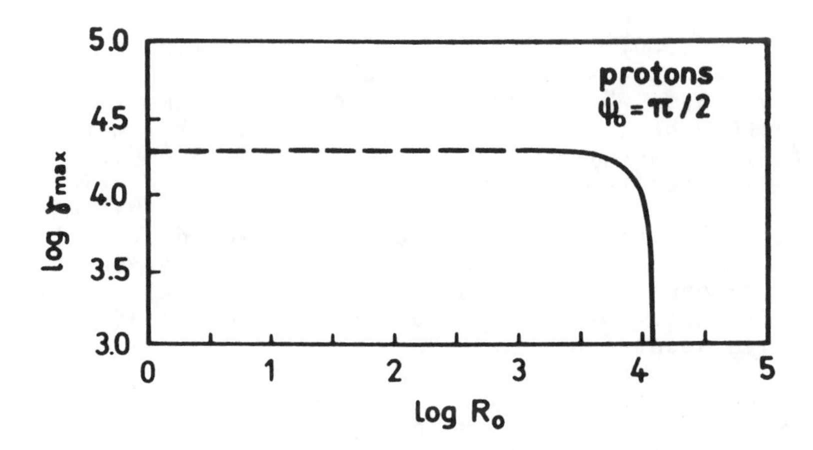

Fig. 8 Demonstration of the existence of the acceleration boundary for protons as calculated on the basis of the standard model

POSSIBLE INJECTION MECHANISMS

The results of the present work demonstrate the existence of a certain region of space surrounding the spinning magnetic dipole in which the electromagnetic forces exerted on a proton e.g. are such that the particle is propagated into the interstellar space with very high energy. To the inside this region of influence is bounded by the "critical surface" with touches down to certain places very near to the pulsar surface. To the outside this region is limited by the "acceleration boundary".

There are two possible injection mechanisms which are being investigated. One is provided by the possible extraction of charged particles from the pulsar atmosphere in places where this is permitted by the topography of the critical surface. The other mechanism is constituted by the possible immigration of neutral particles from the interstellar medium through the acceleration boundary and their subsequent ionization induced by pulsar radiation. One can also think of neutral particles as neutrons entering the aforementioned region and decaying there.

A WORD OF CAUTION

There is no doubt that the real world is much more complicated than is reflected by the simple model which I presented here. For example in those layers, where the pulsar magnetosphere merges with the interstellar plasma, collective forms of motion are expected to take place, Also, in places very near to the pulsar surface, especially in the polar regions, deviations from classical electrodynamics are expected. These are only two of a number of phenomena which certainly modify the performance of the accelerating machine.

But still one can hope that some of the features discussed here, as e.g. the critical surface, the acceleration boundary and the double maximum structure of particle energy will also be ingredients of a more sophisticated theory of pulsars as high energy cosmic ray particle accelerators.

REFERENCES

1. Hewish,A., Bell,S.J., Pilkington,J.D.H., Scott,P.F. and Collins,R.A.
   Nature (London), 217, 709 (1968)
2. Gold,T.
   Nature (London), 218, 731 (1968)
3. Baade,W. and Zwicky,F.
   Proc. Nat. Acad. Sci. 20, 254 (1934)
4. Hess,V.F.
   Phys. Z. 14, 610 (1913)
5. Kohlhörster,W.
   Phys. Z. 14, 1153 (1913)
6. Ostriker,J.P. and Gunn,J.E.
   Astrophys. J. 157, 1395 (1969)
7. Goldreich,P. and Julian,W.H.
   Astrophys. J. 157, 869 (1969)
8. Thielheim,K.O.
   Proc. 19th Int. Cosmic Ray Conf., La Jolla, USA, 210 (1985)
9. Thielheim,K.O.
   Proc. 2nd ESO-CERN Conf., München, FRG (1986), in press
10. Thielheim,K.O.
   Proc. IAU-Symposium No. 125, Nanjing, China (1986), in press
11. Laue,H. and Thielheim,K.O.
   Astrophys. J. (1986), in press
12. Thielheim,K.O.
   (1986), submitted for publication
13. Leinemann,R. and Thielheim,K.O.
   (1986), submitted for publication

CONDITIONS FOR ACCELERATION OF SUPER-HIGH ENERGY COSMIC RAYS IN ACTIVE GALACTIC NUCLEI

Wilfred H. Sorrell
Astronomy Department
University of Wisconsin-Madison
Wisconsin 53706 U.S.A.

ABSTRACT. Super-high energy ($\sim 10^{20}$ eV) cosmic rays could be accelerated in current-carrying beams associated with black-hole accretion discs in galactic nuclei. The current system might resemble an electrical circuit with a voltage source (black hole and accretion disc) coupled to a load (cosmic-ray beam) and a low impedance element (ambient conducting medium). Beam particles can be accelerated to 10^{20} eV provided that the beam channel medium behaves like an inductive circuit with high impedance. But even if these conditions are satisfied, a major difficulty the model faces is the large energy loss caused by photomeson reactions.

1. INTRODUCTION

Although extensive air showers reveal the presence of primary cosmic rays at energies above 10^{19} eV, the composition of these particles is still uncertain. It is likely that a significant flux of protons is present because of the extent to which the air showers fluctuate in their height of development (Walker and Watson 1982). Charged dust grains are unlikely candidates because the depth of shower development constrains the masses of primaries to be much smaller than 10^4 amu (Linsley 1981).

The nature of the cosmic-ray composition strongly influences ideas about the origin of 10^{20} eV particles and about magnetic confinement of particles to our galaxy. As the interstellar magnetic field (~ 3 μG) is unable to confine protons with Larmor radii $\sim$ galactic dimensions (~ 10 kpc), it is often argued that 10^{20} eV protons must originate from nearby extragalactic sources (Virgo supercluster). But Hillas (1983, 1984) pointed out that 10^{20} eV cosmic rays could originate from sources in our galaxy provided that (1) the particles are predominately heavy nuclei (carbon or iron) and (2) the magnetic halo of our galaxy extends to scale heights $\gtrsim 10$ kpc consistent with halo radio observations discussed by Phillips et al. (1981).

It is difficult to decide whether the super-high energy particles originate from galactic or extragalactic sources on grounds of

M. M. Shapiro and J. P. Wefel (eds.), Genesis and Propagation of Cosmic Rays, 227–233.

composition considerations alone. Silberberg and Shapiro (1983) have argued in favour of an extragalactic origin based on the abrupt change in the cosmic-ray spectrum above 10^{19} eV and the preferential arrival directions near the north galactic pole. These authors concluded that the nuclei of Seyfert galaxies in the Virgo supercluster are the principal sources of 10^{20} eV particles. Because this interpretation is the most likely hypothesis at present, it is of interest to consider problems associated with the likely acceleration mechanism and possible ways around the difficulties.

Cavallo (1978) and Hillas (1984) have already summarized the problems associated with particle acceleration by Fermi mechanisms, galactic shocks, and pulsar magnetospheres. None of these proposed mechanisms appear capable of accelerating protons to 10^{20} eV, and at the same time, satisfy source size constraints and avoid particle energy losses from photomeson production, curvature radiation, or e^+e^- pair creation. The present discussion will focus attention on cosmic-ray acceleration by electric dynamos generating relativistic current beams (see Lovelace 1976; Benford 1978). This mechanism has two advantages over others at the outset: First, the cosmic rays are expected to accelerate along magnetic field lines and acquire negligible kinetic energy of motion transverse to field lines. This implies that the particle orbits have zero pitch angles, so the Hillas (1983) constraints on Larmor radii are automatically satisfied. Secondly, as the particles acquire negligible transverse kinetic energy of motion, synchrontron radiation losses are likewise negligible (Sturrock 1971).

2. CONDITIONS FOR COSMIC-RAY ACCELERATION

Black-hole accretion disc with an axisymmetric magnetic field system is expected to behave like an electric voltage source (Lovelace 1976). At distance r from the disc axis of rotation, the poloidal magnetic field $B_p(r)$ threading the disc in a direction anti-parallel to the angular momentum vector is given by

$$B_p(r) = B_S \, (a/r)^{1/2} \quad , \quad a = 6 \, GMc^{-2} \quad , \tag{1}$$

where M is the mass of the black hole, a is the inner disc radius, and B_S is a reference magnetic field strength. As the accretion disc is a perfect conductor, an observer in a non-rotating reference frame will see a radial electric field $E_r(r) = u_\phi(r) B_p/c$ on the disc (top and bottom) surfaces, where $u_\phi(r) = (GM/r)^{1/2}$ is the Keplerian velocity of the disc. The electric voltage drop across the disc is given by

$$V_S = \int_a^b E_r(r) dr = 6^{1/2} \frac{GM}{c^2} B_S \, \ell n \, \frac{b}{a} \quad , \tag{2}$$

where b is the outer radius of the disc, which is determined, in principle, by the specific angular momentum of the ambient infalling plasma. For present purposes we choose the relation $b/a = \lambda \gtrsim 20$ in accordance with standard accretion disc theory (see Novikov and Thorne 1972).

The magnetic flux through the accretion disc is given by

$$\phi_{disc} = (4\pi/3)\, B_S\, a^2\, \lambda^{3/2} \quad , \tag{3}$$

This flux should equal the magnetic flux ϕ_0 threading the galactic nucleus. Thus magnetic flux conservation yields

$$B_S = 3\Phi_0/4\pi\lambda^{3/2}\, a^2 \quad , \tag{4}$$

Hence the disc voltage source has a potential drop given by

$$V_S = \frac{3\, \ln\, \lambda}{2\pi(6\lambda)^{3/2}} \left(\frac{\Phi_0 c^2}{2\, GM} \right) \tag{5}$$

An electric field E_z is expected to be produced in the space above and below the disc. This field would drive an ion current beam from the disc surfaces and along magnetic field lines coupled to the disc. The suggestion that an electrostatic field is present along magnetic field lines might cause many astrophysicists to wince because the frozon-in field theorem says such a phenomenon cannot occur. Nevertheless, this theorem applies to a low impedance (high conductivity) medium, whereas the current beam would be subjected to plasma instabilities that would greatly lower the electrical conductivity and thereby increase the impedance of the medium.

Lake and Pudritz (1985) suggested that instabilities would make the current beam behave like an electrical circuit with a load resistance R_0 and an inductance L_0. The accretion disc acts like a voltage source driving a current impeded by the resistance of space charge effects. The current beam is expected to generate a self-toroidal field that pinches the beam against the space charge repulsion. Because the system has a large inductance, the Bennett pinch and other instabilities could trigger a discharge voltage drop V_{disch} by converting stored magnetic energy into high-energy particles (cf. Alfven and Carlqvist 1967, Lake and Pudritz 1985). The timescale on which the circuit builds up current and stores magnetic energy in the inductance is given by

$$t_{store} = L_o/R_o \quad . \tag{6}$$

The voltage source drives a current $I = V_S/R_0$ while the magnetic energy of the inductance is consumed through a discharge voltage

$$V_{disch} = L_o \frac{I}{t_{disch}} = (t_{store}/t_{disch})\ V_S \quad . \tag{7}$$

Benford (1978) argues that the circuit is closed by a non-relativistic electron current flowing from the ambient plasma into the voltage source. This return current would also help the disc to restore charge neutrality.

The voltage drop V_{disch} is expected to be generated along magnetic field lines threading the disc. This requires the beam channel medium to suddenly become non-conducting with a very high impedance produced by instabilities on timescales $\sim t_{disch}$. The ratio t_{store}/t_{disch} should depend on non-linear effects of plasma instabilities throughout the discharge region. Lake and Pudritz (1985) suggested $t_{store}/t_{disch} \sim 10^2$, based on timescales $\sim 10^4 - 10^5$ sec for X-ray flux variations in active galactic nuclei.

We might ask what conditions can be imposed on t_{store}/t_{disch} in order to accelerate particles to relativistic energies. To calculate the energy gain per particle of mass m and charge Ze, we consider a simple illustrative model for the electric discharge voltage:

$$V_{disch}(z) = \begin{cases} V_S\ \Psi(z) & , \ 0<r<b \ , \ |z| > 0 \\ 0 & , \ r \geq b \ , \ |z| > 0 \\ V_S & , \ a \leq r \leq b \ , \ |z| = 0 \end{cases} \tag{8}$$

where

$$\Psi(z) \equiv t_{store}/t_{disch}$$

with boundary conditions $\Psi(z) = 1$ at $|z| = 0$ and $\Psi(z) = \Psi_\infty$ = constant at $|z| = \infty$. The accelerating electric field is

$$E_z\ (\Psi) = \begin{cases} V_S \dfrac{d}{dz} \Psi(z) & , \ 0<r<b \ , \ |z| \geq 0 \\ 0 \ , \ \text{otherwise} \end{cases} \tag{9}$$

We shall assume that the particle orbits have beamlike pitch angles so that motion occurs only along magnetic field lines. The equation of motion for the particle velocity $u(\Psi)$ is

$$\frac{d}{dt}\left[\frac{mu}{\sqrt{1-u^2/c^2}}\right] = Ze\ E_z\ (\Psi) \tag{10}$$

The solution is given by

$$u(\Psi) = c \left[1 - \frac{1}{(1+\gamma_0(1-\Psi))^2} \right]^{1/2} \tag{11}$$

where

$$\gamma_0 \equiv Ze\, V_S/mc^2 \tag{12}$$

The kinetic energy is therefore

$$W = mc^2 \left[\frac{1}{\sqrt{1-u(\Psi)^2/c^2}} - 1 \right] \tag{13}$$

which becomes

$$W_\infty = (Ze\, V_S)(1-\Psi_\infty) \tag{14}$$

at infinity. Hence, acceleration to high energies is possible provided that t_{store}/t_{disch} decreases outwards along the beam. Thus the current beam could resemble a spark gap in laboratory vapour discharges (cf. Alfven and Carlqvist 1967).

The condition $t_{store}/t_{disch} \lesssim 1$ for acceleration of cosmic rays disagrees with the suggestion that $t_{store}/t_{disch} \sim 10^2$ (see Lake and Pudritz 1985). The results presented here suggest that both the discharge voltage and the particle energy gain can never exceed the energy supplied by the disc voltage source. Such results are consistent with the model discussed by Lovelace (1976).

3. ESTIMATE OF COSMIC RAY ENERGY

If the above conditions are satisfied, then protons will carry away an amount of energy

$$W_\infty \sim eV_S \sim \frac{3\ \ell n(20)}{16\pi(30)^{3/2}} \left(\frac{e\Phi_0 c^2}{2GM} \right) \qquad (\Psi_\infty = 0) \tag{15}$$

where $\lambda = 20$. As a rough estimate of the magnetic flux, we take $\Phi_0 \sim 2.8 \times 10^{35}$ gauss cm^2. This corresponds to a magnetic field strength $\sim 10^{-6}$ gauss on scales ~ 100 pc. The disc field strength is then $B_S \sim 9.2 \times 10^4$ gauss for a typical black hole mass $M = 10^8\ M_\odot$ (Soltan 1982). These values yield

$$W_\infty \sim 3 \times 10^{21}\ \text{eV} \quad \text{for protons} \tag{16}$$

4. ENERGY LOSS

The above estimate for the cosmic ray energy assumes that particles can gain the full voltage supplied by the accretion disc. Nevertheless, the accelerated particles interact with radiation from the disc, and the interactions could lead to particle energy losses by way of photomeson production and e^+e^- pair creation. Colgate (1983) has discussed this problem for quasars, BL Lacertae Objects, and Seyfert nuclei with radiation luminosity $\sim 10^{46}$ erg s^{-1}. He concludes that cosmic-ray energy gains are attenuated down to about 10^{13} eV on photomeson reaction times << 1 yr. Such severe energy losses occur because quasars and active galactic nuclei have radiation energy densities $\sim 10^{18}$ eV cm^{-3} for luminosities $\sim 10^{46}$ erg s^{-1}. Thus, the high luminosities of active galactic nuclei actually discourage acceleration of cosmic rays to super-high energies. Colgate (1983) finds that photomeson damping effects are the strongest for the highest energy particles.

5. DISCUSSION

Although the accretion disc dynamo can supply an adequate voltage to accelerate cosmic rays to high energies, the high radiation luminosities of active galactic nuclei lead to severe energy losses that keep cosmic ray energies below $\sim 10^{13}$ eV. One possible way around the difficulty is to place the discharge voltage drop at distances $\gtrsim$ 100 pc from the central engine. But as active galactic nuclei contain a large amount of forbidden-line gas, such highly conducting gas might become entrained into the beam channel and short-circuit the current. This would be the same as making $\Psi_\infty \sim 1$ (in equation [14]) by lowering the impedance. Thus, unless entrainment and short-circuiting can be avoided, it would appear unlikely that active galactic nuclei are sources of super-high energy cosmic rays.

I am grateful to Dr. H. Völk for a careful reading of the manuscript.

REFERENCES

Alfven, H. and Carlqvist, P. 1967. Solar Physics, 1, 220.
Benford, G. 1978. Mon. Not. R. astr. soc. 183, 29.
Cavallo, G. 1978. Astron. Astrophys. 65, 415.
Colgate, S.A. 1983. Proc. Int. Conf. Cosmic Rays, 18th, Bangalore, 2, 230.
Hillas, A.M. 1983 in Composition and Origin of Cosmic Rays, ed. M.M. Shapiro, 125: Dordrecht, Reidel.
Hillas, A.M. 1984. Ann. Rev. Astron. Astrophys. 22, 425.
Lake, G. and Pudritz, R.E. Proc. 1985. IAU Symp. 107, 471.
Linsley, J. 1981. IAU Symp. 94, 53.
Lovelace, R.V.E. 1976. Nature 262, 649.

Novikov, I.D. and Thorne, K.S. 1972 in Black Holes, ed. C. De Witt and B. De Witt, New York: Gordon and Breach.
Phillips, S., Kearsey, S., Osborne, J.L., Haslam, C.G.T., and Stoffel, H. 1981. Astron. Astrophys. 103, 405.
Silberberg, R. and Shapiro, M.M. 1983 in Composition and Origin of Cosmic Rays, ed. M.M. Shapiro, 231: Dordrecht, Reidel.
Soltan, A. 1982. Mon. Not. R. astr. soc. 200, 115.
Sturrock, P.A. 1971. Astrophys. J. 164, 529.
Walker, R. and Watson, A.A. 1982. J. Phys. G. 8, 1131.

COSMIC RAYS AND A STABLE HYDROSTATIC EQUILIBRIUM OF THE GALAXY

Hans Bloemen
Astronomy Department
University of California, Berkeley, CA 94720

ABSTRACT. It is argued that the stellar, gaseous, magnetic-field, and cosmic-ray components in the Galaxy may currently be in a large-scale hydrostatic equilibrium that is stable against Parker type instabilities. Equilibrium configurations considered in the past were found to be unstable as a consequence of simplifying assumptions, which are inconsistent with recent observations.

1. INTRODUCTION

Since the classical work of Parker (1966, 1969 and references therein) it is well-known that cosmic rays and magnetic fields play an important role in a possible (quasi) hydrostatic-equilibrium configuration of the Galaxy. In such an equilibrium state the gravitational attraction of the stars and interstellar matter towards the galactic plane is balanced by the pressure of the gas, the magnetic field, and the cosmic-ray (CR) particles. Parker considered the case where the lines of force of the magnetic field are parallel to the galactic plane (a horizontal equilibrium). The observed properties of the galactic magnetic field do not preclude a large-scale horizontal equilibrium. Studies of the interstellar polarization of starlight (e.g. Mathewson and Ford 1970), the polarization of synchrotron emission (e.g. Wilkinson and Smith 1974), and the rotation measures of pulsars (Manchester 1974; Thomson and Nelson 1980) and extragalactic radio sources (e.g. Simard-Normandin and Kronberg 1979; Vallée 1983) all indicate that there is a systematic magnetic-field component, preferentially aligned parallel to the galactic plane. The observed random field component of comparable strength, superimposed on the systematic field, has a stabilizing effect (§2). The equilibrium equation is simple for a plane-parallel configuration:

$$\frac{d}{dz}\Big\{P_g(z) + P_{mf}(z) + P_{cr}(z)\Big\} = -\rho(z)g(z) \qquad \text{or} \qquad P_{tot}(z) = \int_z^{\infty} \rho(x)g(x)dx,$$

where P_{tot} represents the total internal pressure (given by the term in braces) as a function of distance z to the mid plane ($P_{tot} \to 0$ if $z \to \infty$). P_g is the gas pressure (thermal and due to macroscopic motions), P_{mf} is the magnetic-field pressure ($B^2/8\pi$, where B is the strength of the plane-parallel magnetic field), P_{cr} is the CR pressure (equal to 1/3 of the CR energy density), g is the z-component of the gravitational acceleration, and ρ is the mean gas density. Parker solved this equation for the case that the gas is isothermal

M. M. Shapiro and J. P. Wefel (eds.), Genesis and Propagation of Cosmic Rays, 235–240.

assuming that $P_{mf} = \alpha P_g$ and $P_{cr} = \beta P_g$, where α and β are dimensionless constants, and showed that the resulting equilibrium is unstable on a time scale of the order 10^7 years. This instability arises from any perturbation which depresses the magnetic lines of force at one point and raises them at another. The gas tends to slide into the depression and concentrate into clouds or cloud complexes (Shu 1974; Mouschovias *et al.* 1974), further depressing the field there, and the CR pressure (which remains uniform along the MF lines) tends to inflate the raised portions of the field. At the disk boundary, the CR inflation of such a bubble may lead to escape from the Galaxy if the inflation is fast. The rate of inflation is probably governed by the random walk of the magnetic-field lines and the CR production rate. If the inflation is slow, most cosmic rays may return to the disk. For equilibrium studies of the final state of the Parker instability see Mouschovias (1974, 1975).

Subsequent studies of possible horizontal equilibrium configurations for the present state of the Galaxy (Kellman 1972; Wentzel *et al.* 1975; Fuchs *et al.* 1976; Fuchs and Thielheim 1979) also assumed that the gas is isothermal and that the pressure components all decrease in the same way with distance from the mid plane; all these equilibrium models are unstable. Some treated the molecular clouds as a separate dynamical system, not coupled to the magnetic field, but they made the same assumptions for the gas-field system. It is evident now that these assumptions are incompatible with a variety of observations (such as for instance the detections of cold, warm, and hot material; the large scale height of the diffuse radio synchrotron emission; the relatively large CR scale height suggested by ^{10}Be measurements). Badhwar and Stephens (1977) dropped these assumptions and derived horizontal-equilibrium models for the solar vicinity that account for the observed radio synchrotron emission in the direction of the galactic poles, but Lachièze-Rey *et al.* (1980; §2) showed that these models are unstable. Also, the hydrostatic-equilibrium models that have been proposed for the gaseous halo (e.g. Weisheit and Collins 1976; Chevalier and Fransson 1984; Fransson and Chevalier 1985) are unstable. This leads to an obvious question, which is addressed in this paper: can the present state of the Galaxy be described at all by a stable horizontal-equilibrium configuration?

2. THE STABILITY CRITERION

The ISM in the galactic plane is subject to a permanent agitation (supernova explosions – cloud motions), but this may not preclude a large-scale hydrostatic equilibrium, particularly away from the plane. In fact, Zweibel and Kulsrud (1975) showed that a small-scale (much smaller than the scale height) turbulent component of the magnetic field, which they attributed to cloud motions, in a horizontal equilibrium of the Parker type has a stabilizing effect due to magnetic shear stresses. Parker (1975) noted that they ignored the most unstable modes (those with short wavelengths perpendicular to both the gravitational and magnetic field), but Lachièze-Rey *et al.* (1980) showed that the latter modes are completely stabilized by the turbulence in the field. From their results it can be calculated that the energy density of the turbulent field has to be only a minor fraction of that of the systematic field. In fact, the first effect of the onset of a Parker instability is to generate this small-scale turbulence. The stabilizing effect of the turbulence on the two-dimensional instability, considered by Zweibel and Kulsrud, is only significant if the energy density of the turbulent field is comparable to that of the systematic field. There is good observational evidence for such a turbulent field on scales of tens of parsecs, but let us forget about this for a while and consider the possibility of a stable equilibrium without additional stabilizing effects.

The normal-modes technique applied by Parker cannot be extended easily to more complex configurations (such as for instance those with α and β being functions of z), but Lachièze-Rey *et al.* (1980) derived a local instability criterion, based on an energy principle, that applies to any horizontal equilibrium state. Using the equilibrium equation, their instability criterion can be written as

$$\frac{1}{P_g}\frac{dP_{tot}}{dz} < \gamma\frac{1}{\rho}\frac{d\rho}{dz},$$

which shows close resemblence to the well known criterion for convective instability of a gas ($dS/dz < 0$; S is the entropy), which contains P_g instead of P_{tot}. γ is the adiabatic index. A configuration meeting the stability criterion of Lachièze-Rey *et al.* is also not convectively unstable, assuming that the sum of magnetic-field and CR pressures does not increase with z. Basically, this criterion defines the minimum fraction of the total internal pressure that has to be due to gas pressure in order to have a system that is not unstable. For an equilibrium of the Parker type, the criterion reduces again to $\gamma < 1 + \alpha + \beta$. It is assumed here that cosmic rays distribute very rapidly along a field line, which is the most stringent case. In addition, let us impose another stringent requirement, namely $\gamma \lesssim 1$ (as considered by Parker). The value of γ is uncertain and position dependent; for a perfect gas γ is $\leq 5/3$ (the maximum ratio of the specific heats), but radiative cooling and inelastic cloud collisions may reduce γ to values below 1. Zweibel and Kulsrud (1975) considered heating sources and derived values as large as 2.

3. EQUILIBRIUM STATES & OBSERVATIONS

Only for the solar vicinity the information on the stellar, gaseous, cosmic-ray, and magnetic-field components is sufficiently accurate to perform a meaningful analysis of the equilibrium and stability of the galactic disk. Still, some relevant characteristics of the system are not well determined by observations, particularly the density, scale height, and temperature of the gaseous halo and the scale height of the cosmic rays and the magnetic field, but there are some observational constraints.

There is conclusive evidence for highly ionized gas far from the plane, obtained primarily from *IUE* observations of C IV and Si IV absorption lines (Savage and de Boer 1979, 1981; Pettini and West 1982; de Boer and Savage 1983). The absorbing halo gas appears to be concentrated at about 1 – 3 kpc from the plane and, depending on the ionizing mechanism adopted, characteristic temperatures of $10^4 - 10^5$ K and densities of $10^{-3} - 10^{-2}$ cm^{-3} have been derived. A significantly hotter halo component extending to much larger distances from the plane might exist [a 10^6 K halo was suggested by Spitzer (1956)], which would not be observable with *IUE*. There is some evidence from soft X-ray observations that such a halo exists (Nousek *et al.* 1982). Although the halo density is small, it has a major impact on the total pressure that is required to balance the gravitational attraction, because of the large z-integration range. The halo density and scale height have therefore to be taken as free parameters in an equilibrium analysis. Exponential density distributions are considered here: $n_{halo}(z) \propto \exp(-z/h_{halo})$. Figure 1 shows some examples of the total pressure distribution, determined from the equilibrium equation, together with the minimum gas pressure that is required for stability. Near the plane, the gas distribution used can be considered "standard" [including the atomic hydrogen with a large exponential scale height of 400–500 pc found by Lockman (1984)]; details are given by Bloemen (1986). The g-distribution presented by Bahcall

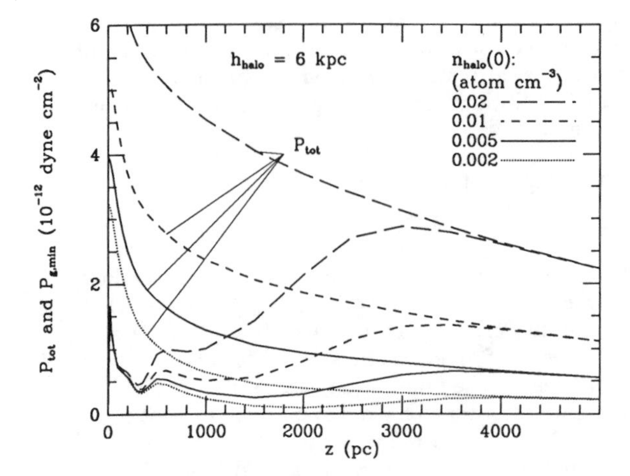

Figure 1: Examples of the distribution of the total internal pressure that is required to balance the gravitational attraction in the solar vicinity and the corresponding distributions of the minimum gas pressure that is required for stability ($\gamma < 1$).

(1984) was used ($z \lesssim 1$ kpc); the dark material is assumed to be distributed as the observed material (stellar and interstellar) and a massive halo, consistent with the rotation curve of the Galaxy, is included ($\rho_{halo}(0)/\rho_{disk}(0) = 0.1$) [see Bahcall 1984]. At $z \gtrsim 1300$ pc, g was taken to be constant ($g_c = 8.3 \times 10^{-9}$ cm s^{-2}). From the examples shown in figure 1, it is clear that the gas pressure is allowed to be significantly smaller than the total internal pressure up to a few kpc from the mid plane, but at larger distances the gas has to be supported by its own (thermal) pressure, requiring a temperature of $\sim 2 \times 10^5 g_c h_{halo}$ (g_c in units of 10^{-8} cm s^{-2} and h_{halo} in kpc).

The residual pressure ($P_{tot} - P_{g,min}$) increases with halo density and is the maximum that can be attributed to magnetic-field and CR pressure. From a comparison with radio continuum observations towards the galactic poles, one can derive lower limits for the density of the halo as a function of its scale height. It turns out that, essentially independent of further assumptions, this relation is approximately given by ($h_{halo} \lesssim 15$ kpc)

$$\left(\frac{n_{halo}(0)}{0.01 \text{ atom cm}^{-3}}\right)\left(\frac{h_{halo}}{1 \text{ kpc}}\right)^2 > \Gamma,$$

with $\Gamma \simeq 17$ if the magnetic-field pressure and the cosmic rays are assumed to have the same (free) z-distribution. So $n_{halo}(0) \gtrsim 0.005$ atom cm^{-3} for the example in figure 1. It was already pointed out by Badhwar and Stephens (1977) that the presence of a gaseous halo is of essential importance to reconcile equilibrium configurations with radio-continuum observations, but the stability criterion requires that its temperature is at least an order of magnitude higher than the value (10^4 K) assumed by Badhwar and Stephens (see below). This high temperature suggests that the maximum halo density is about twice the minimum density given above and that the scale height is probably $\gtrsim 5$ kpc. This is necessary in order (*i*) to have rough pressure equilibrium with the other phases of the ISM, (*ii*) to avoid the situation in which radiative cooling dominates the supernova heat input, and (*iii*) to limit the amount of soft X-rays produced (see Bloemen 1986).

The temperature of the hot medium in the disk and at large distances from the plane ($z \gtrsim 3$ kpc) is found to be typically $\sim 10^6$ K, whereas around $z = 1 - 3$ kpc the temperature is only $(2 - 3) \times 10^5$ K. Such a layer of gas with relatively low temperature is in good agreement with the IUE observations of C IV and Si IV absorption lines from the halo, which seem to originate at these distances from the plane and which require a temperature less than a few times 10^5 K. The results presented here favour collisional ionization, rather than photoionization of a 10^4 K halo as the source of this ionized gas. If the isotropic component of the observed soft X-ray emission in the 0.5 – 1.2 keV MI band (Nousek *et al.* 1982) comes from the halo, it has to originate at $z \gtrsim 3$ kpc in this scenario. The resulting scale height of the sum of cosmic-ray and magnetic-field pressures is only weakly dependent on the scale height of the gaseous halo: the half-equivalent width is 2 – 2.5 kpc for $h_{halo} \gtrsim 5$ kpc. It cannot be excluded that the scale height of the CR nuclei is significantly smaller.

This approach leads to a consistent picture, but it should be kept in mind that the actual parameter values given here are flexible because possible stabilizing effects, as discussed above, have been ignored. In addition, the stability criterion used here is *necessary* (and was applied in a stringent form), but it may not be *sufficient*. However, none of the equilibrium configurations for the present state of the Galaxy studied previously meets this criterion and this exercise shows at least that the simplifying assumptions made in the past may have led to doubtful results.

ACKNOWLEDGEMENTS

I thank E. Bertschinger and C. McKee for valuable comments and gratefully acknowledge receipt of a Miller Fellowship.

REFERENCES

Bahcall, J.N. 1984, *Ap. J.*, **276**, 169.
Badhwar, G.D., and Stephens, S.A. 1977, *Ap. J.*, **212**, 494.
Bloemen, J.B.G.M. 1986, in prep.
Chevalier, R.A., and Fransson, C. 1984, *Ap. J. (Letters)*, **279**, L43.
Cox, D.P., and Smith, B.W. 1974, *Ap. J. (Letters)*, **189**, L105.
Fansson, C., and Chevalier, R.A. 1985, *Ap. J.*, **296**, 35.
Fuchs, B., and Thielheim, K.O. 1979, *Ap. J.*, **227**, 801.
Fuchs, B., Schlickeiser, R., and Thielheim, K.O. 1976, *Ap. J.*, **206**, 589.
Jokipii, J.R. 1976, *Ap. J.*, **208**, 900.
Jones, F.C. 1979, *Ap. J.*, **229**, 747.
Kellman, S.A. 1972, *Ap. J.*, **175**, 353.
Lachièze-Rey, M., Asséo, E., Cesarsky, C.J., and Pellat, R. 1980, *Ap. J.*, **238**, 175.
Lockman, F.J. 1984, *Ap. J.*, **283**, 90.
Manchester, R.N. 1974, *Ap. J.*, **188**, 637.
Mathewson, D.S., and Ford, V.L. 1970, *Mem. Roy. Astron.Soc.*, **74**, 139.
McKee, C.F., and Ostriker, J.P. 1977, *Ap. J.*, **218**, 148.
Mouschovias, T.C. 1974, *Ap. J.*, **192**, 37.
Mouschovias, T.C., Shu, F.H., and Woodward, P.R. 1974, *Astr. Ap.*, **33**, 73.
Mouschovias, T.C. 1975, *Astr. Ap.*, **40**, 191.
Nousek, J.A., Fried, P.M., Sanders, W.T., and Kraushaar, W.L. 1982, *Ap. J.*, **258**, 83.
Parker, E.N. 1966, *Ap. J.*, **145**, 811.
Parker, E.N. 1969, *Space Sci. Rev.*, **9**, 651.
Parker, E.N. 1975, *Ap. J.*, **201**, 74.

Shu, F.H. 1974, *Astr. Ap.*, **33**, 55.
Simard-Normandin, M., and Kronberg, P.P. 1979, *Ap. J.*, **242**, 74.
Spitzer, L. 1956, *Ap. J.*, **124**, 20.
Thomson, R.C., and Nelson, A.H. 1980, *M.N.R.A.S.*, **191**, 863.
Vallée, J.P. 1983, *Astr. Ap.*, **124**, 147.
Webber, W.R. 1983, in *Comp. and Origin of Cosmic Rays*, ed. M. Shapiro, Reidel, 83.
Weisheit, J.C., and Collins, L.A. 1976, *Ap. J.*, **210**, 299.
Wentzel, D.G., Jackson, P.D., Rose, W.K., and Sinha, R.P. 1975, *Ap. J. (Lett.)*, **201**, L5.
Wilkinson, A., and, Smith, F.G. 1974, *M.N.R.A.S.*, **167**, 593.
Zweibel, E.G., and Kulsrud, R.M. 1975, *Ap. J.*, **201**, 63.

VHE AND UHE GAMMA RAY OBSERVATIONS BY GROUND BASED DETECTORS

Wilhelm Stamm
Institut für Reine und Angewandte Kernphysik
University of Kiel
Olshausenstraße 40
D - 2300 Kiel
Federal Republic of Germany

ABSTRACT. Some basic information about gamma ray observations by ground-based techniques are given. In the energy region from 10^{11} eV to 10^{14} eV the Cherenkov light produced by the cascade particles in the earth's atmosphere can be received at ground. For energies greater than 10^{14} eV the particles of the cascade can be measured by detector arrays. The flux sensitivities of typical installations are presented.

1. INTRODUCTION

The domain of gamma ray astronomy starts at photon energies of 10^5 eV and probably ends at photon energies of 10^{20} eV. At low energies (100 keV - 3 MeV) registration is based on photoelectric effect or on Compton scattering. The arrival direction of the impinging photon can be determined by collimators or by the coded mask technique.

At medium energies (3 MeV - 10 MeV) two interactions, namely Compton scattering followed by photoelectric absorption of the same photon, give information about its arrival direction and energy (Compton telescope).

In the HE (High Energy) range (10 MeV - 100 GeV) pair production is the dominant process. With increasing energy electron and positron move more and more exactly in the direction of the incident photon. Their tracks can be measured in a spark chamber for example.

M. M. Shapiro and J. P. Wefel (eds.), Genesis and Propagation of Cosmic Rays, 241–254.

Most of our knowledge about cosmic HE gamma rays is based on the measurements of the SAS-2 and COS-B satellites. More than 200.000 photons in the energy range from 50 MeV to several GeV were detected during the very successful COS-B mission within 6 1/2 years (1975-1982). In addition to a diffuse gamma radiation which can be explained as interaction of cosmic rays with the matter of the galactic disc 25 gamma ray point sources were detected (1). The typical gamma ray flux of these sources amounts to 10^{-6} photons cm^{-2} s^{-1} (E > 100 MeV). Even if the integral spectrum is as flat as $N(>E) \sim E^{-1}$ the 24 x 24 cm^2 spark chamber of the COS-B satellite would have detected only 2 photons of energies greater than 10^{12} eV within one year from such a source. Due to the low flux the energy range beyond 100 GeV will be inaccessible to satellite borne detectors in the near future.

But photons of such high energies can be detected by the cascade showers in the earth's atmosphere. In the VHE (Very High Energy) range 10^{11} eV to 10^{14} eV the atmospheric Cherenkov light of the shower particles can be detected at ground. In the UHE (Ultra High Energy) region $E > 10^{14}$ eV the shower particles can be measured.

2. CASCADE SHOWERS IN THE ATMOSPHERE

Photons of very high energies initiate cascade showers in the earth's atmosphere. By the combined phenomena of pair production and bremsstrahlung the number of particles is increasing exponentially and mainly due to Coulomb scattering they are spread laterally. Based on the well-known electromagnetic interactions there is a lot of theroretical work on cascade development. Besides an analytical treatment (e.g.ref. 2) there exists a large number of Monte Carlo simulations taking all electromagnetic processes into account and giving in addition information about the fluctuations to be expected in experimental data. Again the results of Monte Carlo calculations can be given in the form of analytical expressions (e.g.ref. 3). Fig. 1 shows the number of particles in the electromagnetic cascade as a function of the atmospheric depth. The total number of particles at the observation level is referred as shower size and is related to the primary energy.

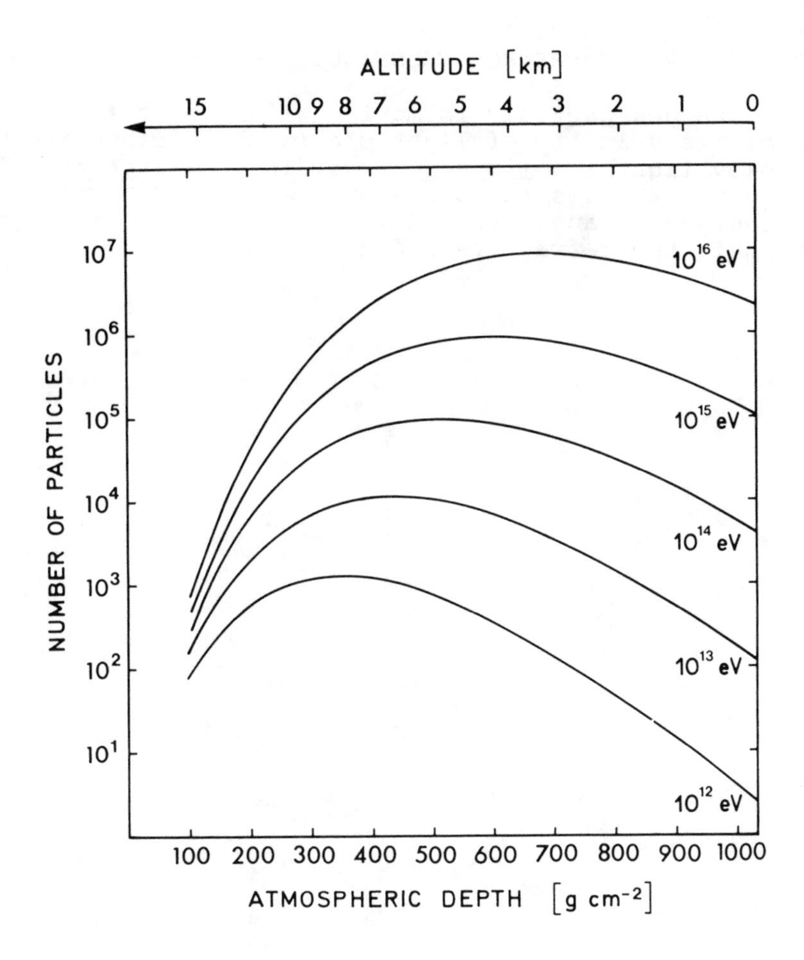

FIGURE 1: Number of particles in photon initiated air showers as a function of atmospheric depth for different energies of the primary photon. Calculations are based on the formula of Hillas (3).

The lateral distribution of particles can be described by the Nishimura Kamata Greisen (NKG)-function (4):

$$f\,(r/R_m) = C\,(s) \bullet (r/R_m)^{s-2} \bullet (r/R_m + 1)^{s-4.5}$$

s age parameter
r distance from shower axis
R_m Molière unit
C(s) normalization factor

3. ATMOSPHERIC CHERENKOV TECHNIQUE

If a charged particle is moving with a velocity v which exceeds the velocity of light c/n in the surrounding medium Cherenkov light is emitted. According to Huyghens priciple the wavefront forms the surface of a cone (see Fig. 2) with the Cherenkov angle θ and it holds $\cos \theta = 1/(n \cdot \beta)$ where n is the refractive index.

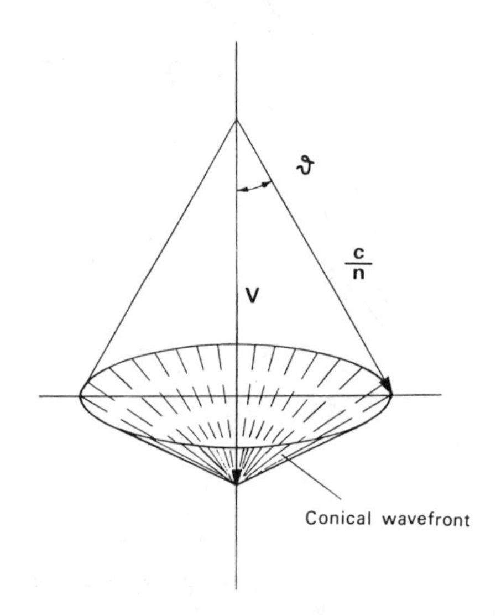

FIGURE 2: Emission of Cherenkov light by a particle moving with the velocity v through a medium of refractive index n.

The refractive index of air is very close to 1. At sea-level the maximum Cherenkov angle is about 1.4° and the Lorentz factor of the charged particle must exceed a threshold of 41.5. The Cherenkov angle becomes smaller with an increase in altitude (Fig. 3).

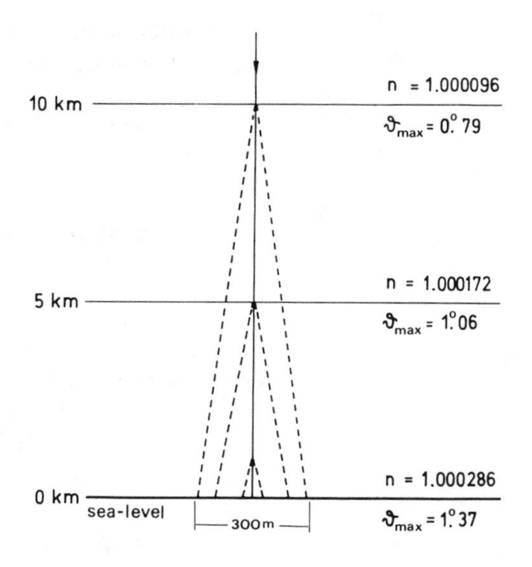

FIGURE 3: Emission of Cherenkov light by a single particle of very high energy moving vertically in the earth's atmosphere.

The Cherenkov light of particles moving parallel through the atmosphere is confined to a light pool of about 250 m diameter at ground. Putting into this area a light collecting reflector with a photo tube in its focus the Cherenkov light flash can be registrated (Fig. 4).

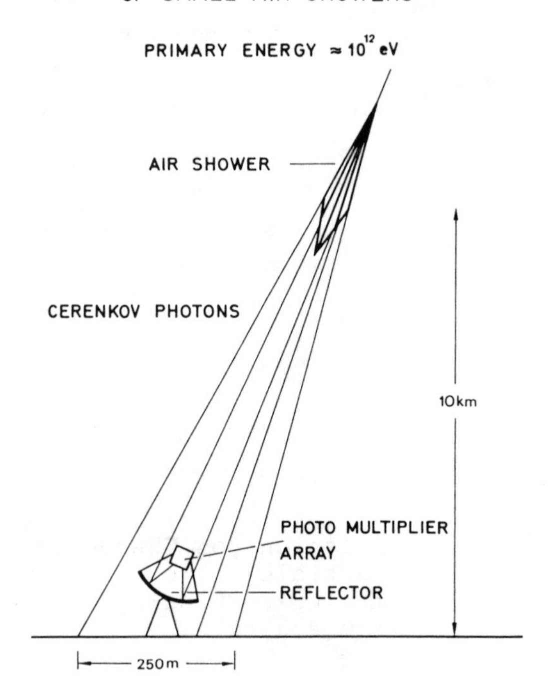

FIGURE 4: Receiver for the Cherenkov light of small air showers.

To predict the photon densities at ground in a more precise approach many effects have to be taken into account. Starting with a VHE photon the angular and energy distributions of the charged particles in the cascade have to be calculated by three-dimensional Monte Carlo simulations. According to the Cherenkov threshold only particles with an energy of more than 21 MeV have to be considered. The angular distribution is affected by the earth's magnetic field. The emission of Cherenkov light varies with the refractive index at different atmospheric heights.

Because the Cherenkov light is emitted mainly at the short wave end scattering and absorption in the atmosphere are of importance (Rayleigh scattering, ozone absorption, and aerosol absorption). Only a half of the light reaches sea-level.

Finally the spectral response of the apparatus has to be taken into account. As results of comprehensive calculations the lateral distribution of Cherenkov photons at the ground is shown in Fig. 5. The photon density is nearly constant up to a distance of 130 m from the shower axis. This corresponds to the sensitive area of a single Cherenkov light receiver for air shower events. For a primary energy of 1 TeV the light intensity amounts to 30 photons per m^2 arriving as a flash of 10^{-8} s.

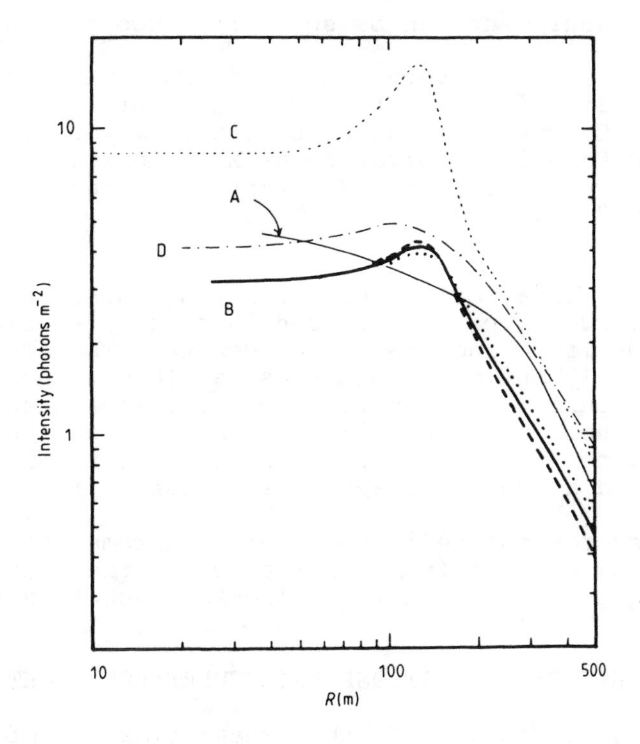

FIGURE 5: Photon densities at sea-level as a function of the distance from the shower axis. Energy of the primary photon is 100 GeV. Results of various comprehensive calculations (A-D) are presented (From Patteson and Hillas, 5).

Because the Cherenkov photons have an angular distribution with respect to the shower axis the image at the celestial sphere is an extended feature of 1^0-2°. It is of circular shape if the shower axis hits the detector and has a comet like shape for larger distances. Recently there are several calculations to find out the differences between hadron and photon induced showers with respect to the Cherenkov light spot.

4. TELESCOPES FOR VHE GAMMA RAY ASTRONOMY

The most simple detector for atmospheric Cherenkov light consists of a reflector with a phototube in its focus. The detector used in the historical experiment of Galbraith and Jelley (6) was of such type.

Modern equipment can be split into two groups:

One using a very large mirror with an array of photomultipliers in its focus in order to record the image of the Cherenkov light spot. The 10 m tessellated reflector of the Whipple Observatory at Mt. Hopkins / Arizona is of such type. It has 37 fast phototubes in its focal plane (7,8).

In the other way an array of telecopes is used. The detector array of the University of Durham at Dugway / Utah was of such type (9). It consists of 3 telescopes separated by 100 metre and one in the centre. They can be operated indepently but moreover it is possible when looking to the same source to measure the time differences which occur when the Cherenkov light disk passes the telescopes. Operating in the latter mode angular resolution can be improved at the expenses of statistics.

Because for the reflectors not a supreme optical quality is required mirrors from solar power stations can be employed at night for atmospheric Cherenkov technique (10).

5. OBSERVATION OF ATMOSPHERIC CHERENKOV LIGHT

The observation of the weak flashes of atmospheric Cherenkov light is possible only in dark and clear nights. So the duty time for a Cherenkov light receiver is small. The site for it has to be selected according to the criteria of optical astronomy.

Gamma ray point sources can only be detected above a high background of showers induced by charged particles. During the time of measurement this background must be known. There are two modes of operation:

At drift scan mode the telescope is fixed. As a result of the earth's rotation the source enters the field of view. At fixed zenith angle a constant background is to be expected. The direct current of the phototubes must be kept constant even if bright stars enter the field of view.

In the tracking mode the telescope follows the source. In this case a background measurement with an off-set in right ascension is necessary covering the same range of zenith angles.

6. EAS ARRAYS FOR UHE GAMMA RAY ASTRONOMY

At energies of the primary photon greater than 10^{14} eV shower measurement can be done by particle detectors at ground. Extensive air shower arrays consist of a number of particle detectors of 1 m^2 typical size separated by some tens of metres. Fitting a lateral distribution function to the sampled particle densities gives information about shower size, core location and age parameter of the shower.

In an air shower all particles are confined in a shower disc moving with the velocity of light through the atmosphere. By measuring the time differences which occur when the shower front hits the detectors the arrival direction of the primary photon or particle can be determined (Fig. 6). This method is based on the work of Bassi et al. (11). Because the shower front is defined well only close to the shower axis the separation of the detectors should not be too large. An angular resolution of about 1^0 can be achieved. According to the cascade development curves there are considerably more particles in the shower disk at high altitude. So the energy threshold of an installation can be lowered. An air shower array for gamma ray astronomy was proposed by Cocconi at the Moscow Conference 1959 (12).

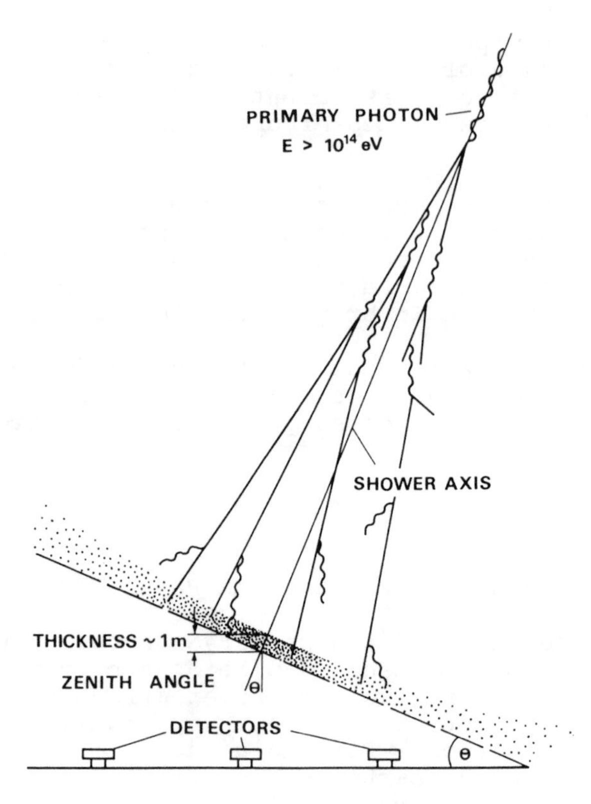

FIGURE 6: Determination of the arrival direction of extensive air showers by fast timing measurements.

7. FLUX INTENSITY OF GROUND BASED INSTALLATIONS

With ground based installations it is generally not possible to identify a primary photon. When measuring only the charged particles of the cascade or the Cherenkov light of these particles photon showers and proton induced showers look quite similar.

Gamma rays from point sources can be detected as an excess of showers from the direction of a sources on the background of cosmic ray showers arriving isotropically.

Due to the limited angular resolutions gamma rays from a point source are spread over an area Ω at the sky. The cosmic ray flux from this area represents the background.

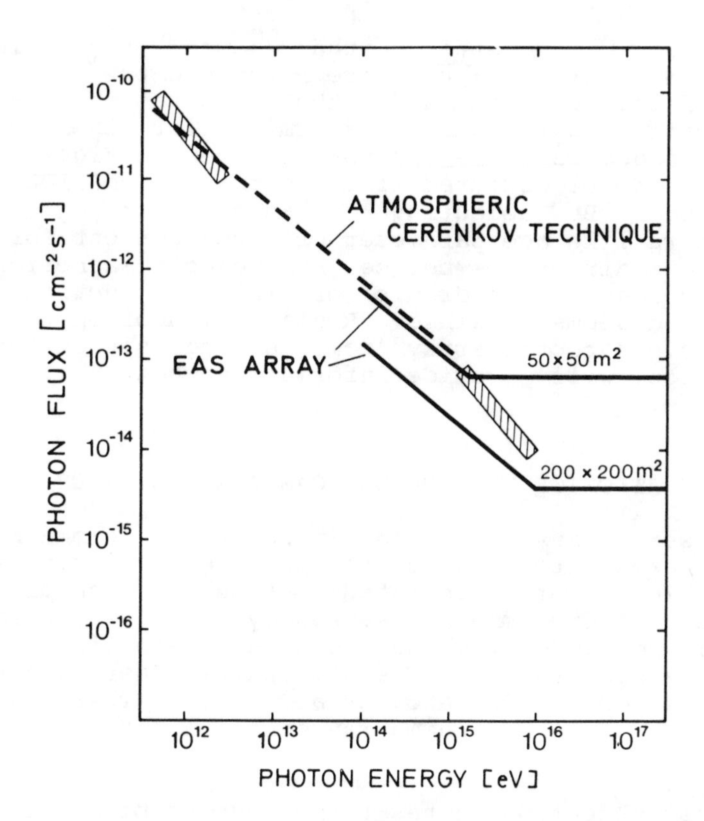

FIGURE 7: Flux sensitivity of ground-based installations for gamma ray astronomy. For Cherenkov light installations the effective area was taken as a circle of 100 m diameter, the observation time as 100 hrs and the background of cosmic rays as from a solid angle of 1 msr.
For the EAS arrays of sensitive area 50 x 50 m^2 and 200 x 200 m^2 the same angular resolution was assumed and the observation time as 1825 hrs. The horizontal lines represent the detection of 10 photons during the observation time. The hatched areas mark measured fluxes from Cygnus X-3.

To find a 3 σ effect the gamma ray flux must exceed $3\sqrt{J_{CR} \cdot \Omega/A \cdot t}$ where A represents the sensitive area of the apparatus and t the registration time and J_{CR} the cosmic ray flux. Flux limits for Cherenkov light detectors and particle arrays are given in Figure 7. Lower fluxes can be detected if the gamma ray emision is time modulated. By selecting cascades according to shower parameters like age parameter or muon content for which is expected a difference between pure electromagnetic cascades and cascades of hadronic origin a reduction of the background seems possible. Registration of photon showers by a big detector array from a strong source with good statistics will provide information what photon showers look like.

8. OBSERVATION OF VHE AND UHE GAMMA RAY SOURCES

There were many activities in observing gamma-rays with ground-based techniques in the early sixties. The observations were stimulated by a publication of Morrison entitled: " On Gamma-Ray Astronomy " (13) and were focused on radio sources for which a concentration of relativistic particles was assumed. The atmospheric Cherenkov technique was employed by Chudakov et al. (14). There were some indications of VHE gamma ray emission but no convincing results.

The first object which has turned out to be a very powerful gamma ray source up to energies of some 10^{16} eV was Cygnus X-3. Using the atmopheric Cherenkov technique Cygnus X-3 was observed from the Crimean Astrophysical Observatory shortly after the giant radio outburst in 1972. An excess (20 % of the background of cosmic ray showers) was detected from the direction of Cygnus X-3 (15). Later it was found out that the VHE gamma ray emission showed the 4.8 hr periodicity well-known from X-ray measurements. In a phase histogram two narrow peaks of VHE gamma ray emission appeared (16). There was a phase shift of the peaks between 1972 and 1975 when using the UHURU-Copernicus-period (p = 0.199622 d) but no shift when taking p = 0.199682 d for the period (17).The latter is very close to that derived later by Parsignault et al. (18) from extended X-ray measurements.

UHE gamma ray emission from Cygnus X-3 was detected by the Kiel extensive air shower experiment (19). Analysing air shower data taken between March 1976 and January 1980 an excess of showers arriving from the direction of Cygnus X-3 was found. When calculating the phase of these showers according to the X-ray ephemeries of Parsignault et al. (18). 13 showers from 31 were found in the phase intervall 0.3-0.4 of a 10 bin phase histogram. The clear excess of events from the direction of Cygnus X-3 together with the peak in the phase plot makes the result highly significant.

Meanwhile there exist a lot of measurements which support the observations. In addition other X-ray sources have been observed as VHE and UHE gamma-ray objects.

9. FINAL REMARKS

With ground-based techniques the most energetic photons from cosmic sources can be observed. This observations may give the answer to the problem of origin of cosmic rays. So big efforts are justified to build large installation. In the past by every extension of the electromagnetic spectrum new surprising astrophysical objects have been revealed.

REFERENCES

(1) Swanenburg, B.N., et al., 1981. Ap. J. Lett. 243, L69.
(2) Nishimura, J., 1967, Handbuch der Physik Bd. XLVI/2 (Springer, Berlin, Heidelberg, New York) p.1.
(3) Hillas, A.M.,1982, J. Phys. G: Nucl. Phys. 8, 1461.
(4) Greisen, K., 1956, Progress in Cosmic Ray Physics, Vol. 3 (North Holland, Amsterdam), p. 3.
(5) Patterson, J.R., and A.M. Hillas, 1983, J. Phys. G: Nucl. Phys. 9, 1433.
(6) Galbraith, W. and J.V. Jelley, 1953, Nature 171, 349.
(7) Cawley, M.F., J. Clear, D.J. Fegan, K. Gibbs, N.A. Porter, T.C. Weekes, 1982, Proc. Int. Workshop on Very High Energy Gamma Ray Astronomy, Ootacamund, TIFR Bombay, p. 292.
(8) Cawley, M.F. et al., 1985, 19th ICRC, La Jolla, 3, 453.
(9) Gibson, A.I., A.B. Harrison, I.W. Kirkman, A.P. Lotts, J.H. Macrae, K.J. Orford, K.E. Turver, and M. Walmsley, 1982, Proc. Int. Workshop on Very High Energy Gamma Ray Astronomy, Ootacamund, TIFR Bombay, p. 97.
(10) Castagnoli, C., G. Navarra, M. Dardo, C. Morello, 1983, Nuovo Cim. 6 C, 327.
(11) Bassi, P., G. Clark, and B. Rossi, 1953, Phys. Rev. 92, 441.
(12) Cocconi, G., 1959, 6th ICRC, Moscow, II, 309.
(13) Morrison, P., 1958, Nuov. Cim. 7, 858.
(14) Chudakov, A.E., V.I. Zatsepin, N.M. Nesterova, and V.L. Dadykin, 1962, J. Phys. Soc. Japan 17, Suppl. A-III, 106.
(15) Vladimirsky, B.M., A.A. Stephanian, and V.P. Fomin, 1973, 13th ICRC, Denver, 1, 456.
(16) Stepanian, A.A., B.M. Vladimirsky, Yu.I. Neshpor, V.P. Fomin, 1975, Astr. Space Sci. 38, 267.
(17) Vladimirsky, B.M., Yu.I. Neshpor, A.A. Stepanian, V.P. Fomin, 1975, 14th ICRC, München, 1, 118.
(18) Parsignault, D.R., E. Schreier, J. Grindlay, and H. Gursky, 1976, Ap. J. 209, L 73.
(19) Samorski, M., and W. Stamm, 1983, Ap. J. 268, L 17.

HADRON AND MUON COMPONENTS IN PHOTON SHOWER AT 10^{15} eV

Ch. P. Vankov and J. N. Stamenov
Institute for Nuclear Research and Nuclear Energy
Sofia 1784
Bulgaria

ABSTRACT. The main average characteristics have been calculated for the hadron and muon components of EAS from primary 10^{15} eV photons. The standard model of nuclear interactions has been used. The obtained results confirm the criteria applied in the analysis of the Tien Shan experimental data for the selection of muon poor and hadron poor showers as gamma-showers. The low average energy of a photonuclear interaction suppresses the generation of high energy muons ($E_\mu > 500$ GeV) in events generated by primary photons of very high energy. The calculated lateral distributions of muons with $E_\mu > 5$ GeV and $E_\mu > 2$ GeV in gamma-initiated showers predict a $\gtrsim 8 \div 10$ times lower density of the muon flux in comparison with the one in showers from primary protons or nuclei.

1. INTRODUCTION

The measurement of extensive air showers (EAS) with unusually low number of hadrons and muons gives a real possibility for the experimental investigation of photons with very high energies in the primary cosmic rays. The reason for this is that the cross section for photoproduction of pions from photons is much lower than the effective cross section for production of pions by hadrons, thus the number of muons and hadrons in a purely electromagnetic shower will be much smaller. For the first time this possibility was pointed out in ref. [1]. The first calculations [2], indeed, showed that the electromagnetic shower are muon "poor". The experimental registration of such showers, until recently, was in practice nonexistant. The existence of a certain number of such muon and hadron poor showers was demonstrated finally in ref. [3], using EAS data from the Tien Shan experimental array. The individual phenomenological characteristics of these showers especially the hadronic component in the core of EAS, unambiguously point to the electromagnetic nature of the registered events. This allows these showers to be interpreted as initiated by a primary photon of energy $\sim 10^{15}$ eV. The progress in this direction is due to the fact that in the analysis of the Tien Shan data a criterion for the amount of the energy flux in the hadron component in

M. M. Shapiro and J. P. Wefel (eds.), Genesis and Propagation of Cosmic Rays, 255–260.

the core of the shower was applied, which selects the primary photon showers more reliably than a muonless criterion only.

Recently the groups in Lodz [4] and Kiel [5] reported registration of EAS with $E_0 > 10^{15}$ eV in the direction of Crab and Cygnus X-3. In fact the "on-source" showers contain unexpectedly large number of muons [6,7], almost as many as nuclear-initiated showers ($\approx 70\%$).

Obviously this situation initiated the recalculation of the characteristics of the muon and hadron components in a pure electromagnetic shower in the atmosphere. The objective of this calculation is directly related to the experiment - first of all to check the validity of the semi-empirical criteria for the selection of gamma-showers applied in the analysis of Tien Shan EAS data, and second - to make more precise some characteristics of the muon component, e.g. the function of the lateral distribution, etc. Our improved knowledge about the characteristics of the elementary act of interaction at high and very high energies and the much more powerful computers compared with the ones in the middle of the 60's give the possibility for a significant improvement in the precision of the new calculations. On the other hand, it is clear, that the new calculations will not alter the main conclusions, namely that the gamma shower has less muons and hadrons (some more general evaluations can be obtained by simply using the formulas of the cascade theory - for example see ref. [8].

2. METHOD

We simulated by Monte Carlo the development of the electromagnetic shower (EMS) in homogeneous atmosphere down to threshold energy $E_{thr} \geq 500$ MeV. The energy of the primary photon was $E_0 = 10^{15}$ eV. The showers were vertical. Because of the large ratio E_0/E_{thr} in the simulation we used statistical weights in order to reduce the computer time. Besides the essential electromagnetic processes at this energies in the modelling we included also the probabilities for photonuclear interaction of photons. We used the cross section proposed in ref. [9]:

$$\sigma_{\gamma p}(E) = 114.3 + 1.647 \ln^2(0.0213 E), \tag{1}$$

where $\sigma_{\gamma p}$ is in μb, and E in GeV.

In these calculations we did not include the process of direct creation of $\mu^+\mu^-$ pairs from the photon, the cross section for which is quite small (11.7 μb at 1000 TeV). As was shown [10] the contribution of this process becomes considerable for the TeV muons.

We assumed that at the point of every photonuclear interaction of the photons begins a nuclear cascade, initiated by a proton with the same energy as interacting photon. To obtain the characteristics of the hadron component at the level of observation we used the differential energy spectra of pions for different depths in the atmosphere obtained by numerically solving of the cascade equations with standard model [11]. The energy spectra for the muons were obtained in the same manner. The lateral distribution of the muons was calculated by the widely used $p_\perp$ - distribution

$$W(p_\perp) \sim \frac{p_\perp}{p_0} \exp\left(-\frac{p_\perp}{p_0}\right); \quad p_0 = 0.24 \text{ GeV/c}$$

where $\langle p_\perp \rangle$ = 0.4 GeV/c is fixed.

To evaluate the energy E_e deposited in the Tien Shan calorimeter from the electron-photon component we use the function of lateral distribution of the average energy per one electron $\varepsilon_{e\gamma}$ [12]. The density of the energy flux is

$$\rho_E(r,s) = \rho_e(r,s) \cdot \varepsilon_{e\gamma}(r,s)$$

where r is the distance to the shower axis, s is the age parameter, $\rho_e(r,s)$ - the function of lateral distribution of electrons. The so called modified NK function was used, i.e.

$$\rho_e(r,s) = \frac{N_e}{(mR_M)^2} f_{NK} \left(\frac{r}{mR_M}, s \right)$$

where N_e is the number of electrons, m = 0.6, R_M = 112 - 6s + 12 s^2. Then

$$E_e = 2\pi \int_{0.2}^{3} dr.r.\rho_E(r,s) = 6.94 \times 10^{13} \text{ eV} \qquad (2)$$

3. RESULTS AND DISCUSSION

The photonuclear interactions in EMS carry off only about 3% of the energy of the primary photon. As only about 0.8% of the primary photon initiate immediately a nuclear shower and because of the fast fractioning of the primary energy in the EMS, the average energy of the photonuclear interaction in the shower is $\bar{E}_{ph}$ = 11.5 GeV ($\bar{E}_{ph}$ = 43 GeV at E_{thr}= 2 GeV).

The average energy of the hadron component at 700 g.cm^{-2} for threshold energy 3.8 GeV (the threshold energy of the Tien Shan data) is

$$\bar{E}_h \ (> 3.8 \text{ GeV}) = 1.62 \times 10^{12} \text{ eV} \qquad (3)$$

If we take into account the results [13] for the lateral distribution of the hadrons with different energies we obtain for the energy of the hadron component, deposited in the Tien Shan calorimeter the value

$$E_h = 3.7 \times 10^{11} \text{ eV} \qquad (4)$$

Then

$$E_h/E_e = 5.3 \times 10^{-3} \qquad (5)$$

Taking into account the total fluctuations of (E_h/E_e) for the Tien Shan experiment [14] we obtain for the threshold value as a criterion for selection of gamma-showers as hadron poor showers

$$(E_h/E_e)_{thr} = 1.5 \times 10^{-2} \qquad (6)$$

which confirms the semiempirical value 1.42 x 10^{-2} used in ref. [14].

The total number of muons with energy E_μ>5 GeV at 700 g.cm^{-2} is 555. Taking into account the dependence of N_μ on E_0[11]we can obtain N_μ for gamma-showers with $E_o \approx 6 \times 10^{14}$ eV (which is the energy of primary photons registered on Tien Shan): $N_\mu = 361$. From here for the ratio of the number of muons in gamma-showers $N_{\mu\gamma}$ and in "normal" showers $N_{\mu A}$ we obtain $N_{\mu\gamma}/N_{\mu A} = 0.047$. With the muon fluctuations for the Tien Shan experiment [14] included we obtain for the threshold value for the selection of muon poor showers

$$N_{\mu\gamma}/N_{\mu A} = 0.15 \tag{7}$$

which is in good agreement with the value 0.11 used in ref. [14].

The analysis of the lateral distribution of the muons with $E_\mu > 5$ GeV in gamma-showers at 700 g.cm^{-2} level (fig.1) shows that shape of the distribution is almost the same as in the case of proton showers. There is a small difference at distances $r < 30$ m where the lateral function for gamma-showers is steeper. The same holds for muons with $E_\mu > 2$ GeV at sea level. The principal difference between the gamma and proton showers is in the values of the muon density $\rho_\mu(r)$ which differ by a factor $\gtrsim 8 \div 10$. The small difference in the form of the distributions for $r < 30$ m cannot lead to a significant change in the total number of muons (only $\approx 20\%$ [15]).

In fig.2 we have compared the lateral distribution obtained by us for muons with $E_\mu > 2$ GeV from a primary gamma with $E_0 = 10^{15}$ eV at sea level with the Kiel's results [6] for muon lateral distribution in showers from the direction of Cygnus X-3 (the normalization is for r = 8 m). The agreement between the shape of the two distributions is good, but one should have in mind that the experimentally obtained differences between "on" and "off" source showers [6] are also small:

$$\rho_\mu \sim r^{-\alpha}, \ \alpha_{on} = 1.11 \pm 0.08, \ \alpha_{off} = 0.94 \pm 0.08 \text{ for } 4 < r < 30 \text{ m.}$$

It is clear that the behaviour of the function of lateral distribution of the muons in gamma-showers cannot explain the large number of muons obtained in ref. [6].

4. CONCLUSIONS

The results obtained from calculations utilizing the standard model of nuclear interactions confirm the criteria applied in the analysis of the Tien Shan experimental results for the selection of muon poor and hadron poor showers as showers from primary photons with $E_0 \approx 10^{15}$ eV.

The low average energy of a photonuclear interaction significantly reduces the possibility for creation of high energy muons ($E_\mu > 500$ GeV) in events initiated from primary photons with very high energies. By the same argument one expects that the results from the calculations will be insensitive to the model of nuclear interactions for energies above the accelerator region.

The shape of the lateral distribution of the muons in gamma-showers does not differ significantly from the one of proton showers and cannot explain the large number of muons registered in showers from the direction of Crab and Cygnus X-3 [6,7].

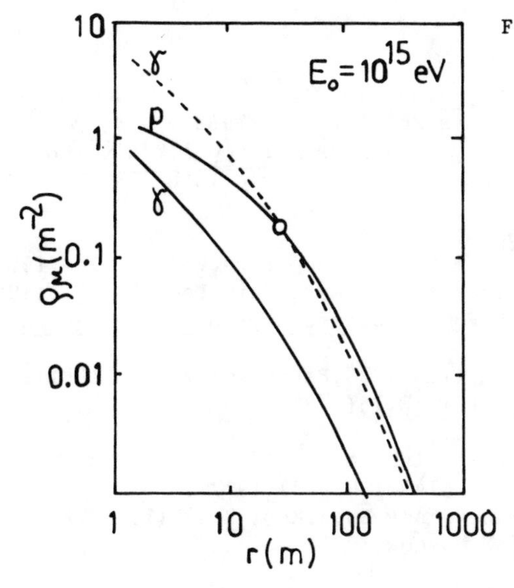

Fig.1. Lateral distribution of muons with $E_\mu > 5$ GeV in showers of primary energy 10^{15} eV at depth 700 $g.cm^{-2}$. Line (p) - primary proton, line (γ) - primary photon (dashed line shows normalized (γ)).

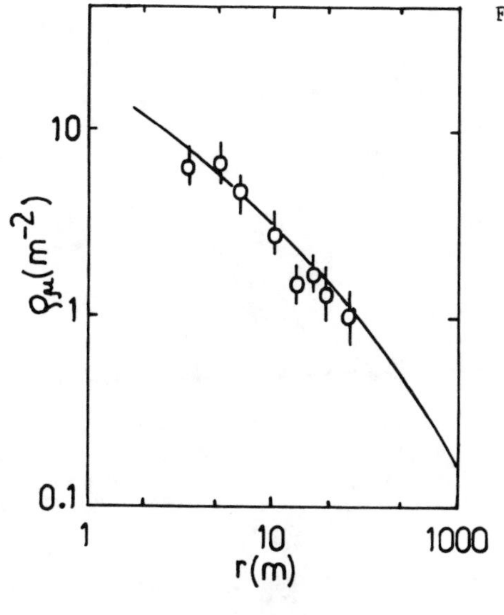

Fig.2. Comparison of the calculated lateral distribution of muons with $E_\mu > 2$ GeV with Cyg X-3 "on" source data from Kiel (the calculated curve is normalized at r = 8 m).

REFERENCES

1. R.Maze and A.Zavadzki,Nuovo Cim.175,625(1960).
2. S.Karakula and J.Wdowczyk,Acta Phys.Polon,24,231(1963).
3. S.Nikolsky, J.Stamenov and S.Ushev, Sov.Phys.JETP 60(1),July 1984.
4. T.Dzikowsky, J.Gavin, B.Grochalska et al., in Proc.18th ICRC, Bangalore,vol.2,p.132(1983).
5. M.Samorski and W.Stamm,Ap.J.268,L17(1983).
6. M.Samorski and W.Stamm, in Proc.18th ICRC,Bangalore,vol.8,p.244(1983)
7. T.Dzikowsky, J.Gawin, B.Grochalska et al.,Acta Univ.Lodz,Fiz.7,5(1984).
8. G.Christiansen, G.Kulikov and U.Fomin, Cosmic Rays of Superhigh Energy (Moscow,1975) (in Russian).
9. L.Bezrukov and E.Bugaev, in Proc.17th ICRC,Paris,vol.7,p.90(1981).
10. T.Stanev and Ch.Vankov, Phys.Lett.158B,75(1985).
11. L.Dedenko,Preprint FIAN A-69 (1965).
12. T.Danilova and A.Erlykin, Preprint FIAN No.15 (1984).
13. V.Romakhin and N.Nesterova, Trudy FIAN,109,p.77(1977).
14. S.Nikolsky, J.Stamenov and S.Ushev, Adv.Spac.Res, 3,131(1984).
15. J.Stamenov, Doct.Sci.Thesis,FIAN,Moscow,1981.

MONOPOLES, MUONS, NEUTRINOS, AND CYGNUS X-3

M.L. Cherry, S. Corbato, D. Kieda, K. Lande, and C.K. Lee
Dept. of Physics, University of Pennsylvania
Philadelphia, PA 19104 USA

ABSTRACT. The deep underground Large Area Scintillation Detector and the surface air shower array at the Homestake Gold Mine are now in operation. Beginning in January 1985, the underground detector has been searching for muons from Cygnus X-3; we have seen no excess signal with the characteristic 4.8 hour period from the direction of Cygnus X-3, with an upper limit below that of the NUSEX result. The surface array has been collecting high energy cosmic ray data, in coincidence with the underground detector, since July of 1985. We describe the initial surface-underground data, and discuss the experiments to search for magnetic monopoles at the level of the Parker limit, neutrinos, and high energy cosmic ray air showers with these instruments and with a new atmospheric Cerenkov detector.

I. INTRODUCTION

The current generation of deep underground detectors runs the gamut from small double beta decay experiments through the 30 m^3 Soudan and NUSEX detectors used to look for energetic muons from Cygnus X-3 to the very large (3000 - 8000 ton) IMB and Kamioka water Cerenkov detectors for proton decay. Depths for the large detectors range from 850 meters of water equivalent at Baksan to 8000 m.w.e. at Kolar. A fairly complete summary of underground experiments is given in the proceedings of the Symposium on Underground Physics, St. Vincent (1985). The experiments at the Homestake Gold Mine in Lead, South Dakota cover a large fraction of the

M. M. Shapiro and J. P. Wefel (eds.), Genesis and Propagation of Cosmic Rays, 261–292.

physics topics requiring large, deep, low-background detectors. They include the 1440 cm^3 ^{76}Ge double beta decay detector of Avignone and colleagues, the 600 ton ^{37}Cl solar neutrino detector of Davis and colleagues, and three new detectors to study high energy cosmic rays and magnetic monopoles. The double beta decay and solar neutrino experiments have been described elsewhere (Cherry, Lande, and Fowler, 1985). Here we will concentrate on the new results from the cosmic ray and monopole detectors.

These new detectors include a deep underground 140-ton Large Area liquid Scintillation Detector (LASD), an extensive air shower array located on the surface above the underground detector, and a Cerenkov telescope to view cosmic ray primary interactions in the upper atmosphere (Fig. 1). The underground detector has a sufficiently low background and ionization threshold to search for massive magnetic monopoles at velocities $\beta \gtrsim 6 \times 10^{-4}$; the area is sufficiently large to reach the Parker limit in three years. In addition, it will be possible to study the zenith angle distribution of neutrino-induced and penetrating muons; search for cosmic point sources of neutrinos, gamma rays, and high-energy cosmic rays (for example, Cygnus X-3) with unparalleled angular resolution (3-10 mrad with the surface-underground telescope); search for low-energy neutrino bursts from stellar collapse events in the Galaxy; measure the multiplicity and transverse momentum distributions of high-energy cosmic ray muons; and study the cosmic ray nuclear composition near 10^{15} eV.

II. LARGE AREA SCINTILLATION DETECTOR -- DESCRIPTION AND INITIAL RESULTS

i) Detector Description

The Homestake Large Area Scintillation Detector is located at the 4850 ft level (4200 meters water equivalent) in the Homestake Gold Mine in Lead, South

Dakota. The chamber was originally excavated in 1965 for Raymond Davis' perchlorethylene solar neutrino detector, and provides an exceptionally low- background environment for experiments to study underground muons, neutrinos, and magnetic monopoles.

The scintillation detector, consisting of 200 scintillator modules with 140 tons of liquid, surrounds the perchlorethylene tank. Each module is a 30 cm x 30 cm x 8 m box arrayed along the surfaces of an 8m x 8m x 16 m volume (Fig. 2). The individual modules are constructed out of 6 mm thick PVC sheets folded and welded into the form of boxes. The boxes are filled with a mineral-oil based liquid scintillator designed for its high transparency, long-term stability, relatively low flammability and toxicity, and low cost. The scintillation light is viewed by two 13 cm diameter hemispherical photomultiplier tubes in each box, one immersed in the oil at each end. The inside of each box is lined with a thin teflon film for total internal reflection. A charged particle passing through a module produces a pulse of light at each of the two phototubes in that module. From the relative amplitudes and arrival times, the position, time, and energy deposit can be determined.

The electronics system for the LASD consists of four components: an amplifier-discriminator board located directly at the phototube; a fast time-digitizing circuit to measure the relative time of firing of each phototube pulse in the system; a pulse shape recorder to measure the amplitude, shape, and duration of each pulse in the detector; and a main trigger and control circuit.

The monopole search has imposed unique requirements on the trigger and pulse shape recording circuits. A slow monopole is characterized by a long (microseconds) pulse in the entering module, a long (tens to hundreds of microseconds) delay time, and another slow (microseconds) pulse in the exiting module. To recognize these slow pulses and the long transit time of the monopole across the detector, we provide each photomultiplier tube with an individual transient pulse recorder, allowing us to digitize individual pulse heights every 10 nsec with 7 bit resolution over a memory span of 300 μsec.

Except for 20 modules on the north face and 10 at the northern end of the top face, the full LASD is now filled with oil and taking data. All of the circuitry except the pulse shape recorder is completed and installed. The final electronic installation is scheduled for this summer. The present data are based on the fast timing electronics only.

Initial calibrations of position vs. relative timing and pulse height have been made using muons in a test module in our surface laboratory in Philadelphia. Since the underground muon rate at Homestake is only 10 muons day^{-1} module^{-1}, these calibrations have been repeated on the operational modules in situ using light-emitting diodes. Underground, we measure a light attenuation length of 3 m, a distance-to-time conversion of 8.5 cm/nsec (which includes the effects of both the index of refraction of the liquid scintillator, n = 1.49, and the internal box optics), and a time resolution of 2.9 nsec. (The distance-to-time conversion relates the time difference measured by the phototubes at either end of the module to the position of the event with respect to the center of the box. Therefore a time delay of 0 nsec corresponds to a particle at the center; a 45 nsec delay corresponds to a particle at one end.)

These in situ measurements are very consistent with the results of the laboratory tests in Philadelphia, and are confirmed by using measured underground muon timing distributions. In Fig. 3a we show a view of the detector looking from the south (the left in Fig. 2). This is a two-muon event, with the two muons passing through two top modules separated by approximately 1.5 m, passing downward through the perchlorethylene tank, and exiting through two bottom modules 25 nsec after they enter the top. In Fig. 3b we show another two-muon event, viewed now from the west. One muon enters through the top and moves down through the west side; the second muon enters undetected through the unfinished northern end and is seen only as it exits through the bottom. "Long-track" events like that in Fig. 3b provide a test of the timing calibrations and reconstruction algorithms. For the sample of "long-track" events passing through both top and bottom modules, we can determine the trajectory trivially by drawing the straight line through the top and bottom modules. For each side wall module, we can then plot the resulting position versus the time difference measured between the two ends of the module. Such a plot is shown for one side-wall box in Fig. 4, where the scatter gives a position resolution of $\pm$17cm. The corresponding angular resolution is 34 cm / 8 m = 2.5^{o}. In Fig. 5 we show the measured angles between the individual particles in vertically downward muon pair events; these measured angles are consistent with the $\pm$1cm position resolution.

To translate from the internal detector coordinates to absolute coordinates on the sky requires that the coordinates and orientation of the underground detector be accurately surveyed with respect to surface markers and celestial coordinates. In principle this is not difficult; in practice, when the sky cannot be viewed from the detector, and when numerous intermediate surveys must be accurately pieced together, this can be a messy operation. We can check our absolute coordinates by comparing the underground directions to directions measured with the surface-underground telescope.

First, we obtain a clean sample of surface-underground events by measuring the time delay between shower and underground events. The muon flight time is 5μsec; a pulse from the underground detector to the surface then takes 12μsec to signal that a coincident underground event has occurred. We show the resulting distribution of time delays in Fig. 6, where we see a uniform background and a sharp peak at 17 μsec; the signal-to-background ratio in the peak bin is 80. The center of gravity of the surface showers can be determined to within 4-5 m for events falling within the boundaries of the array; with a distance of 1600 m to the underground scintillator, and a typical underground lateral dispersion of 2.5 m due to multiple scattering, this corresponds to an angular resolution of approximately 3 mrad. The measured underground directions are consistent with the surface-underground directions to within 2.5^{o}.

A through-going minimum ionizing muon deposits approximately 56 MeV in a single module. In order to obtain the best possible sensitivity to slow, lightly-ionizing massive monopoles, we set the thresholds exceptionally low. By placing a ^{60}Co source (1.2 and 1.3 MeV gamma rays) at a fixed distance of 2 m from each end of each module, and setting the phototube threshold near the upper end of the Compton scattered electron spectrum, we obtain thresholds varying from below 1 MeV near the phototube to approximately 3 MeV at the far end of the module. With this low threshold we become sensitive to the radioactivity background in the rock walls of the room. We are currently analyzing the results of a detailed survey of this rock activity using a large sodium iodide detector in order to better understand the low-energy background observed in the scintillation detector.

ii) Cygnus X-3

Cyg X-3 is an intense and highly variable infrared, radio, X-ray, and high energy gamma ray source observed at energies up to 10^{16} eV. (For recent reviews, see Vladimirskii et al., 1985; Watson, 1985; also Willingale, King, and Pounds, 1985; Johnston et al., 1985.) It is presumed to be a close binary of a neutron star and a massive companion with a 4.8 hour orbital period. Recently, the Soudan (Marshak et al., 1985) and NUSEX (Battistoni et al., 1985) groups have reported observing underground muons from the direction of Cyg X-3 with a period of 4.8 hours and intensity comparable to that of the high energy gamma rays. Since gammas are very poor muon producers, these reports suggest that the transmitting particle is not a gamma ray but rather another, yet unknown, particle. In order to maintain directionality this particle must be neutral. In order to maintain phase coherence over the 30,000 year travel time the particle mass must be very small ($\lesssim$ 5 GeV) compared to its energy. Thus, it is a "gamma ray- like particle" with high muon production probability.

Another unusual characteristic of the observations reported by the two detectors is that the observed muons do not point directly back to the source, but rather only to the general vicinity. The angular deviation is about 5°. There is no explanation for this apparent source smear.

The reported data cover an observing time of about three years. Recently, two new and much larger detectors, at Frejus (Ernwein et al., 1985) and Homestake, have become operational. These new detectors each have about 10 times the aperture of the NUSEX detector (and 15 times that of Soudan) and are located at the same depth as NUSEX. It should thus be possible to repeat and check the NUSEX result with a much higher statistical precision, even over relatively short time intervals.

The Homestake results from 1/15/85 to 9/11/85 have been reported elsewhere (Cherry et al., 1985). Over the period 9/28/85 - 2/23/86 (127.6 days of live time), the Homestake LASD recorded 2.2×10^5 system triggers (1 min^{-1}); when we required single muons and made fiducial cuts, we were left with 2.2×10^4 events; and when we made a muon velocity cut around $\beta = 1$ and accepted only events within 48° of the vertical, we obtained a final sample of 7500 events to examine. A plot of the arrival directions of these events is shown in Fig. 7 for the Cygnus region. Within a $15^o \times 15^o$ bin around Cygnus X-3, we expect 102 background events and see 92 events; within $10^o \times 10^o$ we expect 49 and see 36.

We accumulate background in the same declination band as Cygnus X-3 but trailing the source by 4.8, 9.6, 14.4, and 19.2 hours, so that we look through the same rock depth and reproduce any daily 24-hour background effects in both source and background samples. Phase plots of the data from the $15^{\circ} \times 15^{\circ}$ bin (using the van der Klis and Bonnet-Bidaud (1981) ephemeris) are shown in Fig. 8. We see a 2σ peak in the phase bin 0.8 - 0.9, but a deficiency between 0.4 and 0.7. In the phase bin 0.7 - 0.8, where the NUSEX group see their excess, we see 6.5 $\pm$ 1 background events (compared to an average background of 10.2 events per phase bin). This corresponds to a 1 sigma upper limit of 3.2 events.

A similar phase plot for the period 1/15/85 - 9/11/85 is shown in Fig. 9; again, there is no clear evidence for an enhancement. We have looked at the data with tighter angular cuts, with the Parsignault et al. (1976) ephemeris, with a more recent ephemeris suggested by Mason (1986), and over the entire time interval 1/85 - 2/86; in no case do we see a positive signal similar to that of NUSEX and Soudan.

We have performed Monte Carlo calculations to duplicate the very tight single muon and fiducial cuts. The overall acceptance efficiency is 15%, resulting in a measured vertical muon rate of 1.8×10^{-9} cm^{-2} sec^{-1} sr^{-1}, compared to a rate of 1.6×10^{-9} cm^{-2} sec^{-1} sr^{-1}, based on the earlier Homestake water Cerenkov measurements (Cherry et al., 1982b). Using these calculated efficiencies, and correcting for the detector live times in each phase bin, we find an upper limit to the pulsed flux over the entire interval 1/85 - 2/86, in the bin 0.7 - 0.8, of 5.4×10^{-10} cm^{-2} sec^{-1} sr^{-1}.

Battistoni et al. (1985) do not describe the details of their analysis; to compare our results to the NUSEX results, we must scale our results by the relative detector sizes. With 7 times the exposure time, 0.1 times the area, 0.4 times the angular bin on the sky, and at approximately the same depth, NUSEX reports 13 $\pm$ 0.2 background counts, compared to the Homestake background (scaled to NUSEX) of 6.5 x 7 x 0.1 x 0.4 / 0.15 = 12. Since the background rates are so similiar, we scale our upper limit of 3.2 events in the 0.7 - 0.8 phase bin in the same manner, giving 6 events compared to the NUSEX excess of 19, a factor of 3 below the NUSEX value in just five months.

In addition, we have analyzed our data at the times of the two very large radio flares observed last

October and December. Soudan apparently detected an enhancement of muons near the time of the October flare (Marshak, private communication); we have analyzed our data and see no enhancement in the muon signal.

Finally, it should be emphasized that several other binary pulsars have been detected by the COS-B gamma ray satellite, by ground-based Cerenkov detectors, and by southern hemisphere air shower measurements. These are also candidates for study. We are currently performing the conversion of our event times to barycentric coordinates for the following sources: Hercules X-1, 4U 0115+63, PSR 1953+29, Geminga.

iii) Magnetic Monopoles

The two observable predictions of Grand Unified Theories are the instability of the proton and the existence of massive magnetic monopoles. The search for proton decay has already yielded lower lifetime limits that are in conflict with some versions of GUTs. Since no specific monopole flux predictions are made by GUTs, however, we are guided here only by various cosmological and galactic structure considerations. The stability of the galactic magnetic field and our understanding of its generation place an upper limit on the flux of fast massive magnetic monopoles (the Parker limit; see Turner et al., 1982). This limit does not apply, though, to slow monopoles such as those trapped locally in gravitationally bound orbits. It is also unclear that we understand the generation of the galactic magnetic field sufficiently well to consider this limit on accelerated monopoles as a strict bound (Chernoff et al., 1986). Much stricter bounds come from considerations of neutron stars and white dwarfs. (For a recent review, see Groom, 1986)

The idea that monopoles are gravitationally trapped in orbits in the solar system is particularly attractive because their enormous mass results in a much stronger coupling to gravitational fields than does their magnetic charge to magnetic fields. This gravitational trapping would also involve trapping in collapsed stars, such as neutron stars, giving rise to various effects (such as stellar heating) which have not been observed. However, despite all the astrophysical and particle physics arguments, we are nevertheless left with a search for a particle with a nearly unpredicted flux and unknown velocity spectrum. There are four approaches to monopole searches:

1) Magnetic flux detectors that recognize monopoles by their magnetic coupling to flux loops. These detectors are velocity-independent, but by their nature have so far been limited to relatively small areas.

2) Ionization detectors, either scintillator or gas counters, that detect the monopole by utilizing the effective electric charge resulting from the transformation of moving magnetic charge. Such detectors can be built in large areas, but are sensitive to the monopole velocity and have a lower velocity cutoff. With proper design and sensitivity to low ionization levels, these detectors can put multiple constraints on the detection of a monopole and thus avoid the danger of weakly defined "event candidates", as has occurred in proton decay searches.

3) Material damage detectors, such as mica or CR-39, that search for material damage created by heavily ionizing particles. These detectors have the advantage of extemely large areas and long integration times, but have stringent ionization requirements. The limits they set are based on the pickup and trapping of ionized nuclei by bare monopoles as they pass through the upper layers of the earth's crust.

4) Nucleon decay detectors that look for simultaneous multiple nucleon decays along a linear path. Such catalysis decays would be the result of magnetic monopole stimulation. Catalysis searches impose a dual requirement, however: GUT monopoles and the catalysis effect must both exist, with a sufficiently large product of flux times catalysis cross section.

We have chosen to construct an ionization-sensitive detector utilizing liquid scintillator. In addition to its large size, the LASD, with its low background, sensitivity to low ionization levels, and pulse height and timing capabilities, is

1) insensitive to detailed models of monopole-detector interaction,

2) can impose multiple constraints on the detection of the monopole, and

3) can simultaneously be used for other experiments. The 30 cm thickness of the LASD scintillator insures that a large amount of energy is deposited by a through-going particle. This allows us to set a low ionization limit for such a particle and still be above the signal level associated with radioactivity. In addition, a thick detector allows us to discriminate between fast and slow particles. A relativistic muon will traverse a detector module in 1 nsec, while a slow monopole with $\beta = 10^{-3} - 10^{-4}$ will take between 1 and 10 μsec (Fig. 10).

As discussed above, thresholds are set below 0.1 times minimum ionizing. The large number of photons liberated in traversing a detector element permits a good measure of energy deposition and thus allows us to examine the match between velocity and ionization. Relativistic monopoles are expected to ionize at about 5000 times minimum. For velocities of 10^{-3} -10^{-4} c we expect ionizations between 0.1 and 1 times minimum. The constraints on the monopole can be applied on the entering module, on the exiting module, and on the time-of-flight across the detector. Thus, we have five independent measures: three measures of velocity and two measures of energy loss.

The detector array provides an aperture of 1200 m^2 sr, corresponding to one event in three years at a flux level of 9 x 10^{-16} cm^{-2} sec^{-1} sr^{-1}. This sensitivity limit is compared to the limits obtained with other detectors in Fig. 11. The Homestake sensitivity is shown as a solid line for $\beta \gtrsim 6 \times 10^{-4}$ and a dashed line down to $\beta = 1.5 \times 10^{-4}$ reflecting the uncertainty in the estimates of energy loss at low velocities. The IMB result (assuming a very short pathlength for monopole catalysis) and the Berkeley mica result are the only results at levels comparable to or lower than the projected Homestake sensitivity. The monopole electronics will be installed, and monopole data-taking will begin, by the end of this summer.

iv) High Energy Neutrinos, Discrete Sources, and Neutrino Bursts

Neutrinos detected in underground experiments can be divided into two classes -- low energy neutrinos whose secondaries stop in the detector, and high energy neutrinos whose secondaries escape. For most underground detectors the dividing energy is near 1 GeV. For neutrinos above that energy it is most efficient to utilize the production of charged secondaries in the rock surrounding the detector, thereby amplifying the target mass. A high energy neutrino detector is therefore characterized not by the detector mass but rather by the surface area, which is directly related to the measured flux of muons produced by neutrino interactions in the surrounding rock.

The rock target thickness for the neutrino is related to the range of the secondary muon, and thus linearly related to the neutrino energy E. Since the neutrino cross section also rises linearly with energy, the detection efficiency varies with E^2. Typical cosmic neutrino sources have spectra of the form $E^{-\alpha}$, $\alpha \sim 2 - 3$ depending on the relative contribution of direct and secondary production mechanisms. Thus the flux of secondary muons is roughly independent of energy or decreases slowly with energy.

The main problem for the particle detector is to distinguish between neutrino-induced muons and direct cosmic ray muons. The direct muons have a zenith angle cutoff while the neutrino-induced muon intensity is virtually zenith angle independent. At our depth the direct muons cut off near 60°. This leaves the region between 60° and 180° for the neutrino-induced signal. The time-of-flight measure between faces of the detector permits us to make a clear distinction between upward- and downward-moving particles.

The operative assumption is that the main mechanism for producing high energy neutrinos in stellar sources is by charged pion decay. It is also assumed that the main source of high energy gamma ray production in these sources is by neutral pion decay. From the measured flux of high energy gammas one can then predict the high energy neutrino flux (Fichtel, 1978; Stenger, 1982; Shapiro and Silberberg, 1983; Lee and Bludman, 1985). Unfortunately, this flux is too low to be of interest for any existing or contemplated underground or underwater detector except DUMAND. The models of Eichler and Vestrand (1984) and Hillas (1984), however, suggest neutrino-to-gamma ray flux ratios could be as large as 10^3, resulting in detectable neutrino fluxes.

In addition to these neutrinos from discrete sources, there will also be neutrinos from pion and muon decay in the atmosphere. For this diffuse source, the predictions appear reliable (Gaisser et al., 1983; Dar, 1983). These neutrinos provide a source with which one can investigate the possibility of neutrino oscillation. There are two modes of interest: vacuum oscillation driven by a small mass difference between the various neutrino states, and mass oscillation driven by a difference in interaction potential with matter between the various neutrino species. The flight path available (the earth's diameter) permits an extension of the mass difference range by two orders of magnitude below that available at accelerators or reactors (Ayres et al., 1984).

Low energy neutrino burst events will be produced from stellar collapse events in the Galaxy (Burrows, 1985). The expected neutrino energies will be near 10 MeV, so that a low ionization threshold is essential. Cherry et al. (1982a) have discussed the background levels seen in a search for neutrino bursts with an earlier 150 ton water Cerenkov detector in the Homestake laboratory, and have demonstrated that a detector with the size and depth of the LASD has sufficient sensitivity to detect a stellar collapse liberating 10^{53} ergs of neutrinos at the distance of the Galactic Center.

III. SURFACE EXTENSIVE AIR SHOWER ARRAY

At energies below about 10^{12} eV it is well established that cosmic rays are primarily protons with small amounts of other elements. It has been suggested by various observational and theoretical considerations, however, that near 10^{15} eV the iron content of cosmic rays might become important and possibly even dominant (Yodh, 1981).

Because of the low flux at this energy, direct spectrometer techniques cannot be applied. Indirect mass measurements involve studies of the size, age, height, and lateral distributions of cosmic ray-induced showers, and measures of the muon multiplicity. By combining a large air shower array on the surface with the deep underground detector, and simultaneously detecting the soft electromagnetic part of the shower and the hard hadronic core, we can measure the cosmic ray composition between 10^{14} and 10^{16} eV in a way that essentially depends on energetics. Basically, one compares the total electron number N_e on the surface (i.e., the total energy/nucleus) and the multiplicity of high energy ($E_\mu \gtrsim 2.6$ TeV) muons underground (i.e., the energy/ nucleon). In order to reach our depth, muons must have roughly 2.6 TeV at the surface of the earth. Such muons can be produced by proton primaries with energies in excess of 10^{13} eV or, for example, by iron primaries with energies above a few times 10^{14} eV. A proton generally gives rise to a single high energy muon while an iron, consisting of a superposition of 56 separate nucleons, has a large probability of multiple muon production, particularly above 10^{15} eV. Our data will thus consist of muon multiplicity and separations underground, and shower size at the surface. For small showers ($E \lesssim 10^{15}$ eV) we expect to observe single muons primarily from cosmic ray protons, while for large showers ($E \gtrsim 10^{15}$ eV) we expect a mix of single and multiple muons from protons and heavy (nominally iron) primaries. Monte Carlo simulations illustrating the expected capabilities of the detector system have been described in full detail elsewhere (Cherry et al., 1982b).

The air shower array consists of 27 modules, each 1.2 m by 2.4 m. The modules are filled with 10 cm of liquid scintillator and are viewed by two 13 cm diameter phototubes. Each module contains two high voltage power supplies, amplifiers, discriminators and a coincidence circuit. The modules communicate with cables buried in conduit. The high voltages can be controlled remotely from the electronics trailer. Each module is plateaued at 500 cosmic ray coincidences per second. The liquid has been chosen to have a low paraffin content, and the modules have been carefully insulated to insure that the 10 watts of internal heating are sufficient to keep the scintillator warm during the winter and prevent clouding of the liquid.

Delays of 0 - 2 μsec are added to the cables from each module to compensate for the length of the signal cables. The surface array can then be triggered either alone, in coincidence with the LASD, or in coincidence with the Cerenkov detector. In coincidence with the underground system, we can establish a very low level surface trigger requirement of two modules. When an underground event is seen, we send a pulse up a connecting cable (which follows a tortuous 3 km path, including a 1.6 km ascent up the elevator shaft from the 4850 ft level to the surface). This pulse arrives at the surface electronics about 17 μsec after the surface trigger -- 5 μsec for the muon travel time and a net 12 μsec cable delay, as shown in Fig. 6. The width of the peak is primarily due to the difference in muon travel times for showers striking different points in the array.

Over the period 7/18/85 - 2/22/86, we detected 385 surface-underground coincidences, for a rate of 2.2 day^{-1}. Of a sample of 36 large showers striking inside the dimensions of the surface array, 24 had single muons underground and 12 had multiple muons.

The combined surface array-underground detector also provides an excellent telescope with which to locate cosmic ray point sources. The separation of the two detectors (1.6 km) and the core resolution of the surface array (a few meters) together provide milliradian angular resolution in establishing the source locations. This resolution is at least an order of magnitude better than can be obtained with either a surface array or an underground detector alone. The portion ofthe sky covered by the combined surface-underground telescope is shown in Fig. 12. The latitude of the detectors is $44^{\circ}21'$ N. The surface array operated independently has high efficiency for showers inclined at angles up to 30° from the zenith. The underground detector sees neutrino-induced muons from all directions, but the maximum flux comes from an angular range of half-angle about 30° around the vertical; this is the range of highest atmospheric background but also the range of directions from which the lowest-energy muons can penetrate. A $\pm$ 30° band in the sky is shown in Fig. 12. Several interesting x-ray, γ-ray, and radio point sources are shown, together with the Galactic plane. A line connecting the underground LASD and the current 270' (north-to-south) by 500' (east-to-west) surface array is at an angle of 11° south of vertical. The band from 31° to 35° subtended by the surface array is also shown in Fig. 12. The sources Cyg X-1 and Her X-1 lie within this band; Cyg X-3 and NGC 4151 lie less than 6° to the north, and will be covered by a future expansion of the surface array.

Cygnus X-3 is slightly outside the range of the surface-underground telescope (although it is interesting to note in Fig. 12 that Hercules X-1, another binary pulsar which has been seen by Fly's Eye (Baltrusaitis et al., 1985) at 5×10^{14} eV, is well within the field-of-view). We have therefore constructed a steerable Cerenkov telescope, utilizing a mirror generously given us by the Utah group. The 150 cm mirror collects the Cerenkov radiation emitted in the upper atmosphere by the early interactions of cosmic ray primaries. This Cerenkov radiation is imaged on an array of ten 13 cm diameter phototubes. The signal is read out whenever a muon is observed in the underground detector or an electron shower detected in the surface array. A $10^{\circ} \times 20^{\circ}$ region of the upper atmosphere is viewed by the mirror at any given time. The telescope is steerable so that it can follow Cyg X-3 or any other source through the night.

Funding for the Homestake scintillator experiments is provided by the U.S. Department of Energy. The assistance and generous cooperation of the Homestake Mining Company are deeply appreciated. We are especially indebted to A. Gilles and J. Dunn. In addition, we appreciate the advice and assistance of R. Davis, Jr.

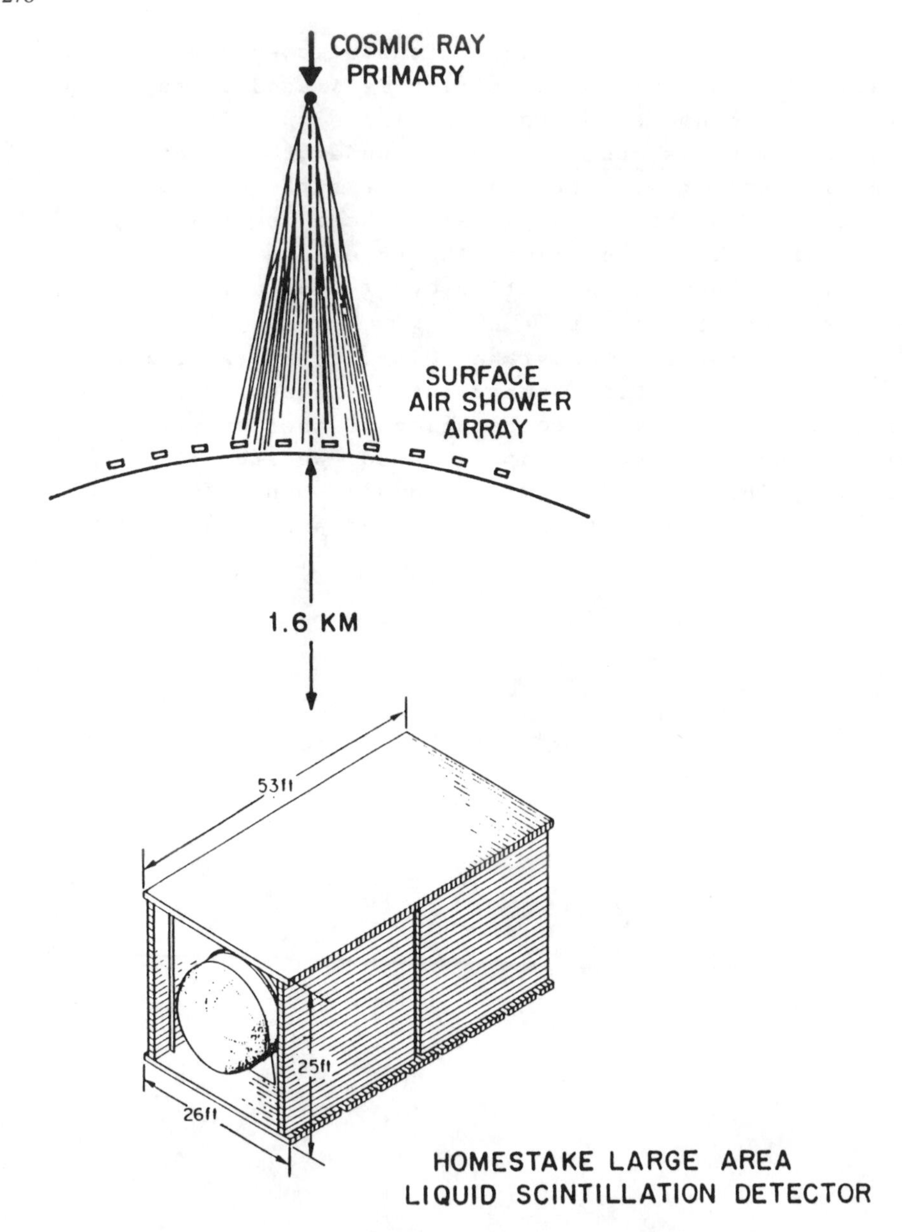

Figure 1. Surface and underground Homestake detectors.

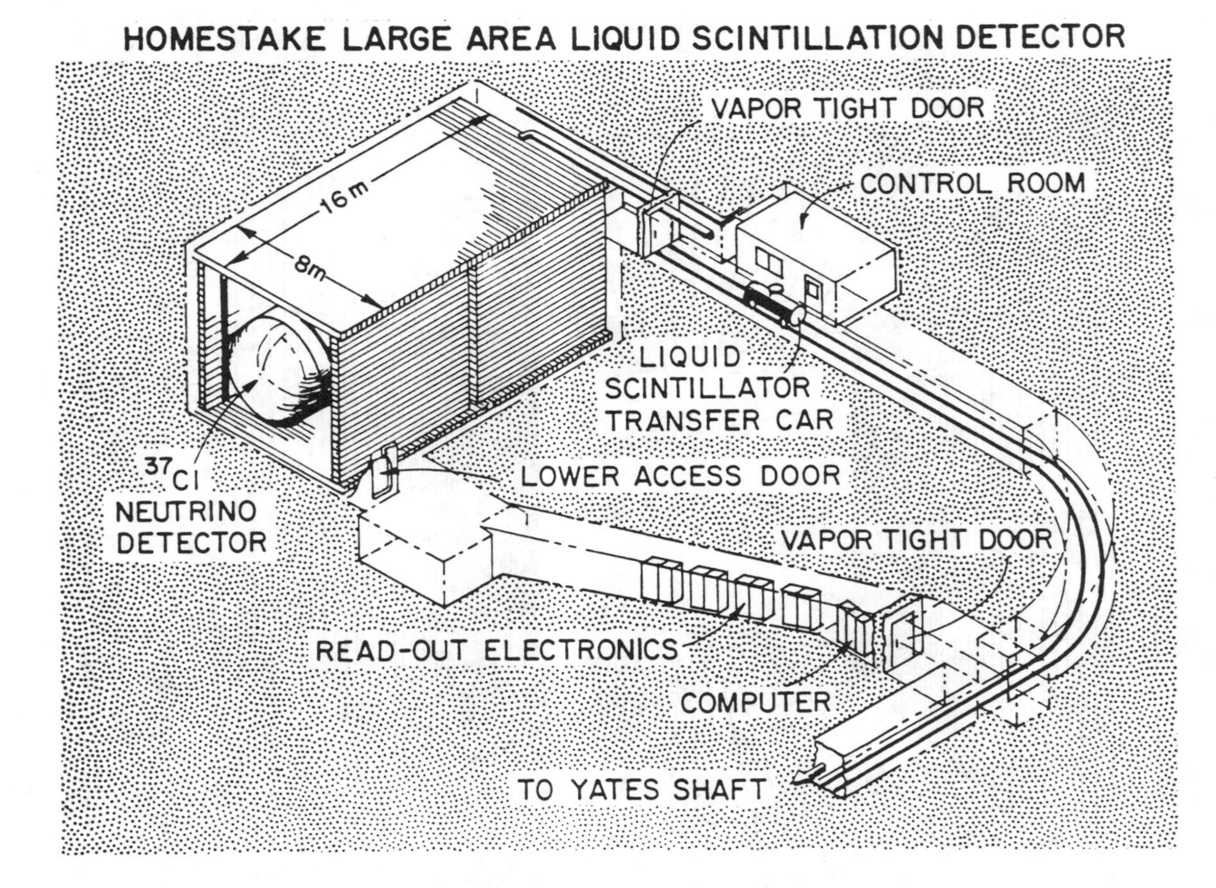

Figure 2. Homestake Large Area Liquid Scintillation Detector.

HOMESTAKE LARGE AREA SCINTILLATION DETECTOR

SOUTH VIEW

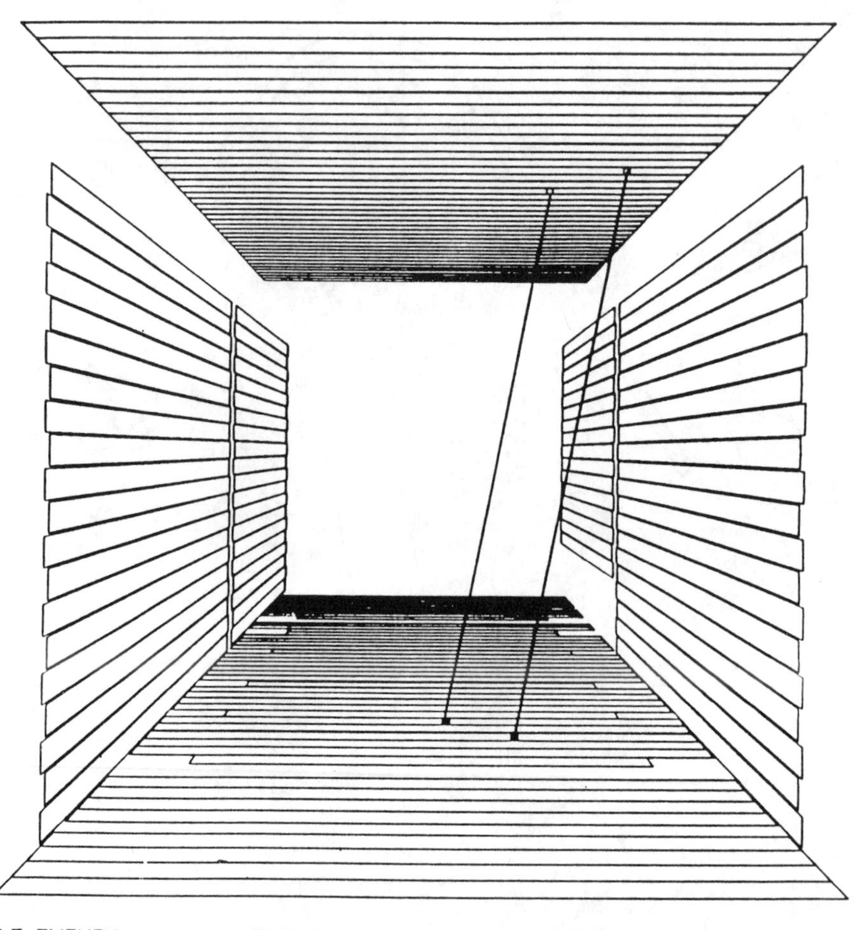

FILE: EVENTY EVENT # 139 12/30/85 19:56:24 UT
MODULES: 14T 17T 12B 14B

Figure 3a. Muon pair events (south view) showing parallel muons.

HOMESTAKE LARGE AREA SCINTILLATION DETECTOR

WEST VIEW

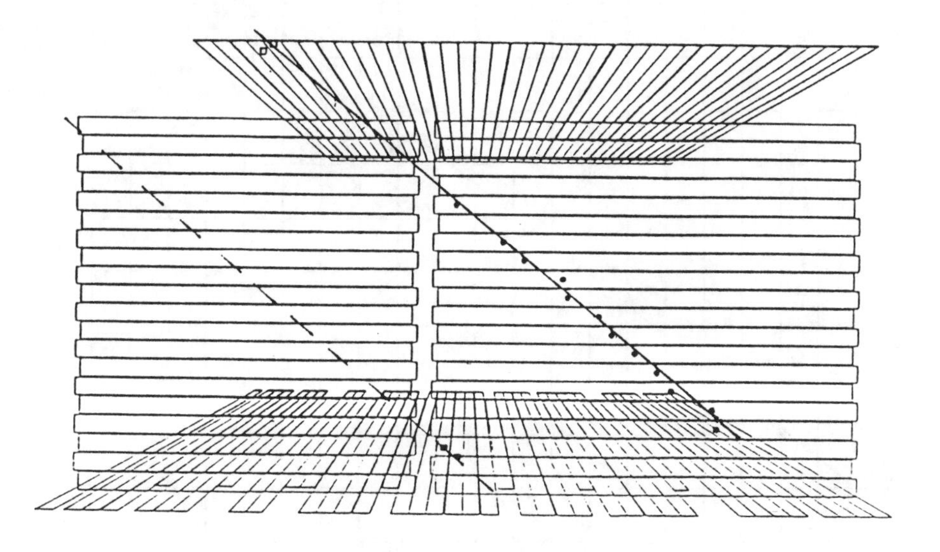

FILE: EVENTB EVENT # 1325 2/16/86 06:44:35 UT

MODULES: 16W 5W 6W 7W 8W 9W 10W 11W 12W 13W 14W 15W 4W
35T 36T 23B 24B

Figure 3b. Muon pair events (west view).

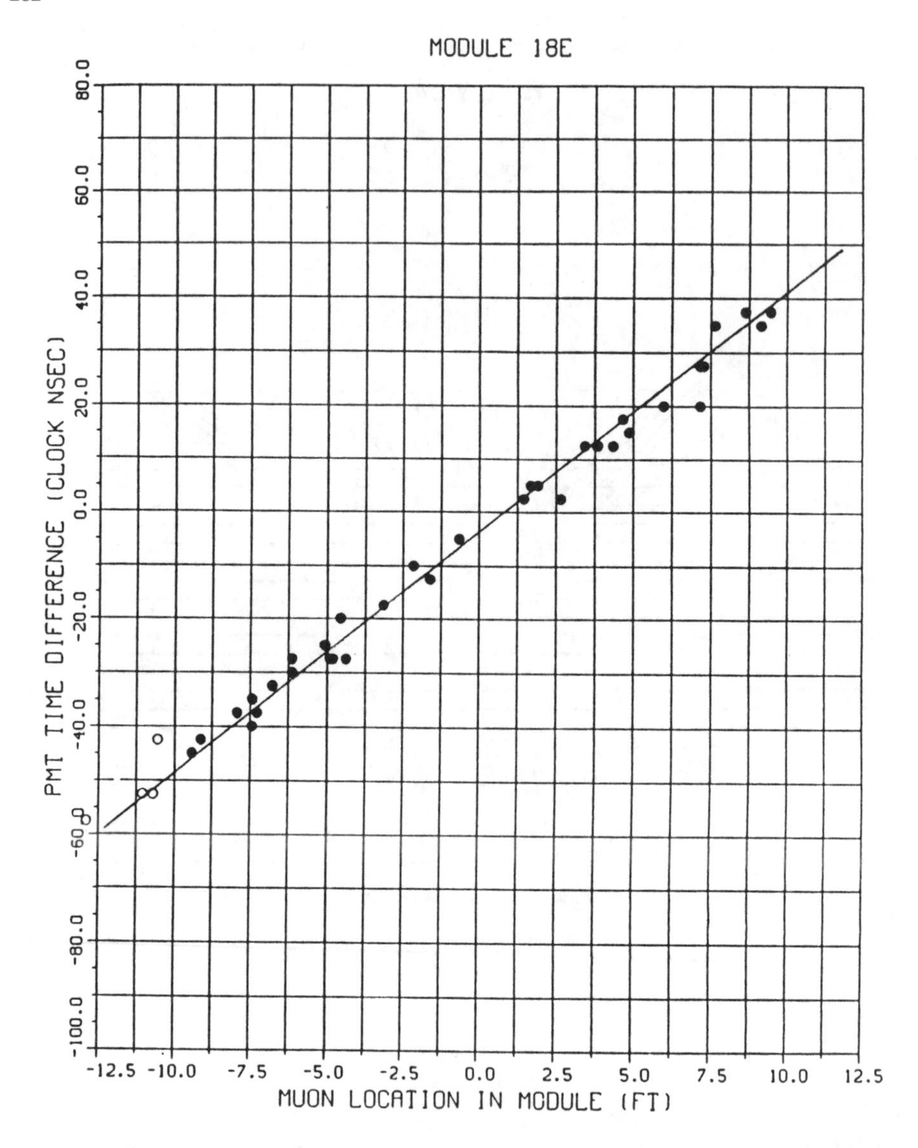

Figure 4. Measured photomultiplier time differences vs. muon location for muon events in a single side wall module.

ANGLES BETWEEN MUON PAIRS

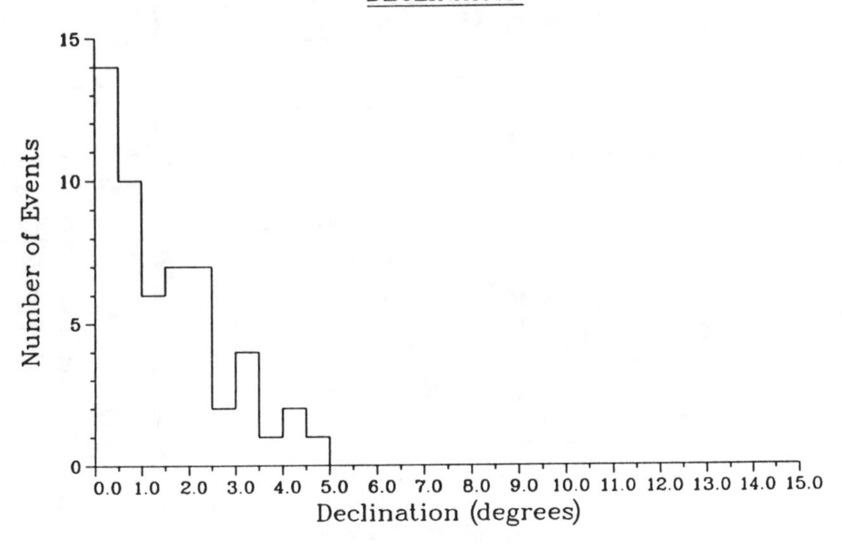

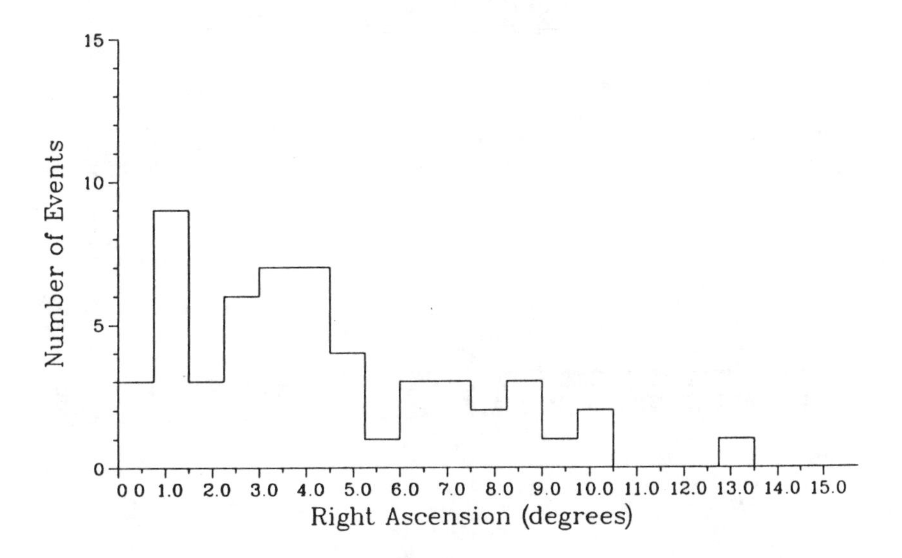

Figure 5. Measured separation between individual muons in two-muon events.

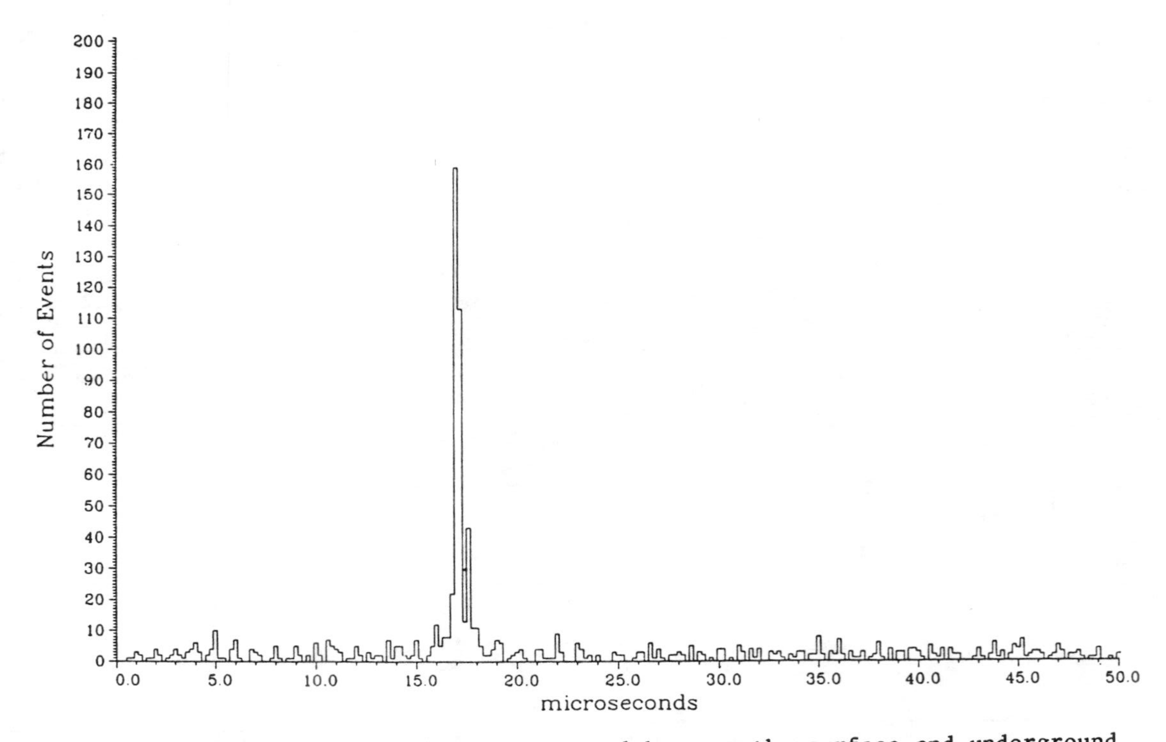

Figure 6. Distribution of delays measured between the surface and underground detectors.

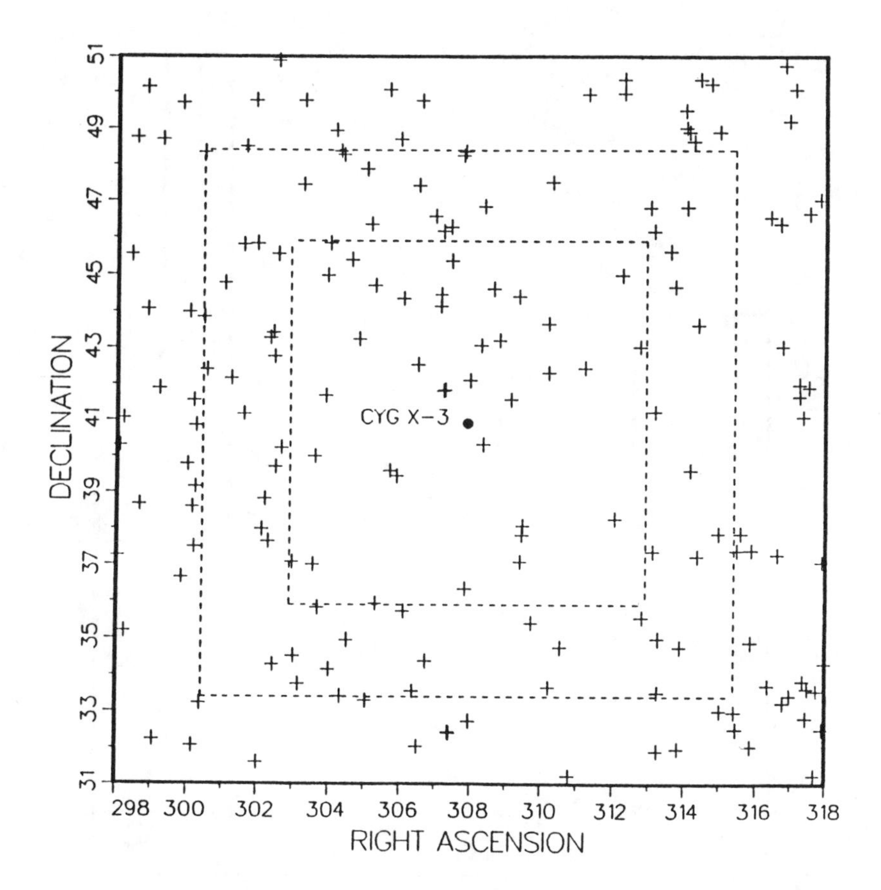

Figure 7. Arrival directions of events on the sky, based on reconstruction of underground events coming from the direction of Cygnus.

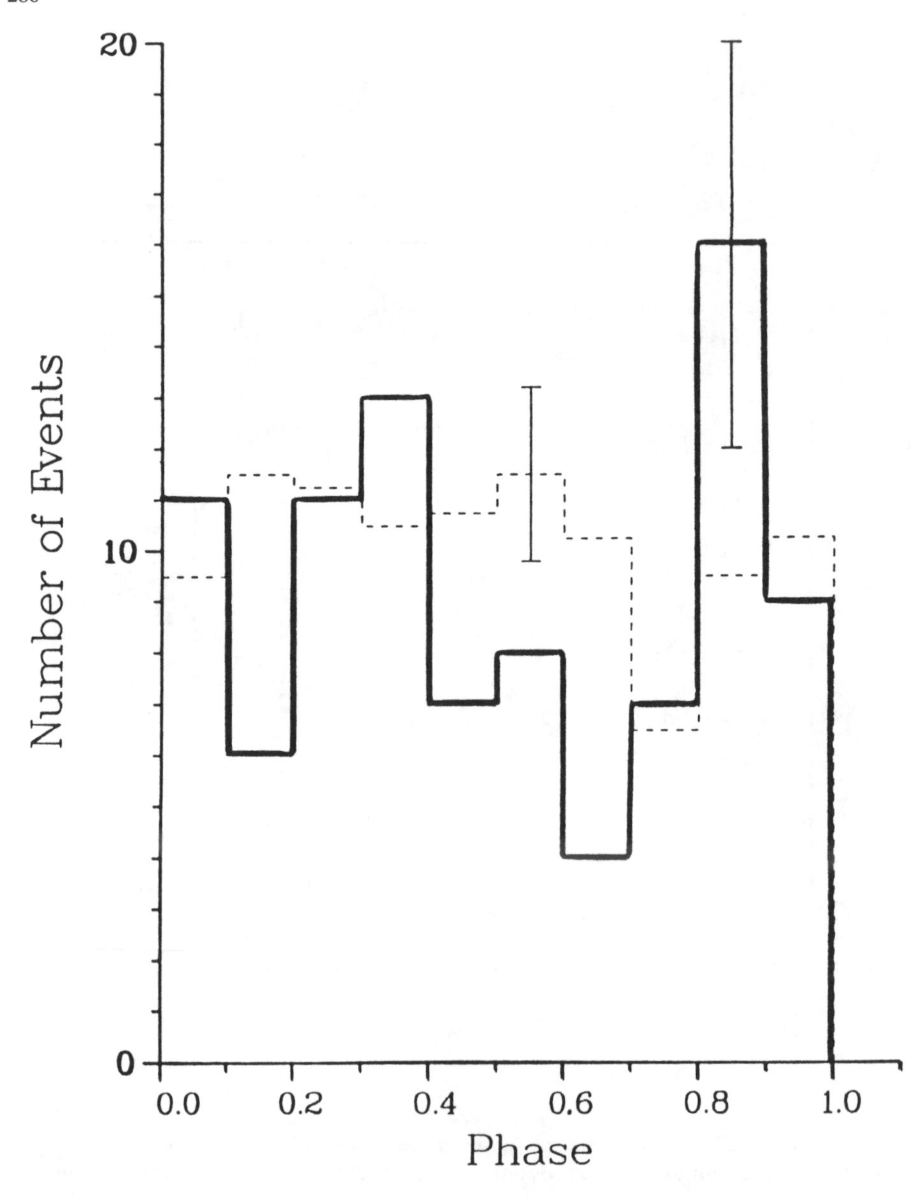

Figure 8. Cygnus X-3 phase plot for the time interval 9/28/85 - 2/23/86. Solid distribution is for events in the 15° x 15° region around the source; dashed distribution is the background measured in the same declination band but for right ascensions following the source by 4.8, 9.6, 14.4, and 19.2 hours.

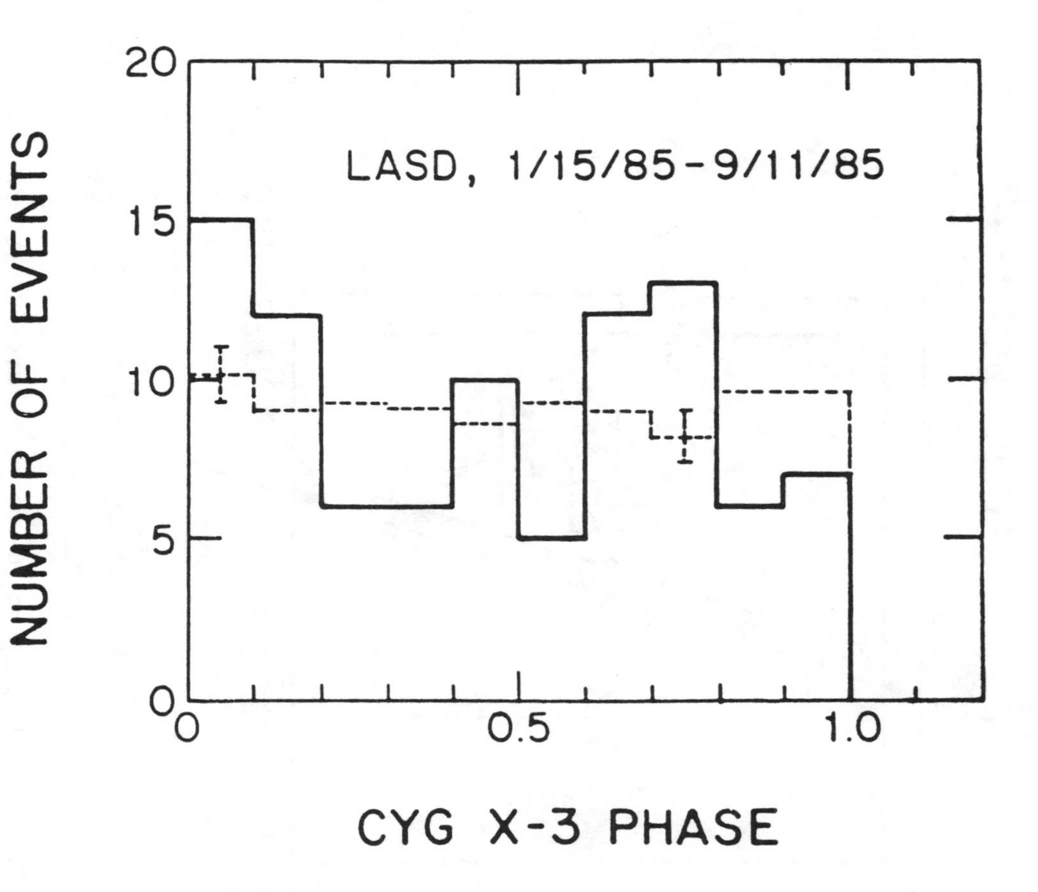

Figure 9. Cygnus X-3 phase plot for the time interval 1/15/85 - 9/11/85 (Cherry et al., 1985b).

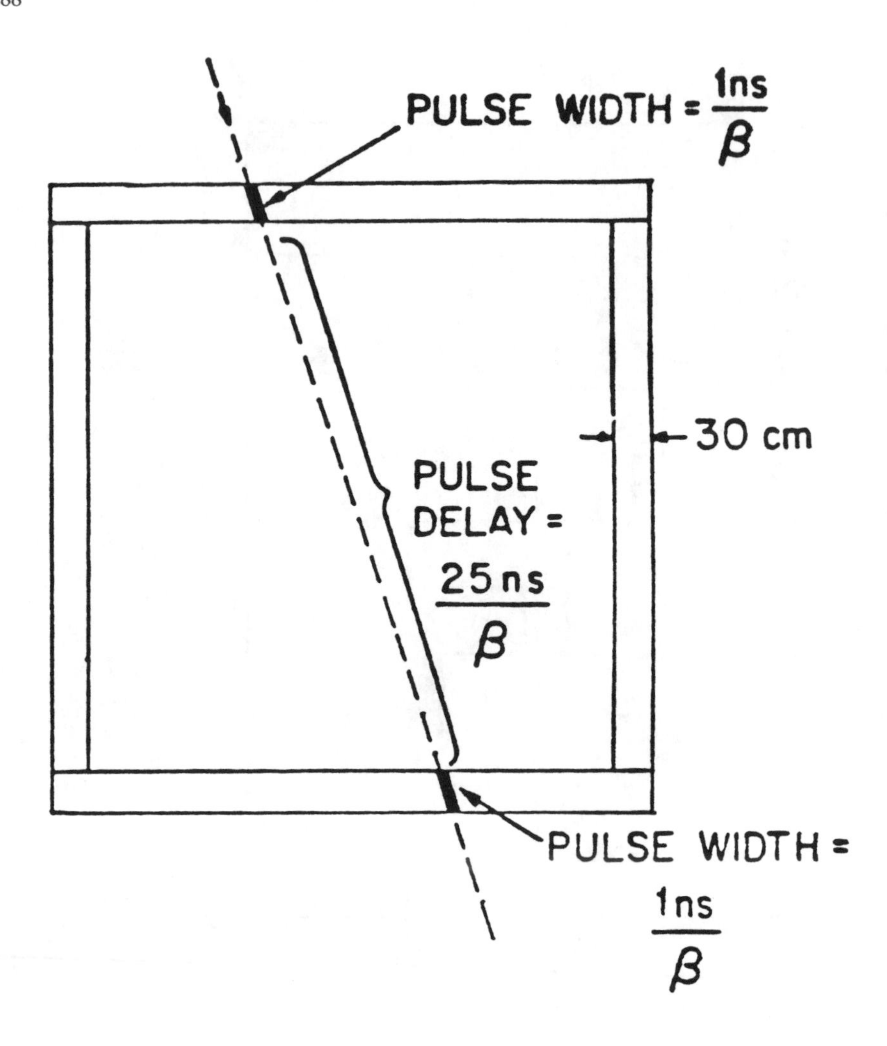

Figure 10. Monopole pulse timing in the Homestake Large Area Liquid Scintillation Detector.

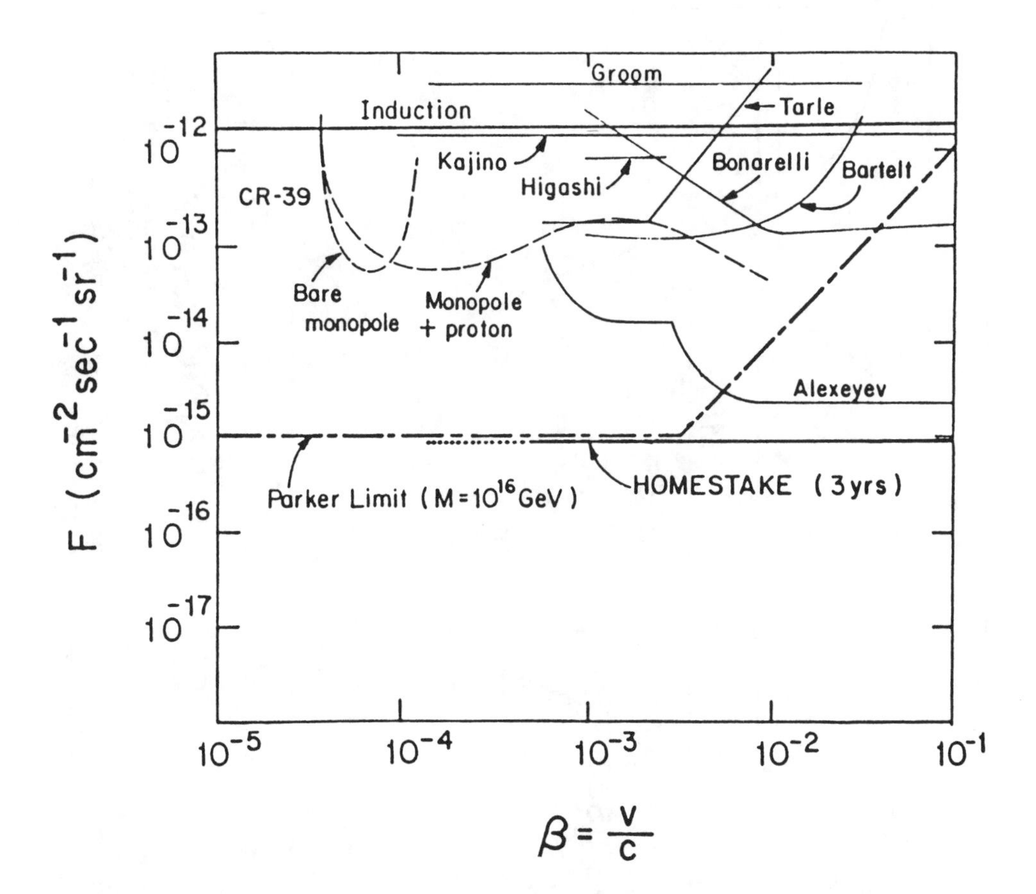

Figure 11. Slow monopole flux limits. The Parker limit is that given for monopoles of $M = 10^{16}$ GeV by Turner et al. (1982). The induction limit is the global limit suggested by Stone (1984). The CR-39 limit is the Price (1984) result; the mica limit comes from Price et al. (1984). The Homestake line is the expected result of running for three years with the LASD. Other limits are references in the citation to Price (1984).

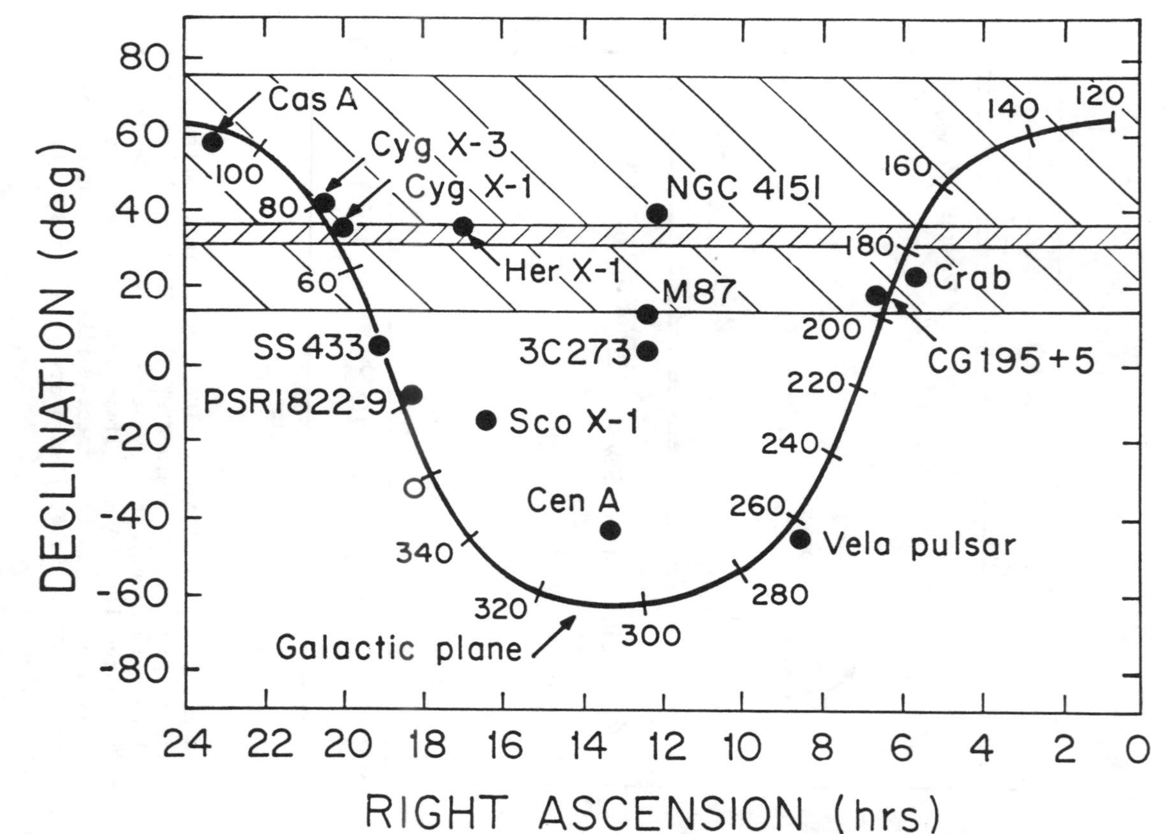

Figure 12. Location of selected x-ray, γ-ray, and radio sources, together with the Homestake field-of-view. The lightly shaded area is the LASD field-of-view, assuming a maximum 30° zenith angle; the heavily shaded area is the region of the sky seen by the surface and underground detectors in coincidence.

REFERENCES

D.S. Ayres et al., Phys. Rev. D29, 902 (1984).

R.M. Baltrusaitis et al., preprint (1985).

G. Battistoni et al., Phys. Letters 155B, 465 (1985).

A. Burrows, in Solar Neutrinos and Neutrino Astronomy, ed. by M.L. Cherry, K. Lande, and W. Fowler, AIP Conf. Proc. No. 126, Am. Inst. of Physics, N.Y., p. 283 (1985).

D.F. Chernoff, S.L. Shapiro, I. Wasserman, Ap. J. 304, 799 (1986).

M.L. Cherry, M. Deakyne, T. Daily, K. Lande, C.K. Lee, R. Steinberg, E.J. Fenyves, J. Phys. G 8, 879 (1982a).

M.L. Cherry, I. Davidson, K. Lande, C.K. Lee, E. M Marshall, R.I. Steinberg, Workshop on Very High Energy Cosmic Ray Interactins, Philadelphia, ed. by M.L. Cherry, K. Lande, R.I. Steinberg, 356 (1982b).

M.L. Cherry, K. Lande, W.A. Fowler, eds., Solar Neutrinos and Neutrino Astronomy, AIP Conf. Proc. No. 126, Am. Inst. of Physics, NY (1985).

M.L. Cherry, S. Corbato, T. Daily, E.J. Fenyves, D. Kieda, K. Lande, C.K. Lee, 19th Intl. Cosmic Ray Conf., paper HE5.1-3, vol. **9**, to be published (1985).

A. Dar, Phys. Rev. Lett. 51, 227 (1983).

D. Eichler and W.T. Vestrand, Nature 307, 613 (1984).

J. Ernwein et al., Proc. 6th Workshop on Grand Unification, Minneapolis (1985).

C.E. Fichtel, DUMAND Summer Workshop on High Energy Neutrino Astronomy, La Jolla (1978).

T.K. Gaisser, T. Stanev, S. Bludman, and H. Lee, Phys. Rev. Lett. 51, 223 (1983).

D. Groom, Revs. Mod. Phys., to be published (1986).

A.M. Hillas, Nature 312, 50 (1984).

K.J. Johnston et al., to be published, Ap. J. (1986).

H. Lee and S.A. Bludman, Ap. J. 290, 28 (1985).

M.L. Marshak et al., Phys. Rev. Lett. 54, 2079 (1985).

K. Mason, Workshop on Cygnus X-3, Naval Research Lab., Washington, unpublished (1986).

D.R. Parsignault et al., Ap. J. Lett. 209, L73 (1976).

P.B. Price, CERN preprint EP/84-28 (1984); D.E. Groom

et al., Phys. Rev. Lett. 50, 573 (1983); F. Kajino et al., 18th Intl. Cosmic Ray Conf., Bangalore 5, 56 (1983); S. Higashi et al., 18th ICRC, Bangalore 5, 69 (1983); J. Bartelt et al., Phys. Rev. Lett. 50, 655 (1983); R. Bonarelli et al., Phys. Lett. 126B, 137 (1983); P.B. Price et al., Phys. Rev. Lett. 52, 1265 (1984); G. Tarle et al., Phys. Rev. Lett. 52, 90 (1984); E.N. Alexeyev, 18th ICRC, Bangalore 5,52 (1983).

M.M. Shapiro and R. Silberberg, Space Sci. Revs. 36, 51 (1983).

V.J. Stenger, Intl. Workshop on Very High Energy Gamma Ray Astronomy, Ootacamund, ed. by P.V. Ramana Murthy and T.C. Weekes, Tata Inst., Bombay (1982).

J. Stone, in Inner Space/Outer Space, Fermilab, to be published (1984).

Symposium on Underground Physics, St. Vincent, to be published, Il Nuovo Cimento (1985).

M. Turner, E.N. Parker, T. Bogdan, Phys. Rev. D26, 1296 (1982).

M. van der Klis and M. Bonnet-Bidaud, Astron. Astrophys. 95, L5 (1981).

B.M. Vladimirskii et al., Sov. Phys. Usp. 28, 153 (1985).

A. Watson, Rapporteur talk, 19th Intl. Conf. Cosmic Rays, La Jolla, vol. 9, to be published (1985).

R. Willingale, A.R. King, and K.A. Pounds, Mon. Not. R. Astr. Soc. 215, 295 (1985).

G. Yodh, Proc. Moriond Astrophysics Mtg. (1981).

MAGNETIC MONOPOLES

W.P. Trower
Physics Department
Virginia Polytechnic Institute and State University
Blacksburg, Virginia 24061 USA

ABSTRACT. The existence of magnetic monopoles would have important consequences for our ideas in particle physics and cosmology. Here we review the status and prospects of the search for these yet unconfirmed objects.

1. INTRODUCTION

The speculation on the existence of free magnetic charge has spanned seven centuries and has spawned much creative theoretical work and considerably fewer experiments [1]. The modern rationale for monopoles was given fifty years ago when Dirac [2] derived a value for its basic magnetic charge, $g = \hbar c/2e = 137/2e$. Later Schwinger [3] gave the magnetic monopole an electric charge and thereby created dyons. The recently fashionable Grand Unification Theories (GUTs) require that monopoles exist and most suggest that they have enormous mass, equal to that of the vector boson divided by α_{GUT}, exceeding that of the proton by at least a hundred thousand [4]. Clearly such monopoles, which can not be produced at accelerators, would be found in the cosmic rays and would have the same velocities as most matter in the galaxy. The monopole size would be just the inverse of the vector boson mass. This complex monopole has been suggested to catalyze nucleon decay at a rate which is highly model dependent[5]. With increasingly less testable theories - the Supersymmetry, Kaluza-Klein and most recently Superstrings - the predictions of monopole properties become increasingly unreliable.

The relevant consequences of GUT theories for

M. M. Shapiro and J. P. Wefel (eds.), Genesis and Propagation of Cosmic Rays, 293–298.

experimentalists are that monopoles can only have been created early in the Big Bang and so will be slow, $\beta\sim.001$; will ionize feebly and so will have great penetration in matter where they will seldom be trapped; and will always produce a current change if they pass through a conduction ring.

All but the most optimistic would now say that no compelling evidence has been seen for the existence of magnetic monopoles in Nature [6]. Here I review the experimental evidence for this pessimism in four detectable processes: induction, catalysis, ionization and trapping.

2. INDUCTION SEARCHES

Alvarez first suggested looking for an electrical current induced by a magnetic monopole threading a closed conducting ring [7]. These tiny induced current changes, which quickly die away in a room temperature ring, will persist in a superconducting ring [8]. Superconducting Quantum Interferometer Devices (SQUIDs) and ambient magnetic field reducing techniques now allow dynamic detectors to be built. Cabrera's was the first of such detectors [9].

The candidate Stanford event [10] was recorded in the single-loop 4-turn 5-cm diameter superconducting niobium wire ring coupled to a SQUID mounted and operated in an ultra-low ambient magnetic field. The event was consistent with the passage of a single Dirac magnetic charge to within 5%, while the baseline noise was typically 1% of such a signal. No similar event has subsequently been seen in either this, or his larger coincidence detector [11]. The previously detected smaller persistient current changes, which could be interpreted as near-miss events, occured in his second detector in only one of the three detector rings, and then about once every two weeks. Although the original signal stands unimpugned,without another in his device it becomes decreasingly credible.

A second candidate event was seen in summer 1985 by a group at Imperial College using independent superconducting loops inside a superconducting shield [12]. Here two horizontal single 2.2 cm^2 rings are threaded through their centers by a 12 cm by 1 m rectangular loop. This design maximized the detector area but the signal for the passage of a monopole no longer had a unique value. After analyzing a number of possible candidates they found one consistient with the passage of a monopole, although its signal was only seen in the window-frame loop.

Three larger, >1 m^2, multiloop detectors with high redundancy are or soon will be, on-line at Stanford, IBM and Chicago. If the techniques on which they are based are proven in their current data runs, it should be possible to extrapolate them into a next generation. Here hundreds of coils providing multiple coincidences and having an area some 100 m^2 could, in principle, be operational by the end of the decade. Currently the sum of all induction detectors place a limit of the cosmic ray monopole flux of ~ 10^{-13}/cm^2/s/sr, a limit that appears to have decreased about a factor of ten per year since 1982.

Magnetic induction detectors rely only on the monopole's magnetic charge and so are independent of all other properties, such as mass, speed, direction, electric charge, ionization, and so on [8]. Thus, induction detectors would be the clear choice for monopole searches if it weren't for their limited areas and high cost.

3. NUCLEON CATALYSIS SEARCHES

A GUT monopole may catalyze baryon nonconserving reaction such as $M + p \rightarrow M + e^+ +$ mesons. This cross section in hydrogen is calculated to be $0.6/\beta$ mb which is smaller than the proton capture cross section ~ $10^{-3}\ /\sqrt{\beta}$ mb for $\beta > 4 \times 10^{-3}$.[5] This means that slow monopoles should first capture a proton before catalyzing the proton's decay. For heavier nuclei the catalysis cross section should be 100 to 1000 times smaller. However, since proton decay experiments have not detected a single nuclear decay event, either monopoles are incredibly rare or they don't catalyze baryon decay.

If catalysis cross section is large, then a passing monopole would trigger a chain of decays along its path in large proton decay detectors like the Kamioka water Cherenkov detector. A search there for two successive decay events yielded null results and the monopole flux limits ~ 10^{-15}/cm^2/s/sr for an assumed catalysis cross section ~ 100 mb and $\beta \sim 10^{-4}$ [13]. The catalysis of proton decays in the sun could produce electron neutrinos with nominal energy of 35 MeV. The Kamioka detector, sensitive to electron energies > 10 MeV, yields a

monopole limit ~ $4 \times 10^{-22} \left(\frac{\beta}{10^{-3}}\right)$/cm^2/s/sr and find a pole

density in the sun of ~ 1 pole/10^{12} grams of solar material.

The catalysis of nucleon decay by monopoles could be an observable energy source for astrophysical bodies without important internal energy sources, such as planets, neutron stars, and white dwarfs. The best limits come from x-ray emission by neutron stars, which are roughly ~ 10^{-22} cm^{-2}/s/sr [15].

4. IONIZATION/EXCITATION EXPERIMENTS

After much uncertainty about the slow monopole energy loss, a new important experiment has provided some confidence that ionization from these particles can be detected. A theory for protons, similar to that for monopoles, was recently tested down to β = .0006 where measurable amounts of light were seen [15]. Comparably slow monopoles should also produce detectable light, so the totality of null results from a variety of large scintillation detectors [6] places a rough limit on the monopole flux ~ 10^{-15}/cm^2/sr/s.

Monopoles have a second ionization energy-loss mechanism in which helium is Zeeman excited into a metastable state[16]. This excitation is transferred by the Penning collisions to a methane molecule which immediately ionizes. The threshold for this process is β ~ .0001. The experimental limit on the monopole flux using this effect is a few 10^{-13}/cm^2/sr/s [17].

Etching of certain materials which can be permanently radiation damaged by the passage of highly ionizing radiation define another class of ionization detectors with which to search for magnetic monopoles[18]. The most sensitive of such detectors is a plastic, CR-39, which is cheaply fabricated into large sheets and can be scanned by semiautomatic means. Although electronic energy loss only occurs for β > .02, diamagnetic effects may make β ~ .0001 searches possible [19].

An indirect monopole search experiment that uses radiation damage looked at ancient mica samples whose age was calibrated by their heavy ion tracks [20]. These mica measurements limits the monopole flux to 10^{-18}/cm^2/sr/s, but under a scenario that has a bare monopole entering the earth, capturing an aluminium nucleus which remains bound and uncatalyzed throughout the mica encounter.

5. TRAPPED MONOPOLE SEARCHES

The earliest search for monopoles trapped in magnetic material attempted to drag them out into a detector using a plused magnetic field and achieved limits of a few tens of monopoles per kilogram of material[21]. In a recent experiment, ancient iron ore was heated above the Curie point allowing any trapped poles to fall under gravity through two waiting superconducting coils. This search using 100,000 tons of ore has produced a limit of a few monopoles per gigagram of magnetite[22].

6. CONCLUSION

Although the magnetic monopole idea has provided a rich source of theoretical speculation, the current two candidate events from two different induction detectors hardly confirm their reality. However, despair should only set in if monopoles are not found in the new generation of large ionization detectors aimed at measuring their abundance at and below the astrophysical bound set by the measured intragalactic magnetic field [23]. This limit, the Parker bound, is ~ 10^{-15}/cm^2/sr/s.

REFERENCES

1. S. Torres and W.P. Trower, Fermilab 85/18 (unpublished) provides a comprehensive biblography of magnetic monopole papers from 1269-1986.

2. P.A.M. Dirac, Proc. R. Soc. London, A133, 60 (1931).

3. J. Schwinger Phys. Rev. 144, 1087 (1966); 173, 1536 (1968); and Science 166, 797 (1969).

4. A.M. Polyakov, JETP Lett. 20, 194 (1974); and G.'t Hooft, Nucl. Phys. B29, 276 (1974).

5. V. Rubakov, JETP Lett. 33, 644 (1981); and C.G. Callan, Phys. Rev. D25, 214 (1982).

6. M. Aguilar-Benitez et al. [Particle Data Group], Phys. Lett. B170, 170 (1986).

7. L.W. Alvarez, Lawrence Radiation Laboratory Physics Note 470, 1963 (unpublished).

8. B. Cabrera and W.P. Trower, Found. Phys. 13, 13 (1983).

9. B. Cabrera Ph.D. Thesis, Stanford University, 1975 (unpublished).

10. B. Cabrera, Phys. Rev. Lett. 48, 1378 (1982).

11. B. Cabrera, M. Taber, M. Gardner, and J. Bourg, Phys. Rev. Lett. 51, 1933 (1983).

12. A.D. Chaplin, M. Hardiman, M. Kortzinos, and J. Schouten, Nature (London) 321 402, (1986).

13. A.M. Koshiba, private communication.

14. E.W. Kolb, S.A. Colgate, and J.A. Harvey, Phys. Rev. Lett. 49, 1373 (1982); and S. Dimopoulos, J. Preskill, and F. Wilczek, Phys. Lett. B119, 320 (1982).

15. S.P. Ahlen, T.M. Liss, C. Lane, Phys. Rev. Lett. 55, 181 (1985); and S.P. Ahlen, private communication.

16. S.D. Drell, N.M. Kroll, M.T. Mueller, S.J. Parke, and M.A. Ruderman, Phys. Rev. Lett. 50, 644 (1983).

17. T. Hara, M. Honda, Y. Ohno, N. Hayashida, K. Kamata, T. Kifune, G. Tanahashi, M. Mori, Y. Matsubara, M. Teshima, M. Kobayashi, T. Konda, K. Nishijima and Y. Totsuka. Phys. Rev. Lett. 56, 553 (1986)

18. R.L. Fleischer, P.B. Price, and R.M. Walker, Nuclear Tracks in Solids (Univ. California Press, Berkeley, 1975).

19. P.B. Price, Phys. Lett. B140, 112 (1984).

20. P.B. Price and M.H. Salamon, Phys. Rev. Lett. 56, 1226 (1986).

21. E. Goto, H.H. Kolm, and K. Ford, Phys. Rev. 132, 387 (1963).

22. T. Ebisu and T. Watanabe, private communication.

23. E.N. Parker, Astrophys. J. 160, 383 (1970); and M.S. Turner, E.N. Parker, and T.J. Bogdan, Phys. Rev. D26, 1296 (1982).

COSMIC RAY BOMBARDMENT OF SOLIDS IN SPACE

Valerio Pirronello
Dipartimento di Fisica, Università di Catania, Catania, Italy
Osservatorio Astrofisico di Catania
Viale A. Doria, 6 - 95125 Catania
Italy

ABSTRACT

Some aspects of the interaction of cosmic rays with solids in space are presented. We first give a short description of experimental results on some effects induced in frozen gases by impinging ions because they constitute the solid base for any reliable astrophysical application. We show the important role played in giant molecular clouds by cosmic rays for the production of molecular hydrogen, the cornerstone of interstellar chemistry. A method to obtain suitable information on the almost unknown spectrum of low energy cosmic rays outside the heliosphere is also presented.

1. INTRODUCTION

Cosmic rays are the only kind of matter entering the inner solar system from the outside of it. Their importance is fundamental because they carry information about their sources, the type and density of matter they have gone through and can help in explaining a lot about high energy phenomena occurring in the Milky Way. They are also the only external agents which release relevant amounts of energy deeply inside interstellar dense and molecular clouds. In this way they heat globally these clouds and determine together with cooling mechanisms their stability.

Most of the effort of cosmic ray physicists has been and is spent, as we have listened here in Erice, into studying the possible astrophysical sources of cosmic radiation, the acceleration mechanisms, especially those responsible for the most energetic rays whose spectrum extends till values greater than 10^{20} eV.

My personal interest is mainly related with the interaction of energetic particles with solids in the interstellar medium, an astrophysical scenario of which cosmic rays are a relevant component because about one third of the total energy density belongs to them. I am mainly interested in cosmic rays having $E < 1$ GeV because they can release the largest amounts of their kinetic energy in the cloud.

M. M. Shapiro and J. P. Wefel (eds.), Genesis and Propagation of Cosmic Rays, 299–313.

The main problem at these energies is that we do not have measurements of fluxes which are representative of those pervading the Galaxy and hence of those bombarding interstellar clouds. This is due to the fact that our measurements are performed deeply inside the heliosphere, that region of the solar system whose electrodynamical properties are due to the presence of the solar wind which carries the magnetic field. It, in turn, acts with an efficient screening effect which prevents most of low energy particles to reach the vicinity of the earth where we can measure them. The edge of the heliosphere is estimated to be between 50 and 100 AU.

In the following, I will first give a short description of laboratory simulations, performed accelerating light ions till MeV energies using a Van de Graaff—simulations which allow to measure the physico chemical modifications induced in solids, mainly ice and mixtures of them, by ion bombardment; then I will present two possible applications of some experimental results. In the first one, using fluxes obtained by Morfill, Völk and Lee (1) on the base of transport theory, I will show how the production of H_2, (a molecule whose importance for interstellar chemistry is such that it is sometime called the seminal molecule) is, at least in same interstellar conditions, due mainly to the cosmic ray bombardment of dust grains covered with "dirty" ice mantles. In the second I will present a method which could allow us to obtain suitable information on the low energy cosmic ray spectrum without going outside the heliosphere.

2. THE SIMULATION

The laboratory simulation of the bombardment of solid material by cosmic rays in space has been performed accelerating light (H^+, He^+,..) ions to MeV energies using electrostatic accelerators and bombarding with them targets of the chosen material at temperatures which are realistic for interstellar space T ~ 10K÷ 100K. Most of my interest has been devoted to molecular "ice" prepared by slow deposition from the gas phase an a cold substrate. They have small binding energies between .5 and .08 eV for species such as H_2O,CO_2, CO, N_2 .... and also they show wide electronic band gaps of many eV which make them good electrical insulators and consequently poor thermal conductors, properties that as we will see later on are important for the type of processes we are going to discuss.

Impinging ions produce several effects such as: crystalline regrowth in amorphous materials, disordering in crystalline ones, erosion and chemical modifications. Such effects depend on several parameters: the charge state and the mass and the energy of the projectile, the composition of the target, and the deposited energy i.e. the energy lost by the particle per unit path length, the so called "stopping power".

The kinetic energy of the ion is progressively transferred to the target atoms via two principal processes:

i) a sort of "direct" collision among the ion and the nuclei partially screened by the most internal electrons (a process which gives rise to the so called "nuclear" contribution to the total stopping $(dE/dx)_n$;

ii) a continuous series of excitations, ionizations and charge excharge reactions produced by the interaction of the ion with the electronic clouds of the striken atoms (the so called electronic stopping $(dE/dx)_{el}$).

Both processes occur "simultaneously" producing a total stopping power

$$\left(\frac{dE}{dx}\right)_{tot} = \left(\frac{dE}{dx}\right)_n + \left(\frac{dE}{dx}\right)_{el}$$

in which the two contributions have different importance according to the energy of the impinging particle; the nuclear contribution is dominant at energies lower than 1keV/amu while at energies higher than this value the electronic one is dominant. It reaches a maximum (whose position for instance for H_2O ice bombardment by light ions is at about 100 keV/amu) and then decreases.

The first important effect due to the energization of target atoms is that some of them can acquire enough energy to overcome the surface barrier and be sputtered from the solid (2-6). Also because of the already mentioned low binding energy of single molecules in these ices, sputtering yields "Y" are very high with respect to those measured for metals and semiconductors (which have binding energies about one order of magnitude higher than those of ices). For this reason sputtering can be a destructive mechanism which can at least under certain conditions affect the presence and the stability of ice mantles on interstellar grains. Some of the results are reported in Table 1.

TABLE 1

Sputtering Yields at 10K by 1.5 MeV He^+

Ice	Binding energy (eV)	Yield
H_2O	.52	8
CO_2	.27	120
CH_4	.18	120
NH_3	.31	140
O_2	,09	120
N_2 [a]	.08	300

[a] estimated

The experimental technique used to measure "in situ" the sputtering yield "Y" (number of ejected molecules per impinging ion) consists simply of measuring the thickness (in mol cm^{-2}) of the deposited layer before and after irradiation with a known dose of projectiles. Thickness measurements are performed using the Rutherford backscattering technique (RBS). It may be worthwhile now to describe this technique because it gives strong experimental evidence for the production, by bombardment, of fluffy layers in the sample.

The experimental procedure is shown schematically in Fig. 1. In the upper part, (a), is shown the silicon substrate on which has been previously deposited a gold marker (~250 Å thick). On top of this marker there is accreted (directly inside the scattering chamber) the ice layer to be irradiated.

In the lower part of the figure, (b), is shown a typical type of energy spectrum of backscattered ions. We see the sharp peak due to ions that bounced off of gold atoms and the broad shoulder due to those ions backscattered by atoms belonging to the "semi-infinite" silicon substrate.

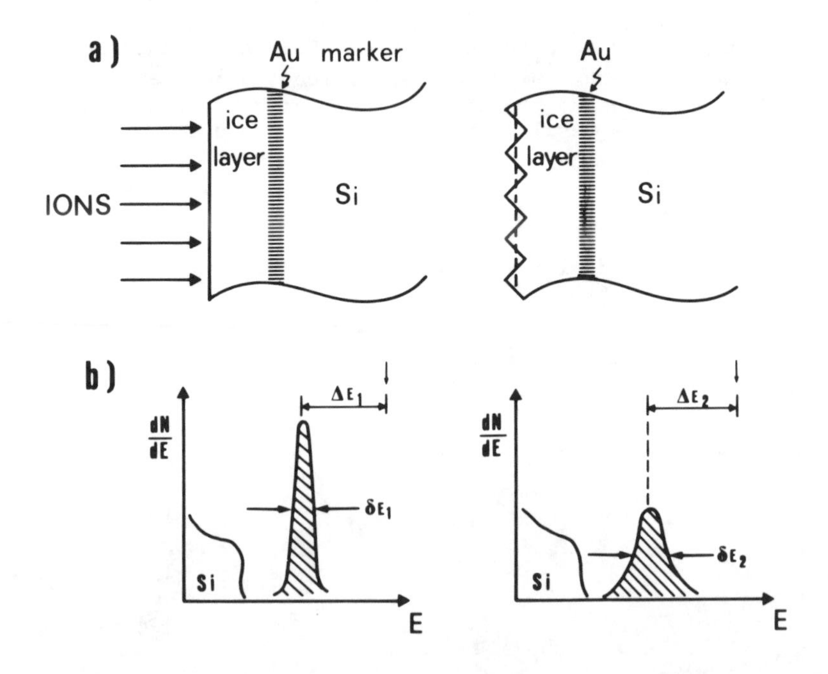

Fig. 1: Schematic of the experimental procedure of RBS measurements.

In this technique the thickness of the ice layer is recorded by the position of the gold marker peak. When initially no ice is accreted on the marker, the position of the Au peak occurs at higher energies (see the arrow). Just after the accretion of ice on the cold substrate the spectrum of backscattered particles presents the Au peak shifted toward lower energies with respect to the original position where the arrow is . The origin of the shift of the marker peak is naturally due to the energy loss of ions in the ice layer both before and after bouncing against the Au atoms. From a quantitative point of view the energy shift "ΔE" is related to the thickness "Δx" of the ice layer through a function "$f(\frac{dE}{dx})$" of the total stopping power; which of course depends on the geometry between the ion beam, the target and the particle detector, i.e.

$$\Delta E = f\left(\frac{dE}{dx}\right)\Delta x \qquad (2)$$

After this first thickness measurement the sample is irradiated with the beam of interest and after a certain controlled fluence another RBS spectrum is taken to remeasure the thickness. After bombardment the shift of the Au marker peak "ΔE_2" is smaller than the previous one because the thickness of the ice layer has diminished. From the difference in thicknesses knowing the dose of particles that irradiated the sample is trivial to evaluate the sputtering yield.

As can be seen from a comparison between the two spectra in the schematic the Full Width at Half Maximum (FWHM) of the peak in the two cases is different; after a fluence of bombarding ions it is wider than before, i.e.:

$$\delta E_1 < \delta E_2 \qquad (3)$$

The explanation of the increase in the FWHM can be found in the upper part of Fig. 1, on the right. Such an increase is due to a change in the surface topography of the bombarded layer. Impinging ions that cross the ice layer where it is thinner lose less energy so in the spectrum they fall in the right wing of the peak, those crossing a thicker layer lose more energy and fall in the left wing of the peak. These fluffy structures are very similar to those observed in Brownlee particles (7), the only sample of dust particles of extraterrestrial origin. The impinging ion loses its energy also breaking some molecules of the ice. Even if such a release of energy is not very large as compared to the nuclear and the electronic one, it has important consequences. It is, in fact, at the base of the chemical modifications induced by the projectiles in these ices and in the same way by cosmic rays in grain mantles in the ISM.

As already mentioned ices are poor thermal conductors; the energy released by the ion remains confined around its track for a time interval long enough ($\sim 10^{-12}$ s) to be thermalized. In this transiently heated high temperature cylindrical region, fragments of molecules are mobile enough to migrate and recombine forming new species. Some of them escape from the solid during the bombardment. This is the case, for instance, for H_2, O_2, H_2O, from H_2O ice irradiated with H^+, He^+, C^+...(4,8) and also the case of CO, CO_2, O_2 from frozen CO_2 (Brown et al., 1981 unpublished). Other species, on the contrary, once formed remain trapped in the frozen layer till when it is allowed to subline. This is the case of formaldehyde, produced in huge amounts (3.7 H_2CO molecules per each 1.5 MeV He^+) in a mixture of H_2O and CO_2 about 1000 monolayers thick by Pirronello et al. (9) and also the case of other polyatomic molecules in several complex mixtures of frozen gases bombarded by light ions (10).

In the following we will apply some of the experimental results just briefly described to the interaction of cosmic rays with icy solids in the ISM.

3. THE FORMATION OF MOLECULAR HYDROGEN

A wide variety of clouds, in which interstellar matter clumps, do exist and wherever the ultraviolet radiation field is screened most of hydrogen is in molecular form. Such a fact gives rise to the problem of the formation of H_2, a problem that has kept the attention of astrophysicists because of the central role played in the development of cloud chemistry, both in dense and diffuse ones, by this molecule that is for this reason sometime called the "seminal" molecule. It has been recognized long ago that radiative association of two hydrogen atoms in the gas phase cannot be in fact responsible for the presence of H_2 because the molecule is formed in a highly excited state and it is not able to emit a photon carrying formation energy "ΔE" before dissociating. Hollenbach and Salpeter (11, 12) proposed a mechanism wich seemed to overcome the problem. It is based on the surface reaction of two adsorbed H atoms on a grain to which is released the formation energy "ΔE". The absorption of ΔE by the grain allows the formed molecule not to dissociate and with the subsequent rise in temperature of the grain (which has a very low thermal capacity) allows the release in gas phase of the formed molecule.
For this mechanism to work an atom has to remain adsorbed on the grain for a time long enough to be reached by another adsorbed atom. If the residence time of adsorbed atoms t_r is longer than the average time interval between adsorptions t_a, i.e. if $t_r > t_a$ with

$$t_r = \frac{1}{\nu} \exp (E_b/KT_g) \qquad (4)$$

where ν = vibrational frequency of H atoms in the adsorption potential
E_b = binding energy
T_g = grain temperature

and $t_a = (\pi a^2 n_H < v > S)^{-1}$

where n_H = number density of H atoms
$< v >$ = mean thermal speed
S = sticking coefficient.

The production rate of H_2 "$dn(H_2)/dt$" per cubic centimeter per second in this case is given by

$$R_{HS} = dn(H_2)/dt = \frac{1}{2}\, n_H \int_{a_{min}}^{a_{min}} n_g(a) < v_H > \pi a^2 S\, \xi\, d_a \qquad (5)$$

where a = grain radius
n = number density of hydrogen atoms
ξ = probability of encounter of two adsorbed H atoms
S = sticking coefficient
n_g= number density of dust grains
$<v>$= mean thermal speed of H atoms

In their original calculation Hollenbach and Salpeter (11, 12) assumed that the grain surface had the structure of a monocrystal. Under this assumption they could show that even at interstellar temperatures at which thermal hopping is inefficient, a very high mobility of adsorbed H atoms is assured by a temperature independent quantum mechanical tunneling from one site to the other. In this way the probability ξ that two adsorbed hydrogen atoms encounter each other giving an H_2 molecule comes to be approximately equal to one.

Taking as usual for the sticking coefficient S $\simeq$.3 one can estimate the production rate of molecular hydrogen R, for instance in giant molecular clouds. We will assume for this type of cloud that the density of molecular hydrogen is $n(H_2) \sim 500\ cm^{-3}$, that of atomic hydrogen is $n_H \sim 1 \div 10\ cm^{-3}$, the kinetic temperature is T ~ 10K and that the size distribution of grains follows that one given by Mathis, Rumple, Nordsieck (13). With these hypotheses one can easily get

$$R_{HS} \sim 5 \times 10^{-16}\ cm^{-3}\ s^{-1}$$

In recent papers Smoluchowski (14, 15, 16) has shown that the original approach to the problem by Hollenbach and Salpeter (11, 12) has to be revisited by taking into proper account the conceivable non-crystalline structure of ice in grain mantles. An amorphous solid can be defined, according to Konnert and Karle(17), as a solid in which structural ordering is at most on the order of 10 Å. The most important difference between the two types of ice, amorphous and crystalline, is not in the number of nearest neighbours, but in the lack of periodicity. Smoluchowski's investigation demonstrates that in the case of armorphous ice for the adsorption sites "there is a broad range of binding energies which is in strong contrast with the four discrete states possible on crystalline ice; about 10% of the sites are quite deep while the remainder are more shallow".

The presence of these few high binding energy sites have an important role in the mobility of H atoms adsorbed on amorphous surfaces. At interstellar temperatures the motion of adsorbed atoms requires a quantum mechanical treatment and is described by the behaviour of its wave function. In the case of a crystalline surface which has a periodic structure the wave packet spreads over almost immediately so an equal probability for every site on the surface to find the H atom exists. On the contrary in the case of an amorphous structure the wave packet describing the atom is quickly localized by tunneling to a site of lowest energy in its vicinity and the probability of encountering another adsorbed H atom by surface migration, in the range of temperatures where thermal hopping is inefficient, becomes exceedingly small.

Such considerations, together with the one that for grains in dense or giant molecular clouds H_2 molecules instead of H atoms can be adsorbed in the usefull sites with a probability that roughly scales according to the ratio between the respective gas phase abundances, allowed Smoluchowski(15) to reevaluate the rate of H_2 formation in interstellar clouds. He obtained a strong temperature dependent rate "R_S" (in fig. 1 of ref. 15) with a maximum at about $T \sim 17K$ which is more than three orders of magnitude lower than the Hollenbach and Salpeter value. It drops below 10^{-27} H_2 formed molecules cm^{-3} s^{-1} at $T \sim 16K$ and decreases exponentially at temperatures higher than about 17K.

At this point it is convenient to evaluate the production of molecular hydrogen with the mechanism I proposed (18) i.e., the cosmic ray bombardment of dust mantles.

For this process the H_2 production rate R_{CR} (H_2) is given by

$$R_{CR}(H_2) = \int_{a_{min}}^{a_{max}} \int_{0}^{E_{max}} Y_{H_2}(E)\ dJ/dE\ n_g(a)\ \pi\ a^2\ dE\ da \qquad (6)$$

where a = grain radius
n_g = number density of grains
$Y_{H_2}(E)$ = H_2 production yield per impinging ion
dJ/dE = differential flux of cosmic rays.

In order to evaluate such a rate we assume that in giant molecular clouds grains are coated by an ice mantle made by H_2O, CH_4, NH_3 in the following per-cent abundances: 52%, 38%, 10%, that the cosmic ray differential flux, bombarding the cloud, is that evaluated by Morfill, Völk and Lee (1) and the grain size distribution is that given by Mathis et al.(13).

With these assumptions I obtain for the production rate of molecular hydrogen induced by cosmic rays

$$R_{CR}\ (H_2) \sim 1.8 \times 10^{-20}\ cm^{-3}\ s^{-1}$$

Such a value is the correct one for $T \lesssim 30K$; for higher temperatures $R_{CR}(H_2)$ increases monotonically because of the monotonic and steep increase in $Y_{H_2}(4)$.

If this rate is now compared with Smoluchowski's results it appears evident that for temperatures below 16K, $R_{CR}(H_2) >> R_S(H_2)$ by orders of magnitude, in the narrow range $16K < T < 18\ K$ it is $R_S \gtrsim R_{CR}$ and at $T > 18K$ again $R_{CR} > R_S$.

After the careful treatment of adsorption and mobility of H atoms on amorphous icy surfaces performed by Smoluchowski I have shown that, at least in giant molecular clouds where grain temperature is around 10K, the direct production of H_2 by cosmic ray bombardment of grain mantles is dominant with respect to the recombination of H atoms on grain surfaces.

4. LOW ENERGY COSMIC RAYS IN THE ISM

As already mentioned in the introduction the spectrum of cosmic rays pervading the Galaxy in the low energy range ($E < 1$ GeV) is almost unknown because of the screening effect offered by the interplanetary magnetic field carried by the solar wind. In order to estimate this flux it is costumary to use the transport theory of ions in magnetic structures (see e.g. ref. 1) together with the measurements of high energy particles which are not much affected by the presence of the heliosphere.

In my opinion (19) an "almost direct" way of getting information on this low energy spectrum of cosmic rays in the ISM exists. It needs some probes that stay outside the heliosphere, record informations on cosmic rays and then, coming in the vicinity of the earth, allow us to obtain this information Fortunately such probes are provided by nature: they are "comets"; either new comets, those that enter for the first time in the inner solar system, or very long period comets.

Comets are considered among the most pristine objects of the solar system, it is infact believed that they formed together with it about 4.6×10^9 years ago. New comets have been for all this time in the so called "Oort cloud", which should extend till 50.000 - 100.000 AU, and in any case well outside the heliosphere. During their residence there cometary nuclei have been bombarded by cosmic rays of all energies, withouth any screening, and they recorded in their external layers, as chemical modifications,information even on low energy particles. When they come close to the Sun the sublimation of external layers of cometary nuclei releases in gas phase species synthesized by ion bombardment; measurements of their production rate "Q" while the comet is approaching will give concentrations of the observed species at various depths in the nucleus (when it was in the Oort cloud) and hence fluxes of ions which reached that depth, i.e. the energy spectrum of particles impinging on the nucleus.

In Fig. 2 we display the energy spectrum of He^+ at varions depth in an ideal cometary nucleus made of equal parts of H_2O and CO_2 if the energy spectrum of cosmic rays impinging on the nucleus were that given in ref. 1.

Using this type of energy spectra one can evaluate the abundance of molecules such as H_2CO in a layer of thickness "dE" and unit area as a function of depth in the nucleus, through the integral

$$\int dt \int Y_{H_2CO}(E')\, dJ'/dE'\, dE' \qquad (7)$$

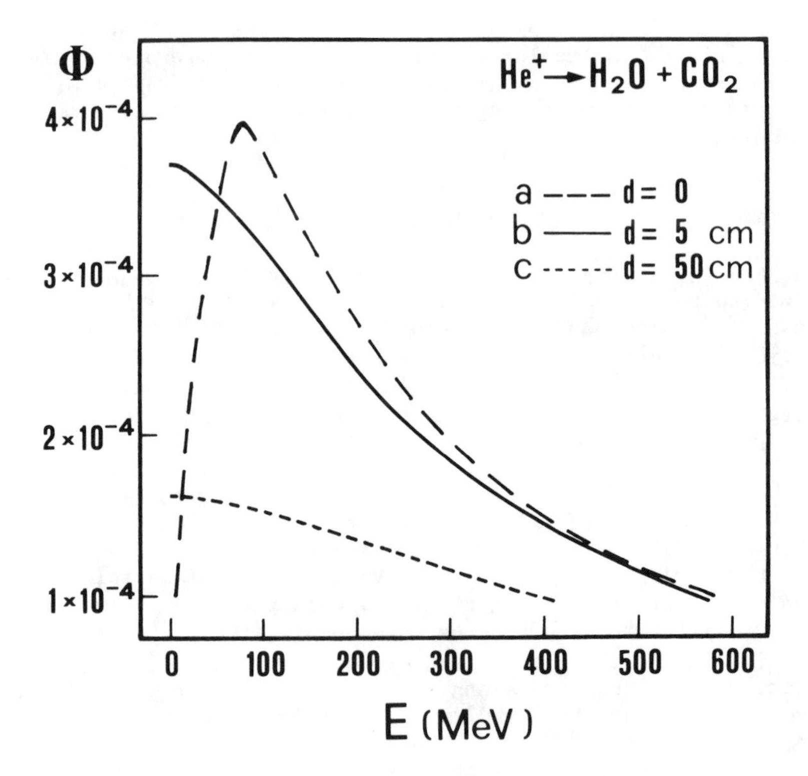

Fig. 2: Differential energy spectra of the low energy helium component of cosmic rays at various depths in an ideal cometary nucleus made of equal parts of water and carbon dioxide ice:

a) differential flux at the surface of the cometary nucleus
b) distorted flux at a depth d = 5 cm inside the nucleus
c) distorted flux at a depth d = 50 cm inside the nucleus

where

$E' = E_{im} - E_{lost}$

E_{im} = energy of impinging ions on the surface of the nucleus

E_{lost} = energy lost to reach the required depth

dJ'/dE' = energy spectrum at the required depth

$Y_{H_2CO}(E')$ = experimental production rate of H_2CO by impinging ions.

The production of H_2CO versus depth inside the nucleus in 10^6 years

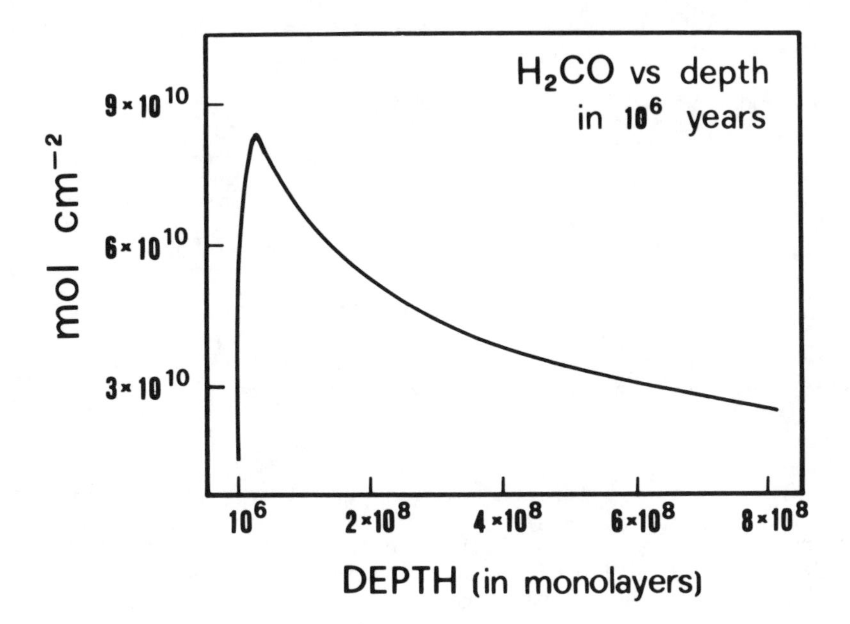

Fig. 3: Production of formaldehyde in an ideal cometary nucleus made of equal parts of water an carbon dioxide by the flux of cosmic rays given in ref. 1 as a function of depth.

via this mechanism is shown in Fig. 3 if the mentioned cosmic ray spectrum is used. In the lack of any dedicated observation of a real comet I will here describe only the method.

Let us now consider the release of the species produced by cosmic ray bombardment, for instance H_2CO, when the comet approaches the sun.

If $N(H_2CO)$ represents the total number of formaldehyde molecules in the coma at any given time, if ν (H_2CO) denotes the photodestruction rate and f (H_2CO) the fractional abundance of H_2CO (that depends on depth) the continuity equation for formaldehyde as the heliocentric distance "r" varies can be written as

$$\dot{r}\,\frac{dN(H_2CO)}{dr} = f\,(H_2CO)\,.\,4\,\pi\,R_n^2\,\dot{Z} - \frac{\nu(H_2CO)}{r^2}\,N\,(H_2CO) \qquad (8)$$

where

R_n= is the radius of the nucleus

$\dot{Z}$ = evaporation rate in mols cm^{-2} s^{-1}

with $\dot{Z}$ obtained integrating the Clausius-Clapeyron equation

$$Z = \frac{N_A P}{(2\pi\ MRT)^{\frac{1}{2}}} \tag{9}$$

where

N_A = Avogadro number

M = mass of the sublimating species

and

$$P = P_o \exp \frac{L}{R} \left(\frac{1}{T_o} - \frac{1}{T}\right) \tag{10}$$

where (P_o, T_o) is a reference point in the vapor pressure-temperature curve and L is chosen equal to that of water because the experiment performed by Pirronello et al. (9) has shown that trapped H_2CO molecules are released only when the sublimation of H_2O occurs.

Together with eq. (7-9) in order to form a complete set of equations we have to add the quasi steady state energy equation

$$\frac{J\ (1-A)}{r^2} = \varepsilon\ \sigma\ T^4 + \frac{L}{N_A}\ \dot{Z} \tag{11}$$

where

J = solar constant
A = bolometric albedo
ε = infrared emissivity
σ = Stefan-Boltzmann constant

and for a comet in a quasi-parabolic orbit

$$|\dot{r}| \simeq (2\ GM_0)^{\frac{1}{2}} \left(\frac{1}{r} - \frac{q}{r^2}\right)^{\frac{1}{2}} \tag{12}$$

where

q = perihelion distance
M_0= Mass of the Sun
G = gravitational constant

Formaldehyde molecules have a lifetime $\tau \sim 3600$ s against photodissociation at 1 AU, so it would be easier to observe their fragments HCO (20). In this case one has to consider together with equation (7) also the following one

$$\dot{r}\ \frac{dN(HCO)}{dr} = \frac{\nu(H_2CO)}{r^2}\ N\ (H_2CO) - \frac{\nu(HCO)}{r^2}\ N\ (HCO) \qquad (13)$$

where N(HCO) is the total number of HCO molecules in the coma.

From the complete set of equations (7-13) it is possible to obtain $N(H_2CO)$, N(HCO), $\dot{Z}$,T as a function of the heliocentric distance r once q, A, ε and R_n are given.

In the simple case of a quasi-steady state, eliminating $N(H_2CO)$ from equation (7) and (13) one easily gets:

$$N(HCO) = \frac{f(H_2CO)}{\nu(HCO)}\ 4\ \pi\ R_n^2\ r^2\ \dot{Z}$$

Observations of N(HCO) as time varies would give $f(H_2CO)$ that would correspond to the concentration of formaldehyde at various depths in the nucleus before the beginning of sublimation. From $f(H_2CO)$ it is them possible to obtain dJ'/dE' and hence dJ/dE the differential flux of cosmic rays impinging on the surface of the nucleus.

5. CONCLUSIONS

Some aspects of the irradiation of icy solids by cosmic rays have been presented. After a brief description of some of the experimental work in which I have been directly involved I have shown that at least under certain interstellar conditions i.e. in giant molecular clouds (but the result can be and will be extended to other type of clouds) the bombardment of dirty ice mantles of dust grains is the dominant mechanism of production of H_2, probably the most important species for explaining the interstellar chemistry. I have also described a possible and perhaps unique method to obtain reliable information on the spectrum of low energy cosmic rays from observations of complex molecules either in new or in very long period comets.

AKNOWLEDGMENTS

I should like to thank: B. Donn; M.M. Shapiro and R. Smoluchowski for helpful discussion, Ms. D. Averna, Mr. G. Lanzafame, and Mr. P. Massimino who are now helping me in improving my earlier calculations, Mr. V. Piparo for his help in carring out some experimental results and Ms. A. D'Amico for typewriting the manuscript.

REFERENCES

1. Morfill, G.E., Völk, H.J. and Lee M.A.: 1976, J. Geophys. Res. 81, p. 5841

2. Brown, W.L., Lanzerotti, L.J., Poate, J.M. and Augustyniak, W.M.: 1978, Phys. Rev. Lett. 40, p. 1027

3. Brown, W.L., Augustyniak, W.M., Brody, E., Cooper, B., Lanzerotti, L.J., Ramirez, A., Evatt, R. and Johnson, R.E.: 1980, Nucl. Instrum. Meth. 170, p. 321

4. Brown, W.L., Augustyniak, W.M., Simmons, E., Marcantonio, K.J., Lanzerotti, L.J., Johnson, R.E., Boring, J.W., Rei,ann, C., Foti, G. and Pirronello, V.: 1982, Nucl. Instr. Meth. 198, p. 1

5. Pirronello, V., Strazzulla, G., Foti, G. and Rimini, E.: 1981, Astron. Astrophys. 96, p. 267

6. Johnson, R.E., Lanzerotti, L.J., Brown, W.L., Augustyniak and Mussil, C.: 1983, Astron. Astrophys. 123, p. 343

7. Brownlee, D.E.: 1978, Cosmic Dust, J.A.M. Mc Donnell ed., John Wiley & Sons Publ. Co., p. 295

8. Ciavola, G., Foti, G., Torrisi, L., Pirronello, V. and Strazzulla G.: 1982, Radiation Effects 65, p. 167

9. Pirronello, V., Brown, W.L., Lanzerotti, L.J., Marcantonio, K.J. and Simmons, E.: 1982, Astrophys. J. 262, p. 636

10. Moore, M.H., Donn, B., Khanna, R. and A'Hearn, M.F.: 1983, Icarus 54, p. 388

11. Hollenbach, D. and Salpeter, E.E.: 1970, J.Chem.Phys., 53, p. 79

12. Hollenbach, D. and Salpeter, E.E.: 1971, Astrophys.J. 163, 155

13. Mathis, J.S., Rumpl, W. and Nordsieck, K.H.: 1977, Astrophys.J. 217, p. 425

14. Smoluchowski, R.: 1979, Astrophys. Space Sci. 65, p. 29

15. Smoluchowski, R.: 1981, Astrophys. Space Sci. 75, p. 353

16. Smoluchowski, R.: 1983, J.Phys.Chem. 87, p. 4229

17. Konnert, J.H. and Kark, J.: 1973, Acta Cryst. A29, p. 702

18. Pirronello, V.: 1986a, IAU Symposium Astrochemistry, Tarafdar ed, Reidel Publ. Co, Dordrecht (in press)

19. Pirronello V.: 1986b, IAU Symposium Astrochemistry, Tarafdar ed, Reidel Publ. Co, Dordrecht

20. Cosmovici, C.B. and Ortolani, S.: 1985, Ices in the Solar System, Klinger J. et al. eds, Reidel Publ. Co., Dordrecht

SOLAR-CYCLE MODULATION OF GALACTIC COSMIC RAYS

John S. Perko
NASA/NRC Resident Research Associate
Code 665
Goddard Space Flight Center
Greenbelt, MD 20771
USA

ABSTRACT. In this paper, I will describe a numerical solution of the spherically-symmetric, time-dependent, cosmic-ray modulation equation, compare it to data taken near the ecliptic, and probe the physical mechanisms responsible for the solar-cycle variations in the cosmic-ray intensity. With a physically defensible diffusion coefficient and other reasonable parameters, I can simulate the spectra of protons and electrons simultaneously over the cycle, the radial intensity gradients for particles greater than about 100 MeV, the overall intensity variation during the cycle, and the time lag in the recovery of low-energy particles behind high-energy ones, known commonly as the "hysteresis." The results suggest that cosmic-ray variations near the ecliptic are dominated by turbulent scattering regions in the heliosphere, overwhelming any effects of gradient and curvature drifts.

1. INTRODUCTION

Since galactic cosmic rays are the only direct sampling of the interstellar medium available, information about them is obviously important. In interplanetary space, however, this sampling is complicated by a redistribution in energy and space that we call modulation. For practical purposes, modulation should be considered for cosmic rays less than 20-30 GeV/nucleon, an important energy regime for production and propagation. So, to understand the behavior of cosmic rays in interstellar space, we need to know how their behavior is affected by the Sun.

Evidence for modulation goes back to the mid-1930's, when Forbush [1938] noted sudden intensity decreases in cosmic-ray measuring stations throughout the world. For over thirty years, neutron monitors on the ground have kept track of cosmic rays of about 5-10 GeV. The decreases they have measured in cosmic-ray intensity at solar maximum and subsequent recovery at solar minimum attest to the Sun's influence over the particles throughout the cycle. [See, for

M. M. Shapiro and J. P. Wefel (eds.), Genesis and Propagation of Cosmic Rays, 315–323.

example, Lockwood et al., 1986.]

The instruments of modulation are the solar wind and the interplanetary magnetic field (IMF). The solar wind is the continuous radial outflow of plasma from the base of the Sun's corona. The IMF is the Sun's magnetic field, frozen into and stretched out by the wind plasma. It is twisted by the Sun's rotation and permeates the heliosphere in all directions. When the solar wind speed is constant, it forms an Archimedean spiral.

Consider here a model of the IMF, in which the constant mean field is superposed by fluctuations, which have scale sizes comparable to the cosmic-ray gyroradius. They could be Alfven waves, magnetosonic waves, or sudden shifts in the field due to turbulence. These irregularities disrupt the normal helical path of the cosmic rays along the field lines and scatter them elastically in pitch angle, the angle between the mean magnetic field and the instantaneous velocity of the particle. We usually assume this scattering to be a diffusive process: that is, the streaming **S** (number of particles crossing unit area in unit time per unit energy) is proportional to the gradient of the cosmic-ray differential number density "U":

$$\mathbf{S} = - \mathbf{K} \cdot \nabla U \,, \tag{1}$$

where **K** is the proportionality factor known as the diffusion tensor, proportional to the mean free path length between scatterings.

In the spherically-symmetric case I will use here, the solar wind carries the magnetic irregularities outward, convecting particles outward as well, and reducing their intensity in the inner heliosphere.

Since the cosmic rays are scattered by irregularities which move in the solar wind, they also suffer energy changes. As seen by the solar wind, the cosmic rays are a mobile gas exerting pressure. Since this pressure has a positive gradient, due to the increasing number of cosmic rays towards the interstellar medium, the solar wind does work against them and increases their energy. However, the cosmic rays find themselves in a spherically expanding medium as seen in the frame of the wind. Thus they are also adiabatically cooled by being confined to this expanding volume. The resolution to this paradox is that some particles do not enter the heliosphere at all, but are scattered back out into interstellar space. They bounce off the outer boundary, a distance of 50-100 AU, and gain energy in the process. The energy they gain, added to the energy lost by cosmic rays in the expanding medium, equals the total work done on cosmic rays by the solar wind, and so energy is conserved. In most neighborhoods where we measure cosmic rays, only the energy loss in the expanding wind concerns us.

Finally, we must speak of drifts, movement of cosmic-ray particles primarily in latitude due to gradients and curvature in the IMF. Theoretically, this should be manifest in various effects due to charge difference among cosmic-ray species and magnetic polarity differences, both of which determine the direction that these drifts take (generally, either toward or away from the ecliptic).

Current experimental evidence shows a slight charge-dependent effect at the last two solar magnetic polarity reversals, in the form of sudden shifts in positive-to-negative particle ratios [Garcia-Munoz et al., 1986]. However, drift theory itself, elaborated primarily by Jokipii and his co-workers (see Isenberg & Jokipii [1979] and references therein) provides predictions over the 22-year magnetic cycle which either do not match the data or can be duplicated by other, rigidity-dependent effects. The theory is basically sound; drifts should be easily noticeable in the data, yet they are difficult to measure in the ecliptic, where most of the data we have today originated. Why this is so remains an important theoretical question.

The effects of modulation are collected in the following equation [Parker, 1965; Gleeson & Axford, 1967]:

$$\frac{\partial U}{\partial t} = \nabla\cdot(\mathbf{K}\cdot\nabla U) - \mathbf{V}_D\cdot\nabla U - \nabla\cdot(\mathbf{V}U) + \frac{\nabla\cdot\mathbf{V}}{3}\frac{\partial}{\partial E}(\alpha E U) \qquad (2)$$

where "U" is the differential cosmic-ray number density, "E" is the kinetic energy, $\mathbf{V}$ is the solar wind velocity, $\alpha=(E+2E_o)/(E+E_o)$, where "E_o" is the particle rest energy, and $\mathbf{K}$ is the diffusion tensor introduced in equation (1). This equation assumes near-isotropic diffusion. The first term on the right-hand side results from diffusion, the third term from convection, and the last term from adiabatic energy loss. The second term accounts for the drift motion of the cosmic rays in the large-scale magnetic field. $\mathbf{V}_D$ is the mean drift velocity. We will ignore it from here on, and then see how the results fare.

Fisk [1980] gives an introductory review of the theory of solar modulation. A more detailed review is in Fisk [1979]. The most comprehensive review is in Gleeson & Webb [1980].

2. TIME-DEPENDENT MODEL

The first numeric, time-dependent solution of this equation was published by Perko & Fisk [1983]. Our technique allowed us to model qualitatively the solar-cycle variation of the cosmic rays, and resulted in a reasonable simulation of particle intensity variations, radial intensity gradients, and, most significantly, the so-called "hysteresis" effect.

This is a phenomenon in which the recovery of low-energy cosmic rays after solar maximum lags behind the recovery of the high-energy ones. This effect is usually shown in a regression plot, with a characteristic "hysteresis" shape, as seen in Burger & Swanenburg [1973]. (See also Figure 4 of this paper.)

The model for this paper is the same as Perko & Fisk, but here I refined the parameters more carefully for comparison to data.

Equation (2) was solved numerically in a spherical coordinate system, in radius only. A continuous cosmic-ray source spectrum was placed at the outer boundary of the modulation region (r=R), beyond which we assume is unfettered interstellar space. The proton source spectrum was just a power law in total energy, with a spectral index of -2.6, the current consensus. I chose the electron source spectrum which Cummings _et al._ [1973] calculated using radio synchrotron data. The inner boundary condition is just S=0, to preclude sources and sinks of cosmic rays at r=0. Finally, V=400 km/s, is everywhere and at all times constant, and R=50 AU.

The form of the diffusion coefficient needs special attention, since it not only provides the pivotal physics for all modulation effects, but also its form and values are not well-known. Although the scattering tensor **K** is, in general, not isotropic, we can try the assumption that particles remain on magnetic field lines. We can also, as shown by Ng [1972], do the actual integration of equation (3) along magnetic field lines. (Further details of the above assumptions can be found in Perko & Fisk [1983].) All this, finally, collapses the tensor to a scalar, "K_r." For an overall review of this and other topics in the propagation of cosmic rays in the solar wind, see Jokipii [1971].

Based on quasi-linear theory at high energies [Jokipii, 1971] and empirical studies of solar particle data at low energies, I can use the following form for the quiet-time "K_r", originally used by Urch & Gleeson [1972]:

$$\begin{aligned} K_r &= A\,\beta\,(P_1P_2)^{\frac{1}{2}}\,, && P < P_2 \\ K_r &= A\,\beta\,(P_1P)^{\frac{1}{2}}\,, && P_1 > P > P_2 \qquad (3) \\ K_r &= A\,\beta\,P\,, && P > P_1 \end{aligned}$$

where P is particle rigidity, β is the speed of the particle over the speed of light, and "A", "P_1" and "P_2" are constants. I used values of "P_1" and "P_2" picked by Urch & Gleeson to match cosmic-ray spectra in 1965 and 1970, solar minimum and maximum respectively. I kept "A" constant throughout, with a value to match the 1965 spectrum of electrons.

Instead of varying "A," as did Urch & Gleeson, I used a system of single, sine-wave scattering disturbances. These are regions of large decreases in the diffusion coefficient, down to 5% of its quiet-time value, which is equation (3). Think of them as thin, turbulent shells propagating out from the origin. I sent these 2 AU-wide disturbances out into the heliosphere at 400 km/s, the solar wind speed. Since "K_r" is directly related to the mean free scattering length of the particles, the disturbances scatter cosmic rays as they go, causing sudden declines in intensity and slow recoveries. These scattering shells were inspired by data from _Pioneer 10_ detectors [McDonald _et al._, 1981], which see the effects of transient regions, originating in active regions of the Sun, passing through the solar system. Deep space probes have tracked single scattering regions which successively depress the intensity at 1 AU and then at the spacecraft.

To simulate a solar cycle, these regions are sent out at an increasing rate going from solar minimum to maximum, and at a decreasing rate from maximum to minimum. That is, an observer at a fixed spot would see one go by about every four months at solar minimum increasing to about one every week at solar maximum. Just the reverse occurs during the declining phase of the solar cycle.

3. SIMULATION RESULTS

Figure 1 shows a sample result of the calculation: 1 GeV proton differential intensity at 25 AU over one solar cycle. The travelling

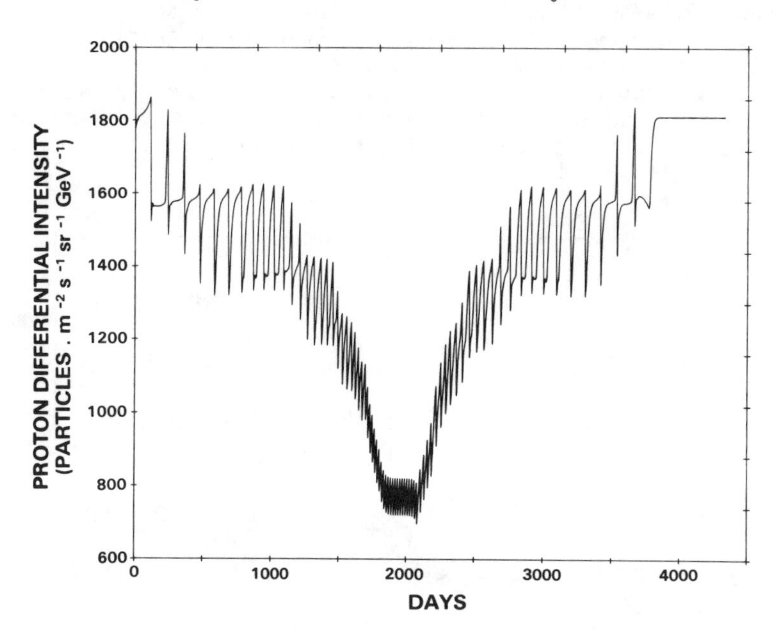

Figure 1. Differential intensity of 1 GeV protons at 25 AU over one solar cycle.

disturbances described above caused the sharp, short-term drops in intensity. The large step decreases, such as the ones at about 1100 and 1500 days, are also seen in the *Pioneer 10* data. The decrease in intensity from solar minimum in 1971 to solar maximum in 1980-1981, as measured by spacecraft occurred in 2 or 3 large steps [McDonald *et al.*, 1985]. Inspection of the numerical simulation reveals that these long-scale decreases are due to the presence of a sine-wave scattering disturbance at the very edge of the heliosphere, where they are still at full strength, but cover a much larger surface area. At this radius, they effectively block the diffusion inward from the cosmic-ray interstellar source. A similar process could arise in interplanetary space, with a buildup of turbulent regions in the outer heliosphere.

Figure 2 is a graph of radial intensity gradients for protons. The horizontal axis is the distance from the Sun of an imaginary spacecraft moving at the speed of Pioneer 10, taking samplings of the simulation results as it goes along. The vertical axis is "G_r" (in percent per AU), the integral radial gradient according to the equation $j_r = j_1 \exp[G_r(r-1)]$, where "j_r" and "j_1" are the intensities at the "spacecraft" (radius "r") and at 1 AU respectively. This is how most gradient data are reported. The boxes are for protons > 95 MeV. Webber & Lockwood [1985] found an almost constant gradient of 3%/AU at these energies and in this time interval with excursions no larger than ±0.5%/AU. The crosses are protons 138-238 MeV. Scattered data in McDonald et al. [1985] show the 140-240 MeV gradient varying between 2 and 6%/AU, with most of them hovering around 4-6%/AU. These data are consistent with the values in Figure 2. The slight peak at about 30 AU is a spatial hysteresis effect in which the intensity at outer distances seems to recover slightly faster than at 1 AU. This behavior does not seem to appear in the data.

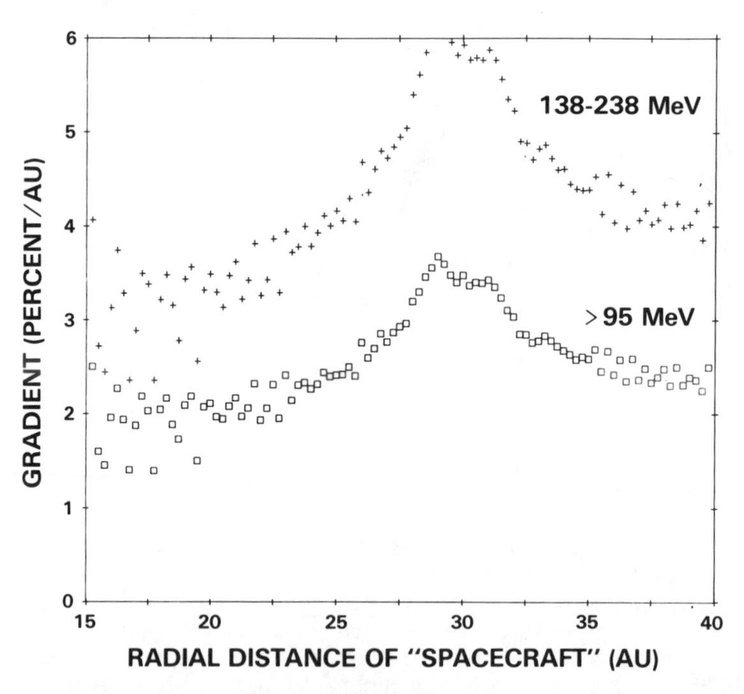

Figure 2. Integral radial intensity gradients calculated by moving at spacecraft speed through simulation results; gradient is in percent per AU between radius on x-axis and 1 AU; boxes are protons > 95 MeV; crosses are 138-238 MeV.

Figure 3 is a group of electron spectra at solar minimum (middle curve) and solar maximum (lower curve). The interstellar spectrum is the topmost curve. The data points are from 1965 (boxes) and 1970 (circles) [Webber & Chotkowski, 1967; Fulks, 1975;] Proton spectra from this model's maximum and minimum (not shown) also fit the 1965 and 1970 proton data. The proton spectra are much easier to fit due to their short rigidity range in the energy range computed, compared to the electrons.

Finally, Figure 4 is a regression plot similar to those used by Burger & Swanenburg [1973], with 10 GeV proton intensity versus 1 GV proton intensity. Time goes from upper right to lower left, then from

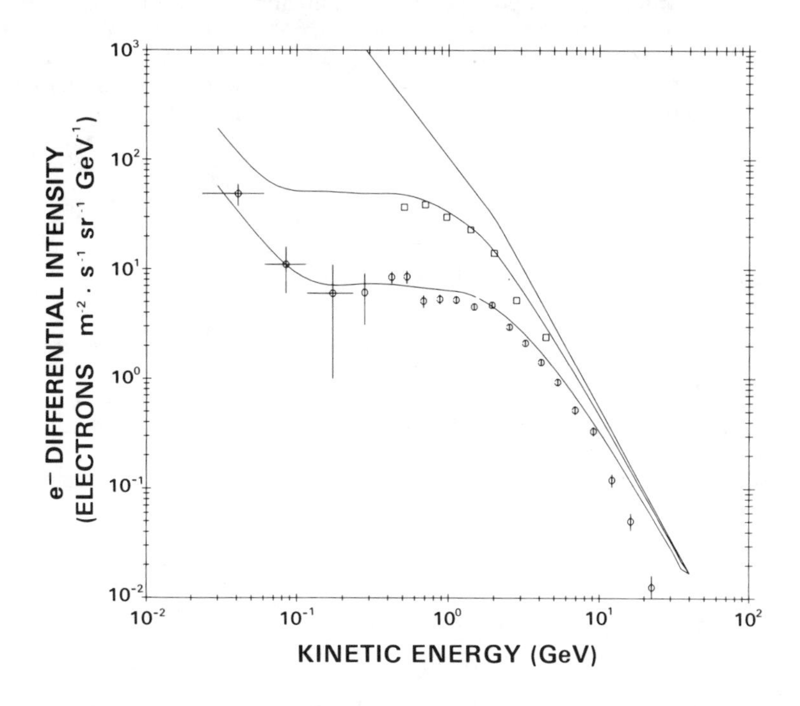

Figure 3. Electron differential energy spectra; interstellar (top), solar minimum (middle) with 1965 data, and solar maximum (bottom) with 1970 data.

lower left back to upper right. The upper line of the curve follows the increasing solar activity and the lower line follows it on the wane. This graph covers the full 11-year cycle. A hysteresis effect between low- and high-energy protons is evident. Inspection of the tabulated simulation results shows that 10 GeV particles returned to their initial solar minimum intensity about 280 days before the 1 GV particles. Burger & Swanenburg [1973] measure a value of about 260 days at these energies. The criss-crossing at the high-intensity end results from a general flattening of the simulated time-vs.-intensity graphs near the end of the cycle.

4. CONCLUSIONS

The original, time-dependent transport equation for cosmic rays in the heliosphere holds over a full solar cycle. Although small charge-dependent effects show up in the 22-year cycle, all significant effects observed in the ecliptic are well-simulated in a spherically-symmetric model. The spectra fit both the solar minimum and maximum data; further, these contemporaneous proton and electron spectra have identical modulation parameters. The gradients for particles above about 100 MeV are consistent with recent data analyses. Finally I have demonstrated a hysteresis, or time lag, between low- and high-energy cosmic rays at about the right magnitude and in the right sense.

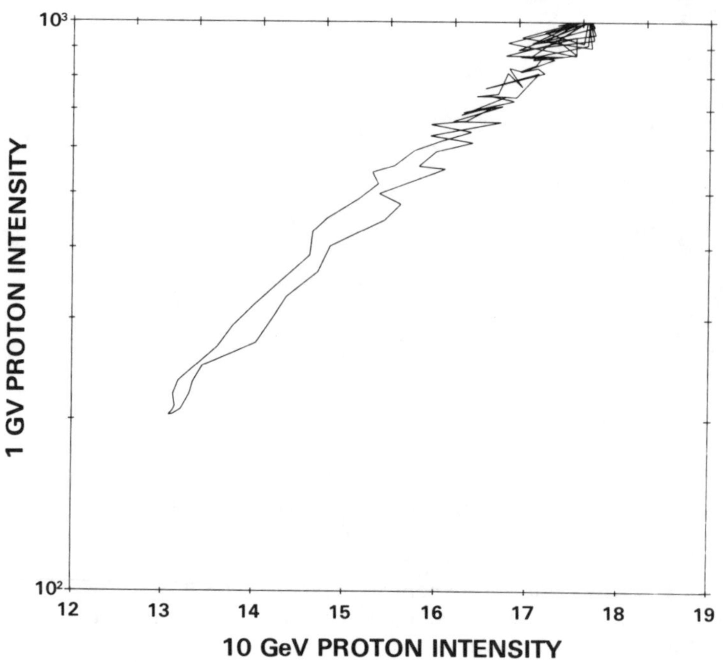

Figure 4. Regression plot of 10 GeV vs. 1 GV proton intensities over one solar cycle; time goes from upper right to lower left(along upper line) and back to upper right, from one solar minimum to the next.

At higher heliocentric latitudes, we must assume different behavior for cosmic rays, due to the IMF structure change with latitude. Even so, it appears that strong turbulent interaction regions, such as those studied by Burlaga *et al.* [1985], dominate cosmic-ray modulation near the ecliptic, with drifts apparently relegated to small effects near Earth, such as changes in the geomagnetic aa index [e.g., Shea & Smart, 1985] and cosmic-ray diurnal variations [e.g., Swinson, 1971]. Indeed, space probe data in the outer heliosphere tell us that modulation out there is much dissimilar to that inside a few AU, possibly making even these secondary phenomena the exception to the rule over the scale of the entire heliosphere.

5. REFERENCES

Balasubrahmanyan, V. K., et al., 9th Intl. Cosmic Ray Conf. (London), **1**, 427, 1965.

Burger, J. J., and Swanenburg, B. N., J. Geophys. Res.(Space), **78**, 292, 1973.

Burlaga, L., et al., J. Geophys. Res.(Space), **90**, 12027, 1985.

Chih, P.-P., and Lee, M. A., J. Geophys. Res.(Space), **91**, 2903, 1986.

Cummings, A. C., et al., 13th Intl. Cosmic Ray Conf. (Denver), **1**, 335, 1973.

Fisk, L. A., in Proc. Conf. Ancient Sun, ed. Pepin et al., p103, 1980.

Fisk, L. A., in Solar System Plasma Physics, Volume I, ed. Parker et al., Chapter I.2.2, p180, 1979.

Forbush, S., Phys. Rev., **54**, 975, 1938.

Fulks, Gordon, J. Geophys. Res.(Space), **80**, 1701, 1975.

Garcia-Munoz, M., et al., Astrophys. J., **202**, 265, 1975.

Garcia-Munoz, M., et al., J. Geophys. Res.(Space), **91**, 2858, 1986.

Gleeson, L. J., Planet. Space Sci., **17**, 31, 1969.

Gleeson, L. J., and Axford, W. I., Astrophys. J., **149**, L115, 1967.

Gleeson, L. J., and Webb, G. M., Fund. Cosmic Physics, **6**, 187, 1980.

Isenberg, P. A., and Jokipii, J. R., Astrophys. J., **234**, 746, 1979.

Jokipii, J. R., Rev. Geophys. Space Phys., **9**, 27, 1971.

Lockwood, J. A., et al., J. Geophys. Res.(Space), **91**, 2851, 1986.

McDonald, F. B., et al., Astrophys. J., **246**, L165, 1981.

McDonald, F. B., et al., 19th Intl. Cosmic Ray Conf. (La Jolla), **5**, 193, 1985.

Ng, C. K., Ph.D. Thesis, Monash University, Melbourne, Australia, 1972.

Ormes, J., and Webber, W. R., J. Geophys. Res.(Space), **73**, 4231, 1968.

Parker, E. N., Planet. Space Sci., **13**, 9, 1965.

Perko, J. S., and Fisk, L. A., J. Geophys. Res.(Space), **88**, 9033 1983.

Rygg, T. A., et al., J. Geophys. Res.(Space), **79**, 4127, 1974.

Shea, M. A., and Smart, D. F., 19th Intl. Cosmic Ray Conf. (La Jolla), **4**, 501, 1985.

Smith, L. H., et al., Astrophys. J., **180**, 987, 1973.

Swinson, D. B., J. Geophys. Res.(Space), **76**, 4217, 1971.

Urch, I. H., and Gleeson, L. J., Astrophys. Space Sci., **17**, 426, 1972.

Webber, W. R., and Chotkowski, C., J. Geophys. Res.(Space), **72**, 2783, 1967.

Webber, W. R., and Lockwood, J. A., 19th Intl. Cosmic Ray Conf. (La Jolla), **5**, 185, 1985.

PARTICLE PROPAGATION AND ACCELERATION IN THE HELIOSPHERE

J.F. Valdes-Galicia, Instituto de Geofisica, UNAM,
04510 Mexico D.F., Mexico.

J. J. Quenby, Blackett Laboratory, Imperial College,
London SW7, U.K.

X. Moussas, Astrophysics Lab., University of Athens,
Panepistimiopolis, Athens 621, Greece.

ABSTRACT. A realistic model of Interplanetary Magnetic Field perturbations has been constructed based on data taken on board spacecraft. The model has been used to study numerically pitch angle scattering suffered by energetic particles (1-100MeV) as they propagate in the Heliosphere. These numerical experiments allow the determination of the pitch angle diffusion coefficent $D_{\mu\mu}$ and the associated mean free path λ. $D_{\mu\mu}$ is found to be always smaller than implied by quasi linear theory, leading to radial mean free paths ($\lambda_r \simeq 0.015$AU) that are at least 3 times larger. Inclusion of Solar Wind velocity measurements in the model producing VxB random electric fields permits the study of stochastic acceleration caused by these fields. Initial results show that these processes might be able to overcome the effects of adiabatic cooling caused by the expansion of the Solar Wind and thus be of some influence in Cosmic Ray acceleration when extrapolated to other astrophysical environments.

1. Introduction

Studies of propagation and acceleration are necessary to have an adequate knowledge of the dynamics of energetic particles within the heliosphere. Propagation involves diffusion parallel and perpendicular to the mean field direction, drift motion and possible scatter free propagation. Acceleration can be at a shock, with or without the help of moving magnetic scattering centres, or by statistical processes in turbulent fields, especially in Corotatung Interaction Regions (CIRs) or behind shocks.

Pitch angle scattering caused by the fluctuating field is one of the fundamental processes governing Cosmic Ray propagation in the interplanetary medium. The Quasi Linear Theory (QLT) approach to the problem (Jokipii, 1966, Hasselman & Wibberenz, 1968) relates diffusion in pitch angle to spectral properties of the measured magnetic

M. M. Shapiro and J. P. Wefel (eds.), Genesis and Propagation of Cosmic Rays, 325–338.

field. The pitch angle diffusion coefficient can then be used to estimate a mean free path λ_{11} along the average magnetic field (Hasselman and Wibberenz, 1970).

It is well known that the QLT calculations of λ_{11} differ from determinations of the same parameter based on solutions of the Fokker-Planck equation produced to fit time intensity and anisotropies during energetic solar proton events (e.g. Ng and Gleeson, 1971, Webb and Quenby, 1973, Zwickl and Webber, 1977, 1978, Hamilton, 1977, Hiseh and Richter, 1981). The difference amounts to a factor of 10.

Several attempts have been made to solve the discrepancy. Earl (1976, 1981) called attention to the Fokker-Planck equation in pitch angle space including the effects of focussing due to the divergent IMF. Kota et al (1982) have produced numerical solutions of the Fokker-Planck equations both for intensity as a function of pitch angle and space coordinates and for the directionally averaged intensity dependant only on space and found agreement in the profiles produced by the two approaches. Hence we must look to causes different than focussing to remove the λ_{11} discrepancies.

Many analytical non-linear corrections to QLT have been produced (e.g. Quenby et al., 1970, Jones et al., 1978, Volk, 1975, Goldstein, 1976) especially to take into account magnetic mirroring due to changes in $|B|$ and not included in the purely resonant interactions assumed by QLT. Some non-linear corrections include also other causes, for example resonance broadening. Goldstein (1976) noted that all these improvements decrease λ_{11} thus making wider the difference with solar proton event determinations. Goldstein (1980) and Goldstein and Mattaeus (1981) have produced two new ideas trying to reduce the gap. The first is to take into account only the 6% power in IMF due to changes in $|\underline{B}|$ caused by non-Alfénic fluctutations. The second is to consider magnetic helicity in the IMF which can cause, in some cases, considerable reduction in resonant scattering. Both of these approaches lead to a $\lambda \simeq 0.3$AU more in accordance with phenomenological determinations. Yet another effort in the same direction has been made by Morfill et al. (1976) who find λ_{11} larger if the $\hat{\underline{k}}$ wave vectors tend to align with the radial direction as predicted by WKB theory, but it is most probable that the propagation vectors tend to align along $\underline{B}$ rather than $\underline{r}$ (Denskat and Burlaga, 1977).

Numerical integration of test particle trajectories into random fields obeying Gaussian statistics were performed by Kaiser et al (1978) and into a field defined by real magnetometer measurements by Moussas et al. (1978). Mirroring near 90°, non-helical trajectories and specific helicity are all automatically taken into account in this last approach.

Here we will be reporting previous (Moussas et al., 1982b, Valdes-Galicia et al., 1984) and new results obtained with pitch angle scattering numerical experiments performed with field models constructed with magnetometer data taken on board Helios 1 and 2 (0.3-0.7AU from the Sun), IMP-G and HEOS(1AU) and Pioneer 10 (5AU).

2. A Numerical Model of the Interplanetary Field

To simulate particle propagation in the IMF we have produced a model which we have called the "Layer model". It is based on measurements of all 3 components of the field taken on board satellites of different heliodistances. Although it has certain differences to adjust to a particular set of data the model could be described as follows:

A string of data taken by a spacecraft (typically 24 hour) is supposed to consist of a series of layers in space. Each of these layers has a width given by the sampling time δt (8 sec for Helios, 2.5 sec for IMP, 0.4 sec for Pioneer) times the solar wind speed(V_{SW}). For this purpose V_{SW} is assumed to be radial and constant. To each of these layers a value of the corresponding measurement of the magnetic vector is assigned. There are no field variations inside a layer. We use a coordinate system whose X axis points in the radial direction away from the Sun, the Z axis is directed towards the solar North pole and the y axis is in the equatorial plane completing a right handed system; i.e. the ith layer has a field:

$$B_i(\underline{x}, \underline{y}, z) = \hat{e}_x B_{xi} + \hat{e}_y B_{yi} + \hat{e}_z B_{zi}$$

where B_{xi}, B_{yi}, B_{zi} are actual measurements taken by a spacecraft magnetometer.

Our method is based upon the assumption that, on a statistical average, the IMF sample taken by the spacecraft would measure the same sort of fluctuations that would be obtained if it could follow a flux tube. In this sense we can refer the method as an approximation to a smooth line by a string of straight lines (see Figure 1). The numerical routine constructed to follow particle trajectories assumes helical path elements and it has been extensively tested with a number of field configurations (Moussas et al., 1982a). We want to emphasize that this is a full 3 dimensional representation of the IMF where not only transverse power is included but also other types of longitudinal perturbations and the phase dependence of waves and discontinuities is retained. For these reasons the layer model should not be confused with the commonly used slab model where only perturbations perpendicular to a mean homogeneous field exist (c.f. Jones et al., 1978, Gombosi and Owens, 1979, Kato et al., 1981). Criticism of our method has been discussed in Valdes-Galicia et al. (1984).

3. Calculation of Diffusion Coefficients

The mathematical description of any diffusive problem involves two coefficients. If s represents any component of the phase space then one coefficient would be $\langle \Delta s^2 \rangle$, describing mean square displacements and the other represents the mean displacement $\langle \Delta s \rangle$ (Born, 1949, Chandrasekar, 1960). However, Dungey (1965) demonstrated that in any case of diffusion where Liouiville's theorem is applicable only one diffusion coefficient is necessary. In the particular case of pitch angle scattering Jokipii (1966) showed that there is in fact a relationship between the first and the second order coefficients. Thus we will be here concerned only with the second order coefficient.

Pitch angle diffusion coefficients were calculated by the technique of Jones et al (1973, 1978) generalized by Moussas et al (1978, 1982a). In the last reference a complete description of the technique is given.

Briefly particles are injected at a single point (S_0) and observed when they reach either of the two preset boundaries "left" (S_ℓ) and "right" (S_γ) of the injection point. A distribution function is constructed weighted by the time step Δt:

$$F(s,\ t + \Delta t) = F(s,\Delta t) + \Delta t(s) \tag{1}$$

where $\Delta t(s)$ is the time spent in each particular bin Δs. After all particles have been lost we arrive at:

$$F(s,\infty) = \sum_{t=0}^{\infty} F(s,t) \tag{2}$$

Because of the absorbing boundaries there a steady state situation is eventually reached and the flux equation corresponding to this situation is (Jones et al., 1978, Moussas et al., 1978):

$$D_{ss} \frac{\delta}{\delta s} < F(s,\infty)> = -J_{L,R} \tag{3}$$

FIG 1 INTERPLANETARY MAGNETIC FIELD LINES

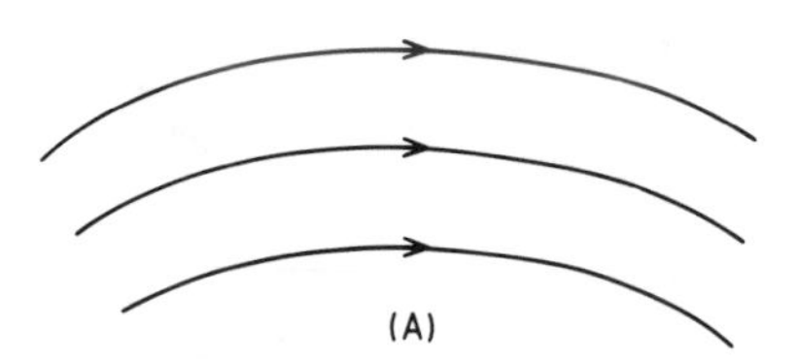

(A)

INTERPLANETARY MAGNETIC FIELD MODEL

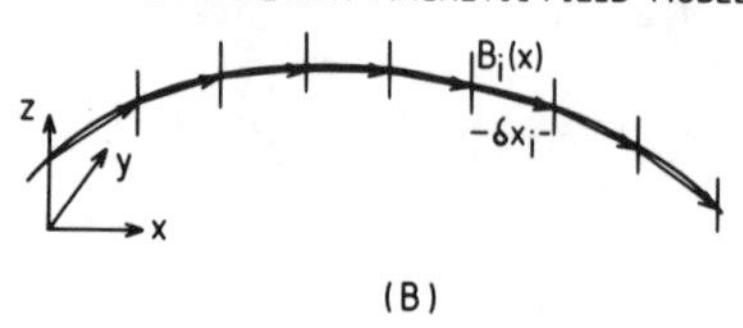

(B)

Figure 1. The IMF model based on actual spacecraft magnetic field measurements. The field line (A) is represented by a polygonal line (B) using successive measurements of the magnetic field vectors ,B(x), spaced at distances $\delta x = V_{SW} \cdot \delta\tau$, where V_{SW} is the solar wind velocity and $\delta\tau$ is the sampling time. This vector model fills up all the space to permit the simulation.

If the fluxes $J_{L,R}$ are measured we can evaluate D_{SS} from the slope of the distribution function. For this case s represents pitch angle cosine which is always measured with respect to the real field direction rather than the average field direction.

The stochastic nature of the scattering process simulated has been demonstrated before (Moussas and Quenby, 1978) and the validity of equation (3) justified.

The parallel diffusion coefficient can be calculated from:

$$K_{11} = \frac{\nu^2}{4} \int_{-1}^{1} \int_{o}^{\mu^1} \frac{1-\mu^2}{D\mu\mu} \quad d\mu \qquad \mu^1 d\mu \qquad (4)$$

where ν is particle velocity (see Hasselman and Wibberenz, 1970). The parallel mean free path is estimated as $\lambda_{11} = 3K_{11}/\nu$.

4. Previous Results

As we mentioned earlier numerical experiments have been performed with IMF data from Helios 1 and 2, IMP-G, Heos and Pioneer 10, covering a range from 0.3 to 5AU heliocentric distance. Results have been presented in two previous publications (Moussas et al., 1982b, Valdes-Galicia et al., 1984). Common features for all the results reported there are the finite level of scattering around $\mu = o$ as opposed to no scattering produced by QLT and the much lower level of dispersion found relative to QLT in the range $|\mu| \gtrsim 0.05$. In terms of λ_{11} this implied values which were at least 3 times larger than those derived following QLT, but still some 2 to 4 times smaller than most determinations of λ_{11} based on solutions of the Fokker-Planck equation. The values obtained for λ_{11} and λ_r , the radial mean free path, are summarised in table I and figure 5. Several points are to be noticed from them:

1) The parallel mean free path is roughly independent of energy in the range 1-100 MeV at 1AU.
2) When the effects of cross field diffusion are included at 5AU, radial transport of particles may be characterised by a constant radial mean free path $\lambda_r \simeq$ 0.01 beyond 1AU, independent of energy. (See also Moussas et al., 1982c).
3) In the range 0.4-1AU, $\lambda_{11} \simeq$ 0.03AU for 100 MeV protons.
4) For heliodistances smaller than 4AU the parallel(λ_{11}) and radial(λ_r) mean free paths decrease with decreasing distance.

The last point is in strong contradiction with the results obtained by Cecchini et al (1980) who found that in order to fit solar proton profile intensities and anisotropies the functional form $\lambda r \propto r^b$ should be fulfilled inside the earth's orbit and b $\simeq$ -2 if $\lambda_r \simeq$ 0.01AU at 1AU. But it could be that the particular set of data employed was not typical of the conditions a solar particle experiences and we decided to repeat these experiments using IMF data taken at the time a solar proton event was observed simultaneously. We will report now on results of numerical integrations performed with data taken on board Helios 2 on day 88 of 1976 (28th March) when the same spacecraft observed solar particles originated at a 1B flare located at S07 E28 in the sun exploding at 1908UT on the same day. The probe was situated at an Helios distance of 0.5AU.

5. Results on the 28 March 1976 Event

In figure 2 the magnitude and orientation of the IMF measured by Helios 2 magnetometer are shown. Strong fluctuations in the orientation of the field with respect to the zenith are noticed and also the azimuthal (PHI) component shows some disturbances. The level of the field fluctuations is illustrated in figure 3 where we present the parallel and transverse power spectra corresponding to the same sample, also shown for comparison are average power spectra obtained by Denskat and Neubauer (1982) using Helios 1 and 2 data over a number of satellite orbits around the Sun. The transverse power (perpendicular to the average field direction) can be considered as typical of 0.5AU fluctuations, whereas the parallel power (parallel to the average field) is perhaps a bit higher in comparison with the Denskat and Neubauer 'standard'.

Particles were injected into a layer model field based on these data and the experiments described in section 3 were carried out using 100MeV particles. Lower energies would require a smaller sampling time for the magnetometer to keep the number of layers per gyroradius sufficiently high as to minimize the effect of discontinuities at every layer end (see Valdes-Galicia et al., 1984).

The final steady state distribution for 975 particles injected at $\mu_o = 0.3$ and removed at $\mu_\ell = -0.6$ and $\mu_R = 0.95$ is shown in figure 4. Care has been taken to inject particles uniformly in phase. A slight change of slope around $\mu = o$ can be seen which corresponds to increased level of scattering as is reflected in the plot of $D_{\mu\mu}$ vs μ which we present in figure 5. There is also drawn for comparison the pitch angle diffusion coefficient predicted by QLT as (Jokipii, 1966, Quenby et al., 1974):

$$D^{QLT}_{\mu\mu} = \frac{1-\mu^2}{|\mu| v} \frac{e^2 \, Vs\omega}{\gamma^2 m^2 c^2} \quad P\left(f = \frac{V\,\omega_o}{2\pi\mu v}\right) \qquad (5)$$

where e is the electronic charge, v is particle velocity, γ is the particle's Lorentz factor, c is the velocity of light and P is the transverse power at the resonant frequency $f = V\,\omega_o/2\pi\mu v$ and ω_o is the gyrofrequency of the particle. We have used $P(f) = Af^{\alpha}$ as is usually assumed and from figure 3 $A = 6.9 \times 10^{-11}$ Gauss^2 and $\alpha = 1.3$ are obtained. It is seen from the figure that $D^{QLT}_{\mu\mu}$ is more than one order of magnitude greater than the numerical simulation value of $D_{\mu\mu}$ for $|\mu| \gtrsim 0.02$. The region $0.3 \lesssim \mu \lesssim 0.45$ where there is no numerical value for $D_{\mu\mu}$ is due to injection effects which invalidate eq(3) (see Kaiser et al., 1978). There is also a slight asymetry in $D_{\mu\mu}$ respect to $\mu = o$. A maximum in pitch angle scattering is found at $\mu \simeq 0.6$ beyond there the scattering tends to zero as μ approaches 1. A minimum in the scattering is observed at $\mu \simeq 0.3$ while at lower values it increases to a peak at $\mu \simeq 0$. If we suppose the $\mu = o$ in $D_{\mu\mu}$ to be

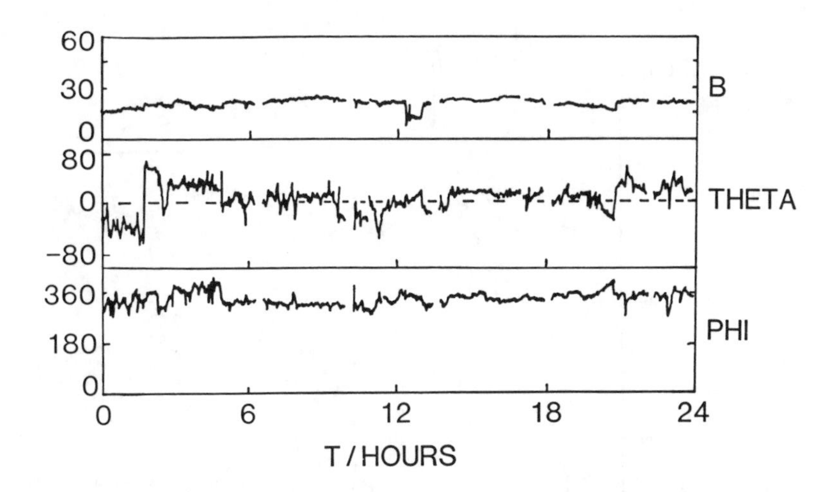

Figure 2. Magnetic field measurements taken on board Helios 2 spacecraft on 28 March 1976 when it was located at 0.5AU from the Sun.

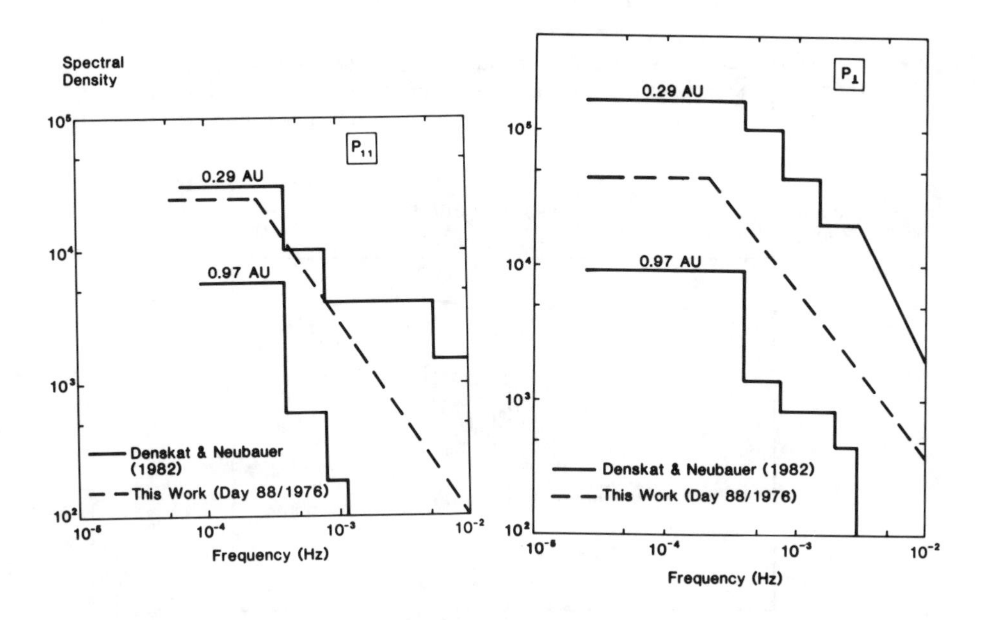

Figure 3. Power spectrum for the field parallel (a) and perpendicular (b) components to the mean field for the field sample shown on figure 2. The corresponding spectra calculated from large samples of Helios data at different heliodistances (Denskat & Neubauer, 1982) are shown for comparison.

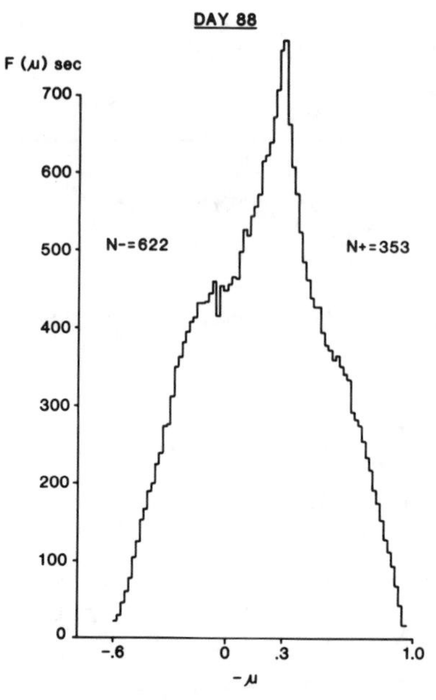

Figure 4. Final steady state distribution for the numerical pitch angle scattering experiment performed on day 88 (28 March) 1976. 975 particles were injected at μ_o = 0.3 and the boundaries set at μ_ℓ = -0.6 and μ = +0.95.

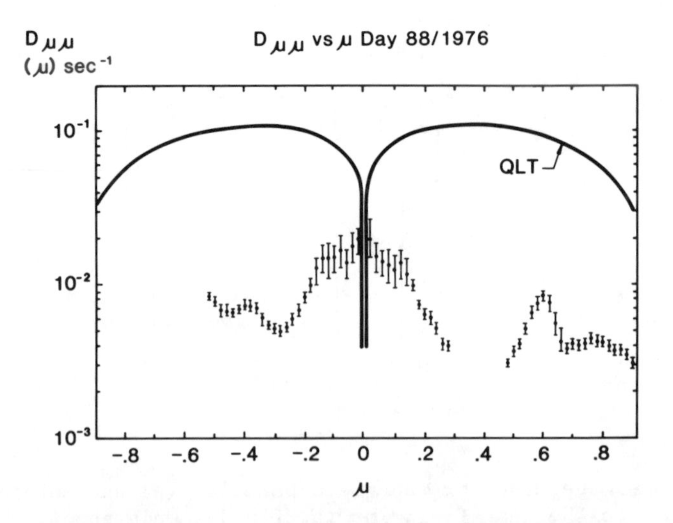

Figure 5. Diffusion coefficient in pitch angle space obtained from the distribution shown in figure 4. Quasi Linear theory predictions for the same magnetic field data used in the experiment are also drawn.

due to particle mirroring caused by the presence of field compressions we can estimate $D_{\mu\mu}$ assuming adiabatic changes in pitch angle (θ) (see Quenby et al., 1970).

$$<d\theta^2> \simeq \frac{\tan^2\theta}{4B^2} <dB^2>$$

Furthermore if we can represent

$$<dB^2> \simeq \int_0^\infty P_M(f)\, d f$$

where $P_M(f)$ is the power spectrum of the magnitude fluctuations of the IMF

$$\int_0^\infty P_M(f)\, d f = 2.97 \times 10^{-10} \text{ Gauss}^2$$

and using

$$d\theta^2 = \frac{d\mu^2}{(1-\mu^2)}$$

if for example $\mu = 0.15$ ($\theta = 81.37^o$) we arrive at: $<d\mu^2> = 7.9 \times 10^{-2}$

Now a characteristic time for the scattering process can be expressed as $t_c = \frac{\alpha}{v\ c}$ where α is the correlation length of field fluctuations and in this case $\alpha = 0.01$AU then

$$D_{\mu\mu}^{MIRROR} = \frac{<d\mu^2>}{t_c} \simeq 6.7 \times 10^{-3} \text{ sec}^{-1}$$

just a factor of 2 below the maximum of $D\mu\mu$ found at $\mu = o$. Thus field compression is a likely explanation of the maximum. The mirroring process rapidly pushes particles across the 90° position. However this will happen only when the particle has been able to get into the region $|\mu| \lesssim 0.2$. Figure 5 implies a simple picture where there is a 'barrier' preventing particles getting into $|\mu| \lesssim 0.2$ region and only a few of them can escape into the high scattering region around $\mu = o$.

Recently, the University of Kiel group, using results from their detectors of solar particles on board the Helios satellites has been able to calculate pitch angle diffusion coefficients based on a propagation model using pitch angle particle distributions (Ng et al., 1983, Green and Schuter, 1985, Beeck and Wibberenz, 1985). In a forthcoming joint publication we will make a detailed comparison of the pitch angle scattering coefficients obtained with our layer model and those obtained via the particle observations.

We have also calculated the parallel and radial mean free paths use of diffusion coefficients detained via equation (4). The values are $\lambda_{11} \simeq 0.07$AU and $\lambda_r = 0.06$AU at 0.5AU. The QLT corresponding values are $\lambda_{11} = 0.008$AU and $\lambda_r = 0.007$AU which are an order of magnitude below the numerically calculated quantitites. Furthermore in figure 6 we present all the values obtained with different numerical experiments we have performed using data from different satellites in a wide range of heliocentric distances. If we take into consideration the λ_r obtained for the 28 March event together with the previous results a much steeper dependence of λ_r on γ will be necessary to fit all values in the region $0.4 \lesssim r \lesssim 1AU$ more in accordance with the findings of Cecchini et al (1980). However, the problem of the absolute magnitude of the small mean free paths inside 0.4AU still is not completely resolved.

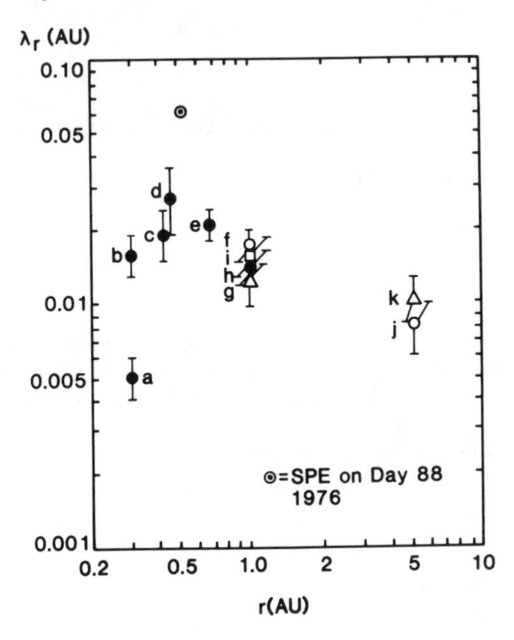

Figure 6. Summary of all the previously obtained radial mean free paths including the result reported here for the 28 March event.

6. Results on Acceleration

Introduction of the actual solar wind velocity measurements into the layer model allows the inclusion of the electric field $\underline{E} = \underline{V}_{SW} \times \underline{B}$ in the trajectory and in such a manner we can account for the energy changes due to this random electric field.

Seen from the rest frame of reference, a particle oscillates regularly in energy during the cyclotron motion due to the mean electric field but also random walks in energy due to wave-particle interactions. It is the latter process that we wish to measure. Working in the rest frame would allow some contribution from the mean $-V \times B$ field to distort the results. In order to reduce to a minimum the unwanted effect we have reduced the data to the solar wind frame of reference by subtracting the average solar wind speed. Then in every layer of the field model we can incorporate an electric field

$$E_i = Vsw_i(x,y,z) \times \underline{Bi}(x,y,z)$$

where $Vsw_i(x,y,z)$ is the residual of the radial component of velocity as measured by the spacecraft. The other two components of the solar wind were not available for the case we have studied but they are nevertheless of relative minor importance.

The numerical code to integrate particle trajectories has been modified to account for the electric field contribution and we have injected 10MeV particles into a field model using Pioneer 11 IMF (0.7sec) and plasma data (5 min) taken when the satellite was located at 2.5AU and saw the passage of a Corotating Interaction Region. At the peak of the turbulence we have been able to calculate energy diffusion coefficients for these particles and we have obtained

$$D_{TT} \lesssim 6 \times 10^{-5}\ \mathrm{MeV}^2\ \mathrm{s}^{-1}$$

Using the energy diffusion equation to calculate characteristic times for acceleration (Dungey, 1965, Moussas et al., 1982a) τ_{ACC} we get: $\tau_{ACC} \simeq 185$ hours if $D_{TT} \propto T^2$ and $\tau_{ACC} \simeq 60$ hours if D_{TT} is

independent of energy for the highest level of acceleration. To be significant in the Heliosphere these processes must be able to overcome the effects of cooling due to the adiabatic expansion of the solar wind; this can be readily calculated from (Gleeson, 1970)

$$\tau_{AD} = \frac{3R}{4V_{sw}}$$

for radial expansion, where R = heliocentric distance. In our case $\tau_{AD} = 195$ hours which is comparable with the acceleration times calculated above and since the characteristic cooling time grows as we go further away from the Sun the stochastic acceleration processes may be an efficient source of accelerated particles in the outer Heliosphere. In fact there is some experimental evidence of stochastic acceleration of particles close to the Earth. Richardson (1985) has analysed a number of particle events seen by ISEE3 situated at the libration point having soft spectra, sunward streaming towards (not away) a shock and particle increases coinciding with an increase in the level of wave activity. These observations could not be explained by the diffusive or scatter free shock acceleration and the only remaining possible source for the particles are the waves present.

If stochastic acceleration is an operative process inside the Heliosphere it might well be also working in other Astrophysical environments where the level of magnetic turbulence is much higher and thus could be one of the manners in which to produce high energy Cosmic Rays.

Table I
MEAN FREE PATHS (RADIAL, λ_r, λ_r^T, AND PARALLEL, λ_{11}) AND DIFFUSION COEFFICIENTS, K_{11} AT DIFFERENT ENERGIES AND HELIODISTANCES

Expt.	Spacecraft	Source	R/AU	T/MeV	λ_{11}/AU	λ_r/AU	λ_r^T/AU*	$K_{11}/(cm^2 s^{-1})$
a	Helios 1		0.31	100	0.006±0.001	0.005±0.001		$(3.5±0.7) \times 10^{20}$
b	Helios 1		0.31	100	0.018±0.003	0.016±0.003		$(1.13±0.2) \times 10^{21}$
c	Helios 2	VMQ84	0.42	100	0.022±0.006	0.019±0.005		$(1.4±0.4) \times 10^{21}$
d	Helios 2		0.45	100	0.032±0.008	0.027±0.008		$(2.08±0.5) \times 10^{21}$
e	Helios 2		0.67	100	0.030±0.006	0.021±0.003		$(1.9±0.3) \times 10^{21}$
f	HEOS 2		1	1	0.033±0.007	0.017±0.004		$(2.3±0.5) \times 10^{20}$
g	and	MQV82b	1	10	0.024±0.005	0.012±0.003		$(5.2±1.) \times 10^{20}$
h	IMP-G		1	100	0.027±0.006	0.014±0.003		$(1.8±0.4) \times 10^{21}$
i	HEOS 2	MQ78	1	117	0.031±0.007	0.015±0.004		$(2.0±0.4) \times 10^{21}$
j	Pioneer 10	MQV82b	5	1	0.083±0.01	0.0031±0.0007	0.008±0.002	$(5.7±0.1) \times 10^{20}$
k			5	10	0.14±0.03	0.0053±0.001	0.01±0.002	$(3.0±0.6) \times 10^{21}$

$\lambda_r^T = \lambda_{11} \cos^2\psi + \lambda \quad \sin^2\psi$ (cf. Moussas et al., 1982b,c)

7. Summary

1) The layer model of IMF produces consistently much lower levels of scattering than predicted by quasi linear theory.
2) The interplanetary scattering processes in the event of day 88 1976 are such that a barrier is impeding particles in their passage into the region $|\mu| \lesssim 0.2$ but when they eventually cross that barrier they are quickly scattered and backscattered across $\mu = o$ by the mirroring process of field compressions.
3) $\lambda_{11} \simeq 0.07$AU at 0.5AU for the event mentioned in 2) and is twice the value obtained previously at a similar distance from the Sun.
4) The radial mean free path obtained $\lambda_r \simeq 0.06$AU if taken together with previous results would produce a steeper dependence of λ_r on r inside 1AU, more in accordance with Solar Proton Event profile fitting requirement.
5) Stochastic acceleration mechanisms could be an important source of energetic particles in the outer Heliosphere and may even become a more efficient process in other astrophysical settings.

ACKNOWLEDGEMENTS

One of us (J.F.V.G.) has benefited from UNAM, British Council and Royal Society of London support for a visit to Imperial College where the paper was finally written. Many thanks are also due to Mr R. Duran and operations staff at UNAM Research Computer Centre where numerical calculations were done. Dr J. A. Otaola kindly provided his code to calculate power spectra and we also had the help of Mr J Arenas to deal with computing.

REFERENCES

Beeck, J. and G. Wibberenz., Preprint, submitted to Astrophys. J., 1985.
Cecchini, S., X. Moussas and J.J. Quenby., Astrophys. Sp. Sci., 69, 425, 1980.
Denskat, K.U. and L.F. Burlaga, J. Geophys. Res., 83, 2215, 1977.
Denskat, K.U. and F.M. Neubauer, J. Geophys. Res., 87, 2215, 1982.
Earl, J. Astrophys. J., 205, 900, 1976.
Earl, J. Astrophys. J., 251, 739, 1981.
Gleeson, L., Astrophys. Sp. Sci., 10, 471, 1971.
Goldstein, M.L., Astrophys. J., 204, 900, 1976.
Goldstein, M.L., J. Geophys. Res., 85, 3033, 1980.
Goldstein, M.L. and W. Matthaeus, 17th Int. C.R. Conf, Paris 3, 294 1981.
Gombosi, T. and A.J. Owens, Proc. 16th Int. C.R. Conf. Kyoto, 3, 25, 1979.
Hamilton, D.C., J. Geophys. Res. 82, 2157, 1977.
Hasselman, K. and G. Wibberenz, Astrophys. J., 162, 1049, 1970.
Hasselman, K. and G. Wibberenz, Z. Geophys. 34, 353, 1968.
Hsieh, K.C., and A.K. Richter, J. Geophys. Res., 86, 7771, 1981.
Jokipii, J.R., Astrophys. J., 146, 480, 1966.
Jones, F., T.B. Kaiser and T.J. Birmingham, Phys. Fluids, 21, 370, 1978.
Kaiser, T.B., T. Birmingham and F.C. Jones, Phys. Fluids, 21, 370, 1978.
Kato, M., T. Sakai and E. Tamai, Proc. 17th Int. C.R. Conf., Paris, 3, 314, 1981.
Kota, J., Merenyi, E., Jokipii, J.R., Kopriva, D.A., Gombosi, J.I., and Owens, J.T., Astrophys. J., 254, 398, 1982.
Morfill, G., H. Volk and M. Lee, J. Geophys. Res., 81, 5481, 1976.
Moussas, X. and J.J. Quenby, Astrophys. Sp. Sci., 56, 483, 1978.
Moussas, X., J.J. Quenby and J.F. Valdes-Galicia, Astrophys. Sp. Sci., 85, 99, 1982a.
Moussas, X., J.J. Quenby and J.F. Valdes-Galicia, Astrophys. Sp. Sci., 86, 185, 1982b.
Ng, C.K. and L.J. Gleeson, Proc. 12th Int. C.R. Conf., Hobart 2, 498, 1971.
Ng, C.K., G. Wibberenz, G. Green, H. Kunow, Proc 18th Int. C.R. Conf., 10, 381, 1983.
Quenby, J.J., A. Balogh, A.R. Engel, H. Elliot, P. Hedgecock, R. Hynds and J.F. Sear, Acta Physica Hungaricae, 29, 445, 1970.
Quenby, J.J., G.E. Morfill and A.C.Durnley, J. Geophys. Res., 79, 9, 1974.
Richardson, I., Planet. Sp. Sci., 33, 147, 1985.
Schluter, W. and G. Green, Preprint, submitted to Astrophys. J., 1985.
Valdes-Galicia, J.F., X. Moussas, J.J. Quenby, F. Neubauer and R. Schwenn, Solar Phys, 91, 399, 1984.
Volk, H.J., Rev. Geophys. Space. Phys., 13, 547, 1975.
Webb, S. and J.J. Quenby, Planet. Sp. Sci., 21, 23, 1973.
Zwickl, R.D. and W.R. Webber, Solar Phys., 54, 457, 1977.
Zwickl, R.D. and W.R. Webber, J. Geophy. Res., 83, 1157, 1978.

PHOENIX-1 OBSERVATIONS OF EQUATORIAL ZONE PARTICLE PRECIPITATION

M. A. Miah, T. G. Guzik, J. W. Mitchell and J. P. Wefel
Department of Physics and Astronomy
Louisiana State University
Baton Rouge, LA 70803-4001

ABSTRACT. The precipitation of magnetospheric particles at low altitude (160 - 300 km) near the geomagnetic equator during moderate geomagnetic conditions was studied by the ONR-602 experiment on board the S81-1 pallet mission in 1982. Significant fluxes of low energy (~1 MeV) protons were observed, with maximum intensity along the line of minimum magnetic field strength. These protons exhibit an altitude dependence that varies as the fifth power of the altitude, and a flux that is higher than that measured in previous missions (circa 1970). The source function, atmospheric loss processes and pitch angle distributions of these particles are investigated.

1. INTRODUCTION

The study of magnetospheric phenomena began with the discovery of the Earth's radiation belts in 1957. This research is important both because the magnetosphere is a major component of our Geospace environment and because it enables us to study basic plasma processes which are known to occur on a larger scale in solar and stellar flares, in galactic magnetic fields, in radio sources, and in the vicinity of neutron stars. Furthermore, the processes occuring in the Earth's magnetosphere may have counterparts in the sources and propagation mechanisms responsible for the galactic cosmic rays.

1.1. The Earth's Magnetosphere

The magnetosphere, sketched in Figure 1, is that part of the Earth's environment in which the geomagnetic field dominates the motion of charged particles. The magnetosphere forms a "tear-drop" shaped cavity balancing the Earth's magnetic field against the pressure of the solar wind which "washes away" geomagnetic field lines in the antisolar direction for hundreds to thousands of earth radii (R_E), forming the geomagnetic tail. Throughout the inner magnetosphere there exist regions known as the radiation belts which are populated by trapped particles of energy from ~10 keV to many MeV. Electrons and nuclei comprise the trapped particles, protons being the dominant

M. M. Shapiro and J. P. Wefel (eds.), Genesis and Propagation of Cosmic Rays, 339–355.

constituent among the nuclei. The radiation belts are commonly divided into the inner and the outer zones based on the low energy electron distirubtions, with the division at about 2 R_E.

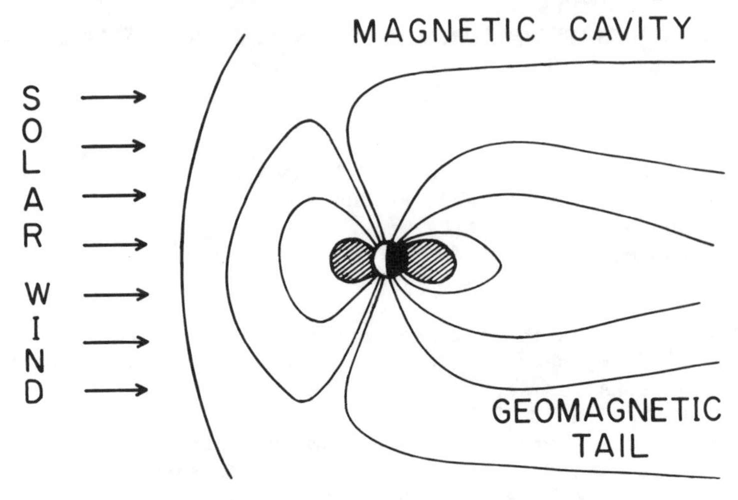

Figure 1. Schematic representation of the Earth's magnetosphere.

The main sources of magnetospheric ions are the galaxy, the sun and the ionosphere. The decay of backscattered neutrons from spallation reactions of galactic cosmic rays in the upper atmosphere contributes protons (10-100 MeV) and electrons (<500 keV). There is also evidence that, beyond the lunar orbit, solar wind plasma gains access into the magnetosphere, while in the polar regions, keV energy ionospheric ions can enter the magnetosphere.

Trapped particles perform a very complex motion in the Earth's magnetic field. They undergo cyclotron motion around magnetic field lines, bounce back and forth between mirror points on either side of the geomagnetic equator, and, finally, drift around the Earth on different L-shells (where L is defined as the equatorial geocentric distance to a particular field line measured in R_E), with electrons drifting to the east and protons to the west. These drifting particles form the ring current, a broad (in latitude), barrel-shaped region of particles whose variation during geomagnetic storms may cause the observed depressions in the horizontal component of the magnetic field. The ring current is a probable source of the particles discussed in this paper.

1.2. Particle Precipitation

Particles are lost from the radiation belts by interactions in the atmosphere and precipitate into the upper atmosphere all over the globe. The worldwide precipitation zones under moderate geomagnetic conditions are: (1) the equatorial zone, (2) the low-latitude zone, (3) the mid-latitude zone, and (4) the auroral zone, as illustrated schematically on Figure 2 (Voss and Smith, 1980). The South Atlantic Anomaly (SAA) region is distinct since in this area the field lines come closer to the Earth's surface than in other areas of the globe. Although these zones are applicable to other energy ranges as well, Figure 2 is based on the precipitation of low energy (keV) electrons and ions, studied mainly with rocket experiments and low-altitude satellites.

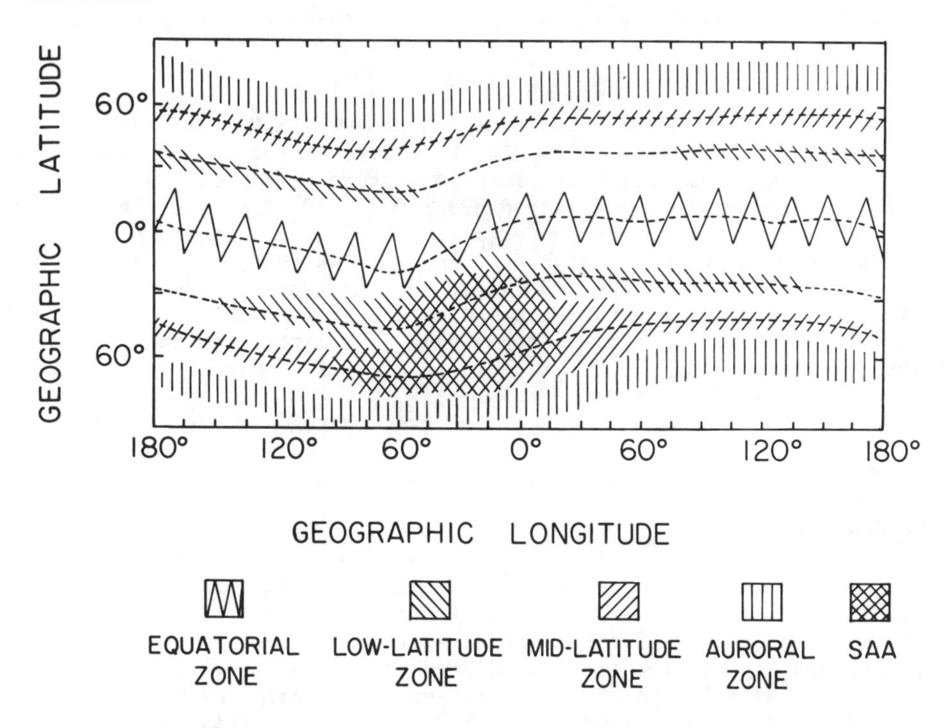

Figure 2. Representation of the global zones of particle precipitation defined by Voss and Smith (1980).

In this report we focus on the equatorial zone and present data on energetic protons (~1 MeV) observed from a low altitude, polar orbiting satellite. The process thought to be responsible for the presence of energetic protons at several hundred kilometers altitude, below the radiation belts, is charge exchange of ring current protons with exospheric hydrogen. In this process, trapped outer belt protons capture an electron from the thermal neutral hydrogen atoms in the exosphere, producing energetic neutrals which leave the source region in the direction of the proton's velocity vector at the time of neutralization. A fraction of these neutrals are directed toward the Earth and enter the atmosphere. At low altitude they are stripped of their electron by collisions with the atmosphere, and the ions so generated become trapped, temporarily, in the Earth's magnetic field forming a low altitude "belt" (Hovestadt et al., 1972; Moritz, 1972; Mizera and Blake, 1973; Scholer et al., 1975). The spatial and temporal behavior of these "quasi-trapped" particles reflects the state of the outer regions of the magnetosphere, the ring current, and the conditions within the upper part of the atmosphere. Thus, these low altitude particles can be used as an indicator of conditions within the magnetosphere.

An estimate of the expected maximum flux of energetic neutral hydrogen at low altitudes originating in the ring current is given by:

$$J_H \simeq 0.3\ \sigma_H(E)\ R_E \int_{2.5}^{3.5} n_H\ (L)\ J_p(L)\ dL \tag{1}$$

where J_p is the outer belt proton flux ($\sim 3.2 \times 10^6$ cm^{-2} s^{-1}sr^{-1} at 1 MeV, Sawyer and Vette, 1976), R_E is the radius of the Earth (6380 km), σ_H is the neutralization or electron capture cross section for protons interacting with neutral hydrogen ($\sim 2 \times 10^{-21}$ cm^2 in the energy range 0.6 - 1 MeV, Toburen et al., 1968), n_H is the exospheric thermal hydrogen density ($\sim 10^5\ R^{-4.25}$ cm^{-3}) and R is the geocentric distance in units of R_E. The 0.3 factor results from the fact that roughly 30% of a given equatorial L shell, in the L range 2.5 - 3.5, can contribute to the neutral hydrogen flux at a particular low altitude point (Moritz, 1972). Equation (1) gives a neutral hydrogen flux $J_H \sim 5.8 \times 10^{-4}$ cm^{-2}s^{-1}sr^{-1} at ~1 MeV.

At low altitudes, these energetic neutral hydrogen atoms will be stripped of their electrons by interactions with the atmosphere. At any altitude, h, the flux of protons is determined by :

$$\frac{dJ_p}{dt} = -vn\sigma\ J_p - vn\sigma_{\Delta E}\ J_p + S(h) \tag{2}$$

where v is the particle velocity, n is the atmospheric density, σ is the electron capture cross section for interactions with the atmosphere, $\sigma_{\Delta E}$ is an "effective" cross section for energy loss by ionization and S(h) is the altitude dependent source term. At these energies the source is the neutral hydrogen which charge exchanges in the atmosphere at this altitude, i.e. $S = vn\sigma' J_H(h)$. This process reaches equilibrium rapidly, and, neglecting energy loss, the steady state solution is:

$$J_p = J_H \left(\frac{\sigma'}{\sigma}\right) \tag{3}$$

where σ' is the electron stripping cross section for interactions with the atmosphere. Using a cross section ratio of $\sim 6 \times 10^3$ at 1 MeV (Toburen et al., 1968), we obtain a rough estimate of the proton flux of ~3.5 protons cm^{-2}-sr^{-1}-s^{-1}, a small but significant value.

Strictly speaking, the estimate given above applies to equatorially mirroring particles, i.e. particles with equatorial pitch angles $\alpha_E = 90^\circ$ (where the pitch angle, α, of a particle performing cyclotron motion around a magnetic field line is defined as the angle between the particle's velocity vector and the magnetic field direction). Ring current particles whose equatorial pitch angles differ from 90° have a geometric factor which is less than the value of 0.3 in Equation (1) and so contribute proportionally less to the observed flux.

The altitude, h_1 (km), at which e^{-1} of the particles have charge exchanged with the atmosphere can be obtained approximately from the relation (Scholer et al., 1975)

$$\int_{h_1}^{\infty} n(h) \frac{dh}{\cos\theta} \simeq 1/\sigma' \tag{4}$$

where θ is the zenith angle of the neutral particles trajectory, and

$$n = n_o e^{-\gamma(h)h} \tag{5}$$

($1/\gamma(h)$ = scale height, n_o = boundary density in cm^{-3}). Taking into account all of the atmospheric constituents (O_2, N_2, O, Ar, He, H), Eq. (4) yields an altitude of 225 km for ~1 MeV neutral hydrogen travelling vertically downward. Some of the protons may undergo additional cycles of neutralization/stripping, with about 7 such cycles being required for a particle of ~1 MeV and of pitch angle 90° to move vertically from 225 km to an altitude of 160 km. During each cycle some of the neutrals will be emitted upward and will contribute to an enhanced source term at greater altitudes or will be emitted in a direction such that they are lost. In addition, during the time the particle is an energetic proton, it will undergo ionization energy loss in the atmosphere which may remove it from the energy range accessible to a given instrument.

Trapped particles, in the course of their complex motion, pass through varying amounts of atmosphere which decreases their energy through atmospheric ionization. If the particle gyroradius is comparable to an atmospheric scale height, then in cyclotron motion the gyrating particle experiences a variation in atmospheric density by a factor of 2.7 between its highest and lowest altitudes. During bounce motion, the farther from 90° in equatorial pitch angle the more variation in density the particle meets. In drift motion, a particle dips down to a very low altitude in the vicinity of the South Atlantic Anomaly and experiences more dense atmosphere than elsewhere in its drift path.

However, for low energy (~1 MeV) particles, the density variation during cyclotron motion is negligible since the cyclotron gyroradius is ~5 km which is much less than the density scale height at altitudes above 160 km. In addition, at low altitudes (160-300 km) a trapped particle's lifetime is only several seconds; so the density variation during drift motion (drift period ~42 minutes) is not important for these particles. Therefore, the effective atmospheric density experienced by low energy, low altitude trapped protons is just the bounce averaged density, which is a function of the equatorial pitch angle of the particle. The bounce average density can be calculated from:

$$\bar{n} = \int (n\, ds/v_{||})/\int (ds/v_{||}) \tag{6}$$

where ds is the element of arc length along the field line upon which the particle is trapped, $v_{||}$ is the component of the particle velocity along the magnetic field direction, n is given by Eq. 5 and the integration is performed between the particle mirror points.

There are two time scales which are important in the motion of low energy protons. The first, the lifetime against electron capture of a proton of a particular equatorial pitch angle trapped on a field line, is given by:

$$t_e = (v \sum_i \bar{n}_i \sigma_i)^{-1} \tag{7}$$

where v is the proton velocity, $\bar{n}_i$ is the bounce average number density of the ith atmospheric constituent and σ_i is the electron capture cross section for interaction with the ith atmospheric constituent. The second time scale is the bounce period, which is the time (independent of the equatorial pitch angle to first order) for a proton to complete one bounce motion. The bounce period in seconds is given by (Lyons and Williams, 1983):

$$t_b = (43.1 \text{ x } L)/\sqrt{K} \tag{8}$$

where K is the proton kinetic energy in keV.

Table I gives some representative values for the bounce period, the bounce average density, the electron capture lifetime and the energy loss for protons of several kinetic energies and equatorial pitch angles at different altitudes. The energy loss is calculated for the bounce period, t_b, or the electron capture lifetime, t_e, whichever is shorter. The bounce periods in Table I are nearly independent of altitude, but t_e is a strong function of altitude, as is the mean energy loss. This is due to the bounce average density which increases rapidly both with decreasing altitude and as the equatorial pitch angle moves away from 90°. There is also a marked energy dependence to the lifetime due to the rapid variation of the electron capture cross section with particle energy.

There are several inferences that can be drawn from Table I. First, at any altitude, the particles with larger $|90° - \alpha_E|$ and the lower energy particles are depleted most rapidly. The increasing amount of interaction with decreasing altitude suggests a possible altitude dependence to the flux. Finally, at low altitudes and energies the particles cannot complete a full bounce motion before capturing an electron so the bounce average density is not applicable in calculating energy loss. In addition, the restripping time at these energies is very short. Thus, once neutralized, the proton is rapidly re-created at a slightly different position (altitude and location along a field line), pitch-angle and at the reduced energy that the proton had (due to ionization loss) before it was neutralized. This leads to a type of spatial/pitch-angle diffusion and spreads the energies of the protons observed at any one time.

Table I.

α_E (deg)	$\bar{n}$ ($g\text{-}cm^{-3}$)	E (MeV)	t_b (sec)	t_e (sec)	ΔE (keV)
h = 300 km					
87.8	7.1×10^{-5}	1.33	2.30	63.0	13
		1.02	2.63	31.0	15
		0.88	2.83	20.0	17
		0.63	3.35	7.5	21
77.3	2.0×10^{-4}	1.33	2.31	22.0	36
		1.02	2.64	11.0	43
		0.88	2.84	7.0	47
		0.63	3.36	2.7	47*
h = 250 km					
87.8	2.4×10^{-4}	1.33	2.29	18.0	43
		1.02	2.61	9.2	52
		0.88	2.81	5.9	57
		0.63	3.33	2.2	71*
77.3	8.4×10^{-4}	1.33	2.29	5.2	151
		1.02	2.62	2.7	179
		0.88	2.82	1.7	122*
		0.63	3.34	0.66	49*
h = 200 km					
87.8	9.3×10^{-4}	1.33	2.27	4.5	167
		1.02	2.59	2.3	178
		0.88	2.79	1.5	118*
		0.63	3.30	0.57	47*
77.3	8.3×10^{-3}	1.33	2.28	0.47	313*
		1.02	2.60	0.35	174*
		0.88	2.80	0.17	116*
		0.63	3.31	0.06	46*
h = 150km					
87.8	7.6×10^{-3}	1.33	2.25	0.54	327*
		1.02	2.57	0.28	181*
		0.88	2.77	0.19	121*
		0.63	3.28	0.07	48*

*Energy loss calculated for t_e, all other for t_b.

2. THE PHONEIX-1 (ONR-602) EXPERIMENT

Our investigation of low altitude energetic particle precipitation was performed with the Phoenix-1 instrumentation on board the S81-1 mission from May through November, 1982. The satellite was in a low altitude, nearly circular polar orbit of inclination ~85° and orbital period ~90 minutes. The orbital plane was 10:30 - 22:30 hours local time. The polar orbit, combined with the long mission duration, yielded many passes at low altitude (160-300 km) over the precipitation zones and provided an opportunity to study these particles in detail. Since the previous measurements were made in the late 1960's or early 1970's, the S81-1 mission provided an opportunity to compare results (particularly intensities) over a 10 year period. The results presented here are restricted to the equatorial zone of Figure 2.

The Phoenix-1 experiment was fabricated by the staff of the Laboratory for Astrophysics and Space Research at The University of Chicago under the direction of Prof. John A. Simpson and was composed of two telescopes: the low energy telescope (LET) and the monitor telescope (MT). The LET was employed to determine the isotopic composition of solar energetic particles at energies above ~10 MeV/nucleon. The MT, shown schematically in Figure 3, was

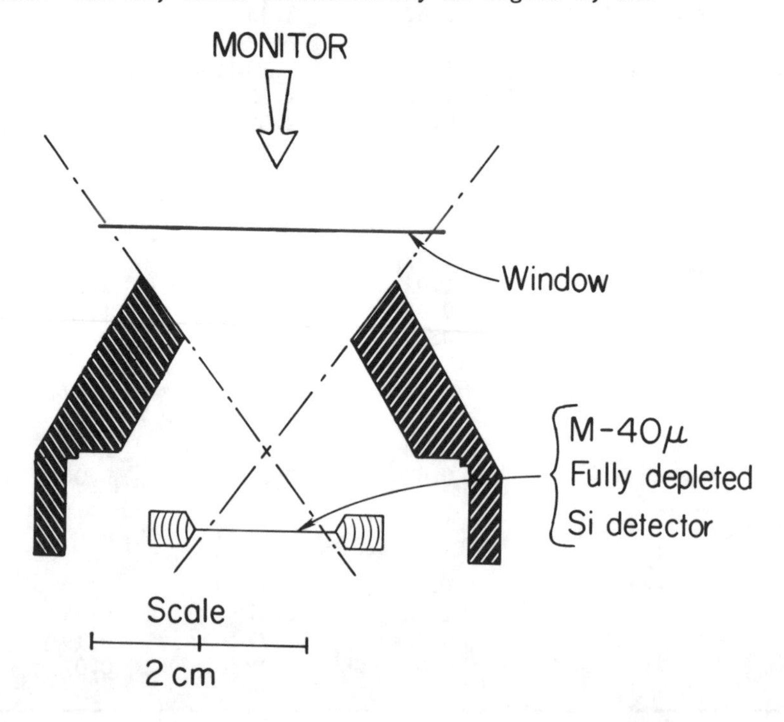

Figure 3. Diagram of the monitor telescope of the Phoenix-1 experiment.

used to study lower energy charged particles of both solar and magnetospheric origin and is the telescope employed for the work described here.

The MT is a single, thin detector, passively shielded instrument which returns three counting rates: ML, MM, and MH. The energy deposit threshold for ML is 0.36 MeV, so it responds to protons in the energy range 0.6 - 9.1 MeV, to alpha particles in the interval ~0.4 - 80 MeV/n and to Z > 3 particles (^{12}C) of energy ≥ 0.7 MeV/n. The threshold for the MM rate is 2.8 MeV, making it immune to protons, and MM responds to alpha particles in the energy range 0.8 to 4.5 Mev/n and to Z ≥ 3 particles in the energy range ~0.5 - 80 MeV/n (^{12}C). The MH threshold at 10.5 MeV responds only to Z > 2 particles in the energy range 1.2 - 11 MeV/n (^{12}C). By comparing the values of the ML, MM, MH counting rates, a measurement of the relative composition of the incident particles is obtained. For a particle population in which the intensities are protons >> alphas >> heavy ions, the three rates measure respectively the proton, alpha and heavy ion components. Each counting rate was accumulated for 4.096 seconds before being read into the telemetry stream.

The passive shielding surrounding the monitor telescope is sufficient to stop normally incident protons of ~40 MeV, making the counting rates relatively immune to background from higher energy particles. The geometrical factor of the monitor telescope for an omnidirectional particle flux is 0.5 cm^2-sr.

The large opening angle, 37.5° half-angle, allows the MT to admit particles with equatorial pitch angles 52.5° - 127.5° at the dipole equator. The instrument was mounted on the three-axis stabilized spacecraft such that the MT axis was tilted at an angle of 2.35° with respect to the local vertical direction, introducing a small asymmetry into the range of pitch angles which depends upon the orbital inclination.

3. PHOENIX-1 OBSERVATIONS AND DISCUSSION

For the analysis reported here the equatorial zone was taken to be the region covering 30° ≥ λ ≥ -30° in geomagnetic latitude and the complete dataset was cut to produce a subset consisting of only this region. This subset was further restricted by removing data which were "contaminated" by the SAA region (see Figure 2) and by eliminating obviously spurious readouts (e.g. negative rates). This produced a clean analysis tape with which to investigate the equatorial particles.

The ML counting rate was found to peak approximately at the equator, but the MM and MH rates showed no such spatial structure. In fact the MM and MH rates are consistent with instrumental background measured by looking at a quiet region of the orbit located between the equatorial and low latitude zones. The MM/ML and MH/ML ratios, for the entire equatorial zone, were about 10^{-3} and 10^{-4} respectively. This indicates that there are, essentially, no helium or heavier nuclei precipitating at the energies observable by the MT. The results found here are consistent with the small amount of data available on these heavy components. At L = 3 and B = 0.19, Spjeldvik and Fritz (1983) quote J_{He}/J_p ~$2 \times 10^{-3} - 10^{-4}$ in the 0.5 - 4 MeV/nucleon energy range. Also, for L = 2.5 - 5, J_{CNO}/J_p is $10^{-5} - 10^{-6}$ in equal energy per nucleon intervals from ~100 to 276 keV/nucleon. Thus, the focus of our analysis of the equatorial zone is on the low energy protons measured by the ML counting rate from the monitor telescope.

3.1. Global Distribution

The extensive coverage of the Earth provided by the S81-1 mission offers the opportunity to study the global distribution of these protons and to locate the center of the equatorial zone precisely. For this analysis the data was collected in 1 degree geomagnetic latitude by 5 degree geomagnetic longitude bins, and the latitude profile in each longitude bin was determined. These profiles showed an approximately Gaussian shaped peak from which the center and the FWHM were determined. The standard deviation was taken as the probable error in the location of the peak flux. Figure 4 (top) shows the

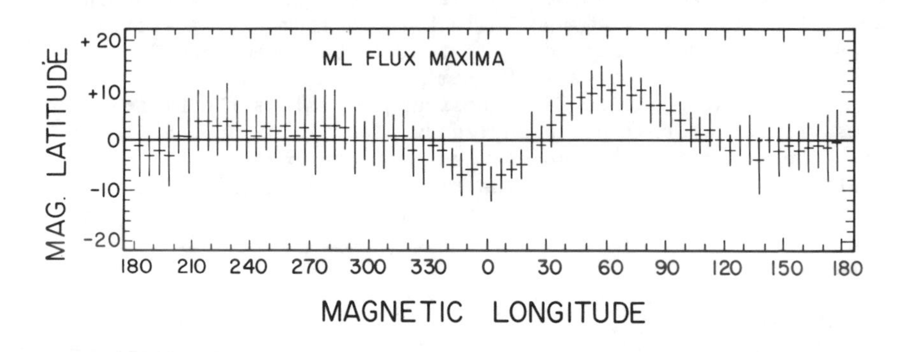

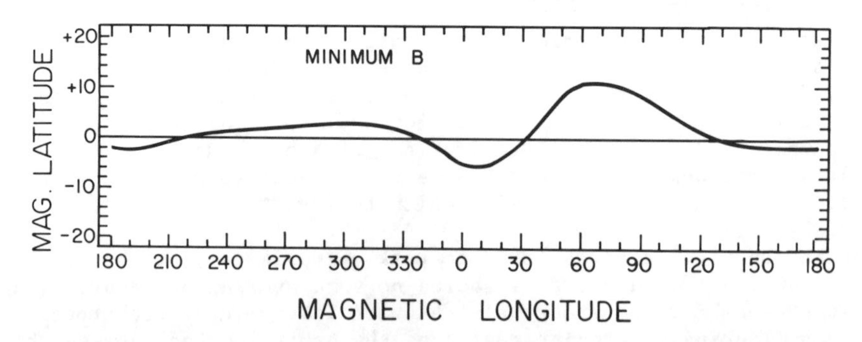

Figure 4. The measured location of the peak ML counting rates (top) compared (bottom) to the location of the line of minimum magnetic field strength (Stassinopoulos, 1970) for the full globe.

locations of the peaks plotted in geomagnetic coordinates, with the lateral bars showing the width of each longitude bin.

The flux maxima do not lie along the geomagnetic equator in this centered dipole model of the Earth's field. This is to be expected since the Earth's field at these altitudes is not a purely dipole field but contains additional components (e.g. from the ring current). The "quasi-trapped" particles which we observe perform bounce motion around the minimum potential point, i.e. the point of minimum magnetic field strength or minimum magnetic field energy density. The global location of this minimum-B line is shown in the lower portion of Figure 4 (Stassinopoulos, 1970), and our experimental results are in good agreement with this location.

For each longitude bin, the peak flux was obtained by dividing the average peak count rate by the geometric factor of the telescope and by the readout time. The uncertainty in the peak flux was determined from the counting rate and the number of readouts included in each peak. A plot of peak flux versus geomagnetic longitude showed no statistically significant deviations from the average flux. Further, the nature of the S81-1 orbit precluded a detailed investigation of local time effects, but for the data available no significant local time variations were observed.

3.2. Altitude Dependence

Over the lifetime of the mission, spacecraft altitudes from 160 to over 280 km were sampled. To study the altitude dependence of the peak flux, the altitude range was binned in 5 or 15 km widths so as to keep comparable numbers of equatorial passes in each bin. Satellite passes with their peak fluxes falling in a given altitude bin were superposed to obtain the average peak flux at that altitude. The resulting altitude dependence is shown on Figure 5 and follows a power law in altitude. A fit to the experimental results gives an exponent for the power law of 5.0 ± 0.2. Plotted on the same figure are the altitude dependent results of Goldberg (1974) for 2-20 keV ions in the equatorial zone, and of Parsignault et al. (1981) and Filz and Holeman (1965) both for ~55 MeV protons in the South Atlantic Anomaly. The slope of our curve correlates well with that of Filz and Holeman (1965) but is steeper than the result reported by Parsignault et al. (1981) whose measurements are confined to a higher altitude range than those of Phoenix-1. The very low energy results of Goldberg (1974), based upon rocket observations, show an exceedingly steep dependence up to 300 km.

Filz and Holeman (1965) interpreted the observed altitude dependence as resulting from ionization energy loss (which removed the particles from the energy window of their instrument) in the increasing amount of atmosphere experienced by the particles at lower altitudes. This explanation suggests a power law dependence of the reciprocal of the mean atmosperic density on altitude. We have calculated, as a function of altitude, the bounce averaged density seen by the protons of various pitch angles, and averaged these densities over the pitch angle distribution given by Heckman and

Nakano (1969). Fitting the reciprocal of these densities to a power law in altitude gives an exponent of 4.9 ± 0.3, sufficient to explain the measurements on Figure 5.

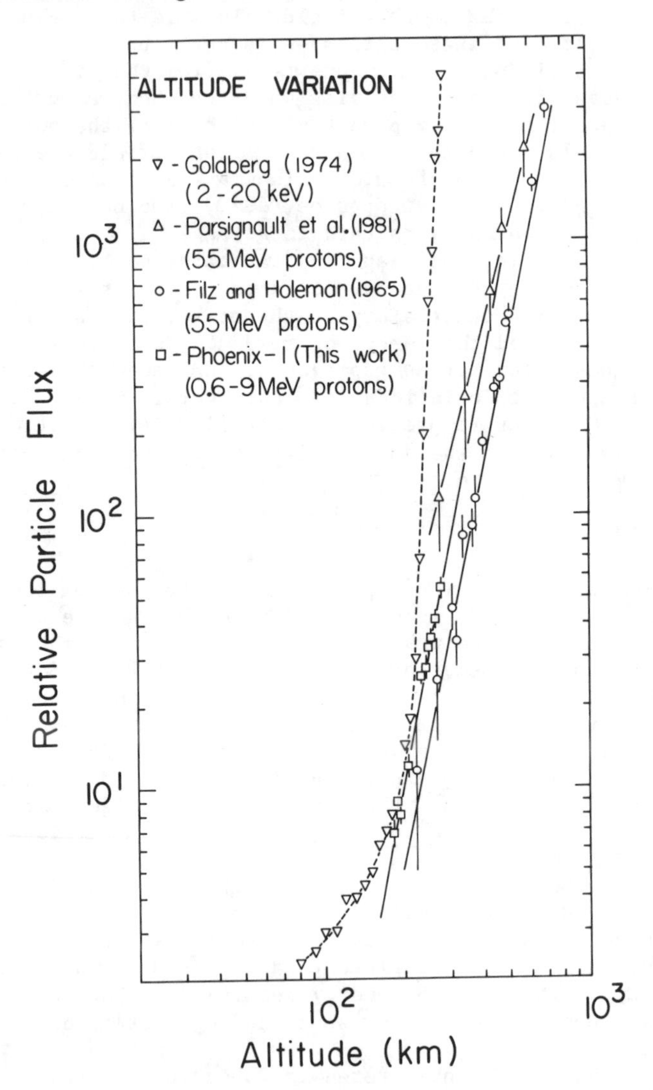

Figure 5. Altitude dependence of the flux measured by Phoenix-1 compared to previous data (see text for details).

The agreement of our measured altitude dependence with that predicted for the atmospheric density and with the dependence measured for 55 MeV protons might be interpreted as implying that atmospheric loss processes dominate the particle behavior in our energy range.

This may, however, be the wrong conclusion since there are significant differences in the physical processes involved at the two energies. The source function for 55 MeV protons is well known since these particles are derived from albedo neutron decay following high energy cosmic ray interactions in the atmosphere. This gives a source that is, to first order, independent of altitude. In addition, at 55 MeV the electron capture cross section is vanishingly small. Equation (2) applied to this case reduces to:

$$\frac{dJ_p}{dt} = S - vn\sigma_{\Delta E} J_p. \tag{9}$$

In a steady state, then, the flux of protons is inversely proportional to the atomspheric density as discussed above, and we would expect the correlation between n^{-1} and altitude to explain the data.

For the ~1 MeV protons studied here, however, the principal source is the neutral hydrogen from charge exchange of protons in the distant magnetosphere, evaluated at the altitude of stripping, times the mean atmospheric density. In the steady state, Equation (3) is obtained which is independent of the atmospheric density. In this model any altitude dependence must derive from the altitude dependence of J_H. The conversion of neutral hydrogen into protons depends upon atmospheric interactions, so that at any altitude, the flux of J_H has been reduced by the conversions that have occured at higher altitudes, and which depend upon the distribution of atmospheric density.

In addition, second order processes must be considered for the ~1 MeV protons. There is a "source" contribution from nearby altitudes due to protons at those altitudes which have neutralized and are directed toward the altitude of interest. Furthermore, the exact geometry of the source must be incorporated, and the energy loss during bounce motion along with the cycles of neutralization and restripping must be investigated in detail. Thus, the complete explanation of the observed altitude dependence requires a detailed model of the interplay of these different physical processes, and the construction of such a model is underway.

3.3. Energy Spectrum

The Phoenix-1 monitor telescope responds to protons over a broad range of energy (0.6 to 9.1 MeV). The spectrum of these protons is expected to be a steeply falling function of energy if the source is indeed the outer radiation belt. The exact energy dependence of the proton flux provides one key test of the model for the origin of these particles. The various observations of the equatorial belt protons are summarized in Figure 6, where we have plotted the peak differential flux versus energy as observed by Moritz (1972), Hovestadt et al. (1972) and Mizera and Blake, (1973). The peak differential flux was calculated by dividing the average peak counting rate by the geometrical factor of the instrument, the readout time, and the energy interval. Both pre-storm, post-storm and average values are presented. At low energies the effects of geomagnetic storms are large, but above ~500 keV the difference is small. The points above 100 keV (prestorm and average data) have been fit to a power law, shown by the solid line, which has an exponent of -2.55 ± 0.11. For this power law the mean energy of the protons observed by Phoenix-1 would be 1.3 MeV.

The power law on Figure 6, calculated from previous data, predicts that Phoenix-1 should observe 38 ± 5 counts per readout for the ML energy range 0.6 - 9.1 MeV. The actual peak rate observed is 9.5 ± 3 counts per readout in the highest altitude bin (centered at 270 km). This is shown as the open square on Figure 6 plotted at 1.3 MeV with

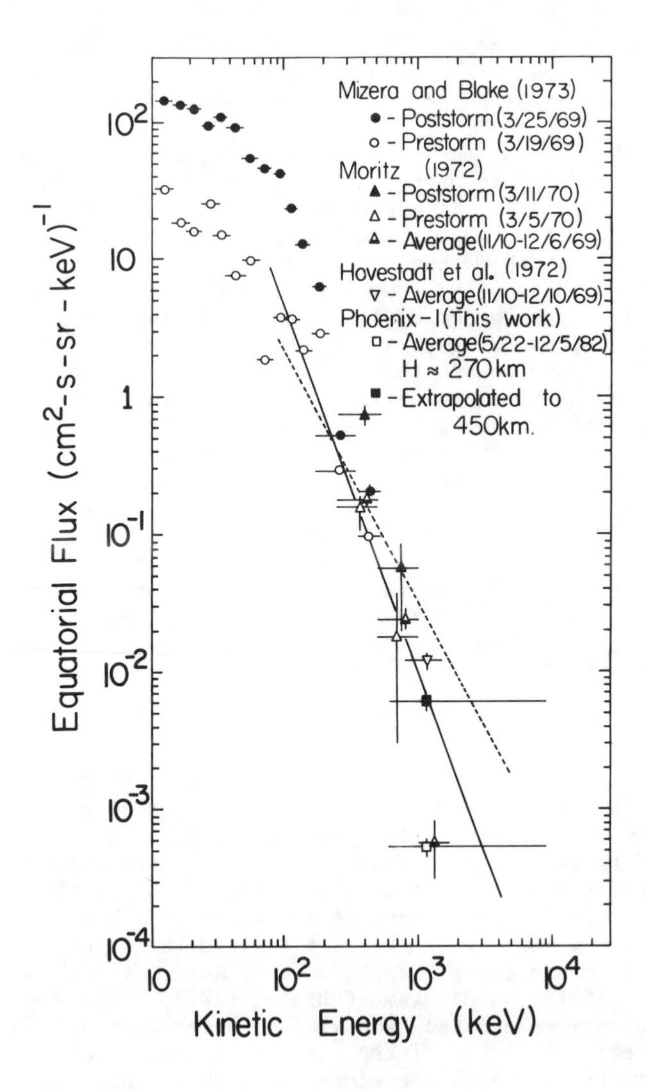

Figure 6. The differential proton energy spectrum.

the full energy range indicated. The previous missions sampled a higher altitude range and, to compare our measurements to theirs, we extrapolated our observed flux to a comparable altitude with other observers (450 km) using the altitude dependence shown on Figure 5. This extrapolation gives a flux of ~6 x 10^{-3} protons $-cm^{-2}-sr^{-1}-s^{-1}-keV^{-1}$ plotted as the filled square. The corresponding peak count rate would be 108 ± 13 counts per readout, a factor of 2.8 larger than predicted by the solid line on Figure 6. The extrapolated flux integrated over our energy interval gives ~50 protons $-cm^{-2}-sr^{-1}-s^{-1}$, significantly above the approximate flux estimate given in the introduction. The dashed line on Figure 6 is drawn to be consistent with the Phoenix-1 extrapolated flux and the previous measurements at several hundred keV and represents a power law exponent of -1.85. The dashed line does not, however, agree with the previous measurements particularly at 0.5 - 1.5 MeV.

An increased proton flux at low altitudes compared to measurements ~1970 could signal a temporal change in the outer belt (ring current) particle population and/or an increase in the exospheric hydrogen density, possibly due to differing solar conditions. However, some of the observed discrepancy may be due to the manner in which the data are compared. Our extrapolated flux employs the measured altitude dependence with the assumption that this dependence is valid up to an altitude of 450 km. This is probably a good approximation but requires checking against model calculations for the altitude dependence.

A more severe problem is the assumption of omni-directional particles which was employed in the scaling for telescope solid angle to obtain the flux values. This assumption is equivalent to a pitch angle distribution proportional to sin α_E. Mizera and Blake (1973) measured the pitch angle distribution for protons above 0.2 MeV with an instrument on the OV1-17 satellite and found a dependence closer to $(\sin\alpha_E)^{10}$. In addition, they calculated the α_E distribution expected if the source of neutrals were assumed to be at low altitudes and isotropic. The resulting distribution is much broader than the data which implies that the source should be anisotropic and at a high altitude relative to their observation point (400-470 km altitude). These observations support the ring current charge exchange hypothesis of Moritz (1972) for the source. Moreover, the observed α_E distribution is in reasonable agreement with that calculated by Moritz (1972) for the charge exchange model and is similar to that calculated by Heckman and Nakano (1969) for high energy particles trapped at low altitudes. Thus, the actual particle α_E distribution is significantly different from that assumed in generating Figure 6.

The consequence of this difference in pitch angle distributions is that the actual flux may be underestimated (or overestimated) depending upon the geometry of the telescope employed. The MT has a detector of 1 cm^2 area and a half opening angle of 37.5° contrasted, for example, to the 0.4 cm^2 with half opening angle of 10.2° telescope used by Moritz (1972) on the Azur satellite. Our MT can observe particles with α_E= 55° while Azur effectively can not. Thus, we see a larger fraction of the pitch angle distribution, but the number of particles actually at the larger $|90° - \alpha_E|$ values are very small, and this reduction is not taken into account properly in the current flux calculation. In addition, the efficiency of a telescope is not the same for all pitch angles. Thus, a careful calculation of the telescope efficiency as a function of pitch angle, taking into account the actual telescope geometry, mounting on the spacecraft, orbital inclination and latitude, longitude, and altitude of the measuring point is required. Such a calculation is in progress and will allow the actual energy spectrum of the radiation and any temporal variability to be determined. This energy spectrum will then be used to test the models for the origin of these low energy protons.

4. CONCLUSIONS

The Phoenix-1 experiment (ONR-602) has provided new measurements of low energy (~1 MeV) "quasi-trapped" protons at low altitudes (160 - 300 km) over the equatorial region. The proton flux peaks at the minimum-B line and, as would be expected from the short lifetime of these particles, shows no detectable longitude dependence. A feature of these measurements, which has not been reported by previous observers, is the observation of a strong altitude dependence (h^5) to the peak proton flux ($160 \leq h(km) \leq 280$). In addition, the flux observed by Phoenix-1, uncorrected for the particle pitch-angle distribution, is considerably larger than that observed previously, which may indicate a temporal variation between 1969 and 1982 in the magnetospheric particle populations. The energy spectrum derived from the data, fit to a power law above ~0.2 MeV, shows an exponent of -2.5, less steep than previous measurements.

The altitude dependence and the energy spectrum provide important tests for the origin of this radiation. Qualitatively, the data supports the hypothesis of charge exchange with ring current ions as the source, but detailed models of the contributing physical processes are still needed before a fully quantitative test of this hypothesis can be undertaken.

5. ACKNOWLEDGEMENTS

We thank the staff of the Laboratory for Astrophysics and Space Research at The University of Chicago for the construction, testing and integration of the instrument and for the initial processing of the data tapes. The S81-1 mission was conducted under the US Air Force Space Test Program who provided integration, launch and telemetry coverage. This work has been supported at LSU by the Office of Naval Research under Contract N-00014-83-K-0365.

6. REFERENCES

Filz, R. S. and E. Holeman, J. Geophys. Res., 70, 5807, (1965).

Goldberg, R. A., J. Geophys. Res., 79, 5299, (1974).

Heckman, H. H. and G. H. Nakano, J. Geophys. Res., 74, 3575 (1969).

Hovestadt, D., B. Hausler and M. Scholer, Phys. Rev. Lett., 28, 1340, (1972).

Lyons, L. R. and D. J. Williams, in Quantitative Aspects of Magnetospheric Physics, (Dordrecht: 1983: D. Reidel Co.) p. 24.

Mizera, P. F. and J. B. Blake, J. Geophys. Res., 78, 1058, (1973).

Moritz, J., Z. Geophys., 38, 701, (1972).

Parsignault, D. R., E. Holeman and R. C. Filz, J. Geophys. Res., 86, 11439, (1981).

Roederer, J. G., Dynamics of Geomagnetically Trapped Radiation, (Heidelberg: 1970: Springer-Verlag) p. 37.

Sawyer, D. M. and J. I. Vette, AP-8 Trapped Proton Environment for Solar Maximum and Solar Minimum, (Greenbelt, MD: 1976: NASA, NSSDC), p. 93.

Scholer, M., D. Hovestadt and G. Morfill, J. Geophys. Res., 80, 80, (1975).

Spjeldvik, W. N. and T. A. Fritz, in Energetic Ion Composition in the Earth's Magnetosphere, ed. R. G. Johnson, (Dordrecht: 1983: D. Reidel Co.), p. 396.

Stassinopoulos, E. G., World Maps of Constant B, L and Flux Contours, (Greenbelt, MD: 1970: NASA), SP-3054, p. 58.

Toburen, L. H., M. Y. Nakai and R. A. Langley, Phys. Rev., 171, 114, (1968).

U.S. Standard Atmosphere, 1976, (Washington, DC: 1976: Government Printing Office), NOAA-S/T 76-1562.

Voss, H. D. and L. G. Smith, J. Atmos. Terr. Phys., 42, 227, (1980).

RECENT IMPROVEMENT OF SPALLATION CROSS SECTION CALCULATIONS, APPLICABLE TO COSMIC RAY PHYSICS

R. Silberberg and C. H. Tsao
E.O. Hulburt Center for Space Research, Code 4154,
Naval Research Laboratory, Washington, D.C. 20375-5000

J.R. Letaw
Severn Communications Corporation, Severna Park, MD 21146

Abstract

About half of the cosmic ray nuclei heavier than helium have suffered nuclear transformations while colliding with the interstellar gas. The semi-empirical calculations we have developed for calculating the still unmeasured cross sections had errors of about 25-50%, depending on the target mass interval. Recent precise measurements (to 10%) have permitted significant improvements in the calculation of cross sections. Several such improvements are discussed. A status report on work-in-progress is presented on a general modification of the spallation equation parameters.

1. INTRODUCTION

Most cosmic ray nuclei with atomic number $Z \geq 6$ have suffered nuclear collisions in the interstellar gas. These collisions alter the elemental and isotopic composition of cosmic rays. The rare nuclides, e.g. the isotopes of Li, Be, and B are enhanced by several orders of magnitude as a result of nuclear spallation of heavier nuclei. The isotopic and elemental source composition is inferred from the observed composition near the earth, with the help of propagation calculations. In these equations, the key terms are loss by nuclear collisions, and production by nuclear spallation, that contain factors for the total and partial inelastic cross sections, respectively. The uncertainty in the inferred source composition is largely due to uncertainties in cross sections, especially for nuclides that have large secondary components. If the secondary component exceeds ~80% of the total, only upper limits can be given for the respective source components. On the other hand, the effects of nuclear spallation with reliable estimates of cross sections, and of the time-dependence of nuclear decay of the radioactive products, provide valuable information on the propagation of cosmic rays. One can then learn about a limited degree of reacceleration of cosmic rays, the confinement time in the Galaxy, the mean path length traversed by cosmic rays, and about the nature of the interstellar medium.

M. M. Shapiro and J. P. Wefel (eds.), Genesis and Propagation of Cosmic Rays, 357–374.

To derive quantitative information from the cosmic ray abundances on the above topics, it is essential to know (or to be able to estimate) reliably the nuclear spallation cross sections and the associated errors. The total inelastic cross sections permit the determination of the fraction of nuclides that survive passage in the interstellar gas. The partial inelastic cross sections give the probability of a given incident nuclide to yield a given product nuclide upon colliding with some target nucleus, e.g. with a proton.

In this paper we shall review the basic features of semi-empirical cross section calculations. We shall then discuss some of the modifications of our cross section calculations, based on recent experimental data. We shall also outline our work-in-progress on a major overall revision of the parameters of the equation for calculating partial cross sections.

2a. SEMI-EMPIRICAL CALCULATIONS OF CROSS SECTIONS

We shall not review the total inelastic cross sections in this paper. These calculations have been described by Letaw, Silberberg, and Tsao (1983), and reviewed at the 1982 Course of the International School of Cosmic Ray Astrophysics (Silberberg, Tsao, and Letaw, 1983). The calculations are accurate to about 2% above energies of 1 GeV and the energy dependence proposed is applicable down to 10 MeV.

The partial inelastic cross sections σ_{ij} (for the production of nuclide i from j) have systematic regularities that permit the design of semi-empirical equations (Rudstam, 1966). Rudstam observed that there are systematic regularities among the relative yields of nuclear spallation reactions that depend on the mass difference of the target and product nuclides and on the neutron-to-proton ratio of the product nuclides. These relationships are described by two factors, illustrated in Figure 1. This illustration shows the spallation cross sections of Fe into various isotopes of argon and vanadium, when iron nuclei are broken up by protons that have energies $E \geq 3$ GeV. The factor $\exp(-P\Delta A)$ describes the diminution of cross sections as the difference of the target and product mass number, ΔA, increases. It is closely related to the distribution of excitation energies discussed by Metropolis et al. (1958) in their Monte Carlo study of nuclear spallation reactions. A large excitation energy results in evaporation of many nucleons, i.e., in a large ΔA. The distribution of excitation energies peaks at small values; correspondingly, the partial cross sections are larger for small values of ΔA. The second factor $\exp(-R|Z - SA + TA^2|^\nu)$, with $\nu \sim 3/2$, describes the distribution of cross sections for the production of various isotopes of an element of atomic number Z. This Gaussian-like distribution is related to the statistical nature of the nuclear evaporation process (Dostrovsky, Rabinowitz, and Bivins 1958). The width of the distribution of cross sections is represented by the parameter R. The parameter S describes the location of the peaks of these distribution curves for small values of the product mass number A. The parameter

T describes the shift of the distribution curves toward greater neutron excess as the atomic number of the product increases. These factors and parameters thus are closely related to nuclear systematics of the prompt intranuclear cascade and nuclear evaporation processes. This is the reason why these relations provide a surprisingly good fit to the experimental partial cross sections. In addition, the numerical values of the parameters are obtained by fitting to thousands of experimental data points. The parameters Rudstam (1966) assigned to his equation are applicable to proton interactions with nuclei heavier than calcium, except when the target-product mass difference is small or large; i.e., it is not applicable for $\Delta A \leq 5$ and $\Delta A \geq 40$.

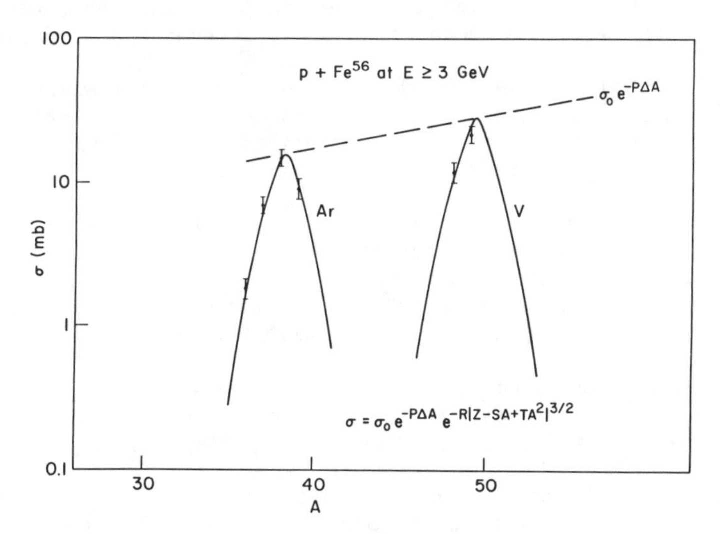

Figure 1: Illustration of the terms of the spallation equation. The example shows the experimental and calculated partial cross sections of iron into isotopes of Ar and V, at energies $\geq$ 3 GeV.

The nuclear reaction systematics of spallation reactions are not applicable to fission and fragmentation reactions, nor to the evaporation of light product nuclei. (Fragmentation is a nuclear breakup process in which a nucleus lighter than a fission product is formed.) Fission products have a higher N/Z ratio than spallation products, because of a lesser degree of nuclear evaporation in fission, and because neutrons are preferentially evaporated from heavy nuclei, since they can penetrate the Coulomb barrier more easily. We (Silberberg and Tsao 1973a, b) have constructed a semiempirical equation resembling Rudstam's equation (1966) with additional parameters, and we have defined regions of target and product mass intervals for fission, fragmentation, and evaporation where corresponding parameters apply.

The equation is applicable for calculating cross sections (in units of mb) of targets that have mass number in the range $9 \leq A_i \leq 209$ and products with $6 < A_j < 200$ at energies >100 MeV/nucleon:

$$\sigma_{ij} = \sigma_o f(A_i) f(E) e^{-P\Delta A} \exp(-R|Z - SA_j + TA_j^2|^{\nu}) \Omega \eta \xi \qquad (1)$$

Silberberg and Tsao 1973a, b; earlier in this section. In equation (1), σ_o is a normalization factor. The factors f(A) and f(E) apply only to products from heavy targets (with atomic number $Z_t > 30$, when A is large, as in the case of fission, fragmentation, and evaporation of light product nuclei. The parameter Ω is related to the nuclear structure and number of particle-stable levels of a product nuclide. The factor η depends on the pairing of protons and neutrons in the product nuclei; it is larger for even-even nuclei. The parameter ξ is introduced to represent the enhancement of light evaporation products.

Eq. 1 is inapplicable to peripheral reactions, that have small values of $\Delta A = A_i - A_j$. For such reactions, a different equation was constructed. A different equation was devised also for the heaviest target elements, the actinides such as Th and U.

New data have permitted the improvement of eq. (1). The improvements have been and are being carried out in two stages: (1) modifications for certain targets, products, and reactions, rather limited in scope, and (2) use of highly precise recent measurements for a major overall revision of the parameters of eq. (1), applicable for a wide range of (target, product) atomic numbers. Stage 1 has been published by Silberberg, Tsao, and Letaw (1985), while Stage 2 is still in progress, with a preliminary report by Tsao, Silberberg, and Letaw (1985). We shall first explore the work during Stage 1.

The improvement of the partial cross section equations of Silberberg and Tsao (1973a, b) has been made possible by several cross section measurements. Lindstrom et al. (1975) measured all the cross sections of ^{12}C and ^{16}O at 2.1 GeV, recently published by Olson et al. (1983). Yiou et al. (1973) and Yiou and Raisbeck (1972) measured the isotopic yields of Be and Li from various light elements. Several isotopic yields from Fe were measured by Perron (1976) and Lagarde-Simonoff et al. (1975); elemental yields from Fe were measured by Webber and Brautigam (1982), Westfall et al. (1979), and Poferl-Kertzman, Freier, and Waddington (1981). The isotopic yields of light elements produced from Ni were measured by Raisbeck et al. (1975). Isotopic yields of ^{197}Au were measured at several energies by Kaufman et al. (1976) and Kaufman and Steinberg (1980). Webber et al. (1983a) measured the isotopic yields of ^{16}O and ^{20}Ne at ~500 MeV per nucleon and Fe at 0.6-1 GeV per nucleon (Webber et al. 1983b). The latter cross sections of Fe are being further adjusted by the authors. Since the work of Stage 1 has been published, we shall not repeat the individual modifications here, but present graphic presentations of the more significant changes, especially those with a major impact on cosmic ray propagation calculations.

We have introduced correction factors that reflect nuclear structures. The "α-particle nuclei" (mass numbers multiples of 4, atomic numbers multiples of 2) ^{12}C, ^{16}O, and ^{20}Ne have (p,pα), (p,2p), and (p,pd) reactions enhanced by 1.8. The effects of these corrections are shown in Figure 2, where the old and new calculated yields at E > 2 GeV are shown, and compared with the data of Olson et al. (1983) and Webber et al. (1983a). Figure 2 also illustrates the effect of increasing the yield of ^{15}N by introducing the factor $\Omega(^{15}N) = 2$. Also the effects of increasing the cross sections of (p,2p) reactions from 21 to 28 mbarns for $6 \leq Z_t \leq 23$, at E > 3 GeV is shown. For $24 \leq Z_t \leq 42$, a value of 33 mbarns is used. These results are illustrated in Figure 3, where the yield of ^{55}Mn from Fe is shown. Also shown in Figure 3 (^{54}Mn, ^{53}Mn) are the increased values of (p,2pxn) peripheral reactions, 27 mbarns at E > 3 GeV, for $A_t \geq 35$ and $Z_t \leq 28$. Also shown in Figure 3 is the reduced yield of ^{54}Fe and ^{53}Fe from ^{56}Fe, based on Webber et al. (1983b); the old calculations gave much weight to the old measurement of Lavrukhina (1963). The new calculation modified eq. (28) of Silberberg and Tsao (1973a), as described by Silberberg, Tsao, and Letaw (1985).

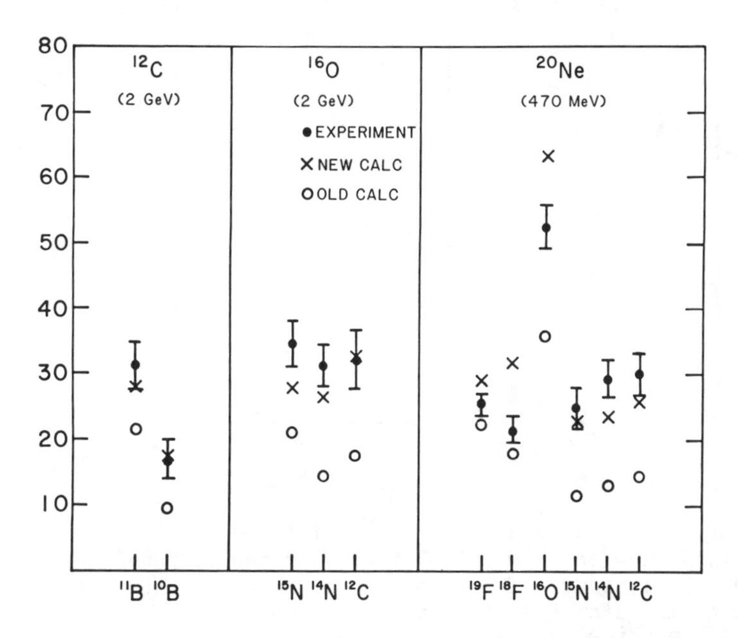

Figure 2: A comparison of some old and new calculated yields of ^{12}C, ^{16}O, and ^{20}Ne with the data of Olson et al. (1983) and Webber et al. (1983a).

Another significant modification affects the ultra-heavy nuclei with $A_t \geq 78$. The measurements of Kaufman and Steinberg (1980) permit the determination of the energy dependence of the partial spallation cross sections of the elements of the Pt-Pb group, with ^{197}Au as a representative target.

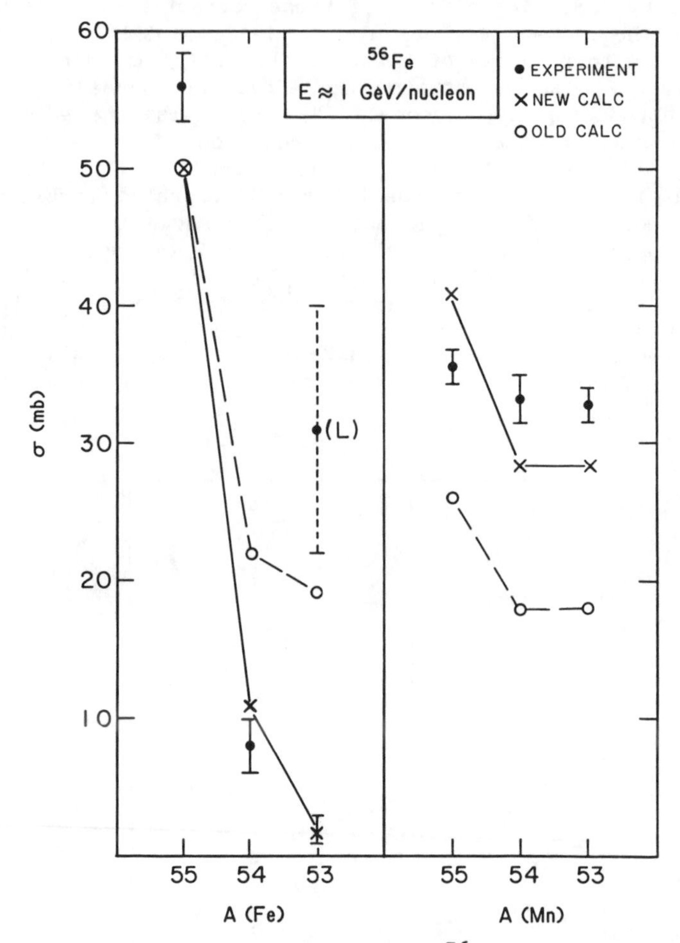

Figure 3: The cross sections of ^{56}Fe into Mn and Fe. The solid lines show the new calculations,and the dashed lines the old calculations. The experimental data of Webber et al.(1983b) and also the old value of Lavrukhina et al. (1963; dashed line) are shown.

Figure 4 shows a comparison of the old calculations based on our 1973 equation parameters, the recent experimental data of Kaufman and Steinberg (1980), and our new stage 1 calculations. We note that

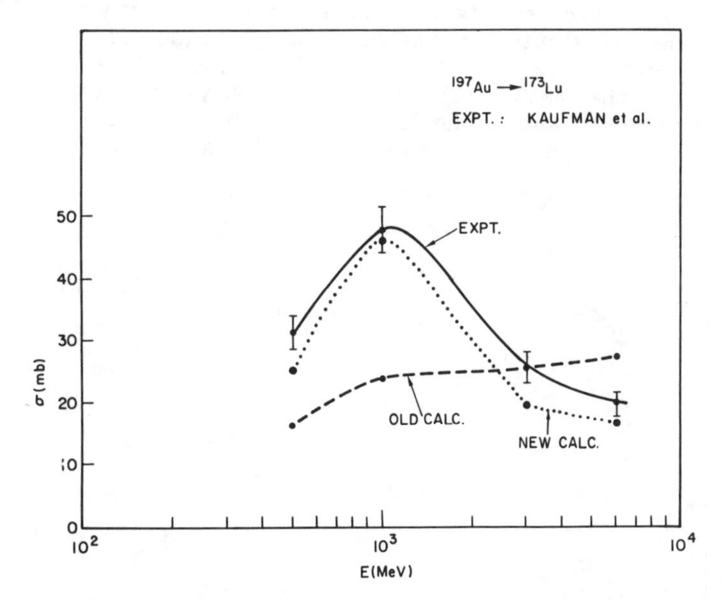

Figure 4: A comparison of the calculated and experimental energy dependence of the spallation cross sections of the Pt-Pb group of elements. The example shown is ^{197}Au into ^{173}Lu; the experimental values are from Kaufman and Steinberg (1980).

the cross section at 1 GeV exceeds that at $E \geq 6$ GeV by more than a factor of 2. This observation applies also to other cross sections of Au for reactions with $10 < \Delta A < 40$ at 1 GeV, as shown in Figure 5. In our stage 1 modifications, (Silberberg, Tsao, and Letaw, 1985), we proposed for the non-peripheral spallation cross sections of elements $76 \leq Z_t \leq 83$ the following correction factors: For $E \leq 1000$ MeV, σ_s is multiplied by

$$f(\Delta A,E) = (\Delta A/28)(E/1000)^{-2/3} \qquad (2)$$

but setting A equal to 15 for A < 15. For $1000 < E < E_o$, (where E_o is the high-energy asymptotic value above which the cross sections are constant), σ_s is multiplied by

$$f(\Delta A,E) = (\Delta A/28)(E/1000)^{-0.06\sqrt{\Delta A-15}} \qquad (3)$$

again setting ΔA equal to 15 for $\Delta A < 15$. Equations (2) and (3) represent special cases; we shall explore in Stage 2 whether a general modification of the parameters of equation (1) could reproduce the experimental data.

A preliminary examination of Figure 8 of Kaufman and Steinberg (1980) suggests that the calculation of the high energy values of spallation cross sections ($E \geq 3$ GeV) in Stage 2 will be easier than at lower energies, because the mass yield no longer obeys the exponential relationship $\exp(-P\Delta A)$, but is nearly constant for $\Delta A \leq 26$ at E = 1 GeV and for $\Delta A \leq 16$ at E = 0.5 GeV, i.e. $P \propto (\Delta A)^{-1}$ for relatively small values of ΔA.

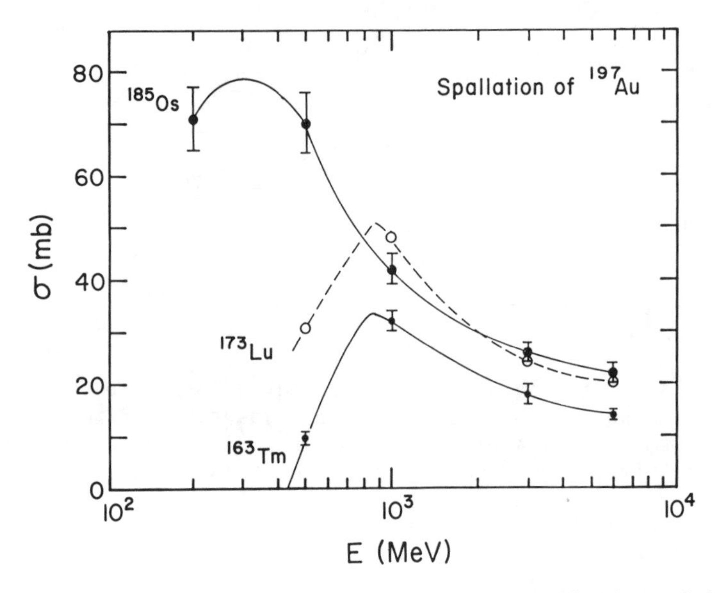

Figure 5: The energy dependence of the spallation of Au into Tm, Lu and Os.

The latter statement is illustrated in Figure 6. The spallation cross sections, as well as the fission peak at 1 GeV and fission + spallation plateau at 11 GeV of Kaufman and Steinberg (1980) are reproduced in Figure 6. This deviation from a simple $\exp(-P\Delta A)$ relationship has to be addressed in our phase 2 calculations. We note that for $E \leq 1$ GeV, the mass yield is nearly independent of ΔA, for small values of ΔA, and then falls off rapidly with increasing values of ΔA.

We note that the cross sections measured at 1 GeV, e.g. by Brewster et al. (1983) and Kertzman et al. (1985) are not representative of the high-energy cross sections, those at $E > 3$ GeV. For $15 \leq \Delta A \leq 30$ the cross sections at 1 GeV are about 2x higher than the high-energy cross sections, while for $60 \leq \Delta A \leq 80$, the cross sections at 1 GeV are about 6x lower.

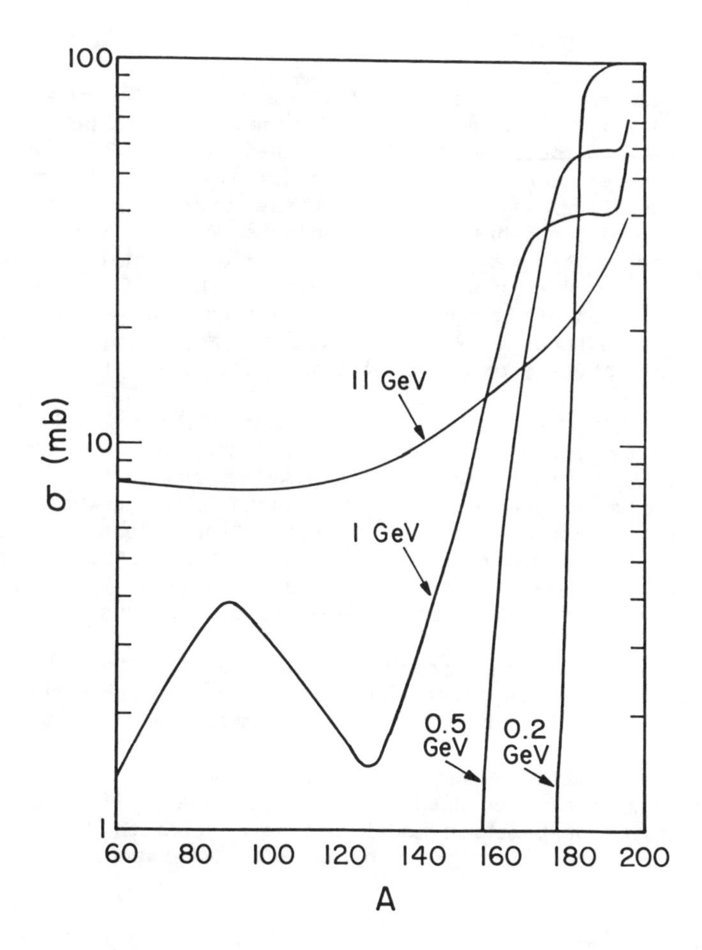

Figure 6: The energy dependence of the mass yields of Au, based on the data of Kaufman and Steinberg (1980).

Another phase 1 modification is the elimination of the discontinuity in calculations between peripheral and spallation cross sections of the ultra-heavy nuclei. Figure 4, of Silberberg, Tsao, and Letaw (1985), illustrates the old and new calculations, with a comparison of the experimental data of Neidhardt and Bächmann (1971a, b).

2b. CROSS SECTION ERRORS

Uncertainty in nuclear interaction cross sections is the most important limitation on the analysis of cosmic ray composition. Cosmic ray primaries, such as C, O, and Ne, have 5-20% source abundance uncertainties due to errors in estimating the contribution of fragmentations of heavier nuclei. These errors have a more profound impact as the contribution of secondaries grows. Thus a more detailed understanding of pre-acceleration atomic selection effects awaits improved source abundances for Na, Al, and Ca. The N source abundance remains doubtful because of the uncertain O --> N cross sections. Source abundances of K, Ti, V, Cr, and Mn, though possibly not negligible, are entirely obscured by cross section errors.

Hinshaw and Wiedenbeck (1983) analyzed the uncertainties involved in computing cosmic ray source composition from observed abundances. Their analysis includes measurement error, total and partial cross section error, and mean pathlength uncertainty. They demonstrated that cross sections are the dominant source of uncertainty. More importantly, they left open the possibility that cross section errors are strongly correlated, meaning only source abundances of nearly pure primaries can be reliably derived from compositional measurements.

In this section we show that cross section correlations in the semi-empirical formulas (Silberberg and Tsao, 1973 and Silberberg et al. 1985) are essentially negligible in cosmic ray propagation calculations concerning targets with $Z < 29$. Our conclusion is based on the excellent agreement between secondary abundances measured by the French-Danish experiment on HEAO-3 (Engelmann et al. 1983) at 3.99 GeV/N and a primitive propagation model. A more detailed discussion of the cross section errors is given by Letaw, Silberberg, and Tsao (1985).

In the error analysis we carried out a standard cosmic ray propagation calculation, (Letaw, Silberberg, and Tsao 1984) with an exponential path length distribution, assuming the errors of the calculated partial cross sections to be 35%, on the basis of Silberberg and Tsao (1973a). The abundances of the secondary elements Be, B, F, P, Cl, Sc, Ti, V, Cr, and Mn were calculated, and compared with the above-mentioned HEAO-3 data.

As shown in Figure 3 of Letaw, Silberberg, and Tsao (1985), and the χ^2 test of that paper, the errors are uncorrelated: The mean deviation is only 12%, even smaller than the value of 18% expected in the case of uncorrelated errors. (The mean deviation has such a small value probably because of updating of cross section calculations since our 1973 paper.)

However, Figures 4 and 5 imply that the energy dependence of the calculated cross sections of the ultra-heavy nuclei $Z \geq 74$, (possibly even $Z > 28$) have (or have had) systematic errors, i.e. correlated errors. E.g., the yields of the group of products of Au with target-product mass difference $10 < \Delta A < 30$ at 1 GeV were underestimated by a factor of two. Thus, in general, the errors are uncorrelated, but cases occur (or have occurred) where the errors are correlated.

2c. STAGE 2 IMPROVEMENTS.

These modifications are still in progress. This section is an interim progress report, outlining remaining systematic discrepancies in spallation cross section calculations.

Several research groups have recently carried out highly precise measurements (to about 10 percent of nuclear spallation cross sections). Our initial emphasis is on high-energy ($E \geq 3$ GeV) spallation reactions of nuclei with atomic numbers $20 \leq Z \leq 83$. The discussion below is confined to these, though in the near future we shall include intermediate energies; there are such data with ~10% precision for $Z \geq 20$, e.g. from Grütter (1982), Heydegger et al. (1972), Orth et al. (1976), and Kaufman and Steinberg (1980).

The high-energy measurements, above 3 GeV, cover a broad range of elements: V, Fe, Cu, Ag, Ta, and Au. Even the small cross sections far off the peak of the isotopic distribution curves have been measured. The semiempirical calculations are compared with the measured values. Preliminary comparisons indicated that the parameters of our spallation relations (Silberberg and Tsao, 1973a) for atomic numbers 20 to 83 need modifications, e.g. a reduced slope of the mass yield distribution, broader isotopic distributions, and a shift of the isotopic distribution toward the neutron-deficient side. The required modifications are relatively small near Fe and Cu, but increase with increasing target mass.

While earlier experimental data were derived from radioactivity measurements after chemical separation of product elements, most of the measurements selected for the present investigation are based on gamma ray line intensity measurements as a function of time. Any systematics introduced by chemical separation are thus avoided.

Table 1 shows the sources of experimental data used in our current analysis. Some of the spallation cross sections are cumulative, i.e. contain the contributions of shorter lived progenitor isotopes. Reactions are omitted in which several isomers are produced, but only one is measured.

Table 1. Sources of Recent High-Energy Experimental Data, $20 < Z < 80$

Author	Target	Energy (GeV)
Husain and Katcoff (1973)	V	3, 30
Asano et al. (1983)	Ti, Fe, Co, Ni, Cu	12
Cumming et al. (1976)	Cu	25
Hudis et al. (1970)	Cu, Ag, Au	3, 29
Porile et al. (1979)	Ag	300
Chu et al. (1974)	Ta	28
Kaufman et al. (1976)	Au	12, 300

Systematic deviations from equation (1) can be explored by comparing the measured and calculated cross sections as a function of $Z-SA+TA^2$ and of ΔA. The former comparison permits a test of systematic deviations as a function of the neutron-richness of product isotopes and of the width of the isotopic spread of the products. The latter permits a test of systematic deviations as a function of the target-product mass difference. After these systematics are corrected, we can explore the smaller systematic differences, as a function of the nuclear pairing factor η, which represents the enhancement of even-even product nuclei and the suppression of the odd-odd products.

Figure 7 shows the ratios of calculated to experimental cross sections of Cu as a function of $Z-SA+TA^2$. A large value of this function implies a small value of A, i.e. a neutron-deficient product. We note a positive slope as a function of $Z-SA+TA^2$. This means that the calculated cross sections of neutron-deficient products, e.g. ^{52}Fe, are overestimated, and neutron-rich ones, e.g. ^{47}Ca, are underestimated.

Figure 8 shows the corresponding data for tantalum. The large negative slope shows that a significant systematic deviation occurs in the calculated cross sections, however, opposite to that for lighter target nuclei like Cu of Figure 7. For Ta, the neutron-rich products are overestimated, instead.

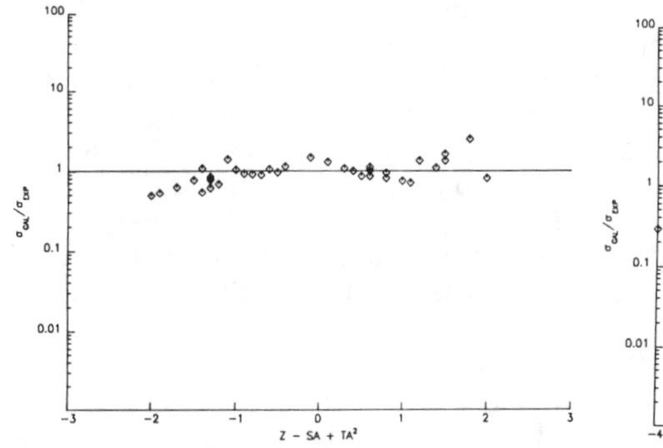

Figure 7: The ratio of calculated to experimental cross sections of Cu, as a function of Z-SA+TA^2, for E > 3 GeV.

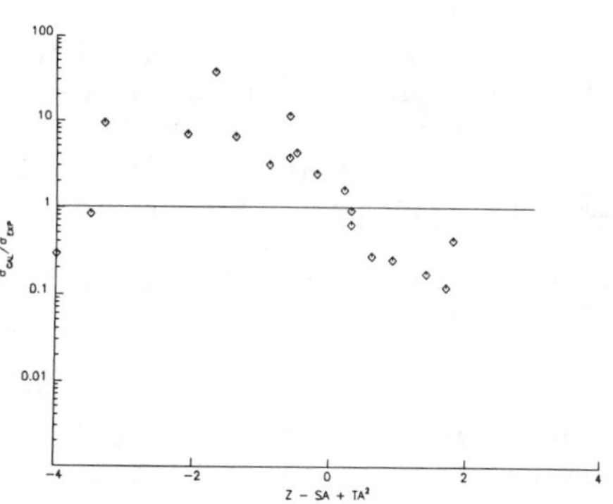

Figure 8: The ratio of calculated to experimental cross sections of Ta, as a function of Z-SA+TA^2, for E > 6 Gev.

Figure 9 compares the calculated and experimental spallation cross sections of Cu as a function of ΔA. We note that for Cu, the systematic deviations are rather small. The largest and smallest values are those near the extreme values of Z-SA+TA^2. After the latter are corrected for, the spread of the ratios about 1 will be very small.

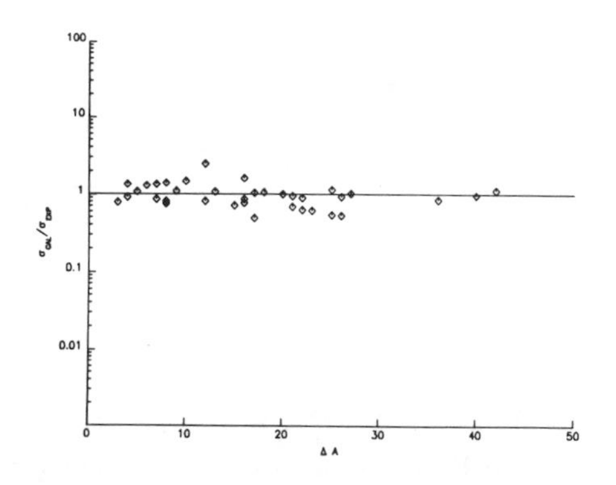

Figure 9: The ratio of calculated to experimental cross sections of Cu, as a function of ΔA, for E > 3 GeV.

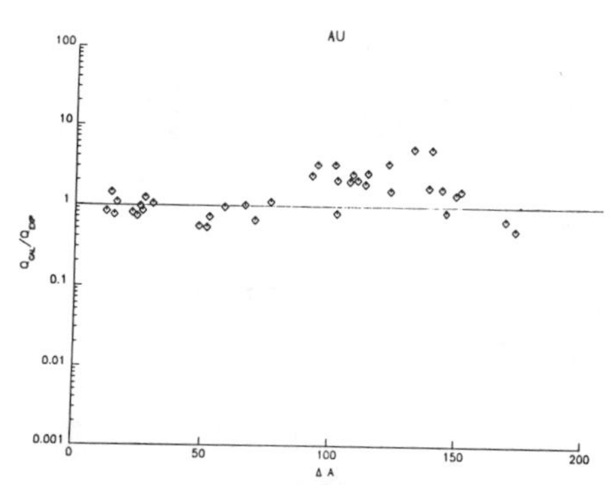

Figure 10: The ratio of calculated to experimental cross sections of Au, as a function of ΔA for E > 6 GeV. The temporary adjustment factors of Tsao et al. (1983b) were applied to the calculation.

Figure 10 compares the calculated and measured cross sections of Au as a function of ΔA. The agreement of the spallation cross sections (i.e. those with ΔA less than approximately 60) is good. However, this agreement was achieved by special correlations for nuclei with $76 \leq Z_t \leq 83$ we proposed (Tsao et al. 1983). Our aim will be to eliminate such special correlations, and adjust the parameters P, R, S, and T so that the whole region $20 \leq Z_t \leq 83$ can be adequately fitted. We note from Figure 8, that for Ta (with $Z_t = 73$), the fit to the data is rather poor. A simultaneous fit to Ta and Au is necessary.

We noted from Figures 7 and 8 that one should increase the calculated cross sections of the n-rich products for targets near $Z_t = 30$, while increasing those of the neutron deficient products near $Z_t = 70$ and 80. This can be accomplished by decreasing S or increasing T in the former case, and by increasing S or decreasing T in the latter case. Since S is associated with A and T with A^2, S is more sensitive for lighter nuclei and T is more sensitive to heavier nuclei. Thus the correction can be accomplished by reducing both S and T, replacing the value 0.486 and 0.00038 given in Table 1D of Silberberg and Tsao by 0.48 and 0.0003 and reducing R to 0.9R. Figure 11 shows how Figure 8 is transformed when the above parameters are used and Figure 12 how Figure 7 is transformed. A correction for $Z-SA+TA^2 < -2$ is still required.

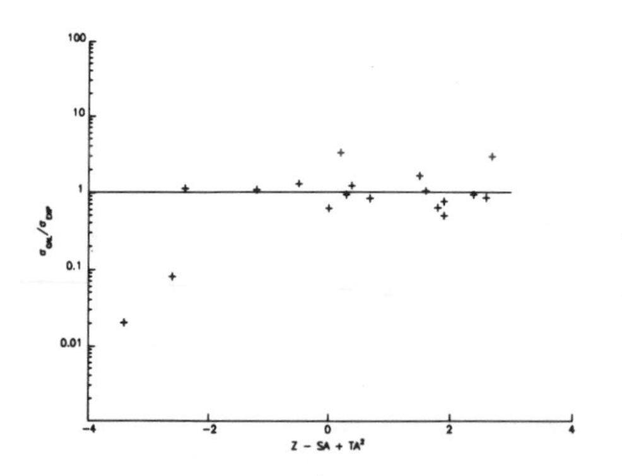

Figure 11: The ratio of calculated to experimental cross sections of Ta as a function of $Z-SA+TA^2$, with the new values of S and T, for E > 6 GeV.

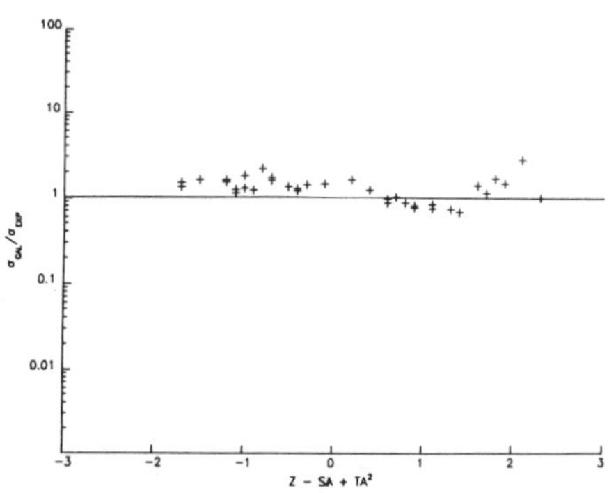

Figure 12: The ratio of calculated to experimental cross sections of Cu, as a function of $Z-SA+TA^2$, with the new values of S and T, for E > 3 GeV.

For the parameters P of equation (1), for energy $E \geq E_o$, and $21 \leq Z \leq 83$ and products $Z \geq 6$, the value $P = 0.77A_t^{-0.67}$ has to be reduced for heavy nuclei. We have not yet carried out a final fit, but an intermediate stage trial function is $P = 2.4A_t^{-0.96}$.

A change in P and R requires a renormalization of the parameter σ_o, with the use of equation 25 of Rudstam (1966). The modifications we plan to introduce involve special complications, because of multiple feed-back loops. An adjustment of P affects the overall normalization factor σ_o, and the energy dependence of the calculations. Adjustments in S and T affect the calculation of fission and fragmentation cross sections, and the parameters f(A) and f(E) of Silberberg and Tsao (1973b) must be re-formulated.

While we have not yet explored the improvement of the calculated energy dependence of cross sections, the change in slope of the mass yields at the right hand side of Figure 6 for $0.2 \leq E \leq 1$ GeV suggests a need to reduce the large calculated cross sections at energy E, by restricting $\sigma(E)$ to values $\sigma(E) \leq [4000/E(\text{MeV})]^{1/2} \sigma(E_o)$, where $\sigma(E_o)$ is the high energy asymptotic value of the cross section, as defined by Silberberg and Tsao (1973a).

We expect that significantly more accurate calculations are possible for targets with atomic numbers $20 \leq Z \leq 83$ after optimizing the parameters P, R, S, and T.

This work is partly supported by the NASA HEAO-3 program.

References:

Asano, T. et al. 1983, Phys. Rev. C28, 1718.

Brewster, N.R., Fickle, R.K., Waddington, C.J., Binns, W.R., Israel, M.H., Jones, M.D., Klarmann, J., Garrard, T.L., Newport, B.J., and Stone, E.C., Proc. 18th Internat. Cosmic Ray Conf. (Bangalore), 9, 259.

Chu, Y.Y., Franz, E.M., and Friedlander, G. 1974, Phys. Rev. C10, 156.

Cumming, J.B., Stoenner, R.W., and Haustein, P.E. 1976, Phys. Rev. C14,1554.

Dostrovsky, I. Rabinowitz, P., and Bivins, R. 1958, Phys. Rev. 111, 1659.

Grutter, A. 1982, Int. J. Appl. Radiat. Isot. 33, 725.

Heydegger, H.R., Garrett, C.K., and Van Ginneken, A. 1972, Phys. Rev. C6, 1235.

Hinshaw, G.F. and Wiedenbeck, M.E. 1983, Proc. 18th Internat. Cosmic Ray Conf. (Bangalore), 9, 263.

Hudis, J., Kirsten, T., Stoenner, R.W.,and Schaeffer, O.A. 1970, Phys. Rev. C1, 2019.

Husain, L. and Katcoff, S. 1973, Phys. Rev. C7, 2452.

Kaufman, S.B. and Steinberg, E.P. 1980, Phys. Rev. C22, 167.

Kaufman, S.B., Weisfield, M.W., Steinberg, E.P. Wilkins, B.D., and Henderson, D. 1976, Phys. Rev. C14, 1121.

Kertzman, M.P. Klarmann, J., Newport, B.J. Stone, E.C., Waddington, C.J., Binns, W.R., Garrard, T.L., and Israel, M.H. 1985, 19th Internat. Cosmic Ray Conf. (La Jolla) 3, 95.

Lagarde-Simonoff, M., Regnier, S., Sauvageon, H., Simonoff, G.N., and Brout, F. 1975, J. Inorganic Nucl. Chem. 37, 627.

Lavrukhina, A.K., Revina, L.D., Malyshev, V.V.,and Satarova, L.M. 1963, J. Exper. Theoret. Phys. 44, 1929 (English trans. Soviet Phys. - JETP, 17, 960).

Letaw, J.R., Silberberg, R., and Tsao, C.H. 1983, Ap. J. Suppl., 51, 271.

_____. 1984. Ap. J. Suppl. 56, 369.

_____. 1985. Proc. 18th Internat. Cosmic Ray Conf. (La Jolla), 3, 46.

Lindstrom, P.J. Greiner, D.R., Heckman, H.H., Cork, B., and Bieser, F.S. 1975 Lawrence Berkeley Laboratory Report LBNL-3650.

Metropolis, N. Bivins, R., Storm, M., Miller, J.M. Friedlander, G., and Turkevich, A. 1958, Phys. Rev. 110, 204.

Neidhardt, B. and Bachmann, K. 1971a, J. Inorganic Nucl. Chem., 33, 2751.

_____. 1971b, J. Inorganic Nucl. Chem., 33, 3227.

Olson, D.L., Berman, B.L., Greiner, D.E., Heckman, H.H., Lindstrom, P.J., and Crawford, H.J. 1983, Phys. Rev. C28, 1602.

Orth, C.J., O'Brien, H.A., Jr., Schillaci, M.E. Dropesky, B.J., Cline, J.E., Nieschmidt, E.B., and Brodzinski, R.L. 1976 J. Inorg. Nucl. Chem., 38, 13.

Perron, C. 1976, Phys. Rev. C14, 1108.

Poferl-Kertzman, M., Freier, P.S., and Waddington, C.J. 1981, Proc. 17th Internat. Cosmic Ray Conf. (Paris), 9, 187.

Porile, N.T., Cole, G.D., and Rudy, C.R. 1979, Phys. Rev. C19, 2288.

Raisbeck, G.M. Boerstling, P., Klapisch,R, and Thomas T.D. 1975, Phys. Rev. C12, 527.

Rudstam, G. 1966, Zs. Naturforschung, 21a, 1027.

Silberberg, R., Tsao, C.H., and Letaw, J.R., 1985 Ap. J. Suppl. 58, 873.

_____. 1983. p. 321 in "Cosmic Radiation in Contemporary Astrophysics, ed. M.M. Shapiro, Reidel Publ. Co., Dordrecht.

Silberberg, R. and Tsao, C.H. 1973a, Ap. J. Suppl. 25, 315.

_____. 1973b, Ap. J. Suppl. 25, 335.

Tsao, C.H., Silberberg, R., and Letaw, J.R. 1983, 18th Internat. Cosmic Ray Conf. (Bangalore), 2, 194.

Tsao, C.H., Silberberg, R., and Letaw, J.R. 1985, 19th Internat. Cosmic Ray Conf. (La Jolla), 3, 103.

Webber, W.R. and Brautigam, D.A. 1982, Ap. J. 260, 894.

Webber, W.R., Brautigam, D.A., Kish, J.C., and Schrier, D.A. 1983a, Proc. 18th Internat. Cosmic Ray Conf. (Bangalore), 2, 202.

Webber, W.R., Brautigam, D.A., Schrier, D.A., and Kish, J.C. 1983b, Proc. 18th Internat. Cosmic Ray conf. (Bangalore), 2, 198.

Westfall, G.D., Wilson, L.W. Lindstrom, P.J., Crawford, H.J., Greiner, D.E., and Heckman, H.H. 1979 Phys. Rev. C19, 1309.

Yiou, F. and Raisbeck, G. 1972, Phys. Rev. Letters, 29, 372.

Yiou, F., Raisbeck, G., Perron, C., and Fontes, P. 1973, Proc. 13th Internat. Cosmic Ray Conf. (Denver), 1, 512.

MEASUREMENTS OF CROSS SECTIONS RELEVANT TO γ-RAY LINE ASTRONOMY

K. T. Lesko, E. B. Norman, R. M. Larimer and S. G. Crane
Nuclear Science Division, Lawrence Berkeley Laboratory
University of California, Berkeley, CA 94720, USA

ABSTRACT: Gamma-ray production cross sections have been measured for the γ-ray lines which are most strongly excited in the proton bombardment of C, O, Mg, Si, and Fe targets of natural isotopic composition. High resolution germanium detectors were used to collect γ-ray spectra at proton bombarding energies of 20, 30, 33, 40 and 50 MeV.

1. INTRODUCTION

Observations of discrete γ-ray lines can provide unique signatures of nuclear reactions occurring in astronomical environments. Because of their highly penetrating nature, γ-rays may provide specific information on astrophysical sites that are opaque to longer wavelength radiation. γ-ray lines have been observed within the solar system, in solar flare events, and in extra-solar system sites such as the galactic center, Centaurus A, SS-433, and perhaps the Crab Nebula. In principle, discrete line γ-ray spectra can be used to obtain the relative abundances and energy spectra of the particles responsible for the γ-ray production. However, in order to make use of such spectra, cross sections for the production of nuclear γ-rays must be known.

There are several mechanisms which result in γ-ray emission: charged particle induced reactions such as inelastic scattering and spallation reactions, radioactive decay to excited states of nuclei, e^+e^- annihilation, and neutron capture. Within the first category, the high cosmic abundances of hydrogen and helium imply that only proton and α-particle induced reactions need be considered.

We present in this paper preliminary results of measurements of γ-ray production cross sections using a proton beam on a variety of targets. In a second paper to be published later we will present results for α-particle bombardments. This work is an extension of the earlier measurements of Dyer et al.[1,2] and Seamster et al.,[3] who determined the γ-ray production cross sections for both proton and α-particle bombardments of a large number of targets from threshold to 23 MeV for proton and from the threshold to 27 MeV for α-particle bombardments.

M. M. Shapiro and J. P. Wefel (eds.), Genesis and Propagation of Cosmic Rays, 375–379.

2. EXPERIMENTAL METHOD

Beams of protons were provided by Lawrence Berkeley Laboratory's 88-Inch Cyclotron over the energy range of 20 to 50 MeV. The beam impinged on targets of natural isotopic composition of C, O, Mg, Si, Fe, and a thick meteoritic sample. The oxygen target was constructed of a thin mylar foil ($C_{10}H_8O_4$). The target thicknesses varied from 200 $\mu g/cm^2$ to 8.16 mg/cm^2 for the various targets, with the exception of the meteoritic target which was a cut and polished slice of the Allende meteorite.

Gamma rays produced by various reactions were observed by two high purity Ge detectors of 110 cm^3 volume. For all but the oxygen target, the detector angles were chosen to be the zeroes of $P_4(\cos\theta)$ to aid in the reduction of the angular distribution data into total cross sections. For the oxygen target, complete angular distributions were determined at each energy. In addition, at one bombarding energy, angular distributions over a wider range of angles were taken for all targets to confirm the multipolarity of the γ-rays.

The energy of the beam was varied in 10-MeV steps from 20 to 50 MeV. Data were also collected at 33 MeV, chosen to coincide with the proposed energy per nucleon of matter in the jet of SS-433. Single parameter energy histograms were collected at each energy and angle combination of all targets with the two detectors. These histograms were stored on magnetic tape for later analysis. In order to determine the system dead-time, the output of the current integrator was used to trigger a pulser that was fed into both detectors preamplifiers and into a scaler.

3. RESULTS

In order to provide a graphic illustration of what the spectra obtained from future γ-ray observatories may look like, we bombarded a thick sample of the Allende meteorite with 33 MeV protons. Allende is a carbonaceous chondrite and contains all but the most volatile elements in roughly their cosmic abundances. Thus, except for the reduction of lines from C, N, O, and Ne, this spectrum is representative of what might be observed from proton interactions with the interstellar medium. As can be seen in Figure 1, γ-ray lines from proton-induced reactions on ^{12}C, ^{16}O, ^{24}Mg, ^{28}Si, and ^{56}Fe are very prominent.

To obtain cross sections from the measurements on the elemental targets, background and deadtime corrected γ-ray yields were determined for the most prominent peaks in each spectrum. These differential cross sections were then used to obtain angle-integrated total cross sections using the technique of Dyer et al.[1]. In order to convert our results into absolute cross sections, we normalized our data obtained at E_p = 20 MeV to those reported by Dyer et al.[1]. The results of this work are summarized in Figures 2-6.

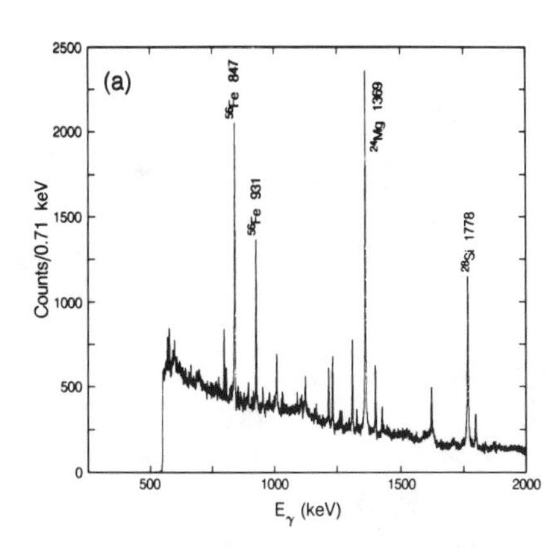

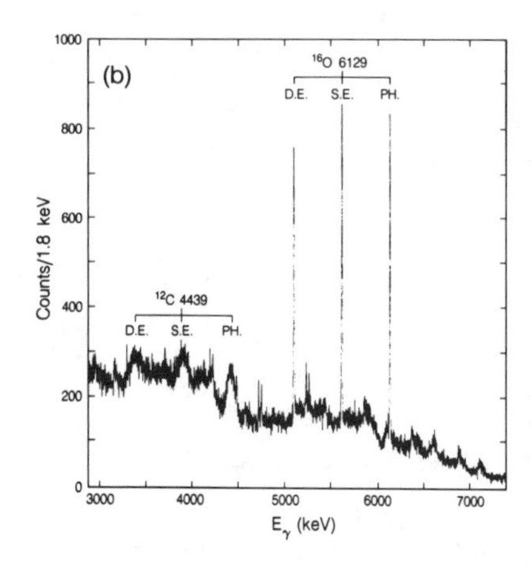

Figure 1. γ-ray spectrum observed during the bombardment of a thick sample of the Allende meteorite with 33 MeV protons. The prominent γ-rays are identified by target isotope. PH, SE, DE stand for, respectively, photo-peak, single escape-peak, and double escape-peak.

In an earlier work, Ramaty et al.[4] used the differential cross sections of Foley et al.[5] and Zobel et al.[6] to deduce total, angle integrated γ-ray production cross sections. This procedure is risky because the measurements were made at a single angle and it is difficult to deduce the accurate total cross sections without angular distribution data. When comparisons between our data and those summarized by Ramaty et al. can be made, we find that the two data sets agree at the 20-50% level. However, for many of the systems we find a much smaller energy dependence of the production cross sections, such that our cross sections at higher energies differ significantly from those of Ref. 4.

It is interesting to compare our measured spectrum from the bombardment of a meteoritic sample to Ramaty et al.'s Monte Carlo simulated γ-ray spectra. In the simulated spectra energetic cosmic ray particles interact with matter in the interstellar medium that contains both small grains and gases with solar-system elemental abundances. With the exception of highly volatile elements, the meteoritic composition is a good approximation to solar-system abundances. While we used a mono-energetic beam of protons and the simulation used a power law distribution for the cosmic ray spectrum, we note striking similarities between the two spectra. The prominent peaks due to Fe, Mg, Si, C and O are all present in both spectra in roughly similar proportions. The γ-ray lines from ^{16}O observed in our spectrum are narrow due to our thick target which assures that most of the excited nuclei will stop before they emit their γ-rays. In the simulation, interactions with the gaseous target components result in substantial Doppler broadening for the ^{16}O nuclei which recoil into vacuum.

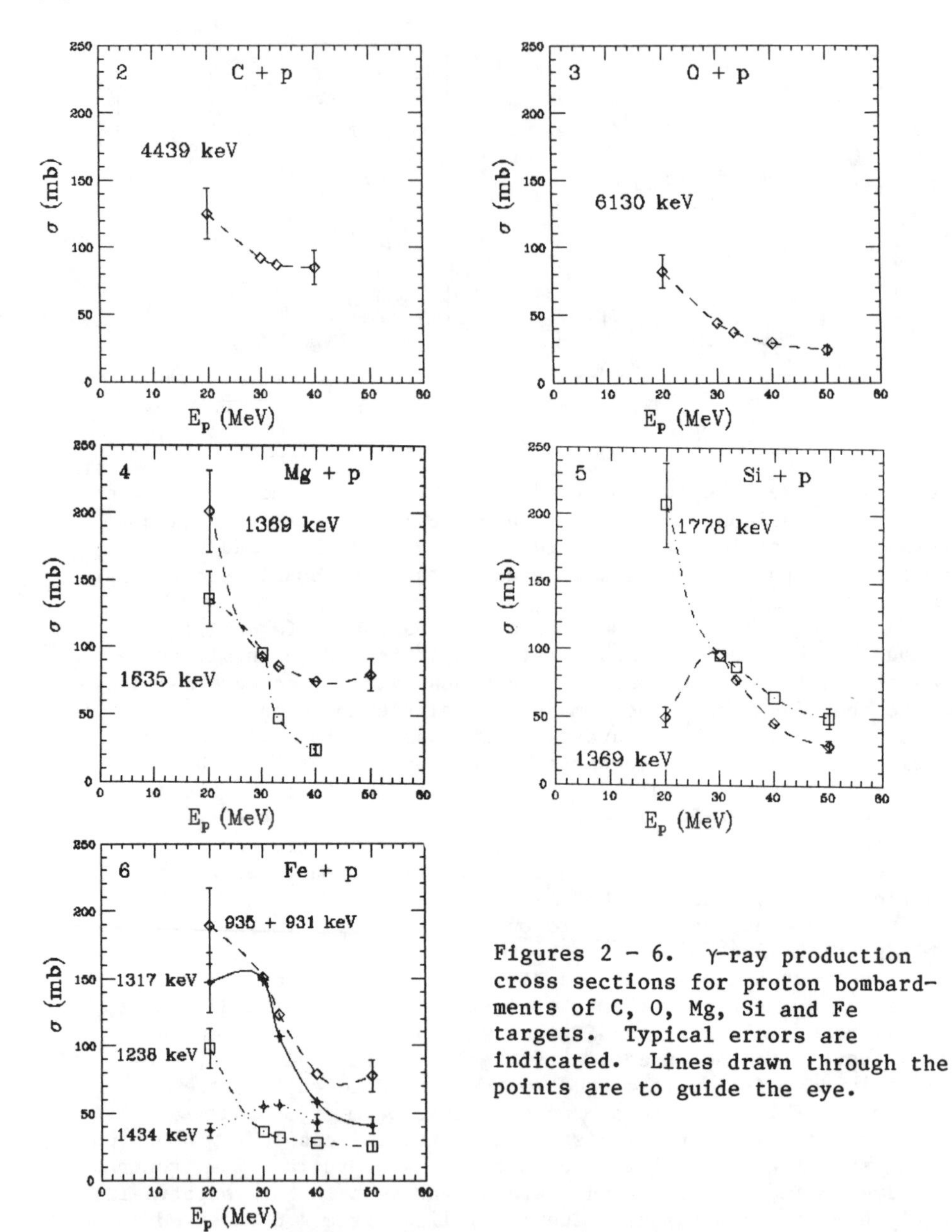

Figures 2 - 6. γ-ray production cross sections for proton bombardments of C, O, Mg, Si and Fe targets. Typical errors are indicated. Lines drawn through the points are to guide the eye.

4. CONCLUSIONS

While the actual number of extra-terrestrial γ-ray sources remains limited to a few, and while the number of γ-ray lines which have been unambiguously identified is even fewer, it is hoped that with additional observations of γ-ray sources in the coming years the γ-ray production cross sections we present here will be of value in determining quantities such as the modes and sites of nucleosynthesis, particle fluxes and nuclear abundances in the interstellar medium, and even the properties of exotic astrophysical environments such as neutron star surfaces.

5. ACKNOWLEDGEMENTS

We would like to express our appreciation to Dr. E. J. Barta for his assistance in target preparation. This work supported by the Director, Office of Energy Research, Division of Nuclear Physics of the Office of High Energy and Nuclear Physics of the U.S. Department of Energy under Contract No. DE-AC03-76SF00098.

6. REFERENCES

1. P. Dyer et al., Phys. Rev. C 23, 1965 (1981).
2. P. Dyer et al., Phys. Rev. C 32, 1973 (1985).
3. A. G. Seamster et al., Phys. Rev. C 29, 394 (1984).
4. R. Ramaty et al., Ap. J. Suppl. 40, 487 (1979).
5. K. J. Foley, et al., Nucl. Phys. 37, 23 (1962).
6. W. Zobel, et al., Nucl. Sci. Eng. 32, 392 (1968).

THE COSMIC RAY NUCLEI EXPERIMENT ON THE SPACELAB-II MISSION

John Mace Grunsfeld
Enrico Fermi Institute and Department of Physics
University of Chicago
Chicago, Illinois 60637, U.S.A.

ABSTRACT

The Cosmic Ray Nuclei Experiment flown on the Spacelab-II mission was designed to measure the elemental composition of individual cosmic ray nuclei (Li to Ni) from 40 Gev/amu to several TeV/amu. The detector utilizes plastic scintillation counters for charge measurement, and gas Cerenkov and transition radiation detectors for energy measurements. The data analysis is in progress and results on the energy spectra are expected in the near future.

INTRODUCTION

The University of Chicago Cosmic Ray Nuclei (CRN) experiment, designed to measure the elemental composition of cosmic ray nuclei at high energies, flew successfully on the Space Shuttle Challenger Spacelab-II mission in July 1985. The scientific group involved with the CRN experiment includes P. Meyer, D. Muller, J. L'Heureux, S.P. Swordy, and J.M. Grunsfeld.

The study of the energy dependence of the cosmic ray composition has proven to be essential in the understanding of the origin, acceleration, and propagation of cosmic rays in the galaxy. Extensive direct observations of cosmic ray nuclei have been performed up to energies of >100 Gev/amu (figure 1, Simpson 1983) with experiments aboard balloons and satellites. At considerably higher energies indirect observations of cosmic ray initiated air showers have yielded the energy spectra of protons and the "all element" spectra for the heavier nuclei (figure 2, Hillas 1984). The cosmic ray nuclei observed at earth contain both primary particles, present at cosmic ray sources, and secondary nuclei produced by the interaction of a primary nucleus with the matter in the interstellar medium. Crucial to the understanding of the acceleration mechanism of the primary nuclei is the shape of the energy spectrum, particularly at high energies where the local effects of solar modulation are negligible. One of the goals of the CRN experiment is to extend the measurements of the primary

M. M. Shapiro and J. P. Wefel (eds.), Genesis and Propagation of Cosmic Rays, 381–390.

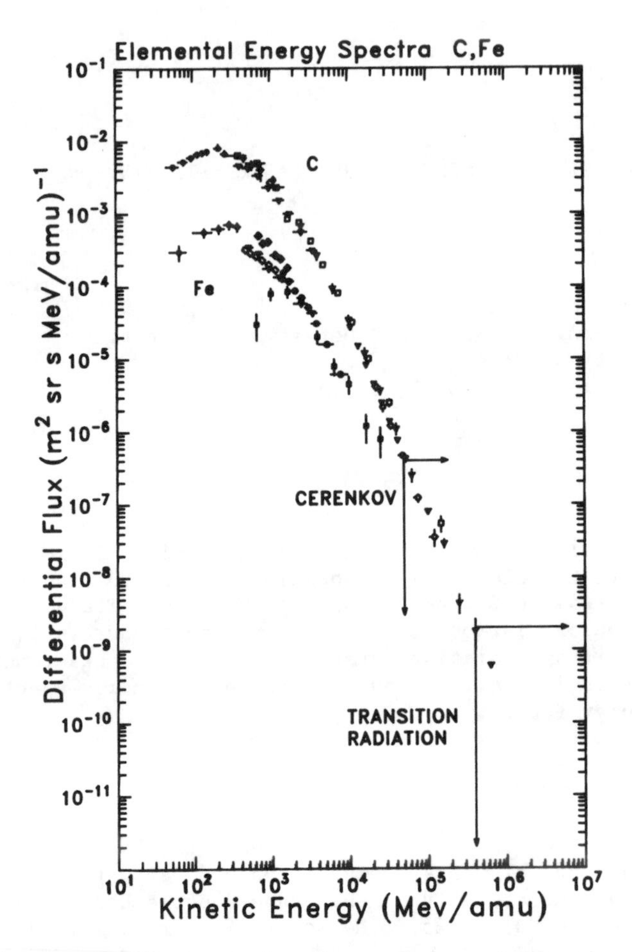

Figure 1. Energy Spectra of carbon and iron indicating the range of coverage of the CRN experiment (Simpson 1983).

elemental spectra beyond the present measurements to several TeV/amu. This new data will also aid in the interpretation of the air shower studies by providing the composition of the primary cosmic ray nuclei in a common energy regime.

The second goal of the CRN experiment is the determination of the energy dependence of the abundance ratio of the secondary to primary nuclei. This ratio gives an indication of the amount of material that the cosmic rays traverse during their propagation in the galaxy. Measurements of this ratio have shown the relative abundances of secondary and primary nuclei to be energy dependent above $\sim$1 Gev/amu. Measurements have been performed up to $\sim$100 Gev/amu indicating a general decrease in the secondary to primary ratio (Caldwell 1977;Simon et al.

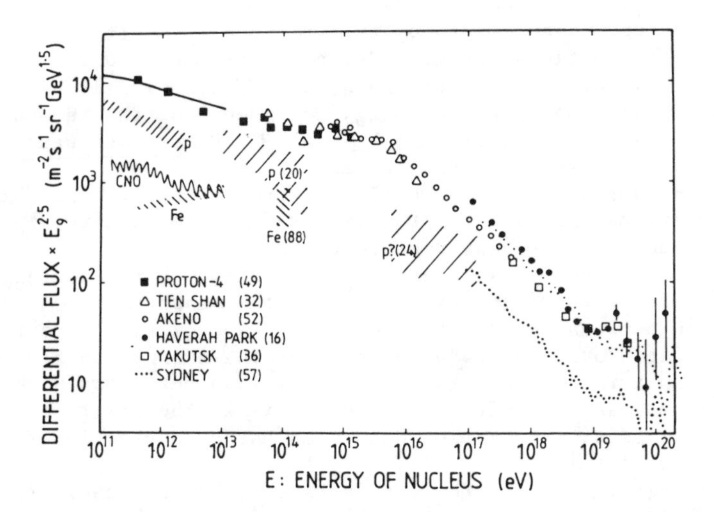

Figure 2. The 'all particle' energy spectrum at high energies. (from Hillas 1984)

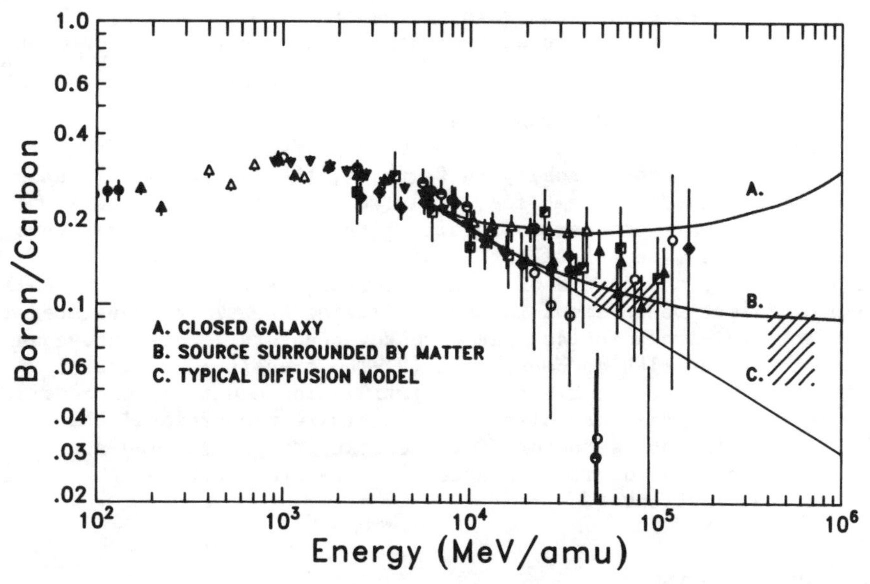

Figure 3. Boron/Carbon ratio as a function of energy. Symbols as in Garcia-Munoz et al. 1984. The curves represent the expected energy dependence for three propagation models: The simple diffusion model, the nested leaky box model (Cowsik and Wilson 1973), and the closed galaxy model (Peters and Westergaard 1977). The shaded areas depict the energy coverage of the CRN measurements.

1980). High energy measurements of this ratio are important, since the interaction cross sections become independent of energy allowing a more straightforward interpretation of the data. Various models have been proposed to explain the observed energy dependence of the secondary to primary abundance ratio (Cowsik and Wilson 1973;Peters and Westergaard 1977). As shown in figure 3, present measurements do not distinguish between the various propagation models. Better statistics and higher energy observations, as will be provided by CRN, are required in order to select the model which best represents the data.

In order to accomplish these objectives two basic challenges had to be addressed. The primary difficulty is the measurement of the energy of each nucleus in this difficult energy regime. To perform this task two gas Cerenkov counters were used for the energy range 50-200 Gev/amu, and an array of six transition radiation detectors were utilized for the range 400-5000 GeV/amu. This is the first time the technique of transition radiation detection has been used for the investigation of cosmic ray nuclei.

The energy spectra of the cosmic rays follow a steeply falling power law resulting in low fluxes at high energies. The second challenge is to accumulate a statistically significant sample of nuclei. This requires both a large collecting area and long exposure of the instrument to the cosmic ray flux. This challenge has been met by maximizing the size of the instrument within the constraints of the Spacelab-II mission and the size of the orbiter payload bay. The instrument is contained in an egg-shaped pressure shell which measures 2.7m in diameter and 3.7m in height. The geometry factors are 5 m^2-sr for the transition radiation detector and 1 m^2-sr for the Cerenkov counters.

As shown schematically in figure 4, the CRN detector consists of three main detector assemblies. (1)-Two plastic scintillators which serve as the main trigger by requiring that each nucleus pass through both counters. The scintillation signal provides an accurate measurement of the charge of each traversing particle. Additionally a time of flight measurement is made utilizing the signal generated in the plastic. (2)-Two identical gas Cerenkov counters located above and below the scintillation counters. (3)-The transition radiation detectors are placed between the scintillation counters and consist of six layers of radiators followed by multiwire proportional chambers (MWPC) used as both detectors for the transition radiation and as position sensitive devices to determine the trajectory of the particle through the instrument.

THE CRN INSTRUMENT

<u>(1)- Charge and time of flight measurement</u>: The charge of each traversing cosmic ray nucleus is measured by analyzing the energy loss ($dE/dx \propto Z^2$) in two plastic scintillators spaced 1.1m apart (for the elemental range lithium to nickel, $3<Z<28$). The material is composed of NE-110 plastic of dimension 2m x 2m, and machined to 1 cm thickness

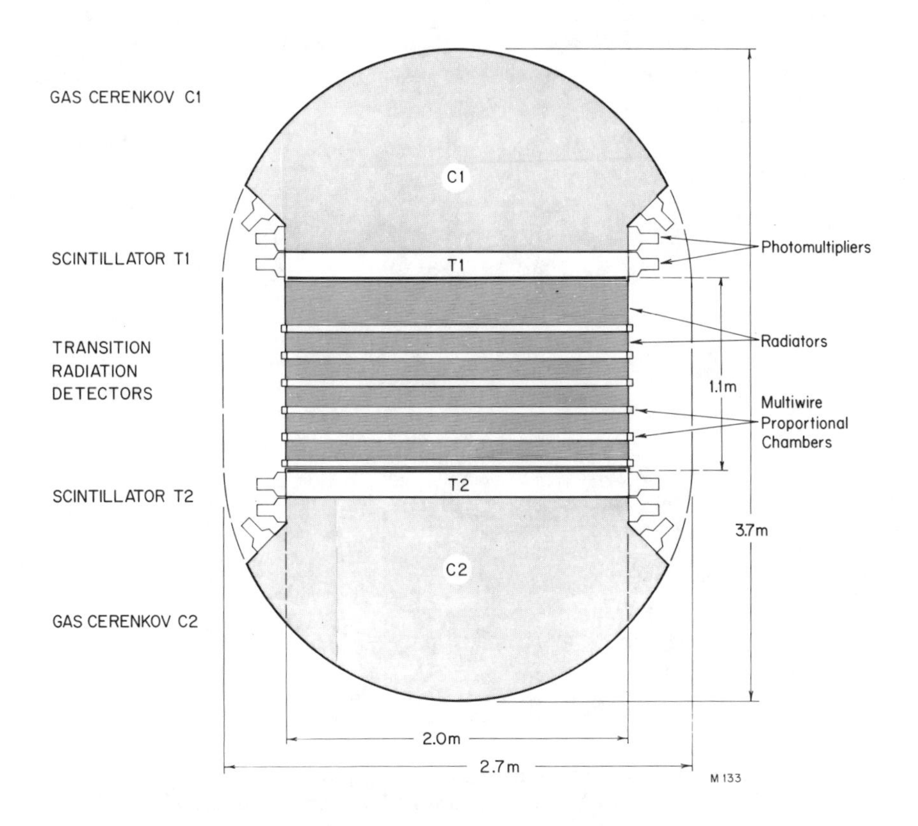

Figure 4. Schematic cross section of the University of Chicago CRN instrument.

with a tolerance of 50 μm. The plastic is viewed, in a light integration box painted with a highly reflective white paint, by 16 5" photomultiplier tubes. With the trajectory information provided by the MWPC's, correction can be made to the scintillation signals for the differing pathlengths in the plastic and for light collection non-uniformities. Additionally each counter is viewed by 16 2" fast photomultiplier tubes which are used to generate the time of flight of each traversing particle. Two measurements, a left and right pair, are made of the time of flight in each counter in order to cancel out the propagation effects within one counter. Examples of the time of flight distribution from flight data are shown in figure 5, demonstrating that adequate separation in time has been achieved to determine the direction of traversal for each particle.

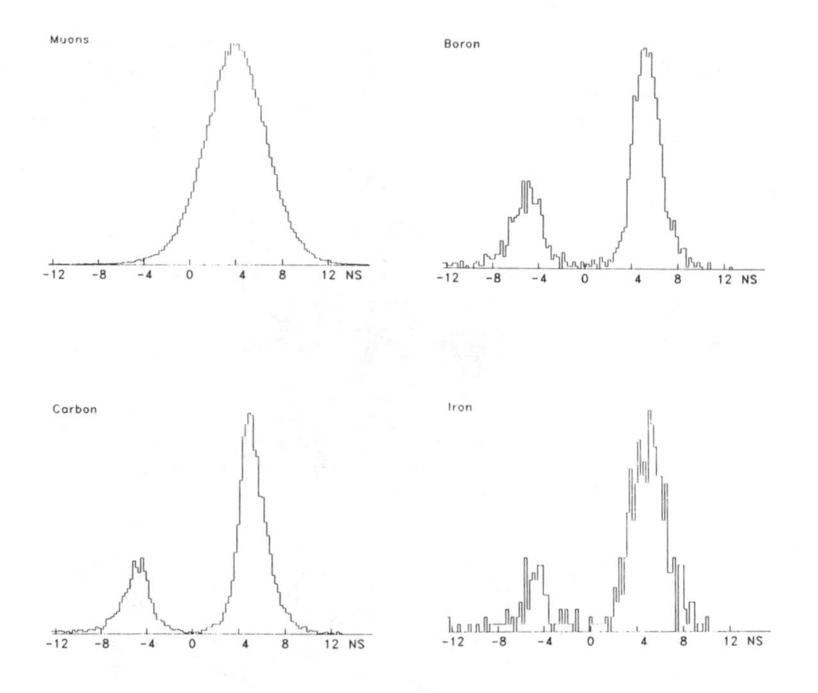

Figure 5. Time of flight distribution from flight data. The time scale reads in nanoseconds showing downward (positive) and upward (negative) traversing nuclei, and muons measured at Chicago in the laboratory.

<u>(2)- Gas Cerenkov Counter</u>: The Cerenkov counters are filled with a N_2-CO_2 gas mixture at atmospheric pressure to achieve a threshold $\gamma_T \sim 40$, allowing energy measurements in the range 40-200 GeV/amu. The CO_2 is utilized to suppress residual scintillation in the gas. The light integration hemisphere is viewed by 48 5" photomultiplier tubes with UV transparent windows. The light collection efficiency is $\sim$16% at 425 nm with the result that a fully saturated, β=1, particle yields approximately 2 photo-electrons for a singly charged particle. Light generated from the passage of the particle through the white paint accounts for approximately 15% of the fully saturated signal. The two independent Cerenkov measurements significantly help to reduce the background.

<u>(3)-Transition Radiation Detectors</u>: X-ray transition radiation results when a relativistic charged particle passes through an interface between two media with differing dielectric properties. The intensity of the radiation emitted from a single interface may be very weak requiring that many interfaces be present for generation of a useful

signal (Cherry 1978). For the CRN experiment, accelerator calibrations of many materials were performed (Swordy et al.1982, 1986, and ref. therein) in order to optimize the performance of the radiators and MWPC's for high energy nuclei. The radiator material consists of polyester and polyolefin batting with fibers of mean diameter 21 and 4.5 μm respectively. The transition radiation is generated in the interface between the fiber and the gas in the space between the fibers. The mixture of the two fibers aids in extending the dynamic energy range of the radiation over that which could be obtained by the use of a single radiator material. Most of the energy of the transition radiation is generated as soft x-ray's. The transition radiation is detected by the MWPC's located below each radiator module. Each MWPC has 200 wires, (2m long, 50 μm diameter stainless steel), in a chamber 2 cm thick with windows made of aluminized mylar (50μm thick). The MWPC's are filled with a mixture of Xe-He-CH_4 gas in the ratio 25:60:15 at atmospheric pressure. This gas mixture optimizes the detection of the transition radiation x-rays while minimizing the ionization signal of the primary charged particle. The resulting signal is a superposition of the charge collected from the ionization of the particle in the chamber and the transition radiation. The response of the chambers from accelerator calibrations is shown in figure 6 demonstrating the dependence of the transition radiation on the Lorentz factor of the primary particle.

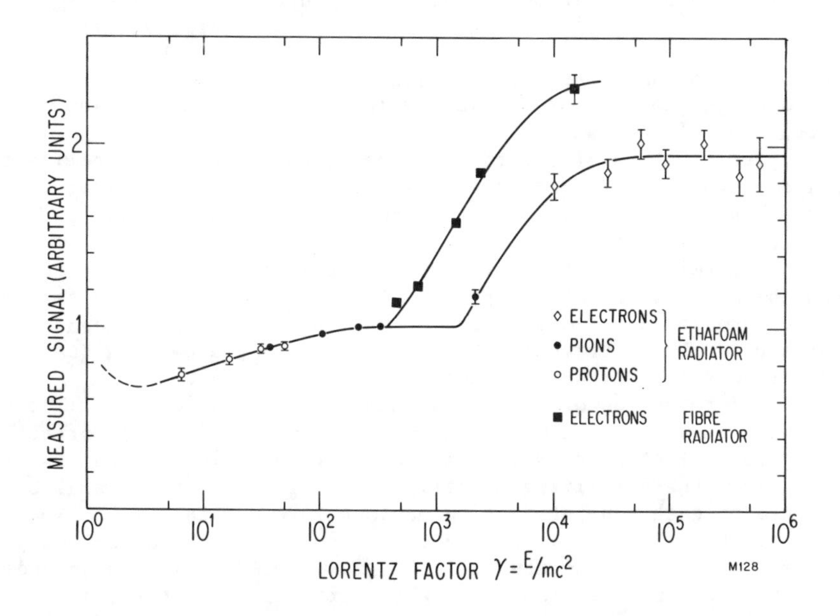

Figure 6. Accelerator calibration results of transition radiation detectors for one radiator-MWPC combination. The CRN experiment uses the fibre radiators. (Swordy et al. 1982)

In addition the signal due to the transition radiation, as well as the ionization signal, scales as Z^2 to higher charges. The transition radiation detectors in CRN should provide a determination of γ in the range $400< \gamma<5000$. The MWPC's also serve as position sensing devices in order to determine the trajectory of the incident particle. Each chamber is electronically organized into 40 positional groups of 5 wires each resulting in a 5 cm positional resolution. To insure a constant gain throughout the flight the pressure between the inside of the MWPC's and the outside had to be maintained to $\leq 10^{-5}$ atm accuracy, while maintaining continuous gas flow for optimum signal resolution.

Each event produced approximately 75 signals which were analyzed to 12 bit accuracy. The formatting of the data as well as instrument housekeeping, systems monitoring, and command operation were performed by a Rolm 1650 computer onboard. The instrument design, including detectors, electronics, and mechanical hardware, and the manufacture, construction and testing of the CRN instrument were all performed at the University of Chicago.

DATA ANALYSIS

The CRN instrument operated for a total of $\sim$180 hours from which the number of cosmic ray events detected was in excess of 40 million. Since less than 0.1% of these detected events are in the energy range of interest to the experiment the analysis of this large volume of data requires a considerable effort. Figure 7 is a flow chart of the overall analysis program. This procedure depicts a program that is being developed and may not be the last word on the subject. Below is a description of each step:

1. Verification of the overall operation of the CRN experiment including the detector systems, the electronics, and the various 'housekeeping' systems. Events which exhibit some peculiarity are rejected at this point.
2. Determination of a straight line trajectory for the particles from the MWPC's. From this process the signals in the detector elements can be normalized to compensate for varying angles of incidence and position in the instrument. Events which do not show a unique straight line trajectory will be rejected at this stage.
3. Determination of a response map of the plastic scintillation counters, as a function of position. This map is used to correct for the light collection nonuniformities versus position in the counter.
4. From the normalized scintillation signals a charge will be assigned to each non-interacting nucleus in the data.
5. Particles which appear to come from the earth, or from any material obstruction in the field of view of CRN will be rejected. This step makes use of the time of flight, particle trajectory, and information regarding the position of other experiments and the orbiter itself.
6. The signals from the Cerenkov counters will be deconvolved in order to obtain an energy spectrum from a pulse height distribution.

7. A response map and gain correction program will be implemented in order to correct for gain nonuniformities in the MWPC"S.
8. The pulse height distribution in the MWPC's will be deconvolved in order to extract the signal from those high energy nuclei producing transition radiation.

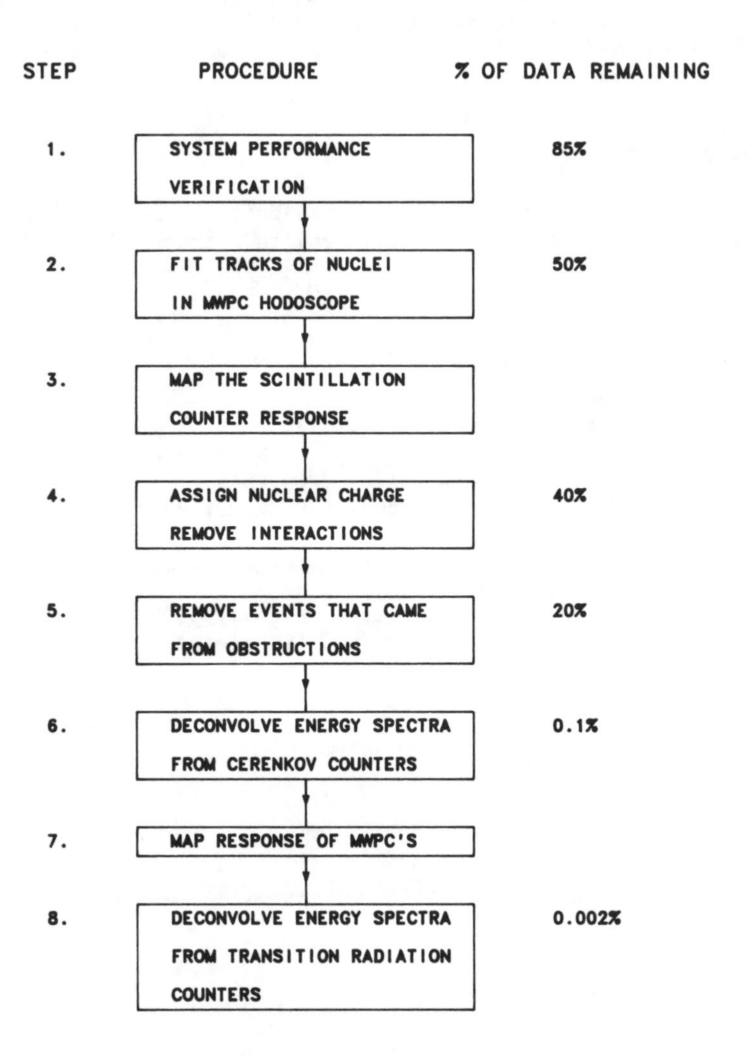

Figure 7. Data Analysis flowchart indicating remaining percentage of data left after each step.

CONCLUSION

At the present time the science team is deep into the development of the data analysis program. The crucial balance that must be maintained throughout the data analysis process is to give sufficient attention to each event, to reject those nuclei not satisfying the various consistency requirements without introducing any charge dependent bias, while proceeding rapidly in the data analysis. The flight has yielded a wealth of high quality data and preliminary results should be appearing soon.

ACKNOWLEDGEMENT

This work was supported, in part, by the United States National Aeronautics and Space Administration under Contracts NAS 8-32828 and NGT 14-020-804.

REFERENCES

Caldwell,J.H., 1977, Ap. J.,218,269
Cherry, M.L., 1978, Phys. Rev. D,17,9,2245
Cowsik, R. and Wilson, L.W.,1973, Proc. ICRC, Denver,1,500
Garcia-Munoz, M. et al., 1984, Ap. J.,280,:L13
Hillas A.M., 1984, Ann. Rev. Astron. Astrophy., 22:425-44
Peters,B. and Westergaard, N.J., 1977, Ap. and Space Sci.,48,21
Simpson, J.A., 1983, Ann. Rev. Nucl. Part. Sci., 33:323-81
Simon, M. et al., 1980, Ap. J.,239,712
Swordy, S.P. et al., 1982, Nucl. Inst. Meth.,193,591
Swordy, S.P. et al., Nucl. Inst. Meth.,(in press)

THE SCINTILLATING OPTICAL FIBER ISOTOPE EXPERIMENT

W. Robert Binns
Department of Physics, and the McDonnell Center for the Space Sciences,
Washington University
St. Louis, Missouri 63130, U.S.A.

ABSTRACT

This paper describes the Scintillating Optical Fiber Isotope Experiment (SOFIE) which is being developed by Washington University and the University of New Hampshire to study the abundances of cosmic ray isotopes in the iron charge region. This detector system is a Cherenkov-Range-dE/dx experiment and utilizes range and trajectory detectors made of scintillating optical fibers, a fused silica Cherenkov counter, and plastic scintillator dE/dx counters to determine the charge and mass of cosmic ray nuclei. A brief description of the balloon flight instrument presently being developed will be given followed by initial results of an engineering model calibration at the LBL Bevalac heavy ion accelerator. In addition a brief discussion of the potential of scintillating fiber trajectory detectors for use in experiments requiring precise trajectory determination such as those being planned for the NASA Particle Astrophysics Magnet Facility (Astromag) program is presented.

1. Introduction

The significance of galactic cosmic ray studies lies in the fact that the cosmic rays are a direct sample of matter from outside the solar system and contain a record of cosmic ray nucleosynthesis and history in the Milky Way galaxy. In addition they are a sample of the interstellar medium which we observe 4.6×10^9 years after the formation of the solar system, and therefore may be different in composition than solar system material owing to the injection of more recently synthesized nuclei, or nuclei with a different history, into the cosmic ray source region. Recent studies of galactic cosmic ray abundances have shown that the GCR elemental source abundances are very similar to solar system abundances, if a first ionization potential (FIP) fractionation enhancement for low FIP elements is applied. A study of the isotopic abundances gives a more powerful, though experimentally more difficult, means of investigating cosmic ray nucleosynthesis and history since isotopes of a particular element are not chemically fractionated, nor are they usually subject to atomic fractionation, as are the elements.

M. M. Shapiro and J. P. Wefel (eds.), Genesis and Propagation of Cosmic Rays, 391–403.

One of the most important cosmic-ray observations of recent years has been the measurement of the isotopic composition of the elements Ne, Mg, and Si, showing that the isotopic composition of the cosmic-ray source is distinctly different from that of the solar system (refs. 1-11). In the cosmic rays the two neutron-rich isotopes of Mg are overabundant relative to the common alpha-particle isotope when compared to the solar system by a factor of about 1.5, and the ratio $^{22}Ne/^{20}Ne$ is higher in the cosmic rays by a factor of three to four. The two neutron-rich isotopes of Si were believed to be overabundant by a factor of about 1.5 in initial investigations, but a recent experiment by Webber *et al.* (ref. 11) is consistent with no enhancement. Meyer discusses and compares these measurements in detail (ref. 12).

Several models have been proposed to explain these results. A model which makes specific predictions for the isotopic composition of other cosmic-ray elements is the supermetalicity model of Woosley and Weaver (ref. 13). This model can explain a neutron-rich excess in Mg and Si, and about half of the excess in Ne as the result of enrichment of the interstellar medium in "metals" (elements with $Z \geq 6$) during the 4.6 billion years since the solar system formed. Nucleosynthesis in massive stars with enhanced metalicity produces an excess of these neutron-rich isotopes. This model predicts enhancements of the ratios $^{34}S/^{32}S$, $^{38}Ar/^{36}Ar$, and $^{54}Fe/^{56}Fe$ by about the same factor of 1.5 as is seen at the heavy Mg and possibly the Si isotopes, but no enhancement for $^{58}Fe/^{56}Fe$. (^{54}Fe is neutron-rich compared with the ^{56}Ni from which ^{56}Fe is the decay product.) A second model which makes specific predictions is the "Wolf-Rayet" model (e.g. refs. 14-17) which invokes enrichment of the cosmic rays by material blown off from Wolf-Rayet stars (helium burning, massive stars). Prantzos *et al.* (ref. 17) have done detailed calculations and conclude that if 1 cosmic ray particle out of 35 originates from a Wolf-Rayet star, the measured ^{22}Ne overabundance of about 3.5 can be accounted for, and predictions of the enrichment of other isotopes relative to solar system material can then be obtained. Specifically this model predicts an enrichment of the neutron rich magnesium isotopes by a factor of 1.5, no silicon, sulphur, calcium or ^{54}Fe enrichment and an ^{58}Fe enrichment of 1.5 (these enrichment factors are relative to a ^{22}Ne enhancement of 3.5, not 4.0 as in Prantzos *et al.*). The predictions of these two models are shown in Fig. 1 along with measured abundances. We see from the figure that a measurement of the iron isotopes, combined with measurements of the lower charge isotopes, can discriminate between these two models.

2. Balloon Flight Experiment

The initial objective of the SOFIE experiment is to measure isotopic abundances in the iron region of the charge spectrum using a Cherenkov-dE/dx-Range balloon borne instrument possessing sufficient mass resolution and geometrical factor to resolve individual isotopes. Our initial emphasis will be to measure the $^{54}Fe/^{56}Fe$

ratio to the precision required to discriminate between the two models discussed above.

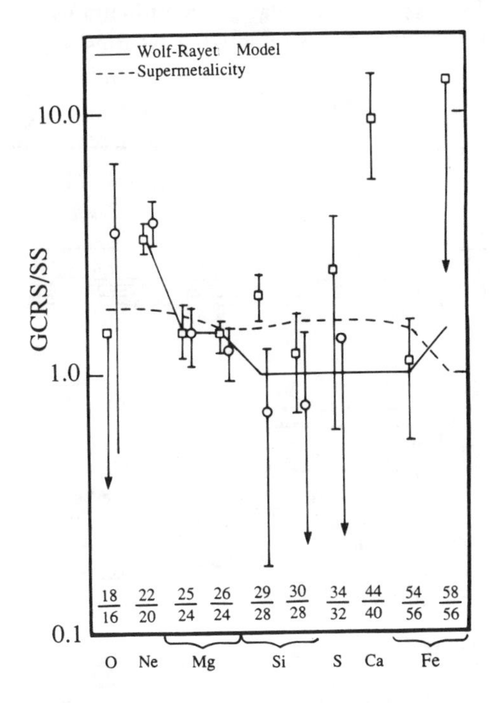

Figure 1. The ratio of the Galactic Cosmic Ray Source abundances to Solar System abundances is plotted for various isotopic ratios. The circle data points are taken from Webber *et al.* (ref. 11) and the square data points from Wiedenbeck (ref. 10). For comparison the predictions of the Wolf-Rayet model (ref. 17) are plotted with a solid line and the Supermetalicity model (ref. 13) predictions are plotted with a dashed line.

A cross-sectional view of the SOFIE instrument is shown in Fig. 2. Because of the nature of the fiber-optic scintillator range detector, the dimensions in this x,z plane are very different from those in a y,z view. The system is 110 cm long in the direction perpendicular to the plane of the figure, so that the range detector has an entry area of 12 cm × 110 cm and depth 4.6 cm. Nuclei entering the instrument acceptance cone are detected by the upper plastic scintillator dE/dx counter, a fused silica Cherenkov counter, a lower plastic scintillator dE/dx counter, a scintillating fiber range detector, and three scintillating fiber hodoscope planes as shown in the figure. The particles of primary interest are those which penetrate through the detector stack and stop in the range detector without interacting. Nuclei which have sufficient energy to penetrate through these counters into the penetration counter located beneath the range counter, or which interact giving fragments which are detected by the penetration counter, will be identified and eliminated from isotope abundance determinations, though they can be useful for detector mapping and detector interaction studies.

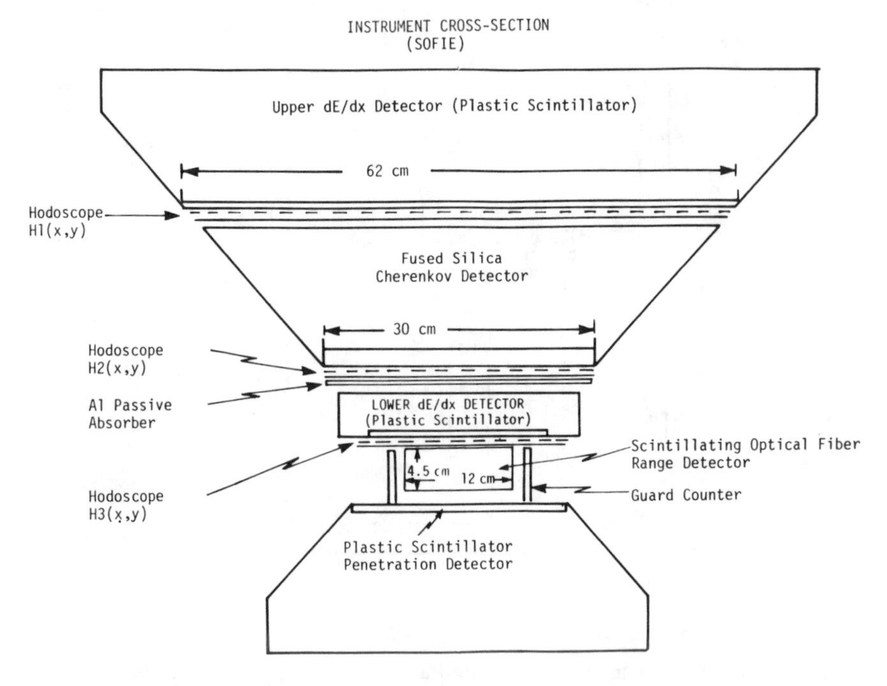

Figure 2. Cross-sectional view of the SOFIE instrument. The length of the instrument in the coordinate not shown in this view is about 1.1 m.

The Cherenkov-Range technique has been used previously by Tarle *et al.* (ref. 18) and Young *et al.* (ref. 19). Our instrument is different in that Tarle and Young both used passive devices for the range detector, etched plastic track detectors or nuclear emulsions, while we use an electronic detector composed of scintillating optical fibers. While our fiber-optic scintillator does not have as fine a spatial resolution as do the plastics or emulsions, the resolution should be adequate for resolving individual isotopes, and it is well matched to the natural range broadening resulting from straggling and multiple coulomb scattering. The advantage of this new range detector is its adapability to spaceflight exposure of very long duration. There is no need for recovery or post-flight correlation of data recorded electronically and tracks accumulated throughout the exposure in the passive detector.

In the absence of nuclear interactions a measurement of Cherenkov and particle range are sufficient to determine the mass of nuclei. Figure 3 is a plot of Cherenkov versus range over the energy interval of about 350-500 MeV/amu at the top of the detector for several iron and manganese isotopes. It shows the clear separation of the iron isotopes with mass from 54 to 58 from the manganese isotopes with mass 53 and 55. The inset shows the resolution in Cherenkov and Range which we expect to obtain, giving an overall mass resolution of better than 0.3 amu. However, as with all Cherenkov-Range or Cherenkov-Total Energy experiments, for high Z nuclei such as iron a large fraction of the primary events will interact before reaching the end of their range. In our experiment we estimate that approximately 65% of the iron nuclei will interact before stopping, thus resulting in a large background of nuclei whose Cherenkov and range are altered.

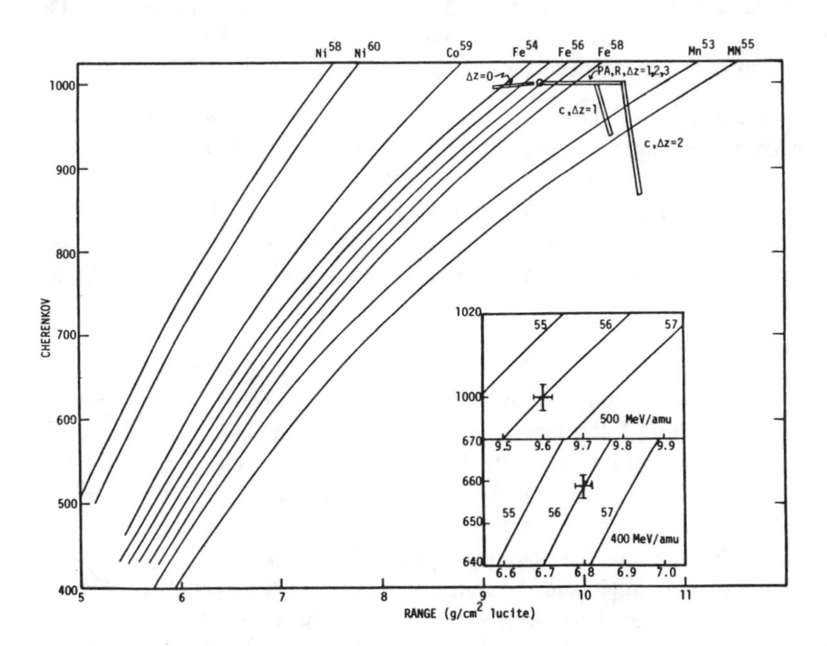

Figure 3. Cherenkov signal (arbitrary units) vs. range for various isotopes. The bars on the main figure indicate events that belong in the circle but interacted in the detector. The labels on the bars PA, R, and C correspond to interactions occurring in the passive absorber or lower dE/dx counter, the range detector, and the Cherenkov counter. ΔZ indicates the charge change in the interaction. Error bars on the insets show rms measurement uncertainties for ^{56}Fe with 500 or 400 MeV/amu at the center of the Cherenkov radiator.

The bars in Fig. 3 show the path of interacted 500 MeV/amu ^{56}Fe nuclei. The bar extending to shorter ranges is for neutron stripping interactions. These interactions can masquerade as ^{55}Fe or ^{54}Fe. Charge changing interactions in general result in extended ranges, and, if the interaction occurs in or above the Cherenkov counter, in reduced Cherenkov emission. These interactions could be misidentified as iron isotopes heavier than ^{56}Fe or as manganese as shown by the bars extending to the right and down. To identify interacted events we have three consistency requirements that will be applied. The first two requirements are that the signals from the upper and lower dE/dx counters must be consistent with the charge estimate obtained from the Cherenkov versus Range plot. This should remove charge changing interactions which occur above the range counter. The third requirement is that the penetration counter which is capable of detecting singly charged particles should not have a signal coincident with the event. This requirement should detect most interactions with the exception of neutron stripping interactions which must be removed by calculated corrections.

The geometry factor of this instrument is about 0.3 m^2sr. When interactions are excluded, we calculate that in a 30 hour balloon flight we will detect about 600 ^{56}Fe nuclei. If the ^{54}Fe to ^{56}Fe ratio is the same as the solar system this will result in a statistical uncertainty of about 17% in this ratio which is sufficient to discriminate between the model predictions described above. In addition we will detect about 30 Ni nuclei, giving a measure of the ^{60}Ni to ^{58}Ni ratio to a statistical accuracy of about 50%.

3. Bevalac Experiment.

An experiment with a test model of the SOFIE instrument has been performed at the Lawrence Berkeley Laboratory Bevalac heavy ion accelerator. This experiment used a Pilot-425 Cherenkov counter with thickness 2.5 cm, a fiber bundle range detector with dimensions 2.5 cm $\times$ 2.5 cm $\times$ 30 cm, and a plastic scintillator penetration counter to do preliminary studies on the performance of such a detector. In addition, a steel plate with thickness 0.79 cm was placed between the Cherenkov counter and the range detector. This was used so that iron nuclei stopping in the range detector would be above the Cherenkov threshold in the Cherenkov counter. Two plastic scintillator paddles provided the coincidence signal for event analysis. The detector is shown in cross-section in Fig. 4. It was exposed to iron nuclei with energies near 500 MeV/amu at the top of the detector.

The range detector shown schematically in Fig. 5 consists of a bundle of scintillating optical fibers which was proximity focused onto the face of a fiber optic reducer and then coupled to an image intensified video camera system on one end and directly coupled to a photomultiplier tube at the other end (not shown in figure). Charged particles traversing the fibers produce scintillation light, about 5% of which is light piped down the fiber to the intensified video system (ref. 20). The intensified video system consists of a G. E. TN-2505 camera which is fiber optically coupled to an ITT-4144, dual microchannel plate, 18mm image intensifier tube. The scintillating fiber bundle was developed jointly by Washington University and Fibre Optics Development Systems, Inc. (ref. 21). The Charge Injection Device (CID) sensor in the camera is a rectangular array of 244×388 pixels, with each pixel having dimensions 23 microns $\times$ 27 microns. The scintillating fiber bundle which is coupled onto this array consists of about 7×10^4 fibers with length 30 cm and a 100 micron square cross-section, joined together into a solid rod with cross-section 2.7×2.7 cm^2. The cladding wall thickness on each fiber was about 7 microns. The attenuation length for the scintillation light for these fibers was poor, being about 11 cm. However more recently we have made fibers with attenuation lengths of about 1 meter. A more detailed description of this detector system is given in reference 22.

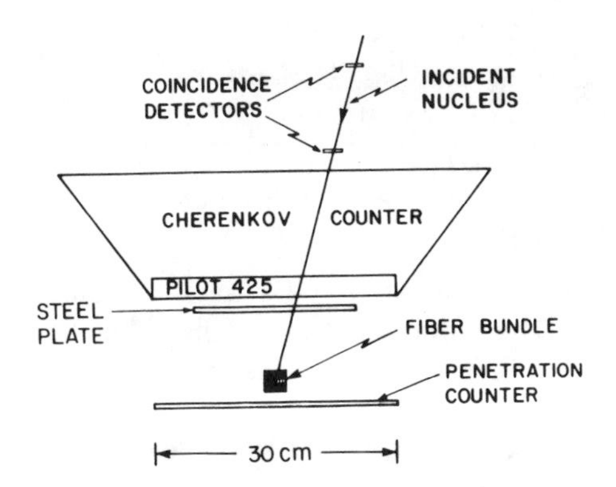

Figure 4. Test model counter exposed to iron nuclei at the LBL Bevalac.

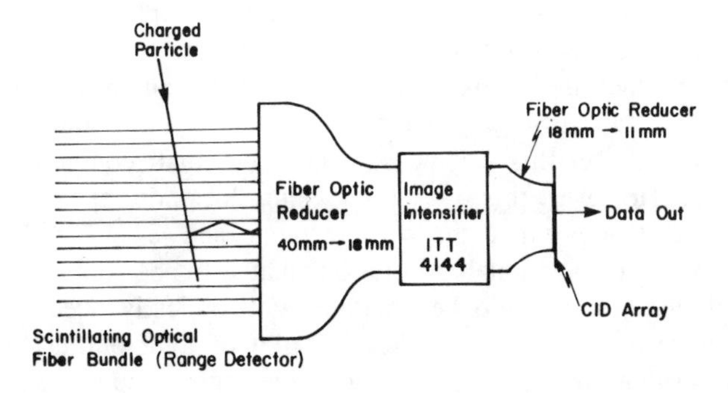

Figure 5. Scintillating optical fiber range detector schematic.

4. Range Detector Results.

A plot of the pixels above threshold for an ^{56}Fe nucleus penetrating into our fiber bundle is shown in Fig. 6. The particle enters the fiber bundle on the right and stops at the left. Each symbol (not resolvable in this figure) represents an 8 bit pulse height for that pixel. One pixel corresponds to roughly one fiber size. This particle had an energy of 529 MeV/amu at the top of the SOFIE instrument and about 200 MeV/amu as it entered the fiber bundle.

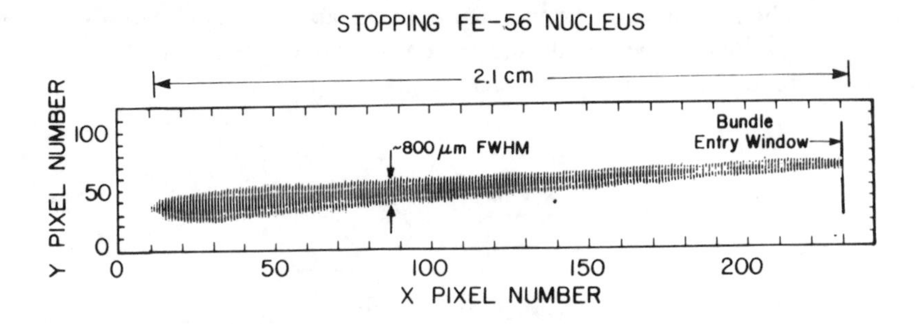

Figure 6. Computer output of ^{56}Fe track stopping in the fiber range detector. The particle enters the fiber bundle on the right and stops at the left. This nucleus penetrated through about 2.1 cm of plastic fibers in addition to the counters located above the fiber detector (see Fig. 4) before stopping.

The breadth of the track corresponds to about 8 fibers (800 microns full width at half maximum) near the end of the particle range. The energy deposition in plastic scintillator should, however, extend over a range of less than 1 fiber width. This track broadening is predominately the result of optical coupling between fibers due to incomplete conversion of primary dye ultraviolet photons emitted within the fiber in which they were produced. Imperfections in the fibers may also contribute to cross talk. Some broadening occurs in the image transfer and intensification process but this can be shown to be small compared to the track breadth observed. However, this track broadening does not appear to significantly degrade our ability to determine the particle range and, as we will show below, actually improves the positional resolution that can be obtained in a fiber hodoscope over that which could be obtained with optically decoupled fibers for heavily ionizing particles. Figure 7 shows a plot of light output from the fiber bundle versus residual particle range where the light output was obtained by summing the light intensity along pixel columns which are nearly perpendicular to the tracks. A running sum of three pixel columns was taken to smooth the data. We see that as the particle slows the light emission increases, in qualitative agreement with the dE/dx Bragg peak, and then drops to zero at the end of the range. However in detail it is evident that there are fluctuations in light output along the track, and these fluctuations occur roughly on the scale of a single multifiber (about 1.5 mm). These fluctuations are believed to be the result of nonuniformities at the multifiber boundaries.

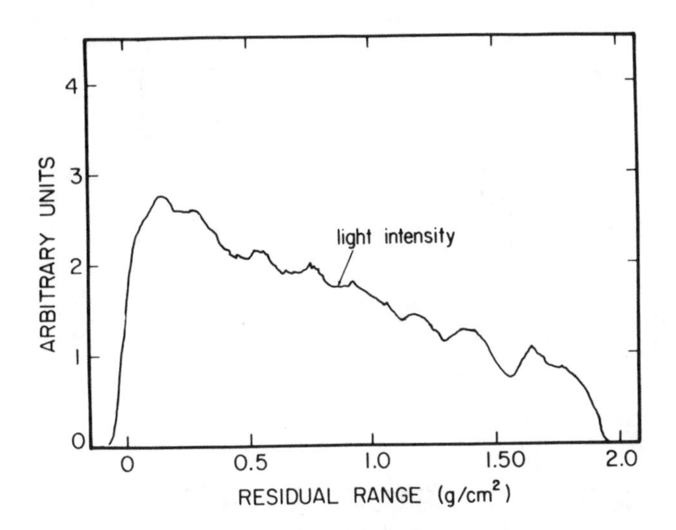

Figure 7. The light intensity is plotted as a function of the residual range of the track shown in Fig. 6.

To obtain the endpoint for stopping particles we identify the half maximum of the light intensity as the end of the range. Figure 8 shows the distributions in range that we obtained for ^{56}Fe nuclei with beam energies 473, 490, and 529 MeV/amu at the top of the detector and having an incidence angle of 10 degrees with respect to the normal to the fiber bundle entry side. These nuclei were selected to have a low penetration counter signal and a Cherenkov signal within about 2 standard deviations of the Cherenkov peak to eliminate interactions; and they have had a first order mapping correction applied to account for variations in the flatness of the entry window into the fiber bundle. From Fig. 8 we obtain a sigma (rms) in the range distributions of 215, 280, and 290 microns for the beam energies 473, 490, and 529 MeV/amu respectively. Calculations show that the combination of multiple coulomb scattering, range straggling, and the finite fiber size will result in a range uncertainty of about 150 microns. The additional broadening which we observe is probably the result of bundle imperfections. We have calculated that for a range measurement uncertainty of 300 microns and using a fused silica Cherenkov counter similar to that developed by Webber *et al.*, (ref. 23) which gives increased photon statistics, hodoscopes which can provide rms resolution of about 250 microns or better, and an Aluminum passive absorber to minimize multiple coulomb scattering, a mass resolution of better than 0.25 amu can be obtained for incident angles less than 30 degrees, and better than 0.30 amu for most particles with incident angle less than 55 degrees. We hope to improve the quality of our fiber bundles which should result in improved range measurement precision in future experiments.

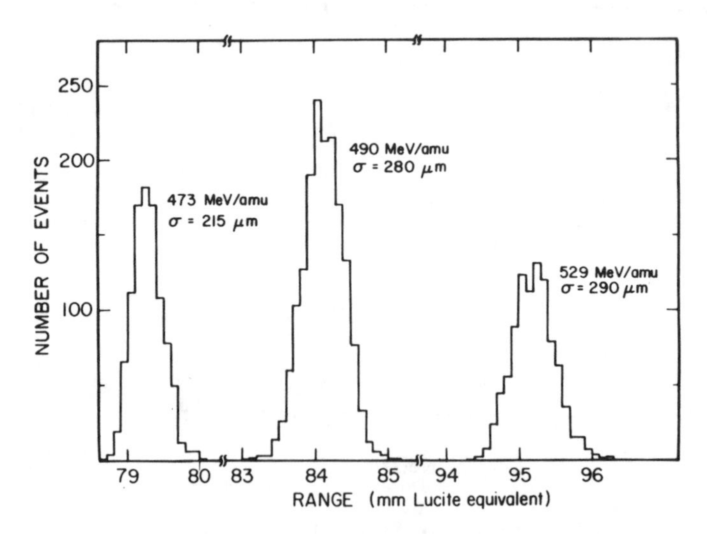

Figure 8. Range distributions for ^{56}Fe nuclei with energies 473, 490, and 529 MeV/amu at the top of the detector, stopping in the fiber bundle.

5. Hodoscope Results.

To study the resolution in position that we could expect to obtain with a hodoscope made of scintillating optical fibers we have calculated, for a single track penetrating through the fiber bundle range detector, the weighted mean position in fiber layers perpendicular to the track direction. These layers were adequately separated to optically decouple them from one another. These points were then fit with a straight line as shown in Fig. 9. The rms deviation of these "individual layer means" from the straight line should then give an indication of the trajectory measurement capability of this technique. Our measurements show a deviation from the best fit line of about 35 microns rms over the entire track length and 10 microns over a limited track segment. It seems clear that the larger deviation over the entire track length (about 200 fiber layers) is the result of systematic nonuniformities in the fiber bundle and that something approaching 10 microns is the true measuring precision which could be obtained using this technique. This measurement precision is actually better than could be obtained if the fibers were optically decoupled from each other. (If light were detected only from the fibers through which the primary nucleus penetrates, then 100 micron fibers would give an rms standard deviation of 30 microns.) This better precision is the result of the additional information contained in the fibers adjacent to the fibers actually traversed by the particle, thus improving our ability to estimate the light intensity distribution mean for the track. The light observed in the adjacent fibers is the result of incomplete conversion of the ultraviolet photons emitted by the primary dye in the plastic scintillator (peak emission at 380 nm) in the fiber traversed by the particle. We have measured the mean attenuation length of these u.v. photons in our more recently polymerized scintillator using a Varian-Carey 219 spectrophotometer and found it to be 430 microns, roughly consistent with the track breadth we observe.

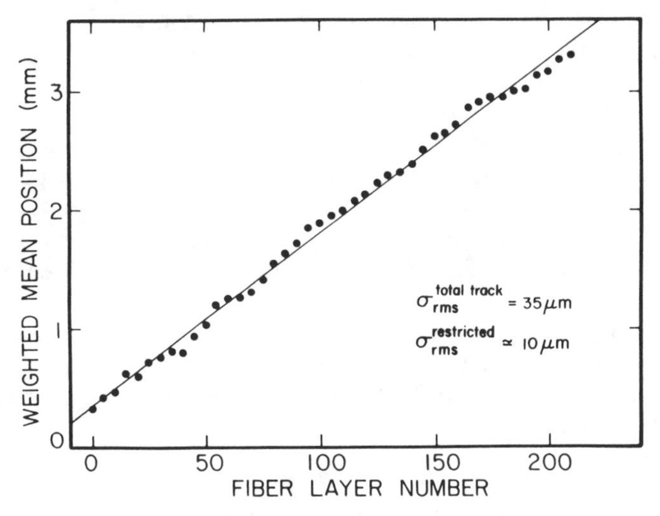

Figure 9: Plot of lateral weighted mean position of the track shown in Fig. 6 versus depth into the fiber bundle.

6. Single Fiber Layer Hodoscopes.

For some Astrophysics experiments such as the planned NASA Particle Astrophysics Magnet Facility (Astromag) there is a need for hodoscopes which can give excellent positional resolution (≤ 50 microns) for large areas (several square meters) with very low detector areal density to minimize multiple coulomb scattering. Scintillating optical fiber trajectory detectors are one of the possibilities for providing such measurements. The low areal density requirement for the high Z Astromag experiments is that a single coordinate measurement should not introduce more than about 150-200 microns (0.0005 radiation lengths) of plastic equivalent into the particle path. This thickness is well matched to a single layer of plastic fibers. However the mean attenuation length of 430 microns for the u.v. photons emitted by the primary dye results in a rather low efficiency of u.v. photon conversion before they exit the fiber layer. For 150 micron fibers we have calculated this conversion efficiency to be about 40%. It would be desirable to increase this conversion efficiency to improve photon statistics, and thus improve the positional resolution which could be obtained. Initial efforts have been made to do this by increasing the concentration of the secondary dye in the scintillator material. For a concentration 3 times that in our standard scintillator we have measured an effective u.v. attenuation length of about 180 microns which gives a u.v. conversion efficiency of about 60%. These fibers exhibit an attenuation length for the visible light (peaked at 425 nm) of about 75 cm. Experiments to improve this efficiency without sacrificing fiber transmission for visible photons are in progress. In addition we have developed fiber ribbons for test to determine experimentally the resolution in position which can be obtained with single fiber layer detectors.

7. Conclusions.

We have developed a new type of electronic detector which is capable of measuring the range of iron nuclei to a precision of 200-300 microns and their transverse position to better than 35 microns. This measurement precision in range, when combined with the Cherenkov and dE/dx measurements described above, is sufficient to resolve individual isotopes in the iron region of the cosmic ray spectrum with a precision of about 0.25 amu for nuclei stopping in the range detector. The excellent resolution in transverse position that was obtained indicates that similar detectors should be very useful as a high precision hodoscope for heavily ionizing particles.

Acknowledgements.

These studies have been done in collaboration with J. J. Connell, M. H. Israel and J. Klarmann at Washington University. Many of the techniques which have made scintillating fibers into a practical detector have been developed by J. W. Epstein at Washington University. This work was supported in part by NASA grants NGR-26-008-001, NAGW-122, and NASA contract NAS5-28657 through Fibre Optics Development Systems, Inc., and in part by the McDonnell Center for the Space Sciences.

References

1. Garcia-Munoz, M., Simpson, J. A., Wefel, J. P., 1979, Ap. J., 232, L95.
2. Mewaldt, R. A., Spalding, J. D., Stone, E. C., Vogt, R. E., 1980, Ap. J., 235, L95.
3. Mewaldt, R. A., Spalding, R. A., Stone, E. C., Vogt, R. E., Proc. 17th Intl. Cos. Ray Conf., 1981, 2, 68.

4. Wiedenbeck, M. E., and Greiner, D. E., 1981, Phys. Rev. Lett., 46, 682.
5. Wiedenbeck, M. E., and Greiner, D. E., 1981, Ap. J. 247, L119.
6. Wiedenbeck, M. E., and Greiner, D. E., 1981, Proc. 17th Intl. Cos. Ray Conf., 2, 76.
7. Freier, P. S., Young, J. S., Waddington, C. J., 1980, Ap. J., 240, L53.
8. Young, J. S., Freier, P. S., Waddington, C. J., Brewster, N. R., Fickle, R. K., 1981, Astrophys. J., 246, 1014.
9. Webber, W. R., 1982, Ap. J., 252, 386.
10. Wiedenbeck, M. E., 1984, Adv. in Space Res., 4, No. 2-3, 15.
11. Webber, W. R., Kish, J. C., and Schrier, D. A., 1985, Proc. 19th Intl. Cos. Ray Conf., 2, 88.
12. Meyer, J. P., 1985, Proc. 19th Intl. Cos. Ray Conf., 9, 141.
13. Woosley, S. E., and Weaver, T. A., 1981, Ap. J., 243, 651.
14. Casse, M. and J. A. Paul, 1982, Ap. J., 258, 130.
15. Maeder, A, 1983, Astron. Astrophys. 120, 130.
16. Blake, J. B., and D. S. P. Dearborn, 1984, Adv. Space Res. 4, 89.
17. Prantzos, N., M. Arnould, J. P. Arcoragi, and M. Casse, 1985, Proc. 19th Intl. Cos. Ray Conf., 3, 167.
18. Tarle, G., S. P. Ahlen, B. G. Cartwright, 1979, Ap. J., 230, 607.
19. Young, J. S., P. S. Freier, C. J. Waddington, N. R. Brewster, R. K. Fickle, 1981, Ap. J., 246, 1014.
20. Binns, W. R., M. H. Israel, and J. Klarmann, 1983, Nucl. Inst. Meth. 216, 475.
21. "Fibre Optics Development Systems, Inc." 427 Olive Street, Santa Barbara, California, 93101, USA.
22. Binns, W. R., J. J. Connell, P. F. Dowkontt, J. W. Epstein, M. H. Israel, and J. Klarmann, 1986, Nucl. Inst. Meth. A251, 402.
23. Webber, W. R., and J. C. Kish, 1983, Proc. 18th Intl. Cosmic Ray Conf. 8, 40.

Cygnus Experiment at Los Alamos

B. L. Dingus, J. A. Goodman, S. K. Gupta, R. L. Talaga,
C. Y. Chang, and G. B. Yodh
University of Maryland, College Park, MD 20742

R. D. Bolton, R. L. Burman, K. B. Butterfield, R. Cady, R. D. Carlini
W. Johnson, D. E. Nagle, V. D. Sandberg, and R. Williams

Los Alamos National Laboratory, Los Alamos, NM 87545
R. W. Ellsworth
George Mason University, Fairfax, VA 22030

J. Linsley
University of New Mexico, Albuquerque, NM 87131

H. H. Chen, and R. C. Allen
University of California, Irvine, CA 92717

Abstract

The Cygnus experiment at Los Alamos National Laboratory has been designed to study, with high angular accuracy, point sources of gamma rays of energy above 10^{14} eV. The experimental detector consists of an air shower array to observe gamma-ray showers and a shielded, large-area track detector to study the muon content of the showers. In this paper we present preliminary data from the array and describe its performance.

Introduction

The study of ultra-high-energy cosmic ray gamma rays provides an exciting new window to explore the origin of high-energy cosmic rays (A. M. Hillas 1984). The interest in this field has grown steadily since the first report of observations of point sources (Stepanian et. al., 1972, Ramanamurthy and Weekes 1982, A. A. Watson 1985). The reports of the Kiel group (Samorsky and Stamm 1983) showed strong evidence for a signal from Cygnus X-3, that showed phase correlation with the orbital period as seen in X-rays. This result also suggested that the observed showers had a muon content much higher than would be expected from gamma-ray-induced showers. Since then, other experiments using both air showers techniques and underground muon detectors have reported results of varying statistical significance. If showers from Cygnus X-3 are muon rich relative to expectation they may indicate the existence of new particle physics phenomena either at the source or in the interactions of high-energy gamma rays with atmospheric nuclei (Barnhill et. al., 1985). None of these experiments (Kiel excepted) has shown a significant signal without the use of phase analysis.

The Cygnus experiment at Los Alamos National Laboratory was designed to search for the presence of point sources of ultra-high-energy gamma rays and to study the muon content of their air showers. This detector was designed to have an angular accuracy of better than 1°, in order to improve the signal-to-background ratio of this experiment. Los Alamos was chosen

M. M. Shapiro and J. P. Wefel (eds.), Genesis and Propagation of Cosmic Rays, 405–409.

as the location of this experiment for several reasons. First, there exists at LAMPF a working, fine-grained track detector (E225), which can be used to detect muons in a clear and unambiguous manner. Second, the detector is located at an altitude of 7000′, which allows this experiment to observe showers produced from lower energy gamma rays than previous air shower experiments done at lower altitudes. Third, the facilities of Los Alamos National Laboratory were available to facilitate the construction and operation of this experiment.

EAS Detector Design

The EAS detector consists of 64 counters placed in an array of radius 60 meters with a typical separation of 14 meters. Each counter contains a scintillator, approximately 1-m^2 by 8-cm thick, with a 2-inch photomultiplier tube positioned 70 cm above the scintillator. Single minimum-ionizing particles selected by small scintillator paddles result in a timing resolution of standard deviation 1.6 ns and produce about 20 photoelectrons in the photomultiplier tube.

EAS Trigger

Every counter is used in making the trigger decision as well as in giving pulse-height and timing information. The basic trigger requires that a given number of counters must fire their individual discriminators within a 300 ns interval, the time for an EAS with a zenith angle of 45° to the array. The discriminators responsible for the trigger are also used to determine the timing, so the threshold is set very low, about 1/10 of a minimum ionizing particle, in order to fire the discriminator on the earliest photoelectron.

A software cut is implemented to eliminate non-analyzable showers before they are récorded on magnetic tape, which reduces our data-taking rate by about a factor of five. Figure 1 shows the online display of a typical event that passes the software criteria.

Muon Detector

Muon information is also recorded for every trigger. The muon detector was designed for studying the elastic scattering of accelerator-produced neutrinos with electrons (E225) and is currently being used for that purpose (Allen, 1985). A multiplexing circuit allows both the neutrino and the air shower experiment to use the detector simultaneously. The detector is shielded above by 1700 g/cm^2 of steel and concrete. Two components of the detector - the multiwire proportional chambers (MWPCs) and the flash chamber calorimeter - are used to determine the muon content and direction in showers. The MWPCs surround the detector with four layers on all six walls except the floor, which has only one layer. Each MWPC is typically 520-cm long by 20-cm wide by 5-cm thick and the horizontal area is 36 m^2. The muon number can be determined exactly for small numbers and systematically for higher densities.

The 208,000 flash chambers cover a volume of 350 by 305 by 348 cm^2 and have sufficient resolution to determine the muon direction and number, as can be seen from the example of Fig. 2. Simulations of a fitting algorithm show that 95% of the tracks can be reconstructed to within 0.5°. However, the flash chambers cannot be triggered for all EASs, which come at a rate of 1.2 Hz.

Smart Trigger

The rate at which it is reasonable to fire the flash chambers in the E225 calorimeter is 0.02 Hz. It was necessary therefore to reduce the rate of triggers to the E225 detector by a factor greater than 50. A computer-controlled hardware trigger was devised to allow only events that come from a specified direction to trigger E225. This smart trigger was needed because the flash chambers must be fired within a microsecond of particles traversing it. This does not allow for timing information to be digitized or processed. The direction of a suspected source moves 1° in four minutes; hence the trigger must be continually updated. This is accomplished by the use of ECLine Camac programmable logic delays. The gates for this trigger are set to a width of 8 ns and the relative delays of each counter are set to zero for showers coming from the source direction. A multiplicity coincidence level is set and only showers coming near the desired direction are accepted. In Fig. 3, we show a plot of the sky (right ascension and declination). The crosses are showers that pass the smart trigger criteria over a 24-hour period in which three sources were being watched, Cygnus X-3, the Crab, and Herc X-1. This trigger selects events in a cone of half angle 9° and gives a trigger every 2 minutes when a source is overhead.

Monte Carlo Calculations

Monte Carlo calculations were performed to study the design and triggering conditions of the array. These calculations included proton- and gamma-ray-induced showers. The primary energies were selected from a spectrum and the cores of these showers were thrown over the area of the detector and the surrounding areas. The hadron component of these showers was simulated fully. Electromagnetic showers were followed down to 500 GeV where Approximation B was used for longitudinal development. The radial distribution of each sub-shower was computed. Muon densities were computed for annular rings for each shower. Differing trigger conditions where imposed and rates for each were computed.

The results of the simulations show that with a trigger requirement of 10 counters, each having more than 2 equivalent minimum ionizing particles, the effective threshold for proton showers is 10^{14}eV, while gamma-ray-induced showers have a threshold of 2 x 10^{14}eV. The muon simulation showed that the E225 detector contained at least one muon for 80% of proton-induced triggers. This number is consistent with the observed number of 75% of showers having one or more muons in data.

The Monte Carlo calculation was also used to study reconstruction algorithms. These simulations showed that with a timing resolution of 2 ns (for large signals) it was possible to obtain a resolution of better than $0.65^{\circ 2}$ for showers that passed our threshold.

Reconstruction and Resolution of Events

The event direction is computed by fitting the shower arrival time distribution. The shower front has a curvature, observed to be (~10ns/60m); this is included in the fits. Counters are weighted in this fit so that counters with larger signals are given more significance. An estimate of the resolution of these counters can be obtained by studying the distribution of the quantity $\chi^2\sigma^2/\nu$, where ν is the number of degrees of freedom, $\chi^2\sigma^2/\nu$ should have the average value of σ^2. For our showers this yields a value of $\sigma = 2$ns for greater than three particles signal.

A test of the random reconstruction error in the array can be obtained from data by the following procedure: counters are divided into two groups - odd and even numbered counters. Each group of counters is used independently to reconstruct the arrival direction. The space angle between these directions is found to have a median value of $< 1^{\circ}$. This predicts a resolution for the combined array of $< 0.75^{\circ 2}$. Work is still being done to improve this resolution.

Three independent tests of pointing accuracy are being undertaken. First, the arrival direction of muons detected in the E225 flash chambers is being compared to the air shower data. Preliminary results indicate reasonable agreement between the two directions, consistent with the expected multiple scattering angle of the muons in the shielding.

Second, a small Cherenkov array has been deployed at the experiment site. This array will be used to determine shower direction and energy in a way that is systematically different from the air-shower method. Tests have been made with these counters and some data have been taken.

Third, we plan to use our existing data to study the shadow of the moon using ordinary hadron data. If our accuracy is greater than ($1^{\circ 2}$), then we should see a substantial reduction of data in bins which contain the moon position. This technique may also work with the sun.

Status

The experiment has been in operation since early March 1986 with more than 40 scintillation counters and with the MWPC information. More than 10^6 events have been recorded. As of July 15, 54 counters are deployed, and by the end of the year nearly 100 detectors will compose an expanded array of radius 90 m. The flash chamber multiplexing scheme is operational, and data are being recorded on a regular basis.

An additional detector (E645) that can give information on muon number and direction, is coming on line this summer to study the oscillations of accelerator-produced neutrinos (Smith, 1985). This detector, consisting of liquid scintillator and drift tubes, has a horizontal area of 56 m^2 and an overburden of 3000 g/cm^2. Liquid scintillator information has been successfully recorded for EAS triggers.

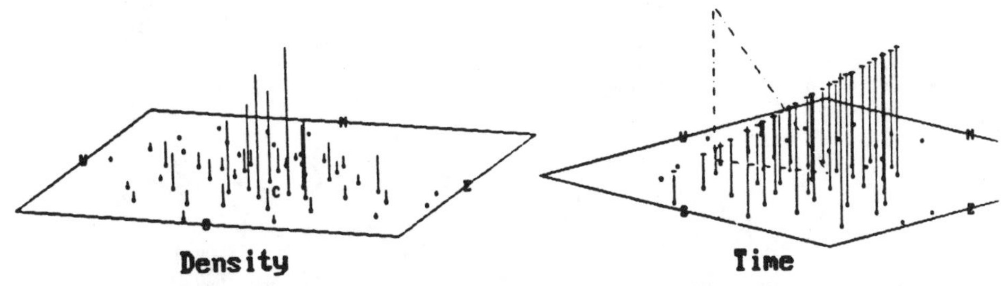

Fig 1. Density profile and arrival time of an EAS event.

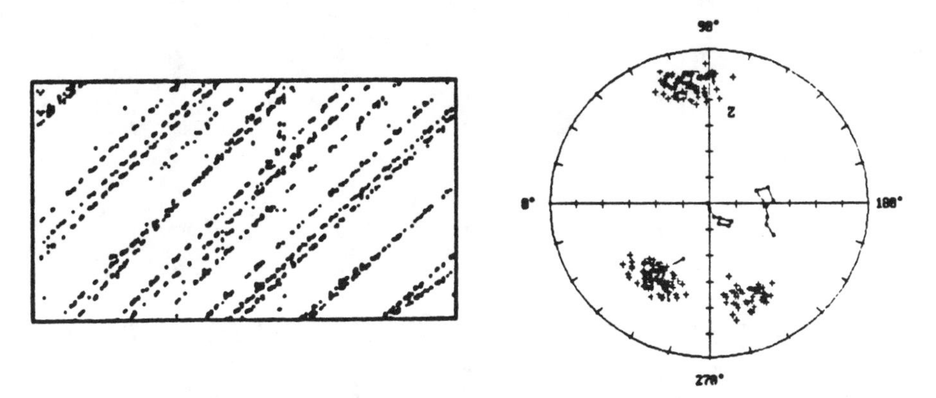

Fig 2. Side view of flash chambers. Fig 3. RA and Dec for smart trigger.

References

A. M. Hillas, Nature, 312, 50, 1984.

A. A. Stepanyan et al., Nature Phys. Sci. 239, 40, 1972.

B. M. Vladirmirski, A. M. Gal'per, B. I. Luchkov and A. A. Stepanyan, Usp. Fiz. Nauk. 145, 255, 1985. Sovet. Phys. Usp. 28, 153 (1985).

P. V. Ramanamurthy and T. C. Weekes, editors. Proc. of the Int. Workshop on Very High Energy Gamma Ray Astronomy, Ootacamand, India, 1982.

A. A. Watson, Rapporteur paper presented at the 19th International Cosmic Ray Conference, La Jolla 1985 (to be published). This paper gives a complete review of UHE gamma ray experiments.

M. Samorski and W. Stamm, Ap. J. Lett. 268, 117, 1983.

M. V. Barnhill, T. K. Gaisser, T. Stanev and F. Halzen, Nature 317, 409, 1985.

R. C. Allen et al., Phys. Rev. Lett. 55, 2401, 1985.

E. S. Smith et al., "Search for Neutrino Oscillations at LAMPF", presented at Moriond Conference on Massive Neutrinos and Astrophysics, Jan. 1985.

MACRO: THE MONOPOLE, ASTROPHYSICS, AND COSMIC-RAY OBSERVATORY

W.P. Trower
Physics Department
Virginia Polytechnic Institute and State University
Blacksburg, Virginia 24061 USA

ABSTRACT: The search for the magnetic monopole has motivated an Italian-American collaboration to construct a large ionization detector in the Grand Sasso Laboratory. This detector is also well suited to observe astrophysical phenomenon such as stellar collapse, particle physics processes such as neutrino oscillation, and cosmic-ray events such as muon bundles. Here we describe the detector and outline some of the accessable physics.

1. INTRODUCTION

An ambitious underground ionization detector, MACRO [1], will serve as an observatory for monopoles, a variety of astrophysical phenomena as well as cosmic rays. Its principal goal is to detect magnetic monopoles [2] by a variety of redundant proven techniques: liquid scintillator counters, helium filled limited streamer tubes, and track-etchable plastic all interleaved with absorber sufficient to range-out charged particles with energies <3GeV. MACRO will also be sensitive to monopoles with and without an attached fermion nucleus (e.g. proton, aluminum, magnesium, etc.) at a flux limit $\sim 10^{-15}/cm^2/sr/s$ per year.

2. THE MACRO DETECTOR

The Italian Grand Sasso Underground Laboratory is a unique facility with three large (17mx110m), well shielded (4km water equivalent), easily accessible experimental galleries supported by shops and computer facilities usually only found at accelerator laboratories. The MACRO detector will be constructed in the center hall beginning in Fall 1987.

MACRO's large area ($\sim 1,300m^2$) and acceptance for isotropic particle flux ($\sim 12,000m^2sr$) makes it ideal for studying rare phenomena in penetrating cosmic-rays. Its modular design (12mx12mx4.6m) allows data taking with the

M. M. Shapiro and J. P. Wefel (eds.), Genesis and Propagation of Cosmic Rays, 411–426.

partial detector to continue while construction of the remaining modules is taking place. The final spatial arrangement of these nine modules is still under discussion.

The deployment of the three detector systems - scintillators, streamer tubes, and plastics - is shown in Fig. 1. Two layers of streamer tube faced scintillation detectors enclose the detector volume in which eight layers of horizontal streamer tubes are interspersed between layers of low radioactivity limestone from the excavation, while six plastic sheets are located in the median plane. The liquid scintillator counters provide dE/dx and timing information, the streamer tubes tracking and ionization data, and the track-etch plastics ionization measurements.

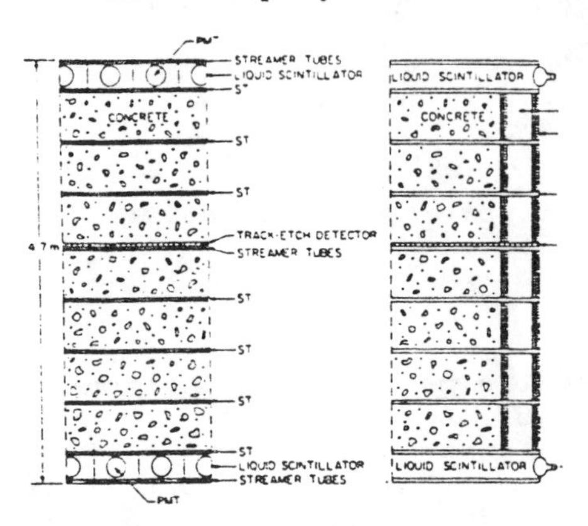

Fig. 1 MACRO detector cross section (a)longitudinal near center and (b)transverse at the edge.

Each of the 572 scintillator counters is a FEP TEFLON (n=1.35) lined PVC box (12mx50cmx25cm) filled with a highly transparent mineral oil based liquid scintillator whose light is collected by two Winstor cone equipped 8-inch phototubes on each end. The associated electronic chains accept a wide dynamic range of pulse heights from each phototube pair.

Each plastic streamer tube cell ($3x3cm^2$) is threaded by a 60μm diameter wire and manufactured in 8-tube units. Lateral and slant readouts allow two-dimensional hit localization. The two units (12mx25cm) cover one scintillator counter.

The detector spatial accuracy from the streamer tubes will be $\Delta x \sim \Delta y \sim \Delta z \sim 1cm$ giving a ten-hit track angular accuracy of $\sim 0.2^{\circ}$ with a time resolution of ~50ns. A single scintillation counter will have a spatial accuracy of ~15cm and timing uncertainty of ~1ns. The ionization loss uncertainty for minimum ionizing particles crossing scintillation counters in both layers is ~5%. The ionization threshold for fully efficient detector triggering by the streamer tubes is 1/100 of a minimum ionizing particle $(\Delta E/\Delta x)_{min}$, while for scintillators, it is 1/10. An individual scintillation counter will detect electrons with good background rejection if $E_e > 10MeV$, while the average minimum energy for muons crossing the whole detector is $E_\mu \sim 3GeV$.

3. THE MAGNETIC MONOPOLE SEARCH

The MACRO will be the first detector capable of searching for super-massive GUT monopoles significantly below the Parker flux limit of 10^{-15}/cm^2/sr/s [3]. Monopoles from Grand Unified Theories (GUTs) move at velocities ~10^{-3}c relative to the Earth. Calculated monopole scintillator response and measured low energy proton scintillation efficiency show that bare and electrically charged monopoles moving with velocities >5×10^{-4}c will produce detectable scintillation signals. Our principal monopole detection scheme, seen in Fig. 2, uses a scintillation counter in each of two layers through which a passing slow monopole produces a long characteristic signal whose waveform is digitized and recorded. Besides continuity of pulse heights, the pulse durations in the two scintillators and time-of-flight between layers, $\Delta t_1:\Delta t_2:T$, must be in the same ratio as the counter thicknesses and separation, d:d:D. The scintillation detectors respond to monopoles of arbitrary electric and magnetic charge with velocities >5×10^{-4}c.

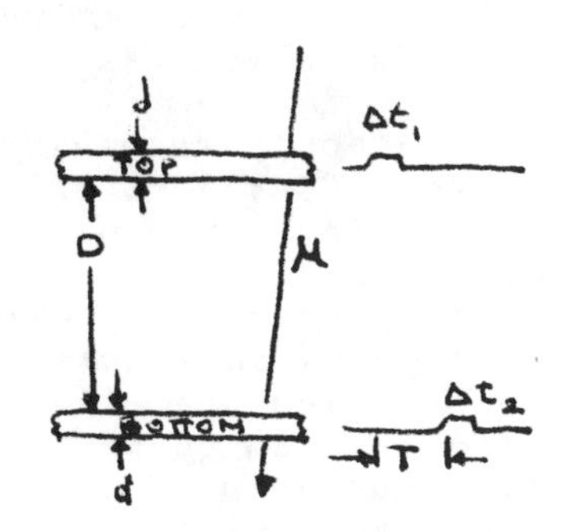

Fig. 2 Schematic of monopole detection strategy.

Helium/n-pentane filled streamer tubes will detect monopoles from level mixing in helium, and resulting collisional excitations which ionize the n-petane. Bare monopoles with velocities >10^{-4}c and electrically charged monopoles with velocities >2×10^{-3}c can be detected [5]. The steamer tubes have a separate trigger.

If the active detectors record a monopole candidate event, the appropriately located plastic detectors (Lexan and CR-39) will be etched and scanned. The CR-39 records bare monopoles for >5×10^{-3}c although diamagnetic effects may reduce this to >5×10^{-5}c [6]. The Lexan detectors record monopole >0.3c and so will only be sensitive if large signals in the electronic detectors are seen.

MACRO sensitivity to monopoles will be ~10% of the Parker limit for a monopole mass of 10^{16}GeV/c^2 in a few years of operation. In this case if no monopole events are detected, then monopoles contribute <4% of the dark matter of the universe.

4. SEARCH FOR STELLAR COLLAPSE

Stars with 8-12 $M_\odot$ evolve as increasingly heavier nuclei are produced and then consumed in a series of exothermic thermonuclear processes [7]. With the formation of a nickel-iron core of ~1.4 $M_\odot$, the electron degeneracy pressure is overwhelmed by the gravitational force of layers outside this core and stellar collapse begins. When the density reaches $\sim 10^{10}$g/cm^3 at $\sim 10^{10}$K heavy nuclei dissociate into free nucleons the protons of which capture electrons further decreasing the degeneracy pressure and accelerating the collapse. The stellar core implosion, which takes a time comparable with gravitational free fall, ~10ms, produces a neutron star or black hole suddenly releasing $E_{tot} \sim 3\times 10^{55}(M/M_\odot)^2(10\ km/R)ergs \sim 3\times 10^{53}$ ergs of the star's gravitational potential energy, most of it in a huge burst of neutrinos and antineutrinos of various species.

The collapse has three distinct phases. In neutronization, ~ms, most of the core protons capture electrons producing neutrons and ν_e. In deleptonization, <1s, the remaining leptons and protons are converted although here a sizeable antineutrino flux can be produced. In cooling, 1-10s, neutrinos of all flavors are emitted with a Fermi-Dirac thermal spectrum attenuated by source absorption [9]. Most of the neutrinos are produced in this phase and they are almost all ν_e which are the most easily detected. MACRO's large mass and high sensitivity should allow us to search effectively for these yet undetected neutrino bursts and so we have made Monte-Carlo calculations in MACRO's ~600 tons of C_nH_{2n} liquid scintillator.

Antineutrinos will be detected by charged-current interactions with free protons while neutrinos will be detected by elastic scattering from electrons. We included module geometry effects, neutrino cross sections, and electron/positron energy loss rates. For the neutrino burst energy distribution we used two sets of collapse parameters [10] which produce positrons: Model I~30,000 e^+ at 1kpc and ~300 e^+ at 10kpc; and Model II ~4,000 e^+ at 1kpc and ~40 e^+ at 10kpc. Since the cross section for antineutrino absorption increases as $E_{\bar{\nu}}^2$, the average positron energy, ~15MeV, is higher than the average of the antineutrinos which produce them. For a detection threshold of 10MeV, we found positrons were detected with efficiencies of 75% (I) and 60% (II). The electron detection from ν_e-e scattering was 25%, since the elastic scattering cross section grows linearly with E_ν and the average electron energy is only ~.5E_ν. Even in the more pessimistic Model II, 40 counts are obtained for a collapse at the galactic center.

Statistical fluctuations of the radioactive background counted in the detector can mimic a neutrino burst. For a 5 year run, 90% confidence level detection of a neutrino burst requires that such spurious bursts be <1/12 year. With this spurious burst rate, Fig. 3 shows the minimum detectable neutrino flux with scintillator background counting rate for three burst durations.

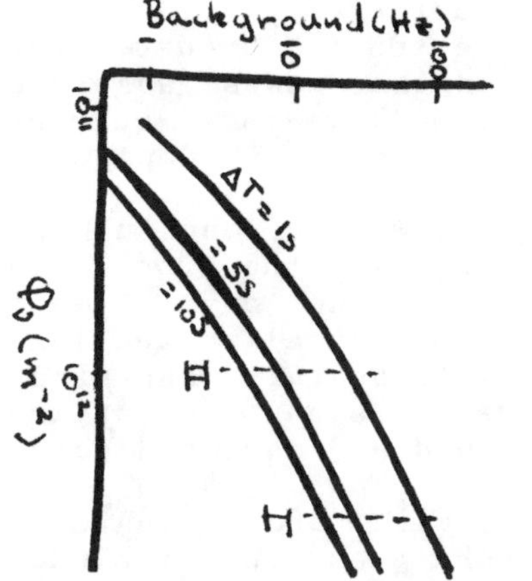

Fig. 3 Minimum detectable neutrino flux with detector background rate.

With an estimated maximum radioactive background rate of 1Hz and a 5s burst duration, the minimum detectable neutrino flux is $\sim 2 \times 10^{11}/cm^2$. Thus, for Model I the maximum distance of a detectable burst is ~40kpc, and for Model II ~15kpc. Thus MACRO should be sensitive to stellar collapse occurring anywhere within our galaxy.

5. SEARCH FOR NEUTRINO OSCILLATION

Atmospheric cosmic-ray neutrinos allow a region of parameter space not accessible to reactor or accelerator experiments to be searched for neutrino oscillations. Although high sensitivity is implied by the long oscillation path (i.e. $D_\oplus \sim 10^4$km) there are several limitations. First, the low statistics make small mixing angles unmeasurable. Second, an average over the neutrino energy spectrum and neutrino directions is much larger than average over source dimensions. Finally, the different refraction indices in matter for ν_μ and ν_e, generated by the charge-current contributions to elastic ν_e-e scattering, produce a decoherence of the ν_μ and ν_e components after a characteristic length of ~9000m in the Earth.

Considering only two neutrino flavours, ν_μ and ν_e, an underground experiment can measure the disappearence of ν_μ or the ν_μ/ν_e ratio. The measured disappearence can only be as well known as the flux or to 10-20%. The Earth's magnetic field and the solar wind add further uncertainties to the low energy spectrum.

The angular distribution of muons produced in the surrounding rock and the ν_μ/ν_e ratio are less sensitive to systematic uncertainties. For neutrinos capable of producing muons in our underground detector $E_\nu \geq 10$GeV so geomagnetic effects can be neglected. Further, the comparison between different path lengths makes absolute knowledge of the flux unnecessary for well measured angular distributions. Finally, the ν_μ/ν_e ratio is less dependent on theoretical uncertainties.

The direct comparison of the downward and upward neutrino fluxes using muons produced in the rock is difficult, since, at the MACRO depth the cosmic muon background is large even near the horizontal direction. Neutrinos that produce muons which will be fully contained in our detector have E<3GeV and their large ν-μ angles decrease the distinction between upward and downward going neutrinos.

We calculated MACRO's sensitivity to neutrino oscillation by Monte Carlo techniques and included generation, transport, and tracking of leptons. We assumed $\Phi(\nu_\mu) = \Phi(\bar{\nu}_\mu)$ and $\Phi(\nu_e) = 0.2 \mathrm{x} \Phi(\nu_\mu)$ [10]. The interaction cross section included quasielastic, Δ and inelastic channels. The ν_μ survival probability was that given by Wolfenstien [11].

A muon must be fully contained in the detector to reduce the background from entering stopping muons, and so its direction of motion cannot be identified. This decreases the flux by as much as 20% and requires that the expected flux be better known than at present. A muon produced in the rock must traverse the detector with a minimum track length to allow its direction to be measured by time-of-flight. Here the median energy neutrino is ~60GeV and the mean ν-μ angle ~3.5°.

The ratio between the measured and expected angular distribution of detected muons (the modulation factor) is shown in Fig. 4 for various Δm^2 at maximum mixing. Matter effects reduce the sensitivity in Δm^2 by about an order of magnitude. In a 3 year exposure a 3σ lower limit of $\Delta m^2 = 5\mathrm{x}10^{-5}$ is achievable. For $\Delta m^2 > 5\mathrm{x}10^{-2}$ the modulation factor again becomes nearly flat establishing the upper limit for MACRO.

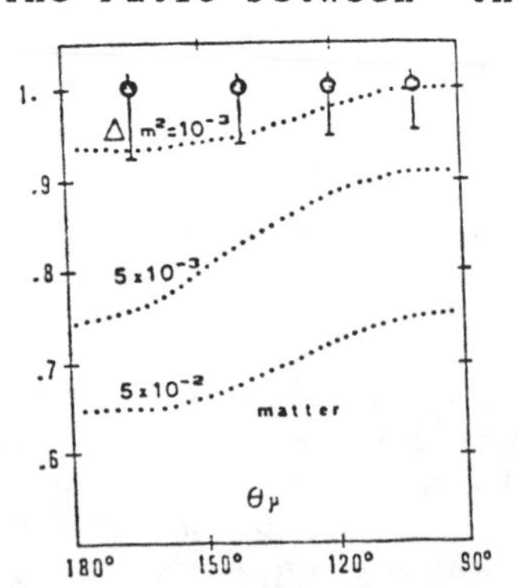

Fig. 4 Modulation factor at MACRO, with matter effects taken into account.

6. HIGH ENERGY NEUTRINO ASTRONOMY WITH MACRO

The MACRO detector will see an omnidirectional muon flux of ~1/m²/hour. With time-of-flight from the scintillation counters, the upward-going muon contamination will be negligible. The background to detecting extraterrestrial neutrino sources is atmospheric neutrino conversion, ~1/day in MACRO.

An underground detector's efficiency for recording extraterrestrial neutrinos rises rapidly with neutrino energy [13]. We estimated MACRO's response to extraterrestrial neutrino sources using Monte Carlo techniques to determine detection efficiency and the angular spread of the muons with respect to the neutrinos direction.

We assumed a power-law spectral distribution for the neutrinos, $F_\nu = K_\nu x E_\nu^{-\gamma}$. Table 1 summarizes our results. The differential neutrino-nucleus cross section includes the effect of the W boson mass and uses the Duke-Owens nucleon structure functions. In passing through the rock we have included only the muon multiple Coulomb scattering. The ν-μ angle distribution is dominanted by a central gaussian core with a tail corresponding to relatively rare large angle scatterings. We therefore quote two angular parameters θ_{RMS} which is Coulomb scattering dominated and θ_{90} which includes 90% of all the muons. The steeper spectra can be easily identified by the larger angular spread.

Table 1. Summary of ν-μ Parameters in MACRO.

γ	E_ν (TeV)	ϵ	F_ν E_T=1GeV	θ_{90}	θ_{RMS}	MDF (ergs/cm²/s)
2.0	23.3	.75	$6x10^{-9}$	$\leq 1^\circ$	0.5°	$2x10^{-8}$
2.2	7.8	.68	10^{-9}	1°	0.6°	10^{-7}
2.6	0.9	.56	$7x10^{-11}$	2.3°	1.2°	10^{-6}
3.0	0.1	.39	10^{-11}	6.0°	2.0°	10^{-6}
3.8	0.01	.14	$3x10^{-12}$	11.5°	4.6°	$2x10^{-4}$

MACRO'S ability to detect extraterrestrial neutrino sources is not limited by the atmospheric background. It's resolution for hard neutrino sources (γ~-2) is <1° with an expected atmospheric background of ~.05/year. So neutrino emission from a celestial object will be seen in clustering of a small number of muons in a ~1° cone around its direction. We require 10 such muons in a five year exposure to define a minimum detectable flux (MDF). Thus, the intrinsic neutrino luminosity of a source at a distance D must be $L_\nu \geq (MDF) 4\pi D^2_{GPC}$ for MACRO to detect it.

MACRO's location at 33°N allows it to survey the southern celestial hemisphere. The time-area dependence on the source celestial declination with the equiexposure contours in the galactic coordinates plot of Fig. 5. The sensitivity corresponds to the maximum exposure in the celestial southern pole direction. So for a given source the MDF of Table I must be scaled by the actual exposed area.

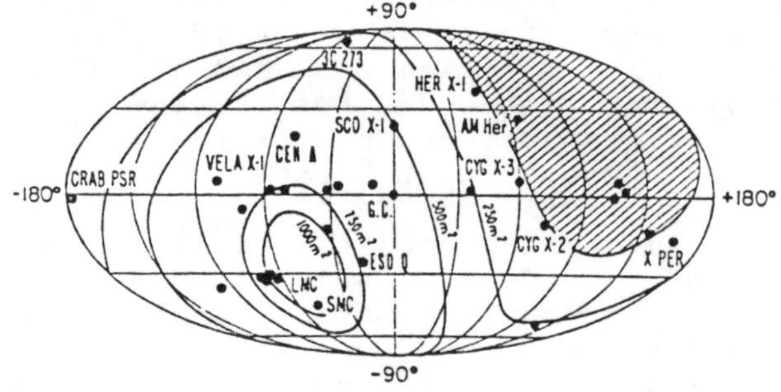

Fig. 5 MACRO field-of-view in galactic coordinates where the shaded region is inaccessible.

With MACRO sensitive to high energy neutrinos, several classes of objects thought to be strong neutrino sources were studied. We discuss briefly the detectability with MACRO of two of these.

Ultra-High Energy γ-rays ($E \geq 10^4$ TeV) have been observed from the X-ray binaries Cyg-X3, Vela-X1 and LMC-X4. Models of these objects suggest that hard (TeV) neutrino emission originates from hadronic interaction of accelerated protons or nuclei on the non-degenerate companion star atmosphere producing mesons which decay into neutrinos. From the observed γ-ray spectrum the neutrino spectral index is estimated to be ≤ 2. Two of these sources, Vela-X1 and LMC-X4, are well positioned for MACRO and the latter is predicted to have a neutrino luminosity easily detectable by MACRO.

Radio and X-ray observations of the compact nuclei of several peculiar galaxies (Type I Seyfert, BL-Lac objects and Quasars) indicate that extreme non-thermal emission gives these objects their huge luminosities. Neutrinos could be associated with this non-thermal emission, especially in the case of continuous reacceleration of particles in the source. From recent high quality X-ray spectra the non-thermal emission has been shown to obey a universal power law distribution implying a particle and neutrino spectrum with $\gamma \sim -2.3$. For MACRO to detect such a source $L_\nu \geq 10^{46} x D^2_{Gpc}$ ergs/cm²/s, far to high for these objects.

7. MULTI-MUON EVENTS

Multimuon events can determine the nuclear composition of the primary cosmic rays in the interesting region >2000 TeV/nucleus. We have parameterized muon production in cosmic ray showers for MACRO by the muon penetration through the earth to the detector in terms of slant depth (X) and mean muon number ($\langle n_\mu \rangle$) [14]. Each nucleon in a primary contributes this same average number of muons and the N_μ distribution is taken to be Poisson. The lateral distribution of muons at the detector was also estimated.

All muons striking MACRO's 12x112m² surface we assumed were detected and have various X values. Because we measure muon direction to ~.2°, when we select on X we change the primary's energy distribution (e.g. events with primaries <2000 TeV/nucleus is ~31% for X=all, but only ~8% for X>7kmwe).

For MACRO to determine the cosmic ray composition >2,000TeV we have studied four models characterized by the primary's composition: all iron nuclei (FE); iron dominating at high energy[3](Md); high energy behaviour same as at 50 GeV/nucleon[3](LE); and all protons(P). In Fig. 6 we show the detected muon multiplicity distributions for these models (solid curves) normalized to the single muon rate. At a given primary energy the proton energy/nucleon is higher than iron and so protons give many more single muon events while iron has a larger multiple-to-single muon ratio and larger lateral spreads (decoherence). The dashed curves show the multiple muons detected in a 10m² detector. For 10 detected muons, the FE yield is a hundred times less. Including single muon events the absolute MACRO yields are ~500K/year for iron and ~13M/year for protons. Depending on primary composition, the number of events with N_μ>5 will be 500 to 20,000!

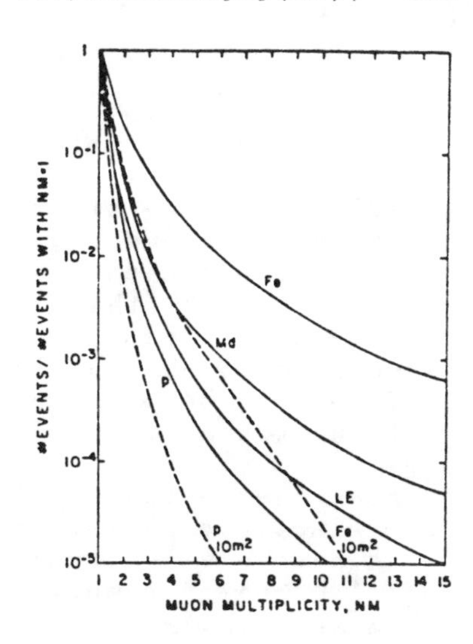

Fig. 6 Muon multiplicity fraction of single muon rate.

One aspect of the muon decoherence which is particularly sensitive to primary composition is the maximum separation of muon pairs in a multiple muon event. In Fig. 7 we show this distribution for the four composition hypotheses and for different primary energy distributions.

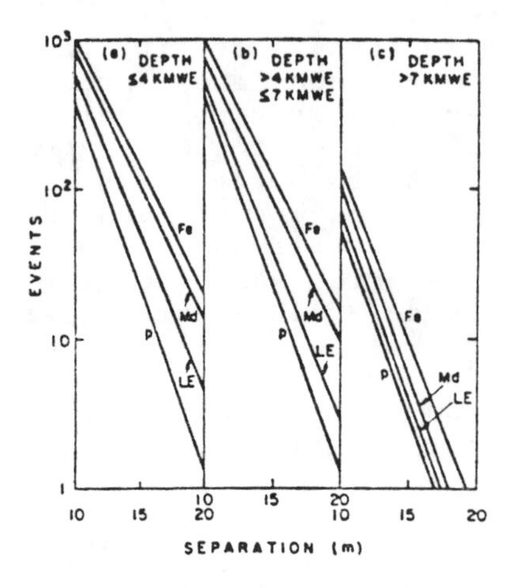

Fig. 7 Maximum muon pair separation in a multiple muon event at three slant depths.

8. SEARCH FOR STRANGE QUARK MATTER

The possible existence of strange quark matter consisting of comparable numbers of u, d and s quarks has been suggested by Witten [16] and discussed by others [17-19]. Such matter could be more tightly bound than nuclear matter because of the Pauli exclusion principal, a three-flavor system should have lower energy/quark than a two flavor system, thus compenstating for the heavier s quark. Such strange quark matter should be absolutely stable, form systems with any number of quarks, be produced in the early universe in the phase nucleon transition and have a slight positive electrical charge.

The flux of these nuclearities reaching the earth [17] is estimated assuming they dominate the galactic dark matter ($\sim 10^{-24}$ g/cm^3 in our neighborhood) and have a velocity of ~.001c characteristic of the solar system's orbital rotation in the galaxy. For a nuclearite mass m_N in grams $F \lesssim 3.9 \times 10^{-18}/m_N/cm^2/s/sr$. Fig. 8 shows this limit with the nuclearite's mass as well as the limits from existing data. For a MACRO $\sim 10^4 m^2 sr$-5year exposure the flux is $< 2 \times 10^{-16}/cm^2/s/sr$.

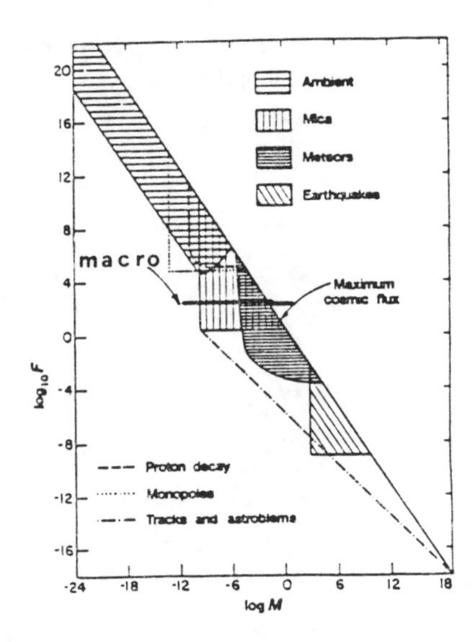

Fig. 8 Nuclearites maximum flux $/km^2/y/2\pi sr$ with mass.

Nuclearites loose energy by elastic or quasielastic interactions with the atoms [18] at a rate of $dE/dx=-\phi\rho v_N^2$, where ρ is the density of the traversed material, v_N is the nuclearite velocity, and ϕ, the cross section of the nuclearite, is $\phi\sim\pi(1Å)^2$ for $m_N<1.5ng$ and $\phi\sim\pi(3m_N/4\pi\rho_N)^{2/3}$ for $m_N>1.5ng$. So $v_N=v_o\exp\{-\phi\rho L/m_N\}$, but if v_N is less than the velocity at which the force with which it penetrates matter is equal to the force with which the material resists its penetration ($v_e\sim(\Sigma/\rho)^{1/2}$, $\Sigma\sim10^9$ erg/cm^3 or a pressure of $\sim10^3$ Atm) this expression looses any meaning. The nuclearite range is $R=(m_N/\phi)\ln(v_o/v_c)$.

Nuclearites with $m_N>4.5\ 10^{-14}g$ reach the earth's surface while those with $m_N>0.1g$ may traverse the Earth's diameter. The MACRO detector may be reached from above if $m_n>3x10^{-11}g$ while those with $m_N>3x10^{-10}g$ which reach MACRO must have galactic velocities.

A velocity v_N nuclearite scattering from an atom gives the latter a velocity $\sim v_N$. A nuclearite passing through a detector produces a temperature increase, part of whose energy is visible light which is transmitted with a luminous efficiency $\eta=(dL/dx)/(dE/dx)$. A lower limit for η is estimated from thermodynamics assuming a black body plasma is created at temperature T by the passage of the nuclearite which expands radially in a cylindrical surface. dL/dx is shown in the right hand scale of Fig. 9 in g/cm^2, assuming $\eta=10^{-5}$ and $v_o=250km/s$. Thus, scintillation detectors are sensitive to nuclearites which produce us scintillation light, but simply because they are transparent to light and are equipped with photomultipliers.

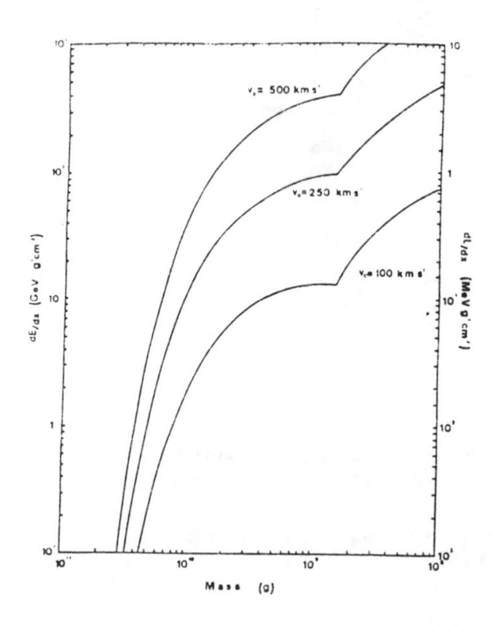

Fig. 9 Nuclearites total luminous/energy loss.

The number of visible photons is estimated as $(dL/dx)/\pi$ where π, the average visible photon energy is 3.leV. For a minimum ionizing muon the energy loss $\sim 2MeV/g/cm^2 x 30 g/cm^2 = 60MeV$ and the number of photons $\sim 3x10^5$, one for every 200eV of energy lost. The number of photoelectrons at the photocatode of each of the two photomultipliers is n_μ(geometrical loss 0.01)(absorption loss 0.33)(photocathode efficiency 0.25)~240. The ratio of nuclearite to minimum ionizing muon photoelectrons, n_N/n_μ, is subject to fewer uncertainties than the absolute value. The nuclearite light produced is over a much long time.

If the scintillator detector is sensitive to particles which give an integrated light yield equal to 1/10 of that of a minimum ionizing muon, the scintillator should be sensitive to nuclearites with $m_N > 3.5x10^{-11} g$ at $\beta_N = 8x10^{-3}$. The trigger must also discriminate on the time distribution of the light similar to that for slow monopoles. The light from the passage of a monopole, $m_M \sim 10^{16} GeV$ and $\beta = 10^{-3}$, can only be discriminated from that of a nuclearite by the precise distribution of the light in two scintillators, the scintillator transit times Δt_1 and Δt_2, and the time-of-flight between the two counters, T.

Table II. Nuclearite, muon and monopole charactertistics in MACRO.

	NUCLEARITE		NUCLEARITE		MUON	MONOPOLE
β_0	0.83×10^{-3}		$1.66\ 10^{-3}$		1	10^{-3}
m_N(g)	5×10^{-11}	10^{-10}	5×10^{-11}	10^{-10}	____	$m_M=10^{16}$GeV
β_{MACRO}	10^{-4}	3×10^{-4}	2×10^{-4}	6×10^{-4}	~1	$\sim10^{-3}$
Δt(s)	7×10^{-6}	3×10^{-6}	4×10^{-6}	10^{-6}	8×10^{-10}	8×10^{-7}
T(s)	10^{-4}	5×10^{-5}	6×10^{-5}	2×10^{-5}	10^{-8}	10^{5}
dL/dx (keV/g/cm^2)	22	160	87	641	31	250
n_N/n_μ	1	5	3	21	1	8

A nuclearite, moving with velocity v_N cannot ionize an atom but may break molecular bounds and indirectly produce ionization from the high energy tail of the emitted light. So the streamer tubes, which are be sensitive to a single electron, would detect nuclearites at some level.

The formation of an etchable track in a plastic detector depends on: REL - the fraction of the total energy loss which remains localized in a small, 0.01μm radius, cylindrical region around the particle trajectory; S_e - electronic energy loss which is equivalent to kinetic energies of δ-rays, <200eV for CR-39. (REL ~90% S_e for not too high energies); and S_n - nuclear energy loss which is only important for low velocity particles; REL$\approx S_n$. Experiments [20] show that nuclear energy loss produce etchable tracks by breaking molecular chains: Mica is sensitive to ions with $10<A<90$ and energy 0.1-3 keV/nucleon, that is with such low velocities that only the $S_n\cong$REL could be at work. The track is formed from atoms ejected from their lattice sites in screened Coulomb collisions. Present REL thresholds are 0.03 GeV/g/cm^2 for CR-39, 1.1 for nitrocellulose, and 2.8 for LEXAN. So nuclearites with an initial velocity v_0=250km/s must have $M_N\sim10^{-10}$g to reach MACRO depths and so CR-39 should be a good detector.

MACRO's scintillation and track-etch detectors should be

sensitive to $m_N > 3x10^{-11}$g nuclearites with v_o=250km/s and in 5 years should establish a limit ~$2x10^{-16}$/cm^2/s/sr.

Acknowledgements: The work presented here was abstracted from other MACRO publications [1, 21, 22].

References

1. The MACRO Collaboration Proposal CALT 68-1224a (1985) (unpublished).

The current collaboration is:
Bari: C. Calicchio, G. Case, C. DeMarzo, O. Erriquez, C Favuzzi, N. Giglietto, E. Nappi, F. Posa, P. Spinelli

Bologna: F. Baldetti, C. Cecchini, G. Giacomelli, F. Grianti, G. Mandrioli, A. Margiotta, L. Patrizli, G. Sanzani, P. Serra M. Spurio

Boston University: S. Ahlen, A. Ciocio, M. Felcini, D. Ficenec, J. Incandela, A. Marin, J.L. Stone, L. Sulak, W. Worstell

Caltech: B. Barish (co-spokesman), R. Batitti, C. Lane, G. Liu, C. Peck

CERN: G. Poulard, H. Sletten

Drexel: S. Cohen, N. Ide, A. Manka, R. Steinberg

Frascati: G. Battistoni, H. Bilokon, C. Bloise, P. Campana, V. Chiarella, A. Grillo, E. Iarocci (co-spokesman), A. Marini, J. Reynoldson, A. Rindi, F. Ronga, L. Satta, M. Spinetti, V. Valente

Indiana: R. Heinz, S. Mufson, J. Petrakis

L'Aguila: P. Monacell

Michigan: D. Crary, M. Longo, J. Musser, C. Smith, G. Tarle

Napoli: M. Ambrosio, G.C. Barbarino, G. Grancagnolo, A. Ounembo,

Pisa: C. Angelini, A. Baldini, C. Bemporad, V. Flaminio, G. Giannini, R. Pazzi,

Roma: G. Auriemma, M. DeVincenzi, E. Lamanna, G. Martellotti, O. Palamara, S. Petrera, L. Petrillo, P. Pistilli, G. Rosa, A. Sciubba, M. Severi

Texas A&M: P. Green, R. Webb

Torino: V. Bisi, P. Giubellino, A. Marzari-Chiesa, L. Ramello

Virginia Tech: B. Laubis, D. Solie, S. Torres, P. Trower

2. See "Magnetic Monopoles" in this volume.

3. E.N. Parker, Astrophys. J. 160, 383 (1970); and M.S. Turner, E.N. Parker, and T.J. Bogdon, Phys. Rev. D26, 1296 (1982).

4. S.D. Drell, N.M. Kroll, M.T. Mueller, S.J. Parke and M.A. Ruderman, Phys. Rev. Lett. 50, 644 (1983).

5. N. Kroll in Monopole '83 edited by J.L. Stone (Plenum, New York, 1984) p.

6. P.B. Price, Phys. Lett. 140B, 112 (1984).

7. J.M. Lattimer, Ann. Rev. Nucl. Part. Sci. 31, 337(1983).

8. S.L. Shapiro, Astrophys. J. 214, 566 (1977).

9. P.V. Kochagin, et al. Proc 17th Int Cosmic Ray Conf. 7, 110 (1981).

10. D.K. Nadyozhin and I.V. Otroshchenko, Sov. Astron. 24, 47 (1980); and A. Burrows and T.L. Mazurek, Nature (London) 301, 315 (1983).

11. L.V. Volkova, Sov. J. Nucl. Part. 31, 784 (1980).

12. L. Wolfenstein, Phys. Rev. D17, 2369 (1978).

13. T. Gaisser and T. Stanev, Phys. Rev. D31, (1985) 2770.

14. T. Gaisser and T. Stanev, Nucl. Instrum. Meth. A235, (1985) p183.

15. G.B. Yohd, T. Stanev and T. Gaisser, in Proc. IOCBAN 1984, Park City, Utah.

16. E. Witten, Phys. Rev. D30, (1984) 272.

17. A. de Rujula and S.L. Glashow Nature (London) 312 (1984) 734.

18. E. Farhi and R.L. Jaffe Phys. Rev. D30, (1984) 2379.

19. A. de Rujula Nucl. Phys. A434, (1985) 605.

20. J. Borg, et al. Rad. Eff. 65, 13 (1982).

21. MACRO contribution to NASA2376 8, 128, 132, 136, 226, 230 (1985).

22. G. Giacomelli, G. Mandrioli, A. Margiotta, L. Patrizii, P. Serra, and M. Spurio. MACRO memo 13/86 (unpublished),

INVISIBLE COSMIC RAYS

A. K. Drukier et al.
Harvard-Smithsonian Center for Astrophysics
60 Garden Street
Cambridge, MA 02138
U.S.A.

ABSTRACT. The PNL/USC ultralow background prototype Ge detector, in the Homestake goldmine, is being applied to searches for dark matter candidates. The low energy data exclude particles with spin independent Z^o exchange interactions, with masses between 20 GeV and 5 TeV, as significant contributions to the cold dark matter of the halo of our galaxy. The existence of stable Dirac neutrinos more massive than 20 GeV is also excluded except for a narrow region around the Z^o resonance.

The use of superconducting micron size grains in the search for the "missing mass" is discussed. The predicted count rates for massive neutrinos and supersymmetric relics from the early universe range from 10^{-1} - 10^{3} counts per kilogram of active detector per day. The earth's motion around the sun produces a significant annual modulation in this signal. Several experimental challenges remain in the development of this detector, but progress is rapid.

1. INTRODUCTION

One of the outstanding problems of astrophysics is that we do not know what comprises 90 % of the mass of the galaxy. In recent years, additional evidence continues to accumulate which confirms the presence of "dark matter" in galaxies. Observations of rotation curves of HI gas in spiral galaxies [1] is perhaps the strongest evidence that the amount of gravitational matter exceeds the amount of luminous matter

M. M. Shapiro and J. P. Wefel (eds.), Genesis and Propagation of Cosmic Rays, 427–443.

observed in stars, gas, and dust. Observations of hot X-ray emitting gas in elliptical galaxies suggests that the dark matter is pervasive and appears in equal abundance in a variety of galaxies[1]. The success of the inflationary scenario in resolving many of the paradoxes of the big bang cosmology also encourages the prejudice that most of the matter in the universe is not only non-luminous but also non-baryonic[2].

The range of possible solutions to the "missing mass" problem is enormous: there is a factor of 10^{67} in mass between the 10^{-5} eV "invisible axions"[3] and 10^{6} solar mass black holes[4]. The GeV range candidates offered by supersymmetric models ("sparticles") lie in middle of this range.

Since this is a conference on Cosmic Rays, I should like to point out that the fluxes of these "cold dark matter candidates" are millions of times larger than the flux of other Cosmic Rays. And yet they are "invisible". If the halo of our galaxy is composed of weakly interacting massive particles (WIMPs), then millions of these particles are streaming through a square centimeter every second. Goodman and Witten[5] realized that this flux could be experimentally detectable. There are, however, two difficulties involved with detecting WIMPs.

A) They don't do much: All of the proposed WIMPs have some conserved quantum number (R-parity of Supersymmetric Theory (SUSY) particles and fourth generation lepton number for massive neutrinos); hence, the end-product of an interaction with a nucleus is at best the deposition of a few keV of energy.

B) They don't do it often: The lowest mass supersymmetric particle (which is stable in many theories) is usually some linear combination of photino and higgsino interaction eigenstates. This Majorana particle has only axial couplings with quarks and thus has a typical elastic nuclear cross-section 10^{-37} cm^2 for WIMPs through its spin-dependent interactions with nuclei. If these sparticles were the galactic missing mass, they would produce 0.1 - 1 counts per day in a kilogram of detector. Scalar neutrinos and massive Dirac neutrinos have vector coupling with quarks, and thus the neutrons in the nucleus constructively interfere to yield a much larger cross-section 10^{-34} cm^2 and a higher count rate 10^3 counts/kg/day. In either case, the search for WIMPs requires a detector with very good background rejection and a low energy threshold.

2. EXPERIMENTAL BOUNDS ON COLD DARK MATTER WITH AN ULTRALOW BACKGROUND Ge DETECTOR

In this part of the paper, I will discuss the use of a germanium diode detector to search for dark matter. This project involved collaborators from not only a range of institutions but also from a wide range of fields including radiochemistry, solid-state physics, particle physics, and astronomy. The collaboration consists of Steve Ahlen (BU), Frank Avignone (Univ. of South Carolina), Ron Brodzinski (PNL), Dave Spergel (Princeton), Graciella Gelmini (Harvard) and myself. A more detailed description of this work will soon be published in Physics Review Letters(6).

It is not uncommon that new technology attracts unanticipated application. Ultralow background Ge detectors, developed specifically for sensitive searches for the ßß-decay of ^{76}Ge, represent a recently developed technology applicable to a number of other fundamental experiments. Th sensitivity required in searches for very weak γ ray lines was first enhanced using the concept of Compton-suppression and live shielding. However, a limitation sensitivity was found due to the radioactive background in the Ge crystal, cryostat and electronic parts, the construction materials of the live shield (including photo-multiplier tubes) and in the bulk shield itself. A detailed study was made of the materials used in commercial low background detectors and a discussion of the improvements achieved by material selection was given in an earlier article(7).

The Ge crystals themselves are virtually free from primordial or induced radioactivity because of extensive zone refinement; however, cosmogenically produced radioisotopes have been observed. In addition, cosmogenic activity in the shielding materials, and materials selected for low background cryostat construction, are also observable(7). The background is greatly reduced by going deep underground with a detector built from selected materials; this diminishes background due to Cosmic Rays and the radioactive background due to nuclei activation by Cosmic muons/neutrons. Figure 1 depicts the results of background reduction over the past four years in terms of gross pulse height spectra and, in different shielding configurations Figure 2 shows the low energy part of the observed spectrum.

While the program of background reduction continues, the present low levels present a new technology which has been applied to a number of interesting searches for rare decay models and exotic particles[8]. They include searches for: $0(\nu)$ and $2(\nu)\beta\beta$-decay of ^{76}Ge, electron decay to γ and ν, dark matter candidates for the hidden mass of the galactic halo and Dine/Fisher/Sreduicki solar axions. Only the search for the dark matter candidates is discussed in this paper.

Because of its low band gap (0.69 eV at 77° K) and high efficiency for converting electronic energy loss to electron-hole (e-h) pairs (2.96 eV per electron-hole at 77°K), germanium detectors are probably the best suited of all existing semiconducting particle detectors for the detection of low velocity recoil nuclei.

In order to interpret the data, the fraction of recoil energy converted to electron-hole pairs must be estimated. At low recoil velocities, much of the nucleus' energy is dissipated through nuclear collisions rather than through electronic collisions. For example, a uranium nucleus with energy 6 MeV incident on a silicon diode produces a signal equivalent to a 2 MeV electron. This effect can be parameterized in terms of a relative efficiency factor (R.E.F.), the ratio of the number of electron-hole (e-h) pairs produced by an incident electron to the number of e-h pairs produced by an incident Ge nucleus with same kinetic energy. The optimum R.E.F. value of 1 is achieved only for $E \gg$ MeV.

The R.E.F. for Ge detectors has been calculated by evaluating the fraction of the primary Ge recoil energy which is lost in electronic collisions. The electronic energy loss of secondary and higher order recoil nuclei has been accounted for in this calculation, and up-to-date electronic and nuclear stopping cross-sections[10] have been used. The calculated R.E.F. of $\approx$ 4 for recoil energy of 15 keV is consistent with experimental data[11].

The observed count rate can be compared to the rate predicted if the halo were comprised of WIMPs. The particles that comprise the halo are assumed to have a velocity distritution function with an r.m.s. of 250 km/s and a maximum of 550 km/s, while the halo, like the galactic spheroid, slowly rotates with a velocity of 80 km/s[12]. This halo model used in the calculation is conservative. If the maximum halo velocity were closer to the local escape velocity of 750 km/s[13], the lower limit of the excluded mass range would drop to 15 GeV.

Since the predicted count rate depends upon the dark matter halo density, the observed count rate can be used to obtain limits on the density of interacting dark matter particles in the halo. Figure 3 shows these limits for particles with spin independent (s.i.) Z^o mediated exchange interactions for a standard isotropic halo model which includes the Sun's motion relative to the galactic halo. Figure 3 can be used for other s.i. vectorial interactions by multiplying the vertical axis by the ratio $(\sigma^{s.i.}_{weak}/\sigma_{s.i.})$.

Since only 7.8 % of the natural isotopic mixture of Ge has a non-zero spin (the isotope ^{73}Ge), our best bounds apply to spin independent (s.i.) interactions. Bounds on dark matter candidates that couple to baryons through Z^o exchange, like stable massive Dirac neutrinos[14] and scalar neutrinos[15], are presented. Our results exclude a halo dominated by particles with scattering cross section $\sigma_{s.i} = \sigma_{weak}$ with masses 20 GeV $\lesssim m \lesssim$ 5 TeV and apply to s.i. reactions in the range of $\sigma^{s.i} = 10^{-1}\sigma_{weak}$ to $\sigma^{s.i} \approx 10^{-28}$ cm^2 for which the dark matter particles would be stopped in the earth's crust before arriving at the detector. Here, σ^{weak} is the scattering cross section for a heavy standard neutrino from a Ge nucleus. This range also includes neutral technibaryons, recently proposed as dark matter candidates[16], having a cross-section = $10\sigma^{s.i.}_{weak}$. They are excluded for masses larger than 20 GeV. The presence in the detector of ^{73}Ge with s = 3/4, allows us to obtain some bounds on particles with spin dependent (s.d.) interactions, in which case our bounds apply to particles in the range

$\sigma = 10^4 \sigma^{s.d.}_{weak}$ to $\sigma \approx 10^{-28}$ cm^2.

The detector background has a smooth contribution from the Compton-scatter of high energy γ emmitters (e.g. ^{40}K) as well as the narrow line components. The low energy peaks are primarily due to the presence of ^{210}Bi in a solder connection used to make electrical contact with the diode and which "sees" the surface of the detector. The solder was removed, and the radioactive shield was upgraded by the use of 448 year old lead in the place of the super-pure copper which has some cosmogenic radioactive contamination. Thus, in fall 1986, the background should be reduced by another factor of ten. The energy threshold was set at 4 keV because of noise at lower energies. The shape of the low energy K absorption edge pairs suggests that $\Delta E/E$(FWHM) $\approx$ 500 eV. The strong increase of noise below E_{th} = 4 keV is largely due to microphonics engendered by mining operations. Hardware and software are being developed to reduce this noise and permit lowering the energy threshold to 1 keV. By the end of 1986, rejection/ detection of the existence of coherently interacting particles of mass $\gtrsim$ 10 GeV should be possible.

It will be difficult to reduce the energy threshold below 1 keV, thus the detection of particles of lower mass will require other detectors. The germanium detector is also not sensitive to particles like the photino that couple through s.d. weak interactions. Because of its low energy threshold and its sensitivity to photinos, the super-heated superconducting colloid detector(17,18) seems like an attractive candidate for extending the limits derived here.

3. DETECTING DARK MATTER WITH SUPERCONDUCTING GRAINS

The limits on weakly interacting halo dark matter from germanium spectrometers, however, only exclude particles more massive than 16 GeV, and it does not seem likely that these limits could be extended below 8 GeV. Unfortunately, the most interesting region for Dirac neutrinos is around 2-3 GeV (the range of the Lee Weinberg limit). Particles with spin-dependent interactions (e.g. photinos, higgsinos, and Majorana neutrinos) also evade detection since most of the abundant isotopes of germanium have zero spin. (They have an even number of both protons and neutrons.) This motivates us to examine the use of new types of detectors in the search of dark matter. I will discuss my work with Dave Spergel (Princeton) and Katie Freese (ITP, Santa Barbara) on the use of superconducting grains to detect the energy deposited in a nuclear recoil[17].

Grains are not a new, but rather an underutilized, detector technology[18,19]. They have been suggested to use in the detection of other rare events such as the elastic scattering of solar neutrinos[20] and the monopoles in the galactic halo[21]. Several groups are now actively working on developing grains and grain detectors: myself, A. Kotlicki and his collaborators at UBC, Vancouver; G. Waysand and his collaborators in Orsay, Saclay and Annency; L. Stodolsky and collaborators at MPI for Physics and Astrophysics, Munich.

In a detector, these grains would be kept in a superheated superconducting state. The deposition of a few keV of energy in an elastic scatter of a WIMP off of a nucleus in the grain heats the grain and flips it from the superconducting to the normal state. The background magnetic field can now permeate the grain. This produces a change in the magnetic flux through a loop surrounding the grains equivalent to the addition of a dipole whose strength is proportional to the product of the grain's cross-sectional area and the background magnetic field. SQUIDs (superconducting quantum interference devices) would be ideal for detecting this small change in flux. The grain could be composed of any type I superconductor. These include gallium and vanadium, both of which are composed mostly of isotopes with an odd number of neutrons, and thus have large cross-sections for particles with spin-dependent couplings.

The energy threshold of a grain, the minimum amount of energy needed to flip the grain, is set by its composition, size, and temperature. Micron size grains are needed to detect photinos of a few GeV.

The grains would probably be coated with a dielectric composed of low Z material. This dielectric separates the grains and suppresses diamagnetic interactions with neighboring grains. More massive photino/ higgsino particles may arise in superstring theories in which large squark masses are predicted. These 10 - 80 GeV sparticles could be detected by large grains. Since the signal is proportional to the cube of the grain radius, larger grains produce a much stronger signal and thus allow the use of less expensive electronics.

The major source of background in the detector is the radioactive decay of trace contaminants in the grains and the surrounding dielectric. Most of these decays produce MeV electrons. Since the dielectric is composed of low Z material, these electrons, which lose energy through Coulomb interactions, will deposit most of their energy in the grains. Thus radioactive decays will flip multiple grains, which will produce a large change of flux in the SQUID loop. The scatter of a WIMP off of a grain will flip only a single grain. Uniform grains, which have similar energy threshold, are needed for this background suppression mechanism to work. Most of the grains must be in the superheated state so that there is little dead material into which the ß particle can deposit its energy.

The problem of grain production is presently the greatest challenge in the development of these detectors. Many of the problems that workers in this field have had with the grains are due to irregularities in grain size and shape. Strong fields near corners of non-spherical grains lead to a spread in energy thresholds. Wide hysteresis curves are a symptom of variations in grain properties. At Vancouver, multiple filtered grains have produced promisingly narrow hysteresis curves which suggest that this process eliminates not only large variations in grain size but also removes irregular grains.

Another challenge in dark matter grain detector development is development of very sensitive electronic read-out. The use of Superconducting Quantum Interferance Devise (SQUID) magnetometers seems necessary. Avoiding background signal due to vibrations producing spurious signals in the SQUID loop requires that the detector is vibrationally well insulated from its environment and may necessitate using multiple SQUID loops as a gradiometer. The SQUID based read-out of Superheated Superconducting Colloid Detector (SSCD) was developed in UBC, Vancouver. Figure 7 shows that the signal due to change of state of a single grain is a linear function of the applied magnetic field. For a grain $R \lesssim 5$ microns placed in a few cubic centimeters read-out loop, a signal to noise ratio of 10 was measured. The collection of grains was irradiated with hard X-ray photons and Quantum Detection Efficiency of 80 % was observed[22].

Since it is possible for both radioactive decays and the elastic scatter of a halo WIMP to flip a single grain, we must find a characteristic of the signal that will allow differentiation from the background. Failing this, any "detection" could be written off as a misunderstanding of the background. Fortunately, the earth's motion around the Sun provides a significant modulation in the background rate.

The Sun is moving around the Galactic Center at a velocity of about 250 km/s. The non-dissipative dark matter in the galactic halo, on the other hand, never collapsed into a disc; thus, it is not rapidly rotating. As a result of the Sun's motion, the detector is moving relative to the galactic halo. Since the grains have an energy threshold, the anisotropy in the velocity distribution alters the predicted count rate.

The earth's motion around the Sun modulates the velocity of the detector relative to the halo. The earth moves around the Sun with a velocity of 30 km/s. Since the ecliptic is inclined at 62° relative to the galactic plane, only a fraction of this velocity is added to the Sun's motion. In January, when the Sun is in Saggitarius (the location of the Galactic Center), the earth is moving at 235 km/s relative to the halo. (The earth's motion around the sun is counter-clockwise, while the Sun's motion around the Galactic Center is clockwise.) In July, more energetic particles will stream through the detector when the earth is moving at 265 km/s relative to the galactic halo. Since the detector has an energy threshold, the flux of more energetic particles produces a higher count rate. In reference (17), we estimated a modulation in the signal of 12 % in a detector sensitive to 20 % of the incident flux. This calculation assumes that the galactic halo has an isothermal distribution of velocities. More detailed models of the galactic halo[23] that include the effect of the growth of the disc on the halo suggest a more radial distribution of velocities. This would imply a higher modulation rate ($\sim 15 - 25$ %).

The existence of this modulation effect does not depend upon details of models of the galactic halo. It only requires that the earth's velocity relative to the rest frame of the halo changes with time. The assumption that the halo is composed of WIMPs implies that it is non-dissiplative; hence, its rest frame should differ from that of the Sun which is composed of baryons which collected in the disc through dissipation. The only other astronomical requirement for the modulation effect is that the earth moves around the Sun. Seeing the modulation effect, however, requires that grains have a narrow distribution of energy thresholds ($\lesssim$ 50 %). This is yet another incentive for developing more uniform grains.

In conclusion, a promising application of grain detectors is their use in the search for dark matter. The earth's motion around the Sun produces a significant modulation which can be used to confirm a detection. While hurdles such as the production of uniform spherical grains remain, recent progress, the rewards of detection, and the powerful limits that could be placed on SUSY theories through non-detection will hopefully continue to motivate experimentalists to surmount these problems.

4. ACKNOWLEDGEMENTS

Names and addresses of the co-authors:

S.P. Ahlen, Department of Physics, Boston University, Cambridge, MA, USA.

F.T. Avignone, Department of Physics, University of South Carolina, Columbia, SC 29208.

R.L. Brodzinski, Pacific Northwest Laboratory, Richland, WA 99352.

K. Freese, Institute for Theoretical Physics, University of California, Santa Barbara, CA.

G. Gelmini, Department of Physics, Harvard University, Cambridge, MA 02138. On leave of absence from Department of Physics, University of Rome II, Via Orzio Raimondo, Rome 00173, Italy.

D.N. Spergel, Institute for Advanced Studies, Princeton, NJ.

I thank my collaborators in UBC, Vancouver (A. Kotlicki, M. le Gros, B. Turrell) for allowing me to describe our results before publication.

5. REFERENCES

(1) Farber, S.M. and Gallager, J.S. Ann. Rev. Ast. Astrophys. 17, 135 (1979). Forman, W., Jones, C., and Tucker, W., Ap.J. 293, 102 (1985).
(2) Hegyi, D.J., Olive, K.A., 'A Case Against Baryons in Galactic Halo', Fermilab preprint Pub-85/26-A.
(3) Dune, M., Fischler, W. and Srednicki, M. Phys. Lett. 104B, 199 (1984).
(4) Freese, K., Price, R. and Schramm, D.N., Ap.J. 275, 405 (1983).
(5) Goodman, M. and Witten, E. Phys. Rev. D 31, 3059 (1985); Wasserman, I., 'On the Possibility of Detecting Heavy Neutral Fermions in the Galaxy', Cornell preprint DE3065 D15-1 (1985).
(6) Ahlen, S.P., Avignone, F.T., Brodzinski, R.L., Drukier, A.K., Gelmini, G., Spergel, D.N. 'Limits on Cold Dark Matter Candidates from the Ultralow Background Germanium Spectrometer', preprint CfA No. 2292, March 1986.
(7) Brodzinski, R.L., Brown, D.P., Evans, J.C., Jr., Hensley, W.K., Reeves, J.H., Woman, N.A., Avignone, F.T., and Miley, H.S., Nucl. Ins. Methods Phys. Res. A239, 207 (1985).
Reese, J.H., Hensley, W.K. and Brodzinski, R.I., IEEE Transactions on Nuclear Science NS-32, 29 (1985).
(8) Avignone, F.T., Ahlen, S.P., Brodzinski, R.L., Dimopoulos, S., Drukier, A.K., Gelmini, G., Lynn, B.W., Miley, H.S., Reeves, J.H., Spergel, D.N., Starkman, G.D. 'Experimental Bounds on ßß-Decay, Cold Dark Matter and Solar Axions with a Ultralow Background Ge Detector', Proc. of Vanderbilt Conference, May 1986.
(9) Avignone, F.T., Brodzinski, R.L., Dimopoulos, S., Drukier, A.K., Gelmini, G., Lynn, B.W., Spergel, D.N., Starkman, G.D. 'Laboratory Limits on Solar Axions from the Ultralow Background Ge Spectrometer', preprint SLAC-PUB-3872, Jan. 1986, submitted to Phys. Rev. Letters.
(10) Wilson, W.D., Haggmark, L.G. and Biersack, J.P., Phys. Rev. B 15, 2458 (1977).
(11) Chasman, C., Jones, K.W. and Ristinen, R.A., Phys. Rev. Lett. 15, 245 (1965).
(12) Bahcall, J. and Casertano, S., 'Kinematics and Density of the Galactic Spheroid', IAS preprint 11 (1985).
(13) Caldwell, H. and Ostriker, J.P., Ap. J. 251, 61 (1981).
(14) Bagger, J., Dimopoulus, S., Masso, E., and Reno, J., Phys. Rev. Lett. 54, 2199 (1985); Kolb, E. and Olive, K., FERMILAB preprint Pub-85/116-A (1985.
(15) Hagelin, J.S., Kane, G.L., and Raby, S., Nucl. Phys. B241, 648 (1984); Ibanez, L.E., Phys. Lett. 137B, 160.
(16) Nussinov, S., Phys. Lett. B 165B, 55 (1985).
(17) Drukier, A.K., Acta Physica Polonica, B17, p. 229-237 (1986); Drukier, A.K., Freese, K. and Spergel, D.N., Phys. Rev. D, 33, p. 3495-3508 (1986).
(18) Drukier, A.K. and Valette, C. NIM 105, 285 (1972); Drukier, A.K., Valette, C., Waysand, G., Yuan, L.C.L., and Peters, F., Lett. Nuovo Cimento 14, 300 (1975); Drukier, A.K. and Yuan, L.C.L., NIM

173, 259 (1980); Drukier, A.K., NIM 201, 77 (1982).
(19) Kouiriti, M., Thesis, Universite Paris VI, Dec. 1984; Seidel, W., Thesis, TU Munich, Jan. 1985; Oberauer, L., Thesis, TU Munich, Feb. 1985.
(20) Drukier, A.K. and Stodolsky, L., Phys. Rev. D 30, 2295 (1984).
(21) Drukier, A.K., 'Search for Magnetic Monopoles with Use of a Superconducting Colloid', preprint MPI-PAE/Pth 58/83; Gonzales-Mestres, L. and Perret-Gallix, D., 'Detection of Magnetic Monopoles with Superheated Type I Superconductors', LAPP-EXP-85-02. For shorter versions of both proposals, see 'Proceedings of 1985 Moriond Conference on Massive Neutrinos in Physics and Astrophysics'.
(22) Kotlicki, A., LeGross, M., Drukier, A.K., Turrell, B., Technical Report, UBC/Vancouver, to be submitted to Cryogenics.
(23) Binney, J., May, A. and Ostriker, J.P., private communication (1986).

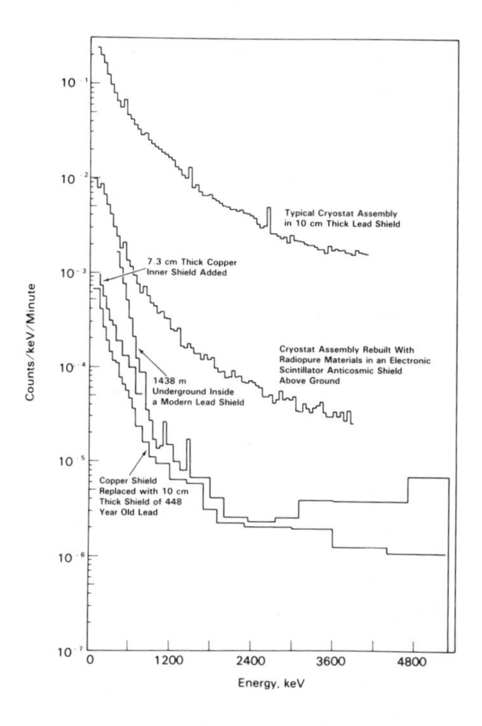

Figure 1. Background spectra of the PNL/USC, 135 cm^3 prototype Ge spectrometer.

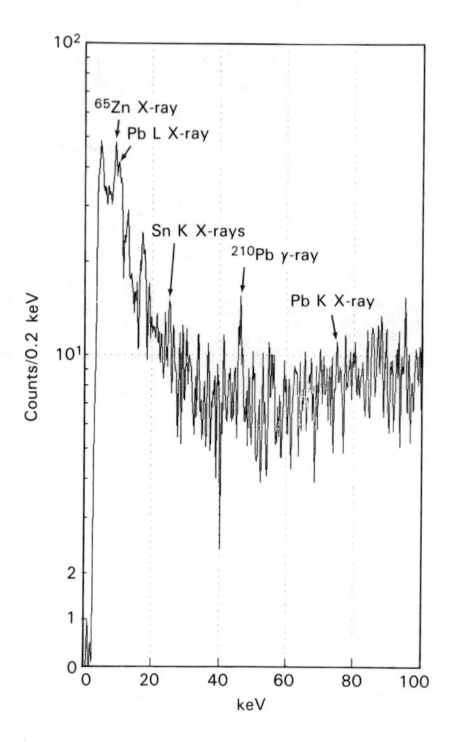

Figure 2. Ten weeks of data from the low energy portion of the Ge detector spectrum.

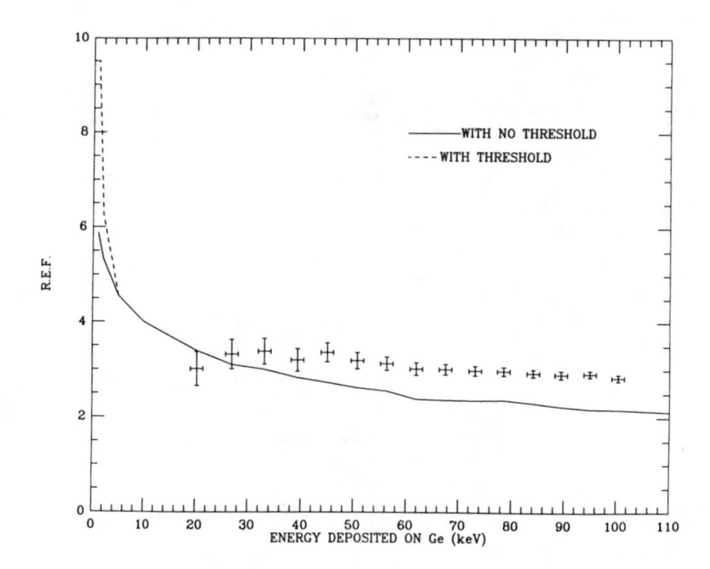

Figure 3. Energy dependence of the Relative Efficiency Factor (REF). The solid curve assumes no kinematic threshold. The dashed curve assumes a threshold of 0.27 keV. The data are from Chasman et al. (Phys. Rev. Lett. 15, 245 (1965).

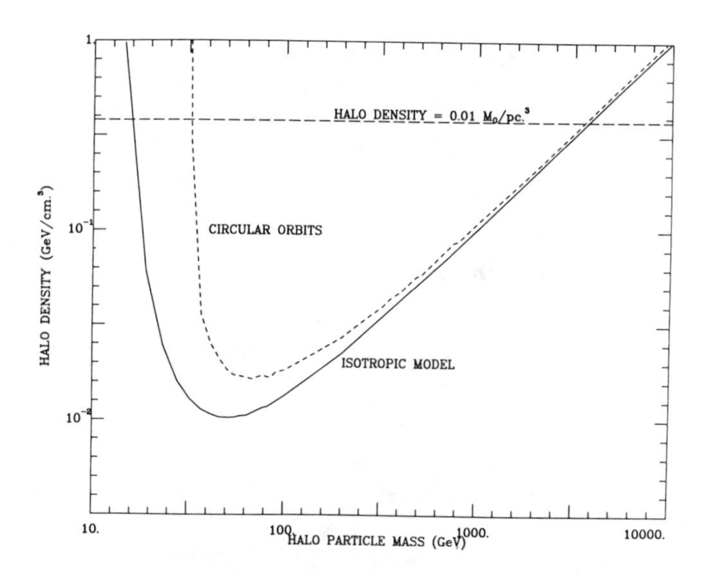

Figure 4. Maximum halo density consistent with the observed count rate.

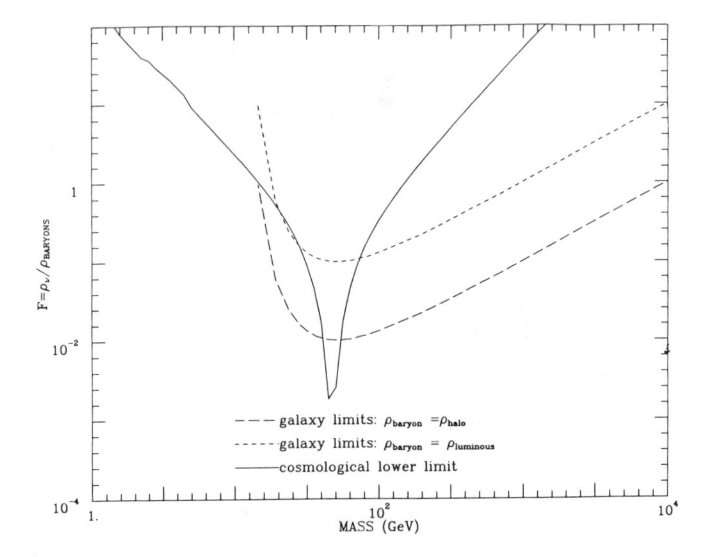

Figure 5. Maximum ratio of total Dirac neutrino mass to total baryonic mass in our galaxy.

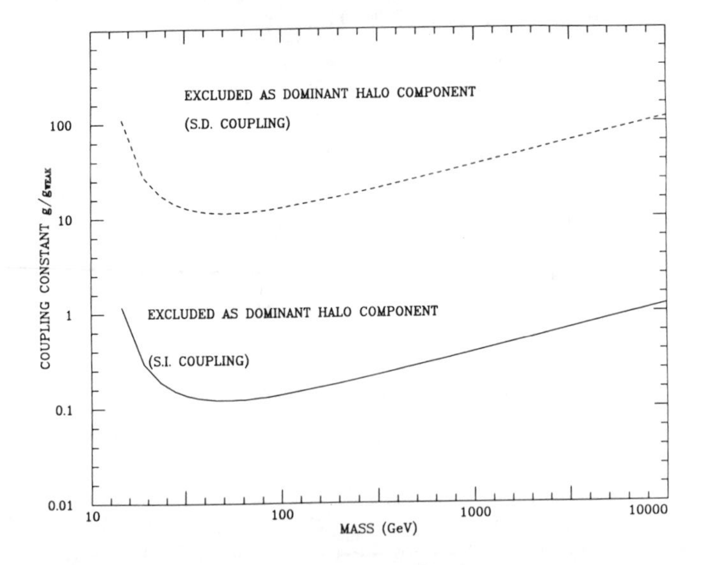

Figure 6. Excluded regions (above the curves) in mass-cross section plane.

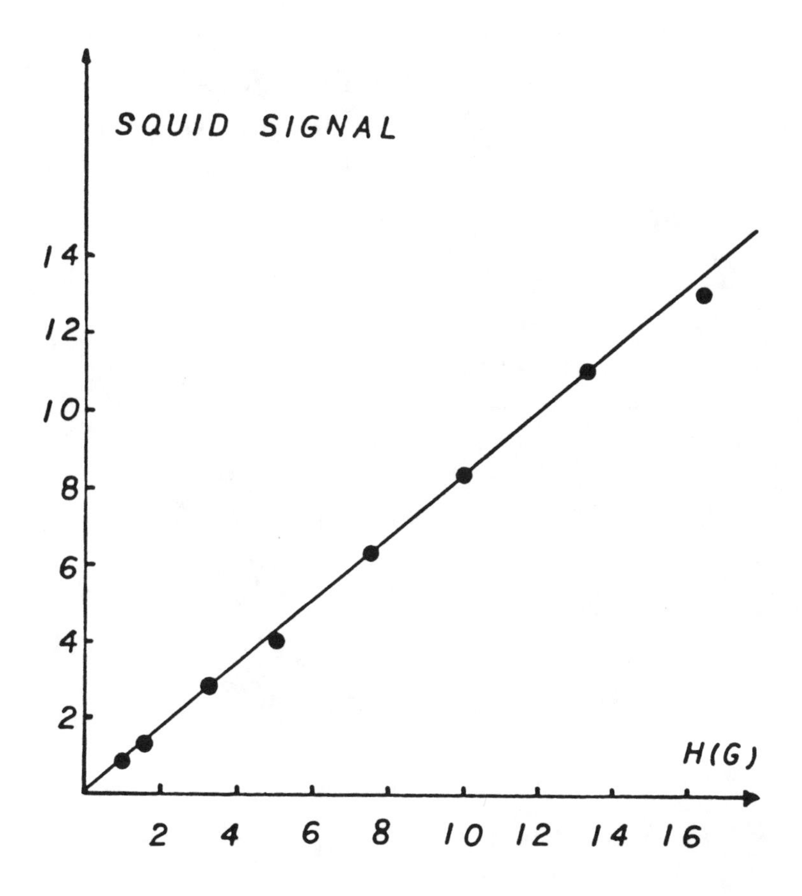

Figure 7. Squid signal from the transition of a single grain as a function of the applied magnetic field.

LIST OF PARTICIPANTS

Mr. Chidi Akujor
Astrophysics Group
Department of Physics and Astronomy
University of Nigeria
Nusukka, NIGERIA

Prof. Fred Ashton
Department of Physics
University of Durham
Science Laboratories, South Road
Durham, DH1, 3LE
ENGLAND

Ms. D. Averna
Osservatorio Astrofisico Di Catania
Citta Universitaria - I
95125 Catania
ITALY

Dr. James J. Beatty
University of Chicago
Laboratory for Astrophysics and
Space Research
933 East 56th Street
Chicago, IL 60637 USA

Miss Claudia Belardi
Consiglio Nazionale Delle Ricerce
Istituto TE.S.R.E.
Via de Castagnoli, 1
40126 Bologna
ITALY

Dr. Robert W. Binns
The McDonnell Center
Department of Physics
Washington University
St. Louis, MO 63130 USA

Dr. Hans Bloemen
Astronomy Department
University of California, Berkeley
Berkeley, CA 94720 USA

Dr. Dieter Breitschwerdt
Max Planck Institut fur Kernphysik
Saupfercheckweg 1,
Postfach 10 39 80
6900 Heidelberg 1
WEST GERMANY

Dr. Carla Cattadori
Piazza Republica 18
Milano, ITALY

Miss Paula Chadwick
Department of Physics
University of Durham
Durham DH1 3RR
ENGLAND

Miss Domitilla de Martino
Osservatorio Astronomico
di Capodimonte - Napoli
Via Moiariello 16
80131 Napoli
ITALY

Ms. Brenda Dingus
Department of Physics and
Astronomy
University of Maryland
College Park, MD 20742 USA

Miss K. M. Doutsi
Department of Astronomy
University of Athens
Panepistimiopolis
GR 15771 Athens
GREECE

Mr. Wolfgang Droge
Max Planck Institut
fur Radioastronomie
Auf dem Hugel, 69
D-5300 Bonn 1
WEST GERMANY

Dr. Andrezj K. Drukier
Harvard-Smithsonian Center
for Astrophysics
60 Garden Street
Cambridge, MA 02138 USA

Miss E. Eleftheriou
Department of Astronomy
University of Athens
Panepistimiopolis
GR 15771 Athens
GREECE

Dr. Jon Engelage
Bldg. 50, Room 248
Lawrence Berkeley Laboratory
Berkeley, CA 94720 USA

Miss Mariella Gottardi
Consiglio Nazionale Delle Ricerce
Istituto TE.S.R.E.
Via de Castagnoli, 1
40126 Bologna
ITALY

Dr. D. A. Green
Cavendish Laboratory Room 930
University of Cambridge
Madingley Road
Cambridge CB3 OHE
ENGLAND

Mr. John M. Grunsfeld
University of Chicago
Laboratory for Astrophysics and
Space Research
933 East 56th Street
Chicago, IL 60637 USA

Mr. Uwe Heinbach
Fachbereich 7, Physik
Universitat Gesamthochschule Siegen
Postfach 101240
D-5900 Siegen
WEST GERMANY

Dr. Davina Innes
Max Planck Institut fur Kernphysik
Saupfercheckweg 1, Postfach 10 39 80
6900 Heidelberg 1
WEST GERMANY

Mr. Joachim Isbert
Fachbereich 7, Physik
Universitat Gesamthochschule
Siegen, Postfach 101240
D-5900 Siegen
WEST GERMANY

Mr. Christoph Koch
Fachbereich 7, Physik
Universitat Gesamthochschule
Siegen, Postfach 101240
D-5900 Siegen
WEST GERMANY

Mr. Giuseppe Lanzafame
Osservatorio Astrofisico
Di Catania
Citta Universitaria - I
95125 Catania
ITALY

Mr. Mark Lawrence
Department of Physics
University of Leeds
Leeds LS2 9JT
ENGLAND

Dr. Kevin Lesko
Bldg. 88
Lawrence Berkeley Laboratory
Berkeley, CA 94720 USA

Mr. Ken Levenson
Space Science Center
University of New Hampshire
Durham, NH 03824 USA

Mr. Mark G. Machin
Imperial College
The Blackett Laboratory
Prince Consort Road
London SW7 2BZ
ENGLAND

Mr. Pier Paolo Maggioli
Department of Physics
University of Southhampton
Southampton SO9 5NH
ENGLAND

Dr. Wojtek Markiewicz
Max Planck Institut fur Kernphysik
Saupfercheckweg 1,
Postfach 10 39 80
6900 Heidelberg 1
WEST GERMANY

Mr. Nicholas Mascarenhas
University of Chicago
Laboratory for Astrophysics and
Space Research
933 East 56th Street
Chicago, IL 60637 USA

Mr. M. A. Miah
Department of Physics and
Astronomy
Louisiana State University
Baton Rouge, LA 70803 USA

Dr. John Mitchell
Department of Physics and
Astronomy
Louisiana State University
Baton Rouge, LA 70803 USA

Mr. Anthony Monk
University of Bristol
H. H. Wills Physics Laboratory
Royal Fort, Tyndall Avenue
Bristol BS8 1 TL
ENGLAND

Dr. Thierry Montmerle
Service D'Astrophysique
CEN-Saclay, DPhG/SAP
91191 Gif sur Yvette, Cedex
FRANCE

Miss Ivana Nastari
via Stazione, 95
40037 Sasso Marconi
Bologna
ITALY

Mr. Klaus Oschlies
Institut fur Reine und
Angewandte Kernphysik
Universitat Kiel
Olshausenstrasse 40/60,
Gebaude No. 20A
2300 Kiel
WEST GERMANY

Miss Eliana Palazzi
Consiglio Nazionale Delle Ricerce
Instituto TE.S.R.E.
Via de Castagnoli, 1
40126 Bologna
ITALY

Dr. John S. Perko
Code 665
NASA-Goddard Space Flight Center
Greenbelt, MD 20771 USA

Dr. Valerio Pirronello
Osservatorio Astrofisico
Citta Universitaria
Vaile Doria
95125 Catania
ITALY

Dr. Graziella Pizzichini
Consiglio Nazionale Delle Ricerce
Instituto TE.S.R.E.
Via de Castagnoli, 1
40126 Bologna
ITALY

Prof. Neil Porter
Physics Department
University college Dublin
Belfield, Stillorgan Road
Dublin 4
IRELAND

Dr. Marius S. Potgieter
Department of Physics
Potchefstroom University
2520 Potchefstroom
SOUTH AFRICA

Dr. Narayan C. Rana
Tata Institute of Fundamental Research
Homi Bhabha Road
Bombay 400 005
INDIA

Mr. M. J. Rogers
University of Durham
Department of Physics
Science Laboratories, South Road
Durham DH1 3LE
ENGLAND

Dr. Manfred Samorski
Institut fur Reine und Angewandte Kernphysik
University of Kiel
Olshausenstrasse 40-60
D-2300 Kiel
WEST GERMANY

Prof. G. Setti
Instituto di Radioastronomia, CNR
University of Bologna
Bologna
ITALY

Prof. Maurice M. Shapiro
Director of the School
Co-Director of the Course
205 Yoakum Parkway, #1720
Alexandria, VA 22304 USA

Dr. Rein Silberberg
Code 4154
Naval Research Laboratory
Washington, DC 20375 USA

Mr. Nigel J. T. Smith
Department of Physics
University of Leeds
Leeds LS2 9JT
ENGLAND

Mr. Wilfred Sorrell
Washburn Observatory
University of Wisconsin-Madison
475 North Charter Street
Madison, WI 53706 USA

Dr. W. Stamm
Institute fur Reine and Angewandte Kernphysik
Olshausenstrasse 40-60
D-2300 Kiel
WEST GERMANY

Dr. C. Sean Sutton
Department of Physics
Smith College
Northampton, MA 01063 USA

Dr. Jacek Szabelski
Institute of Nuclear Studies
Cosmic Ray Laboartory
Ul. Uniwersykecka 5
90-950 Lodz 1, Box 447
POLAND

Prof. K. O. Thielheim
Institut fur Reine und Angewandte
Kernphysik, Mathematische Physik
Universitat Kiel
Olhausenstrasse 40/60,
Gebaude No. 20a,
2300 Kiel
WEST GERMANY

Prof. W. Peter Trower
Department of Physics
Virginia Polytechnic Institute
Blacksburg, VA 24061 USA

Dr. Allan J. Tylka
Code 4154.8
Cosmic Ray Astrophysics
Naval Research Laboratory
Washington, DC 20375-5000 USA

Dr. Jose F. Valdes-Galicia
Institute de Geofisica
Ciudad Universitaria
Circuito Exterior 04510
Mexico D. F., MEXICO

Dr. Christofor P. Vankov
Institute for Nuclear Research and Nuclear Energy
72 Blvd. Lenin
Sofia 1784
BULGARIA

Prof. Heinz Volk
Max Planck Institut fur Kernphysik
Saupfercheckweg 1,
Postfach 10 39 80
6900 Heidelberg 1
WEST GERMANY

Prof. P. J. S. Watson
Department of Physics
Herzberg Laboratories
Carleton University
Ottawa, K1S 5B6
CANADA

Prof. John P. Wefel
Co-Director of the Course
Department of Physics and
Astronomy
Louisiana State University
Baton Rouge, LA 70803-4001 USA

Dr. Kurt W. Weiler
Code 4131
Hulburt Space Center
Naval Research Laboratory
Washington, DC 20375 USA

SUBJECT INDEX

AUTHOR INDEX